THE NEUROBIOLOGY OF ALZHEIMER'S DISEASE

ANNALS OF THE NEW YORK ACADEMY OF SCIENCES
Volume 777

THE NEUROBIOLOGY OF ALZHEIMER'S DISEASE

Edited by Richard J. Wurtman, Suzanne Corkin, John H. Growdon, and Roger M. Nitsch

The New York Academy of Sciences
New York, New York
1996

Cover art: Adolph Gottlieb (American, 1903–1974), *Coalescence*, 1961, oil on canvas, Boris and Sophie Leavitt, on loan to the National Gallery of Art, Washington, D.C. © Board of Trustees.

Library of Congress Cataloging-in-Publication Data

The neurobiology of Alzheimer's disease / edited by Richard J. Wurtman ... [et al.].
p. cm. -- (Annals of the New York Academy of Sciences ; ISSN 0077-8923 ; v. 777)
Papers presented at the Eighth Meeting of the International Study Group on the Pharmacology of Memory Disorders Associated with Aging, held in Züich, Switzerland on February 17-19, 1995.
Includes bibliographical references and index.
ISBN 0-89766-973-8 (cloth : alk. paper). -- ISBN 0-89766-974-6 (paper : alk. paper)
1. Alzheimer's disease--Pathophysiology--Congresses. 2. Molecular neurobiology--Congresses. I. Wurtman, Richard J., 1936- . II. Study Group on the Pharmacology of Memory Disorders Associated with Aging. Meeting (8th : 1995 : Zürich, Switzerland) III. Series.
[DNLM: 1. Alzheimer's Disease--physiopathology--congresses. 2. Alzheimer's Disease--drug therapy--congresses. 3. Diagnostic Imaging--congresses. 4. Amyloid beta-Protein Precursor--physiology--congresses. 5. Apolipoproteins E--physiology--congresses. 6. tau Proteins--physiology--congresses. W1 AN626YL v.777 1996 / WT 155 N494 1996]
Q11.N5 vol. 777
[RC523]
500 s--dc20
[616.8'3107]
DNLM/DLC
for Library of Congress 95-48168
CIP

Second Printing November 1996

MC/PCP
Printed in the United States of America
ISBN 0-89766-973-8 (cloth)
ISBN 0-89766-974-6 (paper)
ISSN 0077-8923

ANNALS OF THE NEW YORK ACADEMY OF SCIENCES

Volume 777

January 17, 1996

THE NEUROBIOLOGY OF ALZHEIMER'S DISEASE[a]

Editors and Conference Organizers

RICHARD J. WURTMAN, SUZANNE CORKIN, JOHN H. GROWDON, AND ROGER M. NITSCH

CONTENTS

[a] The papers in this volume wre presented at the eighth meeting of the International Study Group on the Pharmacology of Memory Disorders Associated with Aging, organized by the center for Brain Sciences and Metabolism Charitable Trust, Cambridge Massachusetts. The meeting, entitled **The Neurobiology of Alzheimer's Disease**, was held on February 17–19, 1995 in Zürich, Switzerland.

Part III. Cytoskeleton and Tau

Part IV. Genetics and Apolipoprotein E

Part V. Novel Therapies for Alzheimer's Disease

Poster Papers: Clinical Features of Alzheimer's Disease

Poster Papers: Apolipoprotein E and Alzheimer's Disease

Poster Papers: Amyloid Precursor Proteins and Aβ

Poster Papers: Neurochemistry and Alzheimer's Disease

Poster Papers: Novel Therapies for Alzheimer's Disease

Poster Papers: Animal Models of Alzheimer's Disease

Financial assistance was received from:

- ALZHEIMER'S ASSOCIATION
- FRENCH FOUNDATION
- HELEN BADER FOUNDATION
- THE JOHN D. AND CATHERINE T. MACARTHUR FOUNDATION

- Alza Corporation
- American Lecithin
- Applied Genetics
- CIBA-Geigy
- Coca-Cola
- Farbenfabriken Bayer
- Ferrer Internacional
- Genentech
- Glaxo Research & Development
- H. Lundbeck A/S
- Hoechst AG
- Hoechst-Roussel Pharmaceuticals
- F. Hoffmann-LaRoche, Ltd.
- Interneuron
- Knoll AG
- Marion Merrell Dow, Inc.
- Merck Research Laboratories
- Miles Pharmaceuticals
- NutraSweet
- Pfizer, Inc.
- Scotia Pharmaceuticals
- SmithKline Beecham Pharmaceuticals
- Synthelabo Recerche
- TEVA Pharmaceuticals
- UCB Pharma
- Warner-Lambert Co./Parke-Davis

Preface

This volume contains the papers and poster abstracts compiled for the eighth meeting of the International Study Group on the Pharmacology of Memory Disorders Associated with Aging (ISG) taking place in Zürich, Switzerland on February 17–19, 1995. The ISG was founded 16 years ago in the belief that the development of effective treatments for Alzheimer's disease (AD) and other dementias would be accelerated by periodic meetings of scientists and physicians from around the world who are actively working on issues related to dementia. There have been seven previous "Zürich Meetings": in 1979, 1981, 1984, 1987, 1989, 1991, and 1993. The proceedings of the second ISG meeting were published by Raven Press in 1982 (*Alzheimer's Disease: A Report of Progress in Research*; Corkin, Davis, Growdon, Usdin, and Wurtman, editors); the proceedings of the fourth ISG meeting were published by Springer-Verlag in 1988 as Supplement 24 to the *Journal of Neural Transmission (Topics in Basic and Clinical Science in Dementia*; Wurtman, Corkin, and Growdon, editors); the proceedings of the fifth ISG meeting were published by Raven Press in 1990 (*Alzheimer's Disease. Advances in Neurology*, Volume 51; Wurtman, Corkin, Growdon, and Ritter-Walker, editors) the proceedings of the sixth ISG meeting were published by the New York Academy of Sciences in 1992 (*Aging and Alzheimer's Disease*, Volume 640; Growdon, Corkin, Ritter-Walker and Wurtman, Editors); and the proceedings of the seventh ISG meeting were published by the New York Academy of Sciences in 1993 (*Alzheimer's Disease Amyloid Precursor Proteins, Signal Transduction, and Neuronal Transplantation*, Volume 695; Nitsch, Growdon, Corkin, and Wurtman, editors).

The overall goal of the ISG meetings is to highlight discoveries that shed light on the potential causes of AD, its pathogenesis, and the biological mechanisms that could underlie its cure. This eighth meeting of the ISG considers five major topics:

- Mapping brain structure and function with neuroimaging
- Amyloid precursor proteins and Aβ
- The cytoskeleton and tau
- Genetics and Apolipoprotein E
- Novel therapies for AD

The first session reviews alterations in brain morphology associated with normal aging and the structural and functional changes that occur during AD. Speakers in this session emphasize changes in brain physiology during mental activation tasks. The second session concerns amyloid and its precursors. Here, speakers consider the potential biological functions of the amyloid precursor protein and its derivatives, and factors that regulate processing of the parent precursor molecule. The third session concerns cytoskeleton pathology in AD. Speakers in this session emphasize the role of tau protein and the molecular genetics of AD. The fourth session concerns the molecular genetics of AD. Speakers in this session discuss apolipoprotein E as a risk factor for developing AD and consider biologic bases for increased frequency of apolipoprotein E4 in AD. The fifth and final session highlights novel therapies proposed for AD. Speakers in this session consider novel approaches to decrease amyloid deposition in brain by muscarinic receptor stimulation and by anti-inflammatory drugs.

The ISG fosters interactions among academic scientists, scientist in industry, clinical investigators, and administrators in public and private agencies concerned with dementia. The eighth ISG meeting continues in the tradition of its predecessors by stimulating research in dementia, and by speeding the transfer of information from the basic sciences to physicians caring for persons with dementia and to the pharmaceutical industry.

The Organizing Committee wishes to thank Mr. David B. Fitzgerald and Ms. Emily Rossie for excellent administrative and editorial assistance, respectively, in the preparation of this volume.

JOHN H. GROWDON

In Vivo Structural Studies of the Hippocampus in Normal Aging and in Incipient Alzheimer's Disease[a]

M.J. DE LEON,[b,g] A. CONVIT,[b] A.E. GEORGE,[c] J. GOLOMB,[b,d] S. DE SANTI,[b] C. TARSHISH,[b] H. RUSINEK,[c] M. BOBINSKI,[f] C. INCE,[b] D. MILLER,[e] AND H. WISNIEWSKI[f]

[b] *Departments of Psychiatry, [c]Radiology, [d]Neurology, and [e]Pathology, New York University Medical Center, New York, New York 10016 USA*

[f] *Institute for Basic Research, Staten Island, New York USA*

ABSTRACT: Population trends indicate that in the near future the size of the elderly population will increase. This will result in a large increment in the numbers of persons suffering mild to severe levels of cognitive impairment. While considerable efforts continue to be made to explain brain changes associated with Alzheimer disease (AD), little is known of the brain changes in aging without dementia or so-called normal aging. Pathologic studies suggest that the medial temporal lobe is informative in the examination of the early brain changes related to AD. However, pathologic studies only offer a single observation and considerable uncertainty exists regarding the likelihood of progression of disease and the development of dementia. Several structural neuroimaging studies have recently investigated this anatomy and recent reports are encouraging for a medial temporal lobe based diagnosis for age-related cognitive impairments. We will present our findings on the MRI anatomy of the hippocampal formation as well as data bearing on the use of hippocampal formation imaging in the diagnosis of AD and as a predictive marker for future dementia. Our findings suggest an anatomically specific relationship between hippocampal volume and secondary memory performance. Because these observations apply to nondemented and normal elderly subjects, we are encouraged that the anatomy of age-related cognitive impairments can be reliably recognized and possibly put to use in therapeutic studies.

INTRODUCTION

By the year 2030 it is estimated that between 17 and 20% or about 50 million of the US population will be over age 65.[1] Dementia affects 1–6%

[a] This work was made possible with grants from the National Institute on Aging: P30 AG08051, RO1 AG12101 and PO1 AG04220; the National Institute of Mental Health RO1 43965 and the New York State Office of Mental Retardation and Developmental Disabilities

[g] *Send correspondence to:* Dr. Mony J. de Leon, Neuroimaging Laboratory, New York University Medical Center, New York, NY 10016, Tel: (212) 263 5805, Fax: (212) 263 6991.

of the population over the age of 65 and 10–20% over the age of 80[2] with its annual incidence approximately doubling for every five years of age between the ages of 75 and 89 years.[3] Other data show that for every demented patient there are approximately 8 nondemented individuals with cognitive deterioration adversely affecting their quality of life.[4] Consequently, a staggering number of elderly with mild to severe cognitive impairments, perhaps 15–20 million, can be expected in the next 35 years, representing a great human and economic toll. While cross-sectional and longitudinal studies commonly observe subtle declines in cognitive functioning associated with aging, the underlying brain mechanisms responsible for age-related declines in cognitive functioning are unclear. The purpose of this chapter is to examine, using structural neuroimaging techniques, the hypothesis that hippocampal formation atrophy occurs early in the natural history of AD and actually precedes neocortical involvement.

Hippocampal Pathology in Normal Aging and AD

The pathological changes in the hippocampus in AD are so common that AD has been characterized as "a hippocampal dementia" (Ball *et al.*[5]). There are many biochemical and structural hippocampal abnormalities documented in AD. The structural pathology, typically observed as part of the neuropathology work-up includes: neuronal loss, neurofibrillary tangles, and neuritic plaques.[6] The common observation of extensive pathology in the hippocampal formation in AD[7] and the recent documentation that the histopathological markers of AD are consistently found in the hippocampus and the parahippocampal gyrus of early cases[8,9] have directed researchers toward examining these structures as possibly the sites of the earliest neurodegenerative changes associated with AD.

Pathologic evidence for the early involvement of the hippocampus in AD is difficult to obtain because the majority of AD patients survive until the later stages of the illness when there are severe diffuse degenerative changes. With these caveats, AD-type histopathological changes concentrated in the hippocampi are often observed in both normal and very mildly impaired elderly and may be suggestive of early brain AD. De la Monte[10] reported that hippocampi (cornu ammonis) both from AD patients and clinically normal elderly with neuropathological evidence of AD showed volume reductions from "pure" (no histologic evidence for AD) elderly control values. Neocortical changes were found only in the demented AD group, suggesting the primacy of the hippocampal changes. Similarly, Price *et al.*[9] found in minimally impaired elderly (not sufficiently impaired to meet criteria for AD), evidence for the neuropathologic changes of AD to be most consistently found in the medial temporal lobe. Nevertheless, it is unknown how prevalent the hippocampal formation changes are in aging populations or if these brain changes are related to the mild memory impairments so often reported in clinical studies of nondemented normal elderly. It is also unknown if long-lived normal elderly individuals with AD-like hippocampal changes will develop neocortical changes and AD-like clinical symptoms.

Neuropsychology

Selective hippocampal lesions in numerous animal models and human case studies are known to produce memory storage deficits while preserving immediate recall, remote recall (historical memories) and intelligence.[11] Some evidence exists that measures that may be more sensitive to hippocampal integrity such as secondary memory tasks emphasizing delayed recall, are somewhat resistant to longitudinal decline over short study intervals in normal aging.[12] These observations have led to the suggested use of delayed recall tasks in the clinical differentiation of normal aging from nondemented patients considered at risk for AD.[13,14] However, efficient memory functioning is not simply based on the integrity of the hippocampus. Many different locations of brain damage have been associated with memory dysfunctions,[15] and generalized atrophic changes occur in normal aged individuals.[16] Consequently, the determination of meaningful brain-behavior relationships in aging humans requires both qualitative and quantitative characterization of the cognitive deficits as well as examination of the anatomic specificity of any statistically significant relationship.

In summary, the pathologic and the neuropsychologic data offer excellent justification for examining the integrity of the hippocampus as a potentially informative anatomy for evaluation of the transition between normal aging and AD-related memory decline.

Neuroimaging Studies

While many cross-sectional structural neuroimaging studies have concentrated on gross measures of atrophy, such as ventricle size, generalized measures of cortical atrophy and severity of white matter disease in order to distinguish between normal elderly and individuals with AD, there has proved to be considerable overlap between the groups and very limited diagnostic utility for these measures. However, the three-year longitudinal study of ventricular size may be of some value as a correlate of clinical decline, as normal elderly when compared with AD patients, do not show appreciable change in short intervals (see Cutler *et al.*[17] and de Leon *et al.*[18] for review). In recent years, several cross-sectional structural studies have examined the region of the hippocampus and suggested that parenchymal volume losses with secondary CSF increases in the perihippocampal fissures may be important in the diagnostic evaluation of AD.[19–21] Moreover, recent studies highlight the value of these assessments in the prediction of future dementia in patients with mild cognitive impairments.[20,22]

Functional or metabolism imaging techniques using modalities such as PET or SPECT have shown some promise in the cross-sectional diagnostic study of AD. Widespread neocortical changes in glucose utilization and cerebral blood flow are typically reported in AD but not in normal aging, and there is a consensus that the temporo-parietal association area deficits are most salient. However, the cross-sectional diagnostic value of neocortical studies appears to be low in less advanced AD cases,[23,24] improving with inclusion of more advanced cases and with longitudinal designs.[25,26] These data suggest the possibility that the diagnostic failure of functional neuroimaging to identify early cases is due to the relative absence of disease in the temporal and parietal association areas in the early stages of AD. For technical reasons, few PET studies have included the region of the medial temporal lobe in their analyses.

In the studies below, we will present both anatomic-pathologic and clinical studies to demonstrate the importance of hippocampal formation imaging in the early detection of age related memory changes and in the prediction of future dementia and a diagnosis of AD.

METHODS AND RESULTS

In Vivo *Anatomy of the Perihippocampal CSF Spaces*

In a series of neuroimaging and neuropathology studies we observed that enlargement of the CSF spaces in the perihippocampal region are indicative of atrophy and volume losses in the hippocampus and the parahippocampal gyrus.[27] These CSF changes involve the transverse fissure of Bichat and its extensions: the choroidal and hippocampal fissures. The lateral aspect of the transverse fissure (LTF) curves around the pulvinar of the thalamus and separates the thalamus from the subicular region of the parahippocampal gyrus[28] (see FIG. 1). On the horizontal (axial) view, (whose plane of acquisition is depicted on the sagittal drawing in FIG. 1) the LTF (arrowheads) is bordered anteriorly by the uncus (see FIG. 2). The lateral boundary of the LTF is the dentate gyrus and the fimbria, medially it communicates with the ambient cistern and cerebral peduncles. The coronal view best permits separation of the choroidal and hippocampal fissure extensions of the LTF (see FIG. 3.) Both the coronal and sagittal planes of section permit identification of the structures at the ventral surface of the transverse fissure (the subiculum of

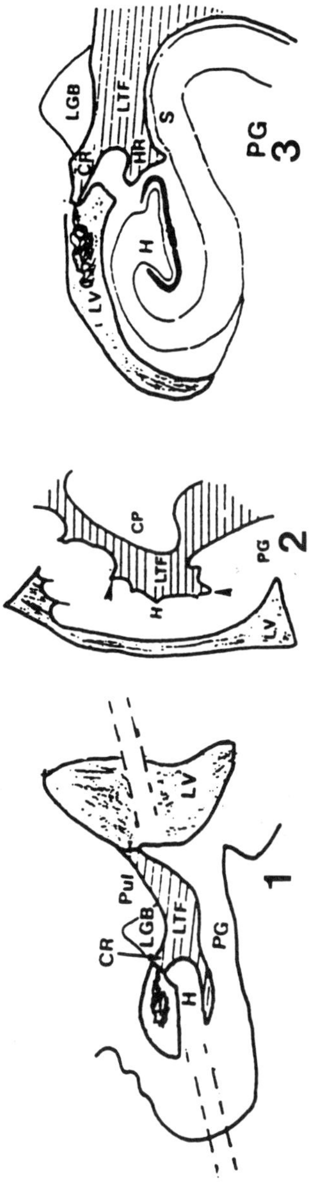

FIGURES 1–3. Schematic representations of sagittal, axial, and coronal view of the right perihippocampal region. Abbreviations; H-hippocampus; LFT-lateral transverse fissure; CR-Choroid fissure recess; HR-hippocampal sulcus recess; LV-lateral ventricle; PG-parahippocampal gyrus; CP-cerebral peduncle; S-subiculum; LGB-lateral geniculate body; Pul-pulvinar.

the parahippocampal gyrus), as well as the identification of thalamic structures at the dorsal surface (the lateral geniculate and the pulvinar).

Cross-Sectional Hippocampal Studies of Normal Aging and AD

Imaging parallel to the long axis of the hippocampus, as is done in our modified axial study, makes it possible to evaluate as a single CSF space the dilatation of the transverse fissure and the associated peri-hippocampal fissure complex. See FIGURE 4 for the validation of the axial CSF location using the coronal view. Images taken in this plane allow a qualitative determination of fissure enlargement to be made from a minimal number of slices. All subjects were evaluated using the same subjective 4-point assessment scale to rate the

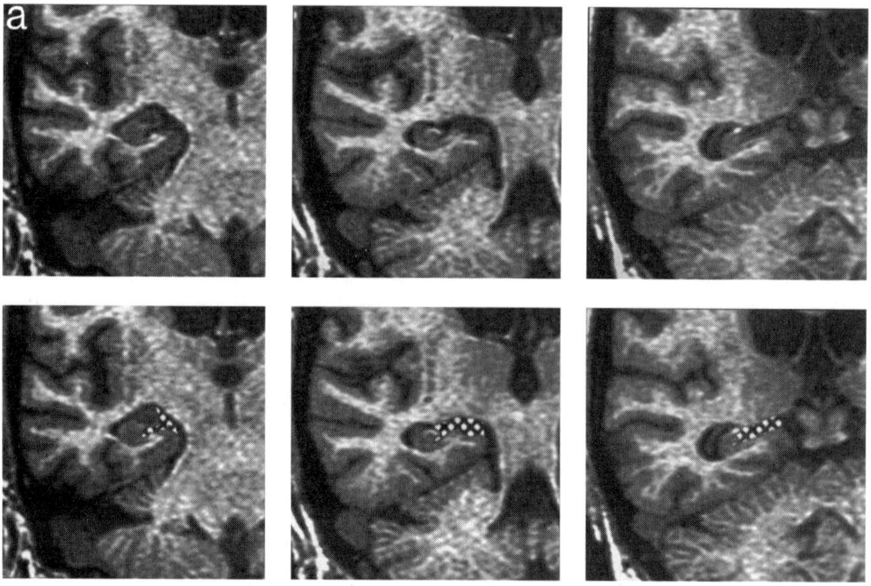

FIGURE 4a-b. The MR scans were obtained on a General Electric Advantage 1.5T imager using a spoiled gradient echo sequence of 35/9/1 (TR/TE/Excitations), a 60° flip angle, an 18 cm field of view and a 256 × 128 acquisition matrix. We obtained 124 contiguous coronal images with a slice thickness of 1.3mm. On 56 of these images (three shown in upper row of (a)), using a twofold image magnification to aid in region drawing (pixel size = 0.35 mm), we outlined a peri-hippocampal fissure region including the transverse fissure of Bichat, the choroidal fissure and the hippocampal fissure. The image intensity values in the region were coded as large open circles, bottom row of (a). All 56 coronal images and the coded region were reformatted into a negative angulation axial plane and transformed (by averaging), into 5 mm contiguous slices (top row of (b)). The reformatted axial images in (b) bottom row, show in large open circles the peri-hippocampal fissure CSF originally drawn on the coronal scan (a). The small open circles surrounding the large open circles represents the partially volumed pixels.

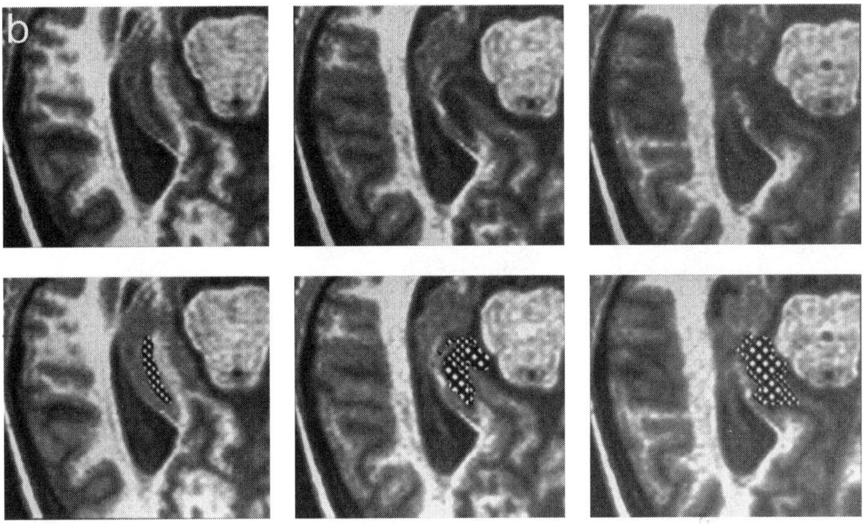

FIGURE 4. *(continued)* The lateral boundary for the peri-hippocampal fissures was the hippocampus and the choroid plexus of the temporal horn of the lateral ventricle. The medial boundary between the transverse fissure of Bichat (lateral transverse fissure) and the blending ambient cistern was drawn as a line that originated at the most medial, horizontal and superior aspect of the subiculum and extended dorsally to the thalamus. Once the medial and lateral margins for the fissures were identified, an ROI was drawn around the area. A signal intensity threshold was used to determine the CSF containing pixels in the ROI. The threshold between the CSF and brain was determined, slice by slice, as the average intensity between "pure" gray matter and CSF based on eight gray matter samples taken from the fronto-temporal cortical ribbon and eight CSF samples taken from the lateral ventricle.

extent of hippocampal atrophy (HA) (0 = no atrophy, 1 = questionable, 2 = mild, and 3 = moderate to severe). The cut-off rating of ≥ 2 in either hemisphere was used to define definite HA.

Using this protocol, we studied 405 patients. In this and all the studies below, the patients received an extensive research protocol of medical, neurologic, psychiatric and neuropsychologic exams. These examinations enabled us to exclude from the study individuals with evidence of conditions that could affect brain functioning or structure and to include only individuals with age or AD-related brain change. Four clinical groups were defined using NINCDS/ADRDA criteria and the GDS. The groups identified included normal elderly (GDS 1 and 2) n = 130; mild AD (GDS 4) n = 73; and moderate to severe AD (GDS 5 and 6) n = 130. The fourth group, the minimal impairment group, n = 72, was defined as having Global Deterioration Scores (GDS) = 3[29] and Mini-Mental State Exam scores (MMSE) > 23.[30]

HA was significantly more prevalent in all clinical groups as compared

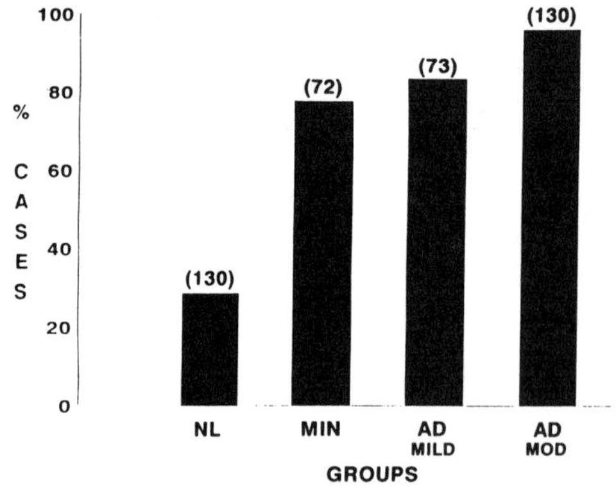

FIGURE 5. The prevalence of hippocampal atrophy in 4 groups (n = 405): normal elderly (NL), minimally impaired elderly (min), and two AD groups with different severity levels of cognitive impairment.

with control. 78% of the minimally impaired (GDS = 3), 84% of the mild AD (GDS = 4) and 96% in the moderate to severe AD (GDS ≥ 5) groups showed HA. Controls showed HA in 29% of subjects (see FIG. 5). Normal controls showed a striking age dependence for HA while the cognitively impaired groups showed prevalence rates independent of age (see FIG. 6).

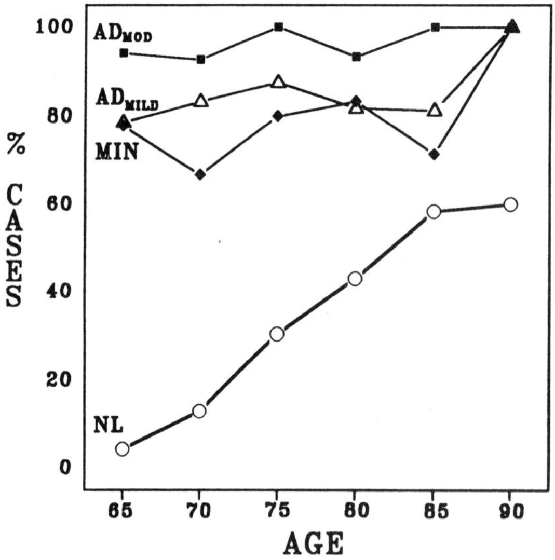

FIGURE 6. The prevalence of hippocampal atrophy for normal, minimal, and two levels of AD as a function of age (n = 405).

Longitudinal Studies Predicting the Development of Dementia

The results of our cross-sectional studies of HA suggested that hippocampal changes occur with a high frequency in minimally affected patients and thereby may be predictive of deterioration to symptoms consistent with the course of AD. In order to assess the early diagnostic value of hippocampal changes, a four-year follow-up CT study was carried out (n = 86) in a normal elderly group (n = 54, aged 70 ± 8 SD) and a minimally impaired group (n = 32, aged 71 ± 9 SD).[22] Over the interval, 23 (72%) of the GDS = 3 group and 2 (4%) of the controls deteriorated to receive the diagnosis of AD (GDS ≥ 4). For the GDS = 3 group, baseline hippocampal ratings ≥ 2 were present in 91% of the declining group and absent in 89% of the non-declining group. The results also showed that the prediction of decline was not related to age as equal numbers of the decliners came from the groups above and below 75 years of age. Nonsignificant differences were found in the percentages of males and females that declined. Both cortical sulcal prominence and ventricular enlargement were related at baseline to the observation of decline. However, as compared with the overall prediction accuracies of 91% for HA, these global measures of brain atrophy yielded overall accuracy scores of 70% and 69% respectively. In summary, the relatively high accuracy of HA for the prediction of decline, and the high rate of preservation in the group not showing HA, suggests the unique potential of HA in the clinical examination of early AD. The results of these studies also point to the need for the continued longitudinal study of the temporal relation between hippocampal volume change, neocortical atrophy and the development of intellectual dysfunction. As only 2/54 of our normal elderly deteriorated to dementia while 8/54 had baseline hippocampal atrophy, in order to carefully evaluate the predictive risk of the hippocampal atrophy the period of observation must be extended beyond the four years and more sensitive indices of hippocampal integrity and neuropsychological performance used. It is of special interest, as expanded below, that normals with HA had significantly lower performances on tests of recent memory with a delay component than normals without HA.

Hippocampal Atrophy in Normal Aging

In order to estimate the neuropsychologic impact of HA in normal aging, from the above cross-sectional study we studied 126 of the elderly normal controls.[31] All the normal elderly selected for the study had MMSE scores ≥ 28, indicating a very high level of functioning. All subjects underwent neuropsychologic tests sensitive to deficits of immediate and recent verbal recall; 29% of these subjects had HA. Using the psychometric measures as set of dependent variables in a MANCOVA design with age and years of education as covariates, significant cognitive differences were observed between the HA-positive and HA-negative groups ($p < .05$). *Post hoc* univariate comparisons demonstrated significant impairments on tests of secondary or delayed verbal memory ($p < .01$) but not tests of immediate memory. This result is consistent with the pattern of memory dysfunction reported in subjects with circumscribed hippocampal lesions. To our knowledge, this observation is the first indication that hippocampal atrophy is associated with decreased cognitive performance among normal elderly. Recently, we replicated this hippocampus-delayed recall correlation using MRI determined hippocampal volumes.[32] In the replication (n = 54), an effort was made to address the anatomic specificity of the observation. The results demonstrated that the volume of the hippocampus but not the superior temporal gyrus was related to the memory performance of the normal elderly.

The Anatomic Specificity of Hippocampal Volume Loss in Patients at Risk for AD

In order to investigate the hypothesis that hippocampal parenchymal volume reductions in the absence of neocortical volume reductions characterized individuals with minimal cognitive impairments, we completed the volumetric MRI and neuropsychological study of 27 normal elderly, 22 elderly GDS = 3 patients and 27 AD patients.[33] As in the above study,[32] we used 4 mm thick T_1-weighted coronal MRI images. We measured five major regions comprising the entire temporal lobe, from the level of the uncus to that of the splenium of the corpus callosum. The regions included the hippocampus, parahippocampal gyrus, fusiform gyrus, the superior temporal gyrus, and the combined middle and inferior temporal gyri. All groups were matched for age, gender and education. The results showed that for the GDS = 3 patients, only the hippocampus showed significant volume loss (14%). In other more recent analyses, these data also confirmed the significant relationship reported above between hippocampus volume and delayed recall performance (in preparation). Medial temporal lobe volume but not lateral temporal lobe volumes were positively related to delayed recall performance. Overall, these data extend the observation of hippocampus change to the GDS = 3 group. Moreover, because the entire temporal lobe was studied, these data also support the potential anatomic specificity of an age-related hippocampal-delayed recall relationship.

The results also clearly showed that lateral temporal lobe volume losses were a necessary condition for the symptoms of dementia. Using logistic regression to classify the minimally impaired and AD patients, the lateral temporal lobe measurements significantly added over and above the medial temporal lobes. The medial temporal lobes correctly classified 67% of patients and inclusion of the lateral regions raised this value to 80%. In the classification of normal and AD patients, the medial temporal lobe correctly classified 87% and with addition of the lateral temporal lobe the classification significantly improved to 91% correct. In the classification of the normal and the minimal groups, the lateral temporal did not increase the classification accuracy.

CONCLUSIONS

In vivo examination of the hippocampal region appears to be of potential diagnostic and predictive value in the clinical study of AD. The data indicate that hippocampal changes appear early in the natural history of AD. When such degenerative changes are present, they are associated with the mild memory and cognitive deficits so often ascribed to individuals at risk for AD. In very mildly impaired individuals, HA appears to be related to the AD process, a conclusion in part supported by both postmortem studies[9] and our neuroimaging studies.[20,22] It is universally observed that hippocampal formation atrophy occurs in AD at postmortem examination, and we now extend these observations to show a remarkably consistent *in vivo* determined relationship. Moreover, the imaging data show the consistent involvement of the lateral temporal lobe in association with dementia. These findings now permit the identification of nondemented patients at increased risk for AD, using a marker relatively free of education and cultural bias.

REFERENCES

1. SCHOENBERG, B. S. 1986. Epidemiology of Alzheimer's disease and other dementing disorders. J Chronic Dis. **39:**1095–1104.
2. CLARK, R. F. & GOATE, A. M. 1993. Molecular genetics of Alzheimer's disease. Arch. Neurology. **50:**1164–1172.
3. PAYKEL, E. S., C. BRAYNE, F. A. HUPPERT, C. GILL, C. BARKLEY, E. GEHLHAAR, L. BEARDSALL, D. M. GIRLING, P. POLLITT & D. O'CONNOR. 1994. Incidence of dementia in a population older than 75 years in the United Kingdom. Arch. General Psychiatry **51:**325–332.
4. LARRABEE, G. J. & T. H. CROOK. 1994. Estimated prevalence of age-associated memory impairment derived from standardized tests of memory function. Int. Psychogeriatrics **6:**95–104.
5. BALL, M. J., V. HACHINSKI, A. FOX, A. J. KIRSHEN, M. FISMAN, W. BLUME, V. A. KRAL & H. FOX. 1985. A new definition of Alzheimer's disease: A hippocampal dementia. Lancet **5:**14–16.
6. KEMPER, T. 1984. Neuroanatomical and neuropathological changes in normal aging and in dementia. *In* Clinical Neurology of Aging. M. L. Albert, Ed. New York: Oxford University Press.
7. HYMAN, B. T., G. W. VAN HOESEN, A. R. DAMASIO & C. L. BARNES. 1984. Alzheimer's disease: Cell-specific pathology isolates the hippocampal formation. Science **225:**1168–1170.

8. BRAAK, H. & E. BRAAK. 1991. Neuropathological staging of Alzheimer related changes. Acta Neuropathological **82:**239–259.
9. PRICE, J. L., P. B. DAVIS, J. C. MORRIS & D. L. WHITE. 1991. The distribution of tangles, plaques and related immunohistochemical markers in healthy aging and Alzheimer's disease. Neurobiol. Aging **12:**295–312.
10. DE LA MONTE, S. M. 1989. Quantitation of cerebral atrophy in preclinical and end-stage Alzheimer's disease. Ann. Neurol. **25:**450–459.
11. SQUIRE, L. R. 1992. Memory and the hippocampus: A synthesis from findings with rats, monkeys, and humans. Psychol. Rev. **99:**195–231.
12. COLSHER, P. L. & R. B. WALLACE. 1991. Longitudinal application of cognitive function measures in a defined population of community-dwelling elders. Ann. Epidemiol. **1:**215–230.
13. PETERSEN, R. C., G. SMITH, E. KOKMEN, R. J. IVNIK & E. G. TANGALOS. 1992. Memory function in normal aging. Neurology **42:**396–401.
14. WELSH, K., N. BUTTERS, J. HUGHES, R. MOHS & A. HEYMAN. 1991. Detection of abnormal memory decline in mild cases of Alzheimer's disease using CERAD neuropsychological measures. Arch. Neurol. **48:**278–281.
15. MESULAM, M. M. 1990. Large-scale neurocognitive networks and distributed processing for attention, language, and memory. Neurol. Prog. **28:**597–613.
16. DE LEON, M. J., A. E. GEORGE, S. H. FERRIS, D. R. CHRISTMAN, J. S. FOWLER, C. GENTES, J. BRODIE, B. REISBERG & A. P. WOLF. 1984. Positron emission tomography and computed tomography assessments of the aging human brain. J. Computer Assisted Tomography **8:**88–94.
17. CUTLER, N. R., R. DUARA, H. CREASEY, C. L. GRADY, J. V. HAXBY, M. B. SCHAPIRO & S. I. RAPOPORT. 1984. Brain imaging: Aging and dementia. Ann. Internal Med. **101:**355–369.
18. DE LEON, M. J., A. E. GEORGE, B. REISBERG, S. H. FERRIS, A. KLUGER, L. A. STYLOPOULOS, J. D. MILLER, M. E. LA REGINA, C. CHEN & J. COHEN. 1989. Alzheimer's disease: Longitudinal CT studies of ventricular change. Am. J. Neuroradiol. **10:**371–376.
19. SEAB, J. P., W. S. JAGUST, S. F. S. WONG, M. S. ROOS, B. R. REED & T. F. BUDINGER. 1988. Quantitative NMR measurements of hippocampal atrophy in Alzheimer's disease. Mag. Resonance in Med. **8:**200–208.
20. DE LEON, M. J., A. E. GEORGE, L. A. STYLOPOULOS, G. SMITH & D. C. MILLER. 1989. Early marker for Alzheimer's disease: The atrophic hippocampus. The Lancet **2:**672–673.
21. JACK, C. R., Jr., R. C. PETERSEN, P. C. O'BRIEN & E. G. TANGALOS. 1992. MR-based hippocampal volumetry in the diagnosis of Alzheimer's disease. Neurology **42:**183–188.
22. DE LEON, M. J., J. GOLOMB, A. E. GEORGE, A. CONVIT, C. Y. TARSHISH, T. MCRAE, S. DE SANTI, G. SMITH, S. H. FERRIS, M. NOZ & H. RUSINEK. 1993. The radiologic prediction of Alzheimer's disease: The atrophic hippocampal formation. Am. J. Neuroradiol. **14:**897–906.
23. REED, B. R., W. J. JAGUST, J. P. SEAB & B. A. OBER. 1989. Memory and regional cerebral blood flow in mildly symptomatic Alzheimer's disease. Neurology **39:**1537–1539.

24. Powers, W. J., J. S. Perlmutter, T. O. Videen, P. Herscovitch, L. K. Griffeth, H. D. Royal, B. A. Siegal, J. C. Morris & L. Berg. 1992. Blinded clinical evaluation of positron emission tomography for diagnosis of probable Alzheimer's disease. Neurology **42:**765–770.
25. Haxby, J. V., C. L. Grady, E. Koss, B. Horwitz, L. Heston, M. Schapiro, R. P. Friedland & S. I. Rapoport. 1990. Longitudinal study of cerebral metabolic asymmetries and associated neuropsychological patterns in early dementia of the Alzheimer type. Arch. Neurol. **47:**753–760.
26. Smith, G. S., M. J. de Leon, A. E. George, A. Kluger, N. D. Volkow, T. McRae, J. Golomb, S. H. Ferris, B. Reisberg, J. Ciaravino & M. A. La Regina. 1992. Topography of cross-sectional and longitudinal glucose metabolic deficits in Alzheimer's disease: Pathophysiologic implications. Arch. Neurol. **49:**1142–1150.
27. Narkiewicz, O., M. J. de Leon, A. Convit, A. E. George, J. Wegiel, J. Moreys, M. Bobinski, J. Golomb, D. C. Miller & H. M. Wisniewski. 1993. Dilatation of the lateral part of the transverse fissure of the brain in Alzheimer's disease. Acta Neurobiologiae Experimentalis **53:**457–465.
28. Duvernoy, H. M. 1988. The Human Hippocampus: An Atlas of Applied Anatomy. Munchen: J. F. Bergmann Verlag.
29. Reisberg, B., S. H. Ferris, M. J. de Leon & T. Crook. 1982. The global deterioration scale for assessment of primary degenerative dementia. Am. J. Psychiatry **139:**1136–1139.
30. Folstein, M. F., S. E. Folstein & P. R. McHugh. 1975. Mini-mental state: A practical method for grading the cognitive state of patients for the clinician. J. Psychiatry Res. **12:**189–198.
31. Golomb, J., M. J. de Leon, A. Kluger, A. E. George, C. Tarshish & S. H. Ferris. 1993. Hippocampal atrophy in normal aging: An association with recent memory impairment. Arch. Neurol. **50:**967–976.
32. Golomb, J., A. Kluger, M. de Leon, S. Ferris, A. Conuit, M. Mittelman, J. Cohen, H. Rusinek, S. de Santi & A. George. 1994. Hippocampal formation size in normal human aging: A correlate of delayed secondary memory performance. Learning & Memory. **1:**45–54.
33. Convit, A., M. de Leon, C. Tarshish, S. de Santi, A. Kluger, H. Rusinek & A. George. 1995. Specific hippocampal atrophy in dementia not yet diagnosable. The Lancet. **345:**266.

Age-related Changes in Cortical Blood Flow Activation during Perception and Memory

C. L. GRADY[a]

Laboratory of Neuroscience, National Institute on Aging, National Institutes of Health, Bethesda, Maryland 20892 USA

ABSTRACT: Although many cognitive functions are affected by age, some are relatively maintained. There also are numerous age-related changes in brain structure, but not much is known about how these changes impact upon the alterations seen in cognition. In order to understand the basis for the reductions and sparings of cognitive function in the aged, experiments were carried out to compare regional cerebral blood flow (rCBF) and performance in young and old subjects on visual perceptual and memory tasks. In the first experiment, which examined the perception of faces and spatial locations, old subjects were as accurate as young subjects, and both groups had rCBF activation in fusiform gyrus during face matching and in superior parietal cortex during location matching. However, old subjects had less activation of prestriate cortex and more activation of frontal and lateral temporal cortex than did young subjects. The second experiment tested recognition memory for faces, a task on which old subjects were impaired. Young subjects had rCBF activation in left prefrontal and inferior temporal cortex and in right hippocampus during encoding of the faces, and in right prefrontal and parietal cortex during recognition. Old subjects showed no significant activation of the areas involved in encoding, but did have increased rCBF in right prefrontal cortex during recognition. These results suggest that the aging brain can demonstrate alterations of the functional systems involved in some cognitive processes, such as perception, that may serve as a compensatory mechanism to maintain performance. Failure to show such compensation, coupled with dysfunction of the areas primarily involved in processing may lead to more marked deficits in performance.

INTRODUCTION

Although much is known about structural and physiological changes that occur in the brain with aging and how cognitive functions are altered with age, the relation between changes in specific brain areas and reductions in cognitive performance is poorly understood. In terms of structural changes,

[a] *Send correspondence to:* Dr. Cheryl L. Grady, NIA/LNS, 10/6C414, 9000 Rockville Pike, Bethesda MD, 20892; Tel: (301)594-7756; Fax: (301)402-0595; Email: grady@alw.nih.gov.

the finding of cerebral atrophy in both gray and white matter with increasing age has been a relatively universal one in the literature, both in autopsy specimens,[1] and structural neuroimaging studies.[2] More important for studies of cognitive aging is the trend towards greater loss of cells in association cortex compared to primary cortex,[3] and neuronal loss in medial temporal regions important for memory, including the hippocampus.[4] Some age-related changes, particularly those in dendrites or synapses, may represent changes to compensate for cell loss. An example is dendritic branching in the hippocampus, which has been found to increase in older brains compared to middle-aged brains, and to decrease in the very old.[5] Similarly, in the monkey, dendrites in prefrontal cortex were found to be increased in old monkeys and to be decreased in very old animals.[6] Some cognitive functions are affected by age, whereas others are relatively maintained or unchanged. For example, some aspects of language processing are reported to be affected by age, such as fluency[7] and discourse,[8] whereas naming is not found consistently to change with age.[9,10] There also are age-related decreases in some types of memory performance, such as episodic and working memory, which are markedly affected by aging.[11] Implicit memory tasks, such as priming, are not affected by age.[12]

The focus of the work presented here is an attempt to elucidate the neurobiological basis for age-related changes in cognition, using both perceptual and memory paradigms. Positron emission tomography (PET), was used to measure regional cerebral blood flow (rCBF) during visual perceptual and visual memory tasks to determine the brain regions involved in these tasks in young and in old subjects, and how the activation patterns differed between groups. The study of various kinds of visual tasks has allowed us to examine functions that are relatively unchanged in older subjects (perception) as well as those that show reductions with age (memory). One hypothesis that we have tested in these experiments is that when old subjects are engaged in performing those types of processing that are relatively spared, they may show reduced rCBF activation in some critical areas but also will show alterations in rCBF activation indicative of compensatory changes. On the other hand, when performing those functions that are impaired, older individuals will demonstrate reduced activation of regions involved in task performance and fail to show compensatory reorganization.

METHODS

These experiments were conducted with $H_2{}^{15}O$ using a tomograph that collects data from 15 planes simultaneously and has 6.5 mm resolution (Scanditronix PC2048-15B). The tasks were begun one minute prior to injection of the isotope and continued throughout the 4-minute scanning period.

Arterial blood samples were collected during each scan for the calculation of CBF in ml/100g/min. In the perception experiment, three tasks were performed by the subjects: a sensorimotor control task, a face matching task, and a location matching task.[13] All tasks were forced-choice, match-to-sample tasks, with sample and choice stimuli presented simultaneously. In the face-matching task the subject was asked to indicate which of two choice faces was a picture of the same person as shown in the sample. One of the choice faces was a distractor and one was a picture of the person in the sample stimulus taken from a different angle or under different lighting conditions. The response consisted of pressing a button with either the right or left thumb, depending on whether the correct stimulus match was on the right or left side of the stimulus array. For the location task the subjects saw three squares which contained a smaller square (filled with a noise pattern) located in different positions within the larger square. The task was to indicate which of the choice squares had the inner object in the same location as that seen in the sample square. The sensorimotor control task consisted of the same noise pattern used in the location task but centered inside the larger square, and the subject was required simply to alternate right and left button presses in response to the presentation of the stimulus. In the memory experiment, subjects performed three tasks in the following order: memorizing a list of faces (encoding), face matching (similar to the task used in the perception experiment), and face recognition. During the encoding task, subjects were shown a set of 32 unfamiliar faces and asked to memorize them. Each face was shown for 4 sec. and the entire set was shown three times in a different order each time. During the face recognition task, subjects were shown two choice faces in each trial and asked to indicate which face had been seen previously in the set of 32 faces. In addition, a sensorimotor control task, identical to the one used in the perception experiment, was performed at the beginning and end of the scanning session.

The serial PET scan design allows within-subject comparisons of the control task and the cognitive tasks of interest, and examination of the interactions of group and task on rCBF activation. Image data were analyzed using Statistical Parametric Mapping.[14] The analysis involves the transformation of each subject's images to correspond to the stereotactic atlas space of Talairach and Tournoux,[15] correction for global rCBF effects, and pixel-by-pixel comparisons (t-tests) among the various tasks to determine the specific areas of activation in the brain that underlie each cognitive process of interest. Between-group comparisons, to assess age-related differences in rCBF activation during encoding and recognition, also were carried out.

RESULTS

In the perception experiment we found that old subjects' performance was relatively maintained; *i.e.*, they performed as accurately as young subjects, but were slower in responding.[16] In young subjects, cortex in the ventral occipitotemporal region (fusiform gyrus) was selectively activated during the perception of faces, and superior parietal cortex was activated during the perception of location.[17] Old subjects showed a similar pattern of activation in these regions of cortex during the two tasks.[16] However, in both tasks, young subjects showed greater activation of prestriate cortex, and old subjects had larger rCBF increases in lateral occipitotemporal cortex and prefrontal cortex (FIG. 1). In addition, during location matching, inferior parietal cortex was more activated in the old subjects.

The memory experiment[18] provides an interesting contrast. Both groups had increased reaction time during the recognition task, compared to the matching task, but there were no age differences in reaction time. However, unlike the perceptual tasks, the old subjects were significantly impaired on

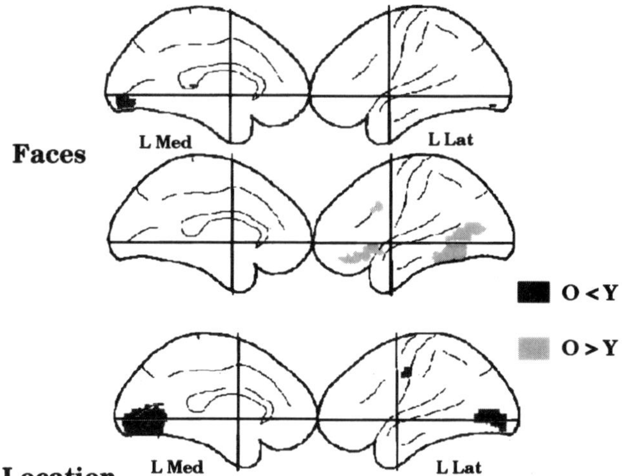

FIGURE 1. Areas of significant differences in rCBF activation ($p < 0.01$) between young and old subjects during the perception of faces and locations, shown on schematics of the lateral and medial surfaces of the left hemisphere (the results were similar for the right hemisphere). The darker shaded areas are those where young subjects had greater rCBF increases during face matching and location matching (compared to the control task) than did older subjects. These areas are primarily in prestriate occipital cortex during both tasks. The lighter shaded areas are those where old subjects had greater rCBF increases during face matching and location matching (compared to the control task) than did young subjects. These areas are in lateral occipitotemporal and prefrontal cortex for both tasks and additionally in inferior parietal cortex during location matching. (Data are from Grady *et al.*[16]).

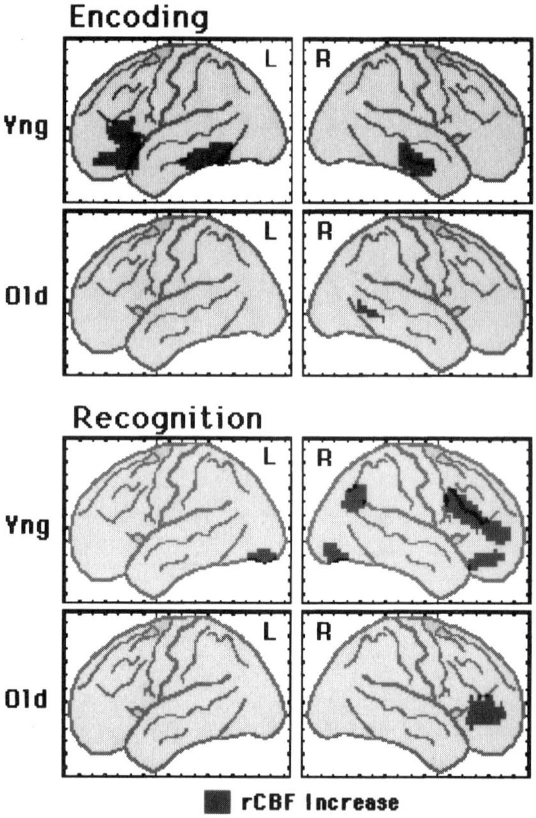

FIGURE 2. Areas of increased rCBF (compared to both the control and face matching tasks; $p < 0.01$) during encoding and recognition of faces in young and old subjects shown on lateral schematics of the left and right hemispheres. During encoding young subjects had activation in left prefrontal and inferior temporal cortex, and in right medial temporal cortex, including the hippocampus. The old subjects had activation only in a region of posterior temporal cortex in the right hemisphere. During recognition, young subjects showed activation in right prefrontal, right parietal and bilateral ventral occipital cortex. Old subjects had activation in right prefrontal cortex.

recognition, showing only 66% correct compared to the 80% correct performance of the young subjects. During encoding, young subjects showed increased rCBF in right hippocampus and in left inferior prefrontal and temporal cortex, and during recognition, they showed activation of right inferior prefrontal and superior parietal cortex, and bilateral ventral occipital cortex (FIG. 2). The old subjects had a strikingly different rCBF activation pattern. They showed no significant activation in those areas that were involved in encoding in young subjects. In addition, although they did show right frontal activation during recognition, similar to that seen in young subjects, they failed to activate the parietal and occipital areas that participate in recognition.

DISCUSSION

The results from these two experiments have important implications for the effects of aging on cognitive function. The results of the perception experiment show that there are separate cortical areas that process object and spatial visual information, and that these two pathways function similarly in both young and old subjects. However, the rCBF patterns were not identical, but showed differences suggestive of more efficient use of occipital visual areas by younger subjects and more reliance by older subjects on different cortical networks, which include areas in frontal cortex, perhaps to compensate for reduced processing efficiency of occipital cortex. Thus, in visuoperceptual tasks, there may be a functional reorganization of cortical networks that allows the maintenance of perceptual performance, although at a cost of reduced response speed.

There also are dissociable neural systems participating in encoding and recognition. During encoding, the active regions are primarily in the left hemisphere, although the right medial temporal region is also involved, and, during recognition, right hemisphere areas are activated. These patterns of rCBF activation are disrupted in older individuals, particularly during encoding. The lack of activation in those areas responsible for encoding suggests that impaired recognition performance in older individuals is due, at least in part, to a failure to encode the faces sufficiently. In addition, there was no activation of additional areas in the old subjects, as there had been in the perception experiment, suggesting that there were no areas activated to compensate for dysfunction of either the encoding or recognition network. Thus, when encoding and retrieving new material, old subjects showed both a functional failure in critical brain areas, and little evidence of a compensatory mechanism to counteract this failure. This finding, together with that of the perceptual experiment, suggests that compensatory changes in cortical networks can be beneficial in maintaining cognitive performance, and that a lack of such changes, coupled with reduced activity of critical areas, will lead to a more marked age-related reduction in cognitive performance.

ACKNOWLEDGMENTS

The work reviewed here represents a collaborative effort with Drs. James Haxby, Barry Horwitz, Leslie Ungerleider, Mortimer Mishkin, and Jose Maisog. The author would also like to thank the staff of the NIH PET Imaging Section for their assistance in conducting the experiments.

REFERENCES

1. ANDERSON, J. M., B. M. HUBBARD, G. R. COGHILL & W. SLIDDERS. 1983. The effect of advanced old age on the neurone content of the cerebral cortex. J. Neurol. Sci. **58:**233–244.

2. COFFEY, C. E., W. E. WILKINSON, I. A. PARASHOS, S. A. SOADY, R. J. SULLIVAN, L. J. PATTERSON, G. S. FIGIEL, M. C. WEBB, C. E. SPRITZEN & W. T. DJANG. 1992. Quantitative cerebral anatomy of the aging human brain: a cross-sectional study using magnetic resonance imaging. Neurology **42:**527–536.
3. KEMPER, T. 1984. Neuroanatomical and neuropathological changes in normal aging and in dementia. *In* Clinical Neurology of Aging. M. Albert, Ed.: 9–52. New York: Oxford University Press.
4. BALL, M. J. 1977. Neuronal loss, neurofibrillary tangles and granulovacuolar degeneration in the hippocampus with aging and dementia. Acta Neuropathol. **37:**111–118.
5. FLOOD, D. G., S. J. BUELL, C. H. DEFIORE, G. J. HORWITZ & P. D. COLEMAN. 1985. Age-related dendritic growth in dentate gyrus of human brain is followed by regression in the 'oldest old'. Brain Res. **345:**366–368.
6. CUPP, C. J. & E. UEMURA. 1980. Age-related changes in prefrontal cortex of macaca mulatta: Quantitative analysis of dendritic branching patterns. Exp. Neurol. **69:**143–163.
7. MCCRAE, R. R., D. ARENBERG & P. T. COSTA. 1987. Decline in divergent thinking with age: Cross-sectional, longitudinal and cross-sequential analyses. Psychol. Aging **2:**130–132.
8. BAYLES, K. A., C. K. TOMOEDA & D. R. BOONE. 1985. A view of age-related changes in language function. Dev. Neuropsychol. **1:**231–264.
9. BOROD, J. D., H. GOODGLASS & E. KAPLAN. 1980. Normative data on the Boston Diagnostic Aphasia Examination, Parietal Lobe Battery, and the Boston Naming Test. J. Clin. Neuropsychol. **2:**209–215.
10. LABARGE, E., D. EDWARDS & J. W. KNESEVICH. 1986. Performance of normal elderly on the Boston Naming Test. Brain Lang. **27:**380–384.
11. CRAIK, F. I. M., R. G. MORRIS & M. L. GICK. 1990. Adult age differences in working memory. *In* G. Vallar & T. Shallice, Eds.: 247–267. Neuropsychological Impairments of Short-term Memory. Cambridge: Cambridge University Press.
12. NEBES, R. D., F. BOLLER & A. HOLLAND. 1986. Use of semantic context by patients with Alzheimer's disease. Psychol. Aging. **1:**261–269.
13. HAXBY, J. V., C. L. GRADY, B. HORWITZ, L. G. UNGERLEIDER, M. MISHKIN, R. E. CARSON, P. HERSCOVITCH, M. B. SCHAPIRO & S. I. RAPOPORT. 1991. Dissociation of object and spatial visual processing pathways in human extrastriate cortex. Proc. Natl. Acad. Sci. USA **88:**1621–1625.
14. FRISTON, K. J., C. D. FRITH, P. F. LIDDLE & R. S. J. FRACKOWIAK. 1991. Comparing functional (PET) images: The assessment of significant change. J. Cereb. Blood Flow Metab. **11:**690–699.
15. TALAIRACH, J. & P. TOURNOUX. 1988. Co-Planar Stereotaxic Atlas of the Human Brain. New York: Thieme Medical Publishers, Inc.
16. GRADY, C. L., J. M. MAISOG, B. HORWITZ, L. G. UNGERLEIDER, M. J. MENTIS, J. A. SALERNO, P. PIETRINI, E. WAGNER & J. V. HAXBY. 1994. Age-related changes in cortical blood flow activation during visual processing of faces and location. J. Neurosci. **14:**1450–1462.
17. HAXBY, J. V., B. HORWITZ, L. G. UNGERLEIDER, J. M. MAISOG, P. PIETRINI

& C. L. GRADY. The functional organization of human extrastriate cortex: A PET-rCBF study of selective attention to faces and locations. J. Neurosci. **14:** 6336–6353.

18. GRADY, C. L., A. R. MCINTOSH, B. HORWITZ, J. M. MAISOG, L. G. UNGERLEIDER, M. J. MENTIS, P. PIETRINI, M. B. SCHAPIRO & J. V. HAXBY. 1995. Age-related reductions in human recognition memory due to impaired encoding. Science **269:**218–221.

Near Infrared Spectroscopy in the Diagnosis of Alzheimer's Disease[a]

C, HOCK,[b,e] K. VILLRINGER,[c] F. MÜLLER-SPAHN,[b]
M. HOFMANN,[b] S. SCHUH-HOFER,[b] H. HEEKEREN,[b]
R. WENZEL,[d] U. DIRNAGL,[d] AND A. VILLRINGER[d]

[b] *Department of Psychiatry, University of Munich, 80336 Munich, Germany*
[c] *Department of Nuclear Medicine, Technical University of Munich, Munich, Germany*
[d] *Department of Neurology, University of Berlin (Charité), Berlin Germany*

ABSTRACT: Near infrared spectroscopy (NIRS) is a new technique that permits noninvasive monitoring of cerebral blood and tissue oxygenation. Recently, we and others have shown that NIRS measurements are sensitive enough to follow changes in cerebral hemoglobin oxygenation due to activation of brain function. Based on these findings we have assessed the influence of aging as well as the influence of neurodegeneration on cerebral hemoglobin oxygenation during mental work. The typical NIRS pattern in young healthy subjects while performing calculation tasks measured in the frontal cortex were increases in oxygenated hemoglobin [HbO_2] and total hemoglobin [HbT] while reduced hemoglobin [HbR] decreased. Elderly healthy subjects showed a significant lower mean increase in [HbO_2] and [HbT] levels. Regression analysis revealed an age-dependent decline in activation-induced local increase of [HbO_2] as well as [HbT]. Furthermore, we monitored changes in cerebral hemoglobin oxygenation in the frontal cortex while patients with probable Alzheimer's disease (AD) performed cognitive tasks. Whereas elderly healthy subjects (as well as patients with major depression, age-associated memory impairments or vascular dementia) again showed clear increases in the local concentrations of [HbO_2] and [HbT] during brain activation, AD patients showed significant decreases compared to the baseline levels in both variables that were most pronounced in the parietal cortex. To clarify whether the different patterns in cerebral hemoglobin oxygenation during cognitive activation were due to an altered functional brain organization in AD or to alterations in the cerebrovascular response to neuronal activation, we are currently performing simultaneous NIRS and (015-H20-)PET measurements during performance of a cognitive task (Stroop test). Our finding of a regional reduced oxygen supply during activation of brain function may be of relevance to the development and the time course of neurodegeneration.

[a] A. V. and U. D. are supported by the Deutsche Forschungsgemeinschaft, Gerhard Hess Programm.

[e] *Send correspondence to:* Dr. Christoph Hock, Department of Psychiatry, University of Basle, Wilhelm-Klein-Str. 27, CH-4025 Basel, Switzerland; Tel: +41-61-325-3111; Fax: +41-61-32508-5514.

INTRODUCTION

Functional neuroimaging techniques, such as positron emission tomography (PET), have increased our knowledge on brain metabolism in normal brain function and neurodegeneration. In addition, there is increasing interest in optical methods that are able to detect signals related to neuronal activity. Near infrared spectroscopy (NIRS) uses the relative transparency of tissue to light in the near infrared (700–900 nm). This near IR "optical window" permits the noninvasive measurement of endogenous chromophore concentrations (*e.g.* hemoglobin or cytochrome aa3) through the intact head and has been applied in sick newborn infants.[1] The NIRS variables are oxygenated hemoglobin [HbO_2], deoxygenated hemoglobin [HbR] and total hemoglobin [HbT] (= [HbO_2] + [HbR]) (for review see ref. 2). Assuming that hematocrit is constant, the changes in HbT are used as an indicator of alterations in cerebral blood volume.[3] Quantitation of another NIRS variable, cytochrome aa3, is a focus of research in several groups at the moment. We and others have recently been able to demonstrate that NIRS, applied as a simple bedside technique, is able to detect changes in hemoglobin oxygenation during brain activation.[4–7] Therefore we were interested in the influence of aging and neurodegeneration on the pattern of cerebral hemoglobin oxygenation during mental work.

PRESENT IMPLEMENTATION OF THE NEAR INFRARED METHODOLOGY

In our investigations we applied the NIRO 500 system (Hamamatsu Photonics K. K.) that uses the difference of absorption spectra of [HbO_2] and [HbR] in the near infrared. The measurements were performed in the reflection mode. The light from the Laser diodes (wavelengths 775, 825, 850, 904 nm) is guided through a fiber optic bundle, the end of which (the optode) is placed strictly at the same site in each subject on the head (left or right forehead (frontal cortex) in the Fp2 or Fp1 position according to the international EEG 10-20 system, or on the left or right parietal region (parietal cortex, superior parietal area) 2 cm rostral to P3 or P4. Another optode leading to the photomultiplier is placed at a distance of 4 cm in a horizontal line lateral to the first optode. The brain volume measured corresponds to a banana-shaped figure beneath the two optodes placed on the head surface.[7] Light source: the laser diodes are pulsed at 1.9 kHz, pulse width: approx. 100 nsec. Light detector: the reflected light is conducted by an optical fiber to a photomultiplier tube, which is connected to a multichannel photon counter, the number of photons at each wavelength is compared with the light output of the lasers.[9] The analysis procedure converts the obtained optical densities (OD) to concentrations of oxygenated and deoxygenated hemoglobin, expressed in millimoles per liter per optical pathlength ($mmol \times l^{-1} \times cm$), using Beer-Lambert's law and an algorithm developed by Wray *et al.*[10] The sampling time for each photon count was 2 seconds. We have estimated the pathlength by multiplying the interoptode distance (4 cm) with the differential pathlength factor (DPF) for the adult head (5.93).[11] All values were averaged and described as mean +/− standard deviation (SD). Assuming that this pathlength estimate is correct, these "arbitrary concentration units" correspond to μM.

AGE-DEPENDENCY OF CHANGES IN CEREBRAL HEMOGLOBIN OXYGENATION DURING BRAIN ACTIVATION

Age-related changes in cerebral oxygen metabolism have been demonstrated in PET studies.[12,13] Our purpose was to evaluate the sensitivity of the NIRS technique to assess differences during normal aging. We hypothesized that the increase of brain tissue oxygenated hemoglobin and total hemoglobin as a response to brain activation was age-dependent (for details see ref. 14).

We examined 29 human subjects, including 12 healthy young volunteers (mean age 28 +/− 4 years, 8 females, 4 males), 17 healthy elderly volunteers (mean age 52 +/− 10 years, 10 females, 7 males). The elderly subjects were carefully examined: The medical screening included a general laboratory screening (including hematocrit), ECG, EEG as well as a physical and neurological examination. All subjects were healthy and free of medication. Informed consent was taken from all subjects before each investigation. Subjects were lying in a convenient position throughout the whole examination. The examination included a period of rest (2 min), a period of cognitive stimulation (subjects had to perform a calculation task, the period of activation lasted 2.75 +/− 1.30 min) and a following period of rest (2 min). Subjects had their eyes closed during the whole experiment. The calculations (such as 100-3, 97-3, *etc.*) were performed aloud, one test series was done per person. The calculation task was chosen because in previous studies we have observed typical changes in NIRS variables ([HbO_2] and [HbT] increase, [HbR] decrease) in young healthy volunteers. Subtracting the mean value for [HbR] as well as for [HbO_2] during the period of rest before the start of the activation period (baseline value) from the mean value for [HbR] as well as for [HbO_2] during the activation period, results in the *change* in [HbR] as well as [HbO_2]. [HbR] and [HbO_2] summed equals [HbT]. We compared the groups on these three variables using two-tailed t-tests.

The typical time course of the NIRS variables ([HbO_2], [HbR] and

[HbT]) during cognitive stimulation in the young and elderly healthy volunteers were an increase in [HbO_2] and [HbT] and a slight decrease in [HbR] shortly after the beginning of the calculation period. The mean increases (given in arbitrary units +/− SD) in the young subjects were: [HbO_2] 2.36 +/− 1.07, [HbR] −0.11 +/− 0.48 and [HbT] 2.24 +/− 1.13. The comparison of young and elderly subjects revealed that the mean increase in [HbO_2] and [HbT] during brain activation was much less pronounced in the elderly ($p < 0.05$). The mean changes in the elderly subjects were [HbO_2] 1.21 +/− 1.38, [HbR] −0.50 +/− 0.72 and [HbT] 0.72 +/− 1.42. Plotting the [HbO_2]-increases and the [HbT]-increases against age (years), significant correlations between age and [HbO_2] ($y = -0.241x + 20.062$: $r = -0.431$, $p < 0.05$) as well as between age and [HbT] ($y = -0.346x + 22.496$: $r = -0.568$, $p < 0.05$) was demonstrated (Pearson type of correlation).

In summary, using NIRS we have shown, that the increase in cerebral hemoglobin oxygenation in response to brain activation declines with physiological aging. Both, the mean increases in oxygenated hemoglobin as well as total hemoglobin during cognitive activation were less pronounced in older than in younger subjects. This age-dependent decline may be due, first, to activation of other brain regions during performance of the calculation task due to an altered functional brain organization, or secondly, to an alteration of coupling between brain cell activity and blood flow. Such an alteration in the coupling may attenuate the (physiological) temporary mismatch between oxygen delivery and consumption that probably is responsible for the increase in [HbO_2] during brain activation.[15]

DECREASE IN CEREBRAL HEMOGLOBIN OXYGENATION DURING MENTAL WORK IN PATIENTS WITH ALZHEIMER'S DISEASE

Several positron emission tomography (PET) studies have shown reduced regional cerebral blood flow, metabolic rate of oxygen and glucose metabolism in AD patients in resting states (for review see ref. 16). We assessed activation-induced changes in cerebral hemoglobin oxygenation measured by NIRS in AD patients compared to elderly healthy controls.

We examined 38 human subjects, including 19 healthy elderly volunteers (mean age 67 +/− 10 years, 14 females, 5 males), 19 patients with probable moderate AD, diagnosis according to the NINCDS criteria (17) (mean age 71 +/− 10 years, 11 females, 8 males, Mini Mental State: 20 +/− 4). Informed consent was taken from all subjects before each investigation.

We again observed the typical time courses of the NIRS variables during cognitive stimulation in the frontal region (prefrontal cortex) and parietal region (superior parietal cortex) of the healthy elderly volunteers. Shortly after the beginning of the activation period the healthy subjects showed an increase in [HbO_2] and [HbT] and a slight decrease in [HbR]. The mean increase in the healthy elderly subjects (n = 19) was: [HbO_2] 0.42 +/− 0.17, [HbT] 1.58 +/− 1.40. In contrast to that, most of the patients with AD showed a decrease in [HbO_2] and [HbT] levels shortly after the beginning of the verbal fluency task. The decreases were more pronounced in the parietal than in the frontal cortex. Compared to healthy elderly controls the AD patients (n = 19) showed a significant decrease in the local concentration (parietal cortex) of oxygenated hemoglobin [HbO_2] (mean (μM) +/− SEM, −0.87 +/− 0.37, $p < 0.05$) and total hemoglobin [HbT] (−1.20 +/− 0.45, $p < 0.05$) while performing a verbal fluency task. The mean increase in the elderly healthy subjects (n = 19) was: [HbO_2] 0.42 +/− 0.17, [HbT] 1.58 +/− 1.40. Interestingly, patients with only mild memory disturbances ("age-associated memory impairments") as well as patients with major depression and also patients with vascular dementia were, at least the limited number measured so far, not different from the healthy elderly controls.

One of the possible interpretations of these data may be that in degenerating brain areas there is a reduction of hemoglobin oxygenation during cognitive tasks in favor of other healthier brain regions due to an altered functional brain organization. Grady *et al.*[18] described an additional frontal activation during performance of an object recognition task in AD patients, possibly reflecting this phenomenon. Secondly, the mechanism of coupling between brain cell activity and blood flow may be altered during neurodegeneration. Both mechanisms may lead to a regional reduced supply with oxygen during activation of brain function. Thirdly, anatomical changes during aging and neurodegeneration may affect the optical properties of the brain (enlargement of the subarachnoid space, brain atrophy, changes in white/grey matter ratio).

LIMITATIONS OF THE PRESENT NIRS METHODOLOGY

The present implementation of the NIRS method is hampered by several limitations. First, the absolute baseline concentration of [HbR] and [HbO_2] cannot be obtained. This would require the exact quantification of the optical pathlength of the near infrared light in brain tissue in each experiment and for each wavelength. Methods to solve this problems are being developed using time domain methods (combination with simultaneous photon time of flight measurements) or frequency domain methods (intensity modulation). Because the absolute baseline value is not known the measurements at the moment refer only to *changes* in absolute concentrations starting from an estimated baseline level. Secondly, the spatial resolution is very poor and restricts the signal to a small brain volume beneath the optodes placed on the head surface.[7] Therefore, only one of the brain regions that is activated during the cognitive task can be detected. However it would be interesting to detect regional specific changes in hemoglobin oxygenation in response to various mental tasks in healthy controls as well as in patients with neurodegenerative disorders in different stages of the disease. This issue may be addressed by multichannel near-infrared optical imaging techniques that are under development.[19] Finally, possible contaminations by chromophore-containing extracranial tissue, such as muscle or skin, have to be considered. The contribution of skin blood flow to the signal can be regarded as minor,[6] but regions with extracranial muscle at the site of measurement have to be avoided.

FUTURE DIRECTIONS

In order to clarify the underlying pathophysiology of the described alterations in cerebral hemoglobin oxygenation in aging and AD, more direct measurements of neuronal brain activity and intracellular metabolism have to be performed simultaneously with the measurements of hemoglobin oxygenation. This goal could be achieved by using electrophysiological techniques (MEG, EEG-brain mapping) and methods to assess intracellular metabolism (MR-spectroscopy). By measuring the intracellular cytochrome-oxidase concentration simultaneously with hemoglobin oxygenation, an improved NIRS method may also be useful to address this issue. In addition, simultaneous measurement of [HbO_2] by NIRS (which has good temporal but poor spatial resolution) and regional cerebral blood flow (rCBF) by PET (which has good spatial but poor temporal resolution) or reduced hemoglobin by functional MRI may clarify the underlying pathophysiological events. In collaboration with the Department of Nuclear Medicine (Director: M. Schwaiger), Technical University of Munich, we are currently performing simultaneous NIRS and ($O15-H_2O$) PET measurement during brain activation. Preliminary results show that in 4 of 5 healthy subjects[20] and 8 of 10 patients with probable AD, PET rCBF changes were paralleled by changes in the concentration of [HbO_2] as measured by NIRS.

In AD a marked reduction of regional cerebral blood flow and cerebral hemoglobin oxygenation may occur during activation of brain function mainly in degenerating brain areas, such as the parietal cortex. This finding of a regional reduced oxygen supply during activation of brain function may be of relevance to the development and the time course of neurodegeneration.

ACKNOWLEDGMENTS

The authors would like to thank L Schleinkofer, Hamamatsu Photonics, Germany, for continuous support as well as R.R. Engel for help in statistical analysis.

REFERENCES

1. WYATT, J. S., M. COPE, D. T. DELPY, S. WRAY & E. O. REYNOLDS. 1986. Quantification of cerebral oxygenation and haemodynamics in sick newborn infants by near infrared spectrophotometry. Lancet **2:**1063–1066.
2. CHANCE, B. 1991. Optical method. Annu. Rev. Biophys. Chem. **20:**1–28.
3. WYATT, J. S., M. COPE, D. T. DELPY, C. E. RICHARDSON, A. D. EDWARDS, S WRAY & E. O. REYNOLDS. 1990. Quantification of cerebral blood volume in human infants by near infrared spectroscopy. J. Appl. Physiol. **68:**1086–1091.
4. CHANCE, B., Z. ZHUANG, C. UNAH, C. ALTER & L. LIPTON. 1993. Cognition-activated low frequency modulation of light absorption in human brain. Proc. Natl. Acad. Sci. USA. **90:**3770–3774.
5. HOSHI, Y. & M. TAMURA. 1993. Detection of dynamic changes in cerebral oxygenation coupled to neuronal function during mental work in man. Neurosci. Lett. **150:**5–8.
6. KATO, T., A. KAMEI, A. S. TAKSHIMA & T. OZAKI. 1993. Human visual cortical function during photic stimulation monitoring by means of near-infrared spectroscopy. J. Cereb. Blood Flow Metabol. **13:**516–520.
7. VILLRINGER A., J. PLANCK, C. HOCK, L. SCHLEINKOFER & U. DIRNAGL. 1993. Near infrared spectroscopy (NIRS), a new tool to study hemodynamic changes during activation of brain function in human adults. Neurosci. Lett. **154(1-2):**101–104.
8. GRATTON, G., J. S. MAIER, M. FABIANI, W. M. MANTULIN & E. GRATTON. 1994. Feasibility of intracranial near-infrared optical scanning. Psychophysiology **31:**211–215.
9. COPE, M. & D. T. DELPY. 1988. System for long term measurement of cerebral blood and tissue oxygenation on newborn infants by near infrared transillumination. Med. Biol. Engl. Comp. **26:**289–294.
10. WRAY, S., M. COPE, D. T. DELPY, J. S. WYATT & E. O. REYNOLDS. 1988. Characterization of the near infrared absorption spectra of cytochrome aa_3 and hemoglobin for the non-invasive monitoring of cerebral oxygenation. Biochem. Biophys. Acta. **933:**184–192.
11. VAN DER ZEE, P., M. COPE, S. R. ARRIDGE, M. ESSENPREIS, L. A. POTTER, A. D. EDWARDS, J. S. WYATT, D. C. MCCORMICK, S. C. ROTH, E. O. R. REYNOLDS & D. T. DELPY. 1992. Experimentally measured optical pathlengths for the adult head, calf and forearm and head of the newborn infant as a function of inter optode spacing. Adv. Exp. Med. Biol. **316:**143–153.
12. LEENDERS, K. L., D. PERANI, A. A. LAMMERTSMA, *et al.* 1990. Cerebral blood flow, blood volume and oxygen utilization. Brain **113:**27–47.
13. MARCHAL, G., P. RIOUX, M. C. PETIT-TABOUÉ, G. SETTE, J. M. TRAVERE, C. LE POEC, P. COURTHEOUX, J. M. DERLON & J. C. BARON. 1992. Regional cerebral oxygen consumption, blood flow and blood volume in healthy human aging. Arch. Neurol. **49:**1013–1020.
14. HOCK, C., F. MÜLLER-SPAHN, S. SCHUH-HOFER, M. HOFMANN, V. DARNAGL & A. VILLRINGER. Age-dependency of changes in cerebral hemoglobin oxygenation during brain activation: A near infrared spectroscopy study. J. Cereb. Blood Flow Metabol. In press.

15. FOX, P. T. & M. E. RAICHLE. 1986. Focal physiological uncoupling of cerebral blood flow and oxidative metabolism during somatosensory stimulation in human subjects. Proc. Natl. Acad. Sci. USA **83:**1140–1144.
16. RAPOPORT, S. I. 1991. Positron emission tomography in Alzheimer's disease in relation to disease pathogenesis: A critical review. Cerebrovasc. Brain Metab. Rev. **3:**297–335.
17. MCKHANN, G., D. DRACHMAN, M. FOLSTEIN, R. KATZMAN, D. PRICE & E. M. STADLAP. 1984. Clinical diagnosis of Alzheimer's disease: Report of the NINCDS- ADRDA Work Group under the auspices of Department of Health and Human Services Task Force on Alzheimer's Disease. Neurology **34:** 939–944.
18. GRADY, C. L., J. V. HAXBY, B. HORWITZ, J. GILLETTE, J. A. SALERNO, A. GONZALEZ- AVILES, R. E. CARSON, P. HERSCOVITCH, M. B. SCHAPIRO & S. I. RAPOPORT. 1993. Activation of cerebral blood flow during a visuoperceptual task in patients with Alzheimer-type dementia. Neurobiol. Aging **14**(1):35–44.
19. HOSHI, Y. & M. TAMURA. 1993. Dynamic multichannel near-infrared optical imaging of human brain activity. J. Appl. Physiol. **75:**1842–1846.
20. VILLRINGER, K., A. VILLRINGER, S. MINOSHIMA, S. ZIEGLER, M. HERZ, S. SCHUH- HOFER, H. OBRIG, C. HOCK, U. DIRNAGL & M. SCHWAIGER. 1994. Frontal brain activation in humans: A combined near infrared spectroscopy and positron emission tomography study. Soc. Neurosci. Abstr.:355.

Functional Imaging Patterns in Alzheimer's Disease

Relationships to Neurobiology[a]

W. J. JAGUST[b]

U.C. Davis Alzheimer's Center, Department of Neurology, University of California, Davis and Center for Functional Imaging, Lawrence Berkeley Laboratory, University of California, Berkeley, Berkeley, California 94720 USA

ABSTRACT: Brain imaging with functional techniques, such as positron emission tomography (PET) and single photon emission computed tomography (SPECT) have been widely applied to the study of dementia. While the pattern of temporal and parietal hypometabolism and hypoperfusion have often been suggested to be of diagnostic utility in ascertaining that a dementia is due to Alzheimer's disease (AD), the exact sensitivity and specificity of this pattern in clinically important situations is unclear. These imaging findings have been of considerable interest, however, in describing the regional patterns of predilection in the disease. Evidence supports the contention that the earliest sites of functional impairment in AD are in the temporal lobes. Surprisingly, however, mesial temporal lobe hypometabolism was difficult to detect in a group of mildly demented AD patients in comparison to a group of healthy older subjects. These results suggest that simple use of mesial temporal lobe metabolic rates as a diagnostic for AD may not be fruitful, and that evaluation of the earliest stages of AD can be most productively studied by investigating healthy older individuals.

IMAGING AND NEUROBIOLOGY IN AD

Brain imaging is one of a limited number of techniques which have been productively applied to the study of living patients with dementia. Structural imaging techniques, such as as x-ray computed tomography (CT) and magnetic resonance imaging (MRI) have demonstrated that patients with dementia, particularly those with clinically diagnosed Alzheimer's disease (AD) show greater cerebral atrophy than comparatively aged nondemented patients.[1] Although the degree of generalized volume loss is related to the degree of cognitive decline of the patient, this generalized atrophy is not specific for Alzheimer's disease and therefore appears to bear an uncertain relationship to

[a] This work was supported by the National Institute on Aging: AG07793 and AG10129.
[b] *Send correspondence to:* Dr. William J. Jagust, Center for Functional Imaging, 55-121, Lawrence Berkeley Laboratory, Berkeley, CA 94720; Tel: (510) 486-6241; Fax: (510) 486-4768.

fundamental neurobiological changes. More specific anatomic changes which have recently been described involve the detection of significant atrophy in mesial temporal lobe structures in AD patients.[2–4] These anatomic changes are of great interest both for reasons of diagnosis, and because they presumably reflect specific pathologic changes which are believed to occur in AD.

Functional brain imaging, using emission tomographic techniques such as single photon emission computed tomography (SPECT) and positron emission tomography (PET) provide clinicians and scientists with three dimensional images of brain function and have also been widely applied to the study of dementia. Most clinical studies using PET have involved quantitative measurements of regional cerebral glucose metabolism (rCMRglc) and regional cerebral blood flow (rCBF), while most SPECT studies have used tracers of rCBF, generally in a semiquantitative ratio of radioactivity in a given brain region normalized to another brain region. Blood flow and metabolism studies have generally shown the same findings in the most common dementia, AD, because in this condition flow and metabolism are coupled.[5] Initial reports using both PET and SPECT identified temporal and parietal association cortices as the brain regions most consistently found to be hypoperfused or hypometabolic in AD.[5,6]

The frequent reports of abnormal function in these cortical regions have prompted a number of investigators and clinicians to examine the utility of these patterns for the diagnosis of the cause of dementia during life. While the question of the clinical utility is still controversial and unresolved, a number of distinct points can be made. First, the pattern of temporal and parietal involvement frequently differentiates AD from other causes of dementia which pathologically or functionally involve the frontal lobe, such as Pick's disease, progressive supranuclear palsy, and frontal lobe dementia.[7] Second, a number of other dementing illnesses may share functional imaging features with AD, specifically vascular dementia and Parkinson's disease with dementia.[8] This overlap may reflect underlying pathological similarities, as both Parkinson's disease and ischemic vascular dementia are frequently associated with Alzheimer neuropathology at autopsy. Finally, neither the sensitivity nor specificity of any pattern for the diagnosis of AD has been well established in a prospective study in which a large sample of unselected dementia cases has been enrolled, followed, and pathologically validated.

In summary, functional images of patients with AD provide considerable information, in the form of patterns of hypometabolism and hypoperfusion, about regional dysfunction. While the diagnostic utility of these techniques is controversial it is clear from the above discussion that a number of interesting parallels between cognitive function, neuropathology, and metabolism or perfusion do exist. For example, patients with frontal lobe dementias display neuropathological abnormalities in frontal association cortex along with behavioral deficits which reflect cognitive functions characteristically associated with damage to these frontal regions.[9] In diagnostic entities which show similar perfusion or metabolic patterns to AD, pathological findings also overlap. Within AD patients as well, good correlations have been described between cognition and regional metabolism or perfusion. Many studies have demonstrated relationships between the regional localization of hypometabolism seen with PET or hypoperfusion seen with SPECT, and the cognitive deficits seen on behavioral testing.[10,11] These sorts of data are consistent with the concept that different regional changes occur in different patients and at different times over the course of AD, and that such changes exert effects on both behavior and functional imaging findings.

Friedland *et al.*[12] suggested in 1985 that the metabolic abnormalities seen in AD patients are a direct reflection of the pattern of regional pathological involvement. Thus, reductions in metabolism in the temporal and parietal lobe are correlated with the localization of the pathology, described by Brun and Englund[33] to severely affect a band through temporal and parietal neocortex and mesial temporal lobes. One hypothesis about the cause of metabolic abnormalities seen in AD, which also relates behavior, pathology, and imaging, is that regional metabolic changes follow regional neuropathological changes. Although there is little direct evidence to support this idea, one study evaluated relationships between neuropathological markers and rCMRglc and found relationships between the density of neurofibrillary tangles and rCMRglc but not between neuritic plaques and glucose metabolism.[14]

One question of particular interest in our laboratory is the determination of which brain regions are initially affected by the functional pathology of AD. This issue has a number of important implications for both clinical medicine and our understanding of the biology of AD. The ability to detect Alzheimer changes in early stages of the disease has important consequences for diagnosis, and as treatments become available, earlier diagnosis becomes increasingly important. A related clinical question involves the determination of which elderly subjects experiencing memory complaints will proceed to develop dementia. Finally, a fundamental concern is related to defining the initial sites of brain involvement in AD. This issue is related to differentiating normal aging from dementia, and assessing key factors in the production of disease symptoms.

Initial studies in our laboratory used SPECT to attempt to define the earliest sites of involvement in AD. We evaluated rCBF in 50 AD patients who were divided into a mildly demented group of 12 (as defined by scores on the mini mental state examination (MMSE) $\geq$ 24) and a moderately demented group of 38 in comparison to 14 control subjects.[15] SPECT data revealed a number of key findings. First, the temporal lobe showed significant reductions in relative rCBF (regional radioactivity values normalized to the values in primary visual cortex) in both hemispheres in both the mildly and moderately demented AD patients in comparison to controls. In parietal cortex, the moderately demented subjects also showed significantly reduced rCBF, but the mildly demented subjects were somewhat intermediate, having significantly reduced flow in only the left hemisphere. In dorsolateral frontal cortex only the moderately demented group showed significant reductions. 75% of mildly demented AD patients had temporal rCBF measurements more than 2 SD below controls, while only 36% of mildly demented patients had similarly low ratios in parietal cortex. For the moderately demented patients, these values were 78% and 61% for temporal and parietal association cortex, respectively. Furthermore, the temporal lobe region was the most sensitive and specific in distinguishing AD patients from controls. Finally, although the temporal lobe region was the most severely involved and the most sensitive in differentiating groups, it was not significantly correlated with dementia severity as measured with the MMSE (Pearson $r = 0.21$). Both dorsolateral frontal cortex and parietal cortex showed significant correlations with the MMSE ($r = 0.51$ and 0.48, respectively). Taken in their entirety, these results suggest that the temporal lobes are the initial brain regions affected by AD, but their use in following the progression of the disease may be limited because of severe involvement of the temporal lobe (*i.e.,* a floor effect) early in the disease process. As the disease advances, the progressive involvement of frontal and parietal cortex provide a reasonably sensitive method of following the course of the disease.

While these findings concerning the temporal lobes are important, the brain regions examined in this study were neocortical. This is a function of the spatial resolution of the particular technique employed. SPECT instrumentation, because of physical limitations related to decreasing count rate sensitivity as resolution increases, is limited in the ability to accurately quantitate radiotracer activity in small brain regions. These limitations are much less significant for PET instrumentation. We employed a tomograph recently constructed in our laboratory, with a resolution of 2.6 mm in plane, to evaluate regions of interest in smaller, limbic brain regions.

It has been known for some time that although the pathology of AD is diffusely distributed in the brain, this distribution is not random but preferentially involves limbic brain regions such as the hippocampal formation and entorhinal cortex. Thus, our investigations using PET were aimed at determining whether hypometabolism in the mesial temporal lobes was the first detectable abnormality in AD. We studied 22 AD patients, 11 mildly demented and 11 moderately demented AD patients (again, using a MMSE score of 24 or greater as evidence of mild dysfunction) and 8 age-matched control subjects.[16] We quantitated absolute metabolic rates in multiple brain regions through the temporal lobes, including mesial temporal cortex (a region actually consisting of hippocampus, amygdala, and entorhinal cortex) anterior temporal cortex (essentially the temporal pole), three neocortical temporal regions (proceeding from a rostral to caudal direction and divided into middle temporal, posterior temporal, and visual association cortex and including most of the temporal neocortex) and primary visual cortex.

While we expected to find that the mildly demented patients showed severe reductions in mesial temporal regions in comparison to controls, in fact the greatest reductions in this group were found in the temporal neocortex.[16] Although the lowest metabolic rates in the mildly demented AD patients were found in the mesial temporal lobes and the temporal poles, these same brain regions also showed the lowest metabolism in the control subjects. In the mildly demented patients, the greatest reductions from controls, and the only statistically significant reductions were found in the neocortex. Widespread reductions, including in mesial temporal lobes, were seen in the moderately demented cases. Interestingly, when the relative involvement of mesial and neocortical temporal lobes was evaluated, there was considerable heterogeneity between AD patients, with some patients showing striking reductions in mesial temporal metabolism relative to neocortical metabolism, and others showing the reverse pattern. The relative involvement of these two brain regions was related to both patient age and cognitive profile, with older patients and more severely amnesic patients showing relatively more severe hypometabolism in the mesial temporal cortex than temporal neocortex.

These PET findings have several important ramifications. First, the implications for clinical practice are that simply imaging mesial temporal lobe function may not provide a useful method of detecting AD in its earliest stages. Although this particular study had several technical shortcomings, most notably the use of axial images rather than coronal images which permit better visualization of mesial temporal lobe structures, these initial results are not encouraging for diagnostic purposes. However, they raise several important questions about the relationship between AD and aging.

Several pathological studies have drawn attention to the prevalence of neurofibrillary tangles in the mesial temporal lobes in older, nondemented individuals, particularly in entorhinal cortex and CA1 of the hippocampus.[17,18] In contrast, these studies have shown that plaque deposition is more often associated with dementia, and even when present in nondemented subjects are usually found in association areas of neocortex. As noted above, DeCarli *et al.*[7] found that metabolic rates were related to tangles, and not plaques. One possible interpretation of these data, therefore, is that the tangles associated with normal aging may reduce brain metabolism in the same regions which are involved early in the course of AD. Our finding, that older AD patients had relatively greater metabolic dysfunction of mesial temporal cortex than neocortex, is also in keeping with this hypothesis, and suggests an interaction between aging and AD in the manifestations of the disease.

In addition to this interpretation of the data, it is important to also keep in mind that regional metabolic abnormalities in a variety of diseases reflect not only local effects, but remote effects such as diaschisis related to deafferentation. Such phenomena may clearly operate in degenerative disease as well as vascular ones[19] so that regional metabolic changes in dementia reflect the totality of local changes and deafferentation. This concept is closely related to the idea that synaptic activity is both a key feature of dementia[20] and a major determinant of glucose utilization. Thus, the metabolic findings in any specific brain region in AD represent complex relationships between local and remote pathology.

In summary, functional brain imaging is capable of demonstrating patterns of abnormal function which may be of considerable importance in determining the initial sites of abnormality in AD. These findings generally bear strong relationships to clinical disease features, and are thus likely to be of key importance in symptom production. While considerable evidence suggests that initial stages of functional pathology severely affect the temporal lobe, relationships between normal aging and AD may make detection of early changes in the mesial temporal lobes of AD patients difficult. This suggests that additional studies of healthy older individuals may be the most productive way of investigating the earliest stages of AD.

REFERENCES

1. JERNIGAN, T. L., D. P. SALMON, N. BUTTERS & J. R. HESSELINK. 1991. Cerebral structure on MRI, Part II: Specific changes in Alzheimer's and Huntington's diseases. Biol. Psychiatry **29:**68–81.
2. DE LEON, M. J., A. E. GEORGE, L. A. STYLOPOULOS, G. SMITH & D. C. MILLER. 1989. Early marker for Alzheimer's disease: the atrophic hippocampus. Lancet **2:**672–673.
3. JACK, C. R., R. C. PETERSEN, P. C. O'BRIEN & E. G. TANGALOS. 1992. MR-based hippocampal volumetry in the diagnosis of Alzheimer's disease. Neurology **42:**183–188.
4. SEAB, J. P., W. J. JAGUST, S. T. S. WONG, M. S. ROOS, B. R. REED & T. F. BUDINGER. 1988. Quantitative NMR measurements of hippocampal atrophy in Alzheimer's disease. Magn Reson. Med. **8:**200–208.
5. FRACKOWIAK, R. S. J., G. POZZILLI, N. J. LEGG, G. H. DUBOULAY, J. MARSHALL, G. L. LENZI & T. JONES. 1981. Regional cerebral oxygen supply and utilization in dementia. Brain **104:**753–778.
6. FOSTER, N. L., T. N. CHASE, L. MANSI, R. BROOKS, P. FEDIO, N. J. PATRONAS & G. DICHIRO. 1984. Cortical abnormalities in Alzheimer's disease. Ann Neurol. **16:**649–654.
7. JAGUST, W. J., B. R. REED, J. P. SEAB, J. H. KRAMER & T. F. BUDINGER. 1989. Clinical-physiologic correlates of Alzheimer's disease and frontal lobe dementia. Am. J. Physiol. Imaging **4:**89–96.
8. DUARA, R., W. BARKER, D. LOEWENSTEIN, S. PASCAL & B. BOWEN. 1989. Sensitivity and specificity of positron emission tomography and magnetic reso-

nance imaging studies in Alzheimer's disease and multi-infarct dementia. Eur. Neurol. **29**(suppl. 3):9–15.
9. MILLER, B. L., J. L. CUMMINGS, J. VILLANEUVA-MEYER, K. BOONE, C. M. MEHRINGER, I. M. LESSER & I. MENA. 1991. Frontal lobe degeneration: clinical, neuropsychological, and SPECT characteristics. Neurology **41**:1374–1382.
10. EBERLING, J. L., B. R. REED, M. G. BAKER & W. J. JAGUST. 1993. Cognitive correlates of regional cerebral blood flow in Alzheimer's disease. Arch Neurol. **50**:761–766.
11. HAXBY, J. V., R. DUARA, C. L. GRADY, N. R. CUTLER & S. I. RAPOPORT. 1985. Relations between neuropsychological and cerebral metabolic asymmetries in early Alzheimer's disease. J. Cerebral Blood Flow Metab. **5**:193–200.
12. FRIEDLAND, R. P., A. BRUN & T. F. BUDINGER. 1985. Pathological and positron emission tomographic correlations in Alzheimer's disease. Lancet **i**:228.
13. BRUN, A. & E. ENGLUND. 1981. Regional pattern of degeneration in Alzheimer's disease: Neuronal loss and histopathological grading. Histopathology **5**:: 549–564.
14. DECARLI, C., J. R. ATACK, M. J. BALL, J. A. KAYE, C. L. GRADY, P. FEWSTER, K. D. PETTIGREW, S. I. RAPOPORT & M. B. SCHAPIRO. 1992. Post-mortem regional neurofibrillary tangle densities but not senile plaque densities are related to regional cerebral metabolic rates for glucose during life in Alzheimer's disease patients. Neurodegeneration **1**:113–121.
15. EBERLING, J. L., W. J. JAGUST, B. R. REED & M. G. BAKER. 1992. Reduced temporal lobe blood flow in Alzheimer's disease. Neurobiol. Aging **13**:: 483–491.
16. JAGUST, W. J., J. L. EBERLING, B. C. RICHARDSON, B. R. REED, M. G. BAKER, T. E. NORDAHL & T. F. BUDINGER. 1993. The cortical topography of temporal lobe hypometabolism in early Alzheimer's disease. Brain Res. **629**:189–198.
17. ARRIAGADA, P. V., B. A. MARZLOFF & B. T. HYMAN. 1992. Distribution of Alzheimer-type pathologic changes in nondemented elderly individuals matches the pattern in Alzheimer's disease. Neurology **42**:1681–1688.
18. PRICE, J. L., P. B. DAVIS, J. C. MORRIS & D. L. WHITE. 1991. The distribution of tangles, plaques, and related immunohistochemical markers in healthy aging and Alzheimer's disease. Neurobiol. Aging **12**:295–312.
19. AKIYAMA, H., R. HARROP, P. L. MCGEER, R. PEPPARD & E. G. MCGEER. 1989. Crossed cerebellar and uncrossed basal ganglia and thalamic diaschisis in Alzheimer's disease. Neurology **39**:541–548.
20. TERRY R. D., E. MASLIAH, D. P. SALMON, N. BUTTERS, R. DETERESA, R. HILL, L. A. HANSEN & R. KATZMAN. 1991. Physical basis of cognitive alterations in Alzheimer's disease. Synapse loss is the major correlate of cognitive impairment. Ann Neurol. **30**:572–580.

Molecular and Functional Magnetic Resonance Neuroimaging for the Study of Dementia[a]

R. GILBERTO GONZÁLEZ[b]

Massachusetts General Hospital NMR Center and Neuroradiology Section, Massachusetts General Hospital, Harvard Medical School, Boston, Massachusetts, USA

ABSTRACT: The living brain's structure, function, and underlying chemistry are increasingly being revealed in sharpened detail by the extraordinary evolution of magnetic resonance (MR) technology. Recent years have ushered in a wealth of new information about neurophysiology and pathological states owing to such technologies as functional MR imaging (fMRI), MR spectroscopy (MRS), and MR spectroscopic imaging (MRSI). These advances are of substantial benefit in the study of the dementias, especially Alzheimer's disease (AD). One primary objective of our laboratory at the Massachusetts General Hospital NMR Center is to utilize these extant and emerging MR technologies to further understanding of the human brain as it undergoes assault by AD. Our approach is guided by the belief that the pathological states observed in the AD brain must be accompanied by structural, chemical and/or physiological changes that can be made visible in an *in vivo* MR study. Quantitative measurements by MRI medical temporal lobe structures have been shown to be abnormal by several groups including our own. This use of MRI will be reviewed by others. In this paper, I will review recent advances in the application of MR for the study of chemical and functional brain abnormalities in dementia, and in particular to the investigation of AD.

MR SPECTROSCOPY

Magnetic resonance spectroscopy (MRS) is a noninvasive method to identify and quantify brain chemistry. The physics underlying MRS is identical to that of magnetic resonance imaging (MRI), and spectroscopic experiments can be performed by using the same instruments as in MRI. In general the molecules amenable to study by *in vivo* MRS are water soluble and of low molecular weight (up to several hundred daltons). The concentrations of MR visible molecules are in the millimolar range. Although not nearly as sensitive

[a] The research described here was supported by NIH grants AG05134 and AG10689.

[b] *Send correspondence to:* Dr. R. Gilberto González, MGH-NMR Center, 149 13th Street, Charlestown, MA 02129; Tel: (617) 726-8628; Fax: (617) 724-3338.

as positron emission tomography, a number of biologically interesting brain chemicals are accessible to measurement by MRS *in vivo.*

In vivo MR spectroscopy of the human brain has been available for almost a decade. Early spectroscopic investigations generally used phosphorus (^{31}P) MRS, which allowed molecules related to bioenergetics, such as ATP and phosphocreatine, to be observed within the brain. In addition, other molecules that contained phosphorus and were related to membrane metabolism could be detected by this technique. Early ^{31}P-MRS investigations of Alzheimer's disease (AD) were encouraging,[1] but subsequent, increasingly quantitative techniques revealed that changes in steady-state level brain metabolites could not be observed by phosphorus MRS.[2–4] One study that coupled PET with ^{31}P-MRS and demonstrated that the concentrations of high energy phosphates were maintained despite abnormally decreased glucose metabolism.[4] The poor sensitivity of ^{31}P-MRS remains the technique's major drawback: At present, large areas must be sampled to achieve a satisfactory signal-to-noise ratio. However, the advent of higher field magnets, with their attendant increase in sensitivity, may allow ^{31}P-MRS to yield information of important use in studies of the dementias.

Proton (^{1}H) MRS has been used more extensively in recent years to investigate the brains of subjects with AD and other dementing syndromes. There are many advantages to ^{1}H-MRS, but chief among them is that the technique exploits what is inherently the most sensitive of all nuclei, the proton, which is ubiquitous component of biomolecules. Moreover, the requirements for changes in clinical MRI instrumentation to use ^{1}H-MRS are minimal. Advances in ^{1}H-MRS were previously hampered by the difficult task of water suppression, but this technical difficulty has been overcome.

A variety of interesting molecules may be monitored by proton MRS. FIGURE 1 shows a typical ^{1}H-MR spectrum obtained from a small region of

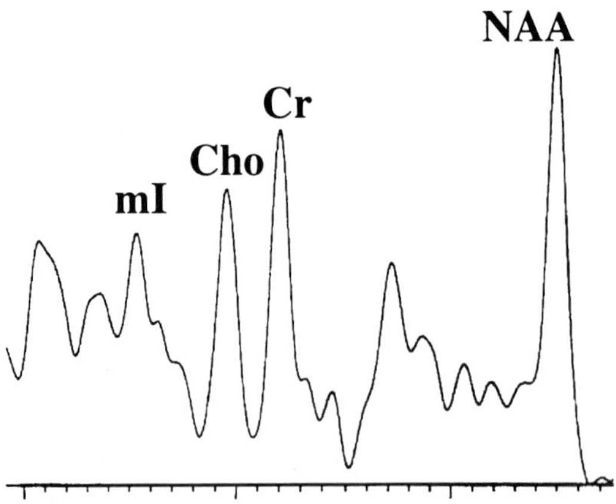

FIGURE 1. Typical proton MR spectrum obtained from a 8 ml voxel localized ot the midline of the parietal lobes in a volunteer. The major resonances are labeled as follows. NAA: n-acetylaspartic acid + other acetyl-containing molecules; Cr: creatine + phosphocreatine: Cho: choline-containing compound; mI: myo-inositol.

the brain from a normal subject. N-acetyl aspartic acid (NAA) is readily observed in the proton spectrum, usually as its largest resonance. Tallan *et al.* first described the high concentration of this form of NAA, which is located exclusively in nervous tissue.[5] Subsequently, Nadler confirmed that NAA is localized to that brain tissue which contains neurons by demonstrating its absence in bovine glial tumors.[6] Immunocytochemical methods have demonstrated unequivocally that NAA is specific to neurons and is uniformly distributed throughout the brain.[7,8] These findings indicate that NAA is a neuronal marker.

The advent of cerebral *in vivo* proton NMR spectroscopy has resulted in an increased interest in NAA. NAA contains three resonances observable by proton NMR. The N-acetyl methyl group possesses three NMR visible protons that resonate as a singlet at 2.0 ppm. This peak is prominent in the *in vitro* and *in vivo* proton spectrum of brain because of the high concentration of NAA (second only to glutamate) in the adult mammalian nervous system. This prominence, in conjunction with evidence for the metabolite's neuronal specificity, suggests that quantitative NMR spectroscopic measures of NAA in neurodegenerative disease should provide a noninvasive measure of neuronal loss. This hypothesis has already been incorporated in several recent studies of patients with neurodegenerative disease.[9–12] While significant results verifying the validity and sensitivity of NAA as a neuronal marker in NMR spectroscopic studies have been obtained, the core of this work has remained *in vitro*. We applied MRSI techniques in quantitative investigations of neurodegeneration in a rat model of excitotoxicity to obtain *in vivo* evidence that NAA is a valid neuronal marker.

Our data demonstrated the ability of MRS to provide quantitative measures of neuronal loss in a striatal lesion rat model. The data indicated that NAA is an excellent *in vivo* and *in vitro* marker for neuronal loss, and that MRS provides an index of neuronal loss comparable to post mortem biochemical assays.[13,14] This animal work supports the incorporation of quantitative proton MR spectroscopy that makes use of NAA as a neuronal marker into routine clinical imaging protocols. Such incorporation could provide a quantitative, periodic assessment of the progression of neuronal loss in patients suffering neurodegenerative diseases.

Another prominent resonance within the proton MR spectrum is creatine (Cr). This resonance is the sum of the creatine and phosphocreatine content of tissue. Phosphocreatine has been investigated by ^{31}P and proton MRS and was found to be normal in AD. The choline (Cho) resonance is complex and consists primarily of phosphocholine, glycerophosphocholine, and a small contribution from choline itself. After studies indicated that GPC is elevated in AD brains, it was hoped that this resonance might be detected *in vivo* by proton MRS. To date, detection has not occurred. It appears that the reported decrease in PC found in AD balances the increase of GPC. In another vein of investigation, myo-inositol (MI) (a sugar found in relatively high concentrations in the mammalian brain that can be observed by proton MRS when short echo-pulse sequences are employed) was reported to be elevated in AD.[12] However, this has not been reproduced in other laboratories including our own.

The majority of proton MR studies of AD have employed the single voxel method. That is, a spectrum is acquired from a single location and a relatively small amount of tissue (on the order of 10 ml) is sampled. Obvious disadvantages exist with this approach. If the disease process is not homogeneously distributed throughout the cerebrum, improper location selection may lead to erroneous conclusions. To avoid this pitfall, a method that can sample multiple areas of brain at the same time must be used. Such a method exists and is known as MR spectroscopic imaging (MRSI). Although it is beyond the scope of this paper to discuss the method in detail, the technique acquires spatial information about metabolites in a manner similar to that used in MR imaging. The capability of this approach to sample large portions of the cerebrum outweighs its most important disadvantage, the relatively long time (30 minutes or more) needed to obtain information from most of the brain.

The great advantage of MRSI is that characteristic spatial patterns of abnormal metabolism may be mapped. For example, it has been observed that the characteristic pathological lesions observed in AD are not uniform across the brain, but occur in a hierarchical manner: The earliest and most pronounced abnormalities occur within the medial temporal-lobe structures, followed by the high order (heteromodal) association cortices.[15–18] We hypothesize that neuronal loss also follows this stereotypical distribution; this hypothesis may be tested by mapping the distribution of NAA in the AD brain. Other metabolic derangements may also prove to display such distributions.

FUNCTIONAL MRI

Cognitive tests may demonstrate evidence of dementia, such as impaired memory and language deficits, even when brain CT or MRI studies appear intact for age. Because of this observation, physiologic measurements of the brain have been devised to improve our assessment of brain function. Studies using positron emission tomography (PET) and single photon emission computed tomography (SPECT) have proven to be superior to routine, anatomic CT and MRI in differentiating AD from other causes of dementia. PET[19–21] and SPECT[22] have been shown to be able to demonstrate regional deficits in perfusion and metabolism in AD. For this reason, they are often employed in the assessment of dementia.

Functional magnetic resonance imaging (fMRI) recently has emerged as a method that may provide information similar to SPECT and PET. Because the MR data needed to generate a functional brain image can be acquired rapidly, it is possible to perform fMRI during otherwise routine MRI studies of patients undergoing evaluation for dementia. There are two general types of fMRI. One known as magnetic susceptibility (MS) fMRI requires the administration of an exogenous MR contrast agent such as Gadolinium DTPA. This method is used to derive dynamic cerebral blood volume images of the brain. The other method is known as blood oxygen level dependent (BOLD) fMRI; this technique will be discussed in the next section.

Rapid imaging is important in MS fMRI because it is necessary to track a bolus of contrast material as it passes through the brain. When the contrast agent arrives in the brain, a local loss of MR signal results from the disturbance in the local magnetic field. This phenomenon is known technically as the magnetic susceptibility effect. The loss of signal is directly proportional to the blood volume within small vessels, such as capillaries and venules, and thus this information may be used to generate dynamic cerebral blood volume images. While conventional MR instruments are capable of acquiring images on the order of one image per second, they generally can do so only for a single, or at most a few, levels of the brain. The technique known as echo-planar imaging (EPI) is capable of rapidly acquiring data from across the entire brain, and is thus preferred in susceptibility contrast MR imaging.

To evaluate the potential use of SC-fMRI for the study of AD, we tested the hypothesis that cerebral blood volume images derived from magnetic susceptibility fMRI provide information similar to 18-fluorodeoxyglucose (18FDG) PET studies in patients under evaluation for dementia. In our preliminary study, we evaluated subjects by PET and fMRI. The clinical diagnoses of our patients included probable AD, possible AD, Pick's disease, and primary progressive aphasia. The imaging studies were compared quantitatively by co-registration of PET and fMRI images, followed by regression analyses of corresponding regions of interest. Qualitative analysis consisted of categorizing each brain into 8 regions, and the classification of each region as normal or abnormal through visual inspection.

We found excellent correlation coefficients (mean $r = 0.58$) between registered fMRI and PET images. Statistically significant correlations were observed in 72 of 74 brain slices, with an average p value of better than 10^{-11}. Qualitatively, the concordance between fMRI and PET was slightly less than 80%, which was highly significant ($p < 0.0001$). The full details of this study will be published elsewhere.[23] FIGURE 2 shows an example of the high correlation between SC-fMRI and PET in a patient with probable AD.

The finding that quantitative and qualitative comparisons of images obtained by MS-fMRI and 18FDG PET were highly correlated supports the hypothesis that magnetic susceptibility fMRI provides functional information similar to 18FDG PET, and that MS-fMRI may be useful for the study of

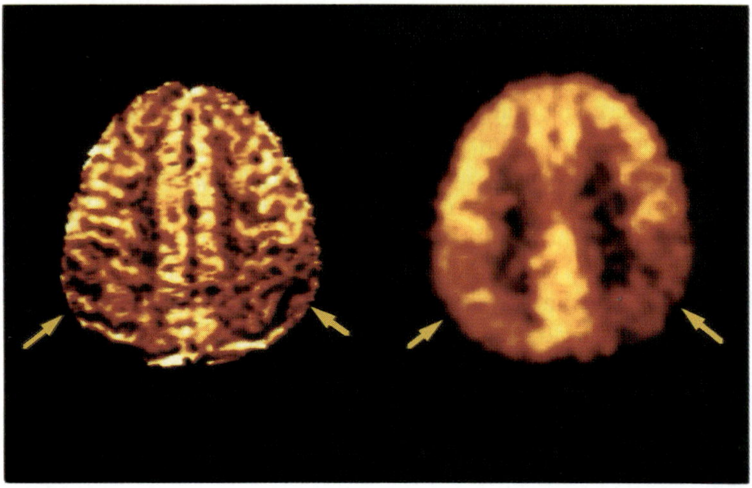

FIGURE 2. Transaxial SC-fMRI dynamic cerebral blood volume *(left)* and 18FDG PET *(right)* images from a patient with probable Alzheimer's disease. The images are from approximately the same axial plane, and were selected from data encompassing the whole brain. Areas of relatively higher blood volume or glucose metabolism are depicted in yellow, and those of relatively lower blood volume or metabolism appear red in these images. Note the corresponding regions of decreased blood volume and metabolism in the parietal lobes bilaterally.

dementia. Although the correlation coefficients between fMRI and PET were very high in the regions of the central cerebrum, we did observe areas with lower correlations. These areas included the posterior fossa and region of the inferior temporal lobes. The most likely explanation for this observation is the substantial amount of artifact that echo planar imaging generates in these regions.

Our results suggest that dynamic cerebral blood volume images derived from MS-fMRI provide information similar to other neuroimaging techniques, such as PET or SPECT, that measure brain function and are thus of use in the study of AD. The clinical value of MS-fMRI in helping to differentiate AD from other dementiias will require further research, which we have commenced. Another related, outstanding question is whether fMRI can measure the progression of AD, and thereby aid in the evaluation of potential AD therapies. Were fMRI to prove useful in tracking disease progression, the technique's usefulness in the study, and potentially in the treatment, of AD would be great. With fMRI, data are acquired rapidly and the procedure can be incorporated easily into a standard brain MRI examination. Thus, magnetic susceptibility fMRI may be an important new technique for investigations of AD and other dementias.

BRAIN ACTIVATION STUDIES USING BOLD-fMRI

The technique known as BOLD-fMRI is the most recent and perhaps the most exciting new method for investigating the brain in action. This type of fMRI has been used to measure brain activity during a variety of conditions, including visual stimulation, motor activity, and even cognitive test performance. The technique exploits the inherent paramagnetic nature of deoxygenated hemoglobin to measure brain activation. Different levels of blood oxygen saturation create differential tissue contrast, which is measurable in the MR image. PET studies have shown that when a part of the brain is activated, there is increased perfusion to that region. It appears that this increase in regional perfusion usually provides more oxygen than is required by the activated brain. This results in a venous oxygen saturation in the activated regions that is higher than in regions of the brain that are not activated. Because blood coursing through activated brain has a lower content of deoxyhemoglobin, a relative increase in the MR signal is observed.

The change in the inherent MR signal produced by brain activation depends on many factors, but usually it does not exceed 1-2% when studies are performed at a magnetic field strength of 1.5 Tesla. This small change in the signal is not observable by direct inspection; therefore, subtraction or statistical methods are required. The experimental approach is simple. First, images are obtained during a baseline state, then images are acquired during the performance of an activation paradigm. The latter images are subtracted from the former, or a statistical analysis is performed to identify regions of the brain that are activated by the paradigm. Because MRI poses no known risk to people, activation studies can be repeated as often as is necessary to answer the question at hand. Unlike PET brain activation studies, BOLD-fMRI can detect patterns of activation in individuals.

For the last 2 years, brain activation studies that employ fMRI techniques have been developed at the MGH-NMR Center by using simple visual, motor, somatosensory, and auditory stimulation. Initial investigations relied on stimulus materials and task paradigms that had been used successfully in PET experiments, *e.g.*, full-field visual stimulation and finger sequencing. Brain activation studies obtained in our laboratory and others with fMRI techniques have focused on the primary visual and motor cortices because these regions have been well characterized in primates and humans.[24–28] In our laboratory, more recent fMRI studies obtained by using the BOLD technique have been used to investigate high-order areas, such as visual function in MT, real vs. imaginary movement, and memory function. Those of us interested in AD have concentrated our efforts on the development of paradigms for the study of brain activation during memory function. Three types of memory brain activation studies have been undertaken, including studies of long-term explicit memory, working memory, and semantic memory. Our progress in these areas is briefly summarized below.

BOLD-fMRI of Long-term Explicit Memory Brain Activation

It has been postulated that during learning, an interaction is established between the medial temporal-lobe memory system and distributed storage sites in the neocortex. Using BOLD-fMRI, we studied young and older normal subjects while they performed picture encoding and picture recognition tasks. Asymmetric spin-echo echo-planar imaging was employed to evaluate signal changes throughout the whole brain. To assess picture encoding, subjects were asked to look at a consecutive series of complex, colored pictures and to remember them so that they could recognize them later. As a control condition, the subjects were scanned while viewing a single repeating picture. Statistical maps (using the Kolmogorov-Smirnov statistic) that compared the encoding with the control tasks revealed significant bilateral increases in signal intensity in the hippocampal and parahippocampal regions and at the temporal-occipital junction. Recognition memory was assessed through the evaluation of statistical maps derived by comparing images obtained during a two-alternative, forced-choice recognition test (in which subjects identified pictures previously viewed) with images obtained during a two-alternative, forced-choice identification test (in which subjects identified indoor vs. outdoor pictures). The statistical maps revealed consistent signal changes that occurred bilaterally in the dorsolateral prefrontal cortex (areas 46 and 10). This BOLD-fMRI experiment provides new evidence about the participation of the medial temporal-lobe system during information encoding, and also about the locus of cortical areas that support picture recognition memory. Our data also indicate that similar task-specific activation patterns occur in young and older normal subjects.

BOLD-fMRI of Working Memory Brain Activation

To date, we have tested 5 young control subjects and 7 older control subjects on a working memory task similar to the one described by Cohen *et al.*[29] A sequential task-activation paradigm that alternated between two conditions was employed. During the experimental condition, subjects viewed a series of letters one at a time and were asked to press a response key whenever they saw a letter that repeated itself with one intervening letter. For example, given the letter string A,C,H,C,L, the subject would press the key on the second presentation of the letter C. During the control condition, subjects viewed a series of letters one at a time and were asked to look at

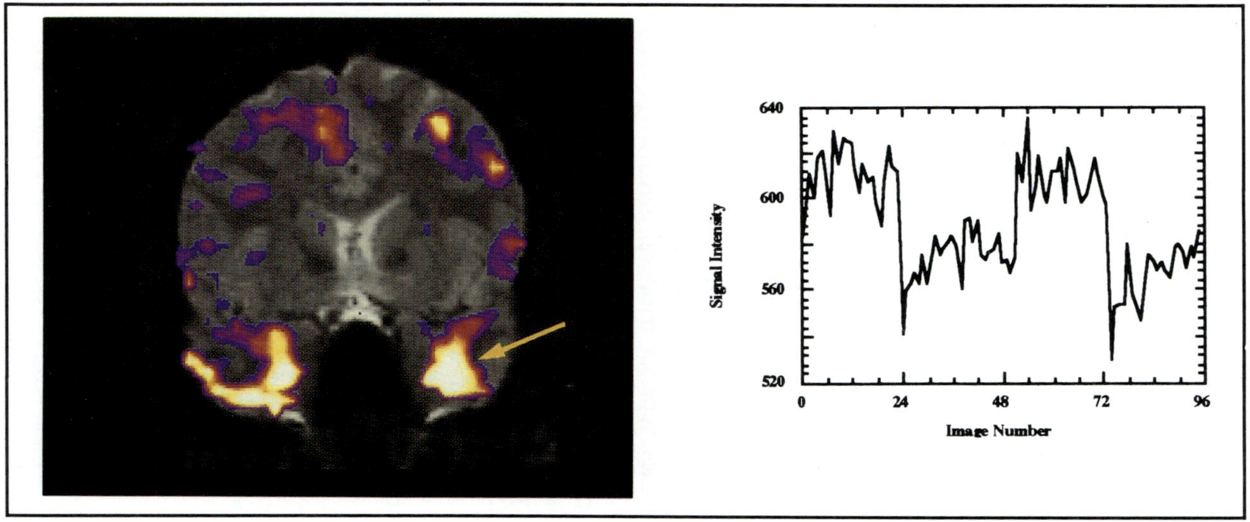

FIGURE 3. BOLD-fMRI detection of brain activation during a working memory task. This figure shows the result from one of the young control subjects who performed the 2-back working memory task. The brain activation statistical map is shown on the left superimposed on T1 image, and the real time course of the MR signal intensity change measured at one location within the left medial temporal lobe. In this case, the signal change was extremely robust with nearly a 10% change in the MR signal intensity at this location between the two tasks.

each letter and press a key whenever they saw the target letter X. We are currently evaluating all of our data, but our preliminary findings show consistent activation of the dorsolateral frontal regions, with less consistent activation of the medial temporal lobes. An example of temporal-lobe activation during this task is shown in FIGURE 3.

BOLD-fMRI of Semantic Memory Brain Activation

Functional MRI language experiments have been carried out in 14 normal control subjects. Paradigms of passive word presentation (visual and acoustic), verb generation to concrete nouns, and a visual object naming task (Boston Naming) have been employed. Some patterns have emerged from the comparison of these different tasks. In general, we see an anterior to posterior gradient of activation, with anterior areas showing greater signal intensity increases. Regions which activate most consistently in the visual object naming task include: occipital and temporo-occipital cortex, dorsolateral prefrontal cortex, pars orbitalis (Brodmann area 47), Broca's area (areas 44 and 45), motor cortex (areas 4 and 6), angular gyrus, cingulate gyrus, and supplementary motor area. Less consistent but frequently activated areas include Wernicke's area (left posterior superior temporal gyrus), supramarginal gyrus, and anterior prefrontal cortex (areas 9 and 10).

We are now embarking on brain activation studies of memory in subjects with a clinical diagnosis of AD. We hope to focus our studies on those in the earliest stages of the disease. The important questions that we hope to answer include the following: Do AD patients demonstrate a decrease or a paradoxical increase in activation in brain areas that normally activate during memory tasks? Does the AD brain activate in different brain locations than controls subjects, perhaps attempting to compensate for failing memory? Neuroimaging investigations of brain activities that are vulnerable to the devastation of AD through the use of fMR techniques should help us to understand many aspects of the course and toll of this pathology. Clinically, modern MR neuroimaging methods may provide a means for earlier diagnosis in AD and an improved capacity to assess the disease's progression.

ACKNOWLEDGMENTS

In this review I took the liberty of presenting data from work involving several close collaborators. I wish to thank Drs. Suzanne Corkin, Chantal Stern, John H. Growdon, Alex Guimaraes, Peggy J. Jennings, Mr. Robert M. Sugiura, and Ms. Cindy Carr. I also wish to thank Ms. Joanne Fordham for editing this manuscript.

REFERENCES

1. Brown, G. G., J. H. Garcia, J. W. Gdowski, S. R. Levine & J. A. Helpern. 1993. Altered brain energy metabolism in demented patients with multiple subcoritcal ischemic lesions. Arch. Neurol. **50:**384–388.
2. Bottomley, P. A., J. P. Cousins, D. L. Pendrey, W. A. Wagle, *et al.* 1992. Alzheimer dementia: Quantification of energy metabolism and mobile phosphoesters with P-31 NMR spectroscopy. Radiology **183:**695–699.
3. Gonzalez, R. G., A. R. Guimaraes, G. J. Moore, A. Crawley, L. A. Cupples & J. H. Growdon. Quantitative *in vivo* 31P magnetic resonance spectroscopy of Alzheimer's disease. Alzheimer Rel. Dis. In press.
4. Murphy, D. G. M., P. A. Bottomley, J. A. Salerno, C. DeCarli, M. J. Mentis, C. L. Grady, D. T. Teichberg, K. R. Giacometti, J. M. Rosenberg, C. J. Hardy, M. B. Schapiro, S. I. Rapoport, J. R. Alger & B. Horwitz. 1993. An *in vivo* study of phosphorus and glucose metabolism in Alzheimer's disease using magnetic resonance spectroscopy and PET. Arch Gen. Psychiat. **50:**341–349.
5. Tallan, H., S. Moore & W. Stein. 1956. N-acetyl-L-aspartic acid in brain. J Biol Chem. **219:**257-264.
6. Nadler, J. & J. Cooper. 1972. N-acetyl-L-aspartic acid content of human neural tumours and bovine peripheral nervous tissues. J. Neurochem. **14:** 551–554.
7. Moffett, J., M. Namboodiri, C. Cangro & J. Neale. 1991. Immunohistochemical localization of N-acetylaspartate in rat brain. Neuroreport **2:**131–134.
8. Simmons, M. L., C. G. Frondoza & J. T. Coyle. 1991. Immunocytochemical localization of N-acetyl-aspartate with monoclonal antibodies. Neuroscience **45:** 37–45.
9. Chong, W., M. Paley, I. Wilkinson, *et al.* 1994. Localised cerebral proton MR spectroscopy in HIV infection and AIDS. AJNR **15:**21–25.
10. Gideon, P. & O. Henriksen. 1992. *In vivo* relaxation of N-acetyl-aspartate, creatine plus phosphocreatine, and choline containing compounds during the course of brain infarction: A proton MRS study. Magn. Reson. Imaging **10:** 983–988.
11. Jarvik, J., R. Lenkinski, R. Grossman, J. Gomori, M. Schnall & I. Frank. 1993. Proton MR spectroscopy of HIV-infected patients: Characterization of abnormalities with imaging and clinical correlation. Radiology **186:**739–744.
12. Miller, B., R Moats, T. Shonk, R. Ernst, S. Woolley & B. Ross. 1992. Alzheimer disease: Depiction of increased cerebral myo-inositol with proton MR spectroscopy. Radiology **187:**433–437.
13. Guimaraes, A., T. Nguyen, D. Kennedy, S. Hyman & R. Gonzalez. 1991. Quantitative magnetic resonance imaging of experimental neurotoxicity. Proc. Soc. Magn. Reson. Med. **1:**297.
14. Guimaraes, A. R., P. Schwartz, M. R. Prakash, C. A. Carr, U. V. Berger, B. G. Jenkins, J. T. Coyle & R. G. Gonzalez. Quantitative *in vivo* ^{1}H NMR spectroscopic imaging of excitotoxicity in rat brain. Submitted.
15. Arnold, S. E., B. T. Hyman, J. Flory, A. R. Damasio & G. W. Van Hoesen. 1991. The topographical and neuroanatomical distribution of neurofibrillary

tangles and neuritic plaques in the cerebral cortex of patients with Alzheimer's disease. Cerebral Cortex **1**:103–116.

16. ARRIAGADA, P. J. GROWDON, J. E. HEDLEY-WHYTE & B. HYMAN. 1992. Neurofibrillary tangles but not senile plaques parallel duration and severity of Alzheimer's disease. Neurology **42**:631–639.
17. HOF, P. R., L. M. BIERER & D. P. PERL. 1992. Evidence for early vulnerability of the medial and inferior aspects of the temporal lobe in 82 year-old patient with preclinical signs of dementia—Regional and laminar distribution of neurofibrillary tangles and senile plaques. Arch Neurol **49**:946–953.
18. HOF, P. & J. MORRISON. 1994. The cellular basis of cortical disconnection in Alzheimer disease and related dementing conditions. *In* Alzheimer Disease. Terry R., R. Katzman & K. Bick, Ed. 197–229. New York Raven Press, Ltd.
19. FRACKOWIAK, R., C. POZZILLI, N. LEGG, *et al.* 1981. Regional cerebral ozygen supply and utilzation in dementia: a clinical and physiological study with oxygen-15 and positron tomography. Brain **104**:753-778.
20. BENSON, D., D. KUHL, R. HAKINS, M. PHELPS, J. CUMMINGS & S. TSAI. 1983. The fluorodeoxyglucose 18F scan in Alzheimer's disease and multi-infarct dementia. Arch. Neurol. **40**:711–714.
21. RAPOPORT, S. 1991. Positron emission tomography in Alzheimer's disease in relation to diseas pathogenesis: A critical review. Cerebrovasc. Brain Metab. Rev. **3**:297-335.
22. JOHNSON, K., B. HOLMAN, T. ROSEN, J. NAGEL, R. ENGLISH & J. GROWDON. 1990. Iofetamine I 123 single photon emission comupted tomography is accurate in the diagnosis of Alzeimer's disease. Arch Intern. Med. **150**:752–756.
23. GONZALEZ, R. G., A. FISCHMAN, A. GUIMARAES, C. A. CARR, C. STERN, E. F. HALPERN, J. H. GROWDON & B. R. ROSEN. Cerebral blood volumes from fMRI parallels cerebral metabolism by 18FDG PET in patients evaluation for dementia. Am. J. Neurorad. In press.
24. KWONG, K., J. BELLIVEAU, D. CHESLER, *et al.* Dynamic magnetic resonance imaging of human brain activity during primary sensory stimulation. Proc. Nat. Acad. Sci. USA **89**:5675-5679.
25. BANDETTINI, P. A., E. C. WONG, R. S. HINKS, R. S. TIKOFSKY & J. S. HYDE 1992. Time course EPI of human brain function during task activation. Magn. Res. Med. **25**:390-397.
26. TURNER, R., J. P. WEN, K. K. KWONG, D. LEBIHAN, T. ZEFFIRO & R. BALABAN. 1993. Functional mapping of the human visual cortex at 4 Tesla and 1.5 Tesla using deoxygenation contrast EPI. Magn. Reson. Med. **29**:277-279.
27. OGAWA, S., T. D. W., R. MENON, J. M. ELLERMANN, *et al.* 1992. Intrinsic signal changes accompanying sensory stimulation: Functional brain mapping using MRI. Proc. Nat. Acad. Sci. USA. **89**:5951-5955.
28. TOOTELL, R. B. H.., K. K. KWONG, J. W. BELLIVEAU, J. R. BAKER, *et al.* 1993. Mapping human visual cortex: Evidence from functional MRI and histology. Inves. Ophth. Vis. Sci. **34**:813.
29. COHEN, J. D., S. D. FORMAN, T. S. BRAVER, B. J. CASEY, D. SERVAN-SCHREIBER & D. C. NOLL. Activation of prefrontal cortex in a non-spatial working memory task with functional MRI. Human Brain Mapping. In press.

Clinical and Neuroimaging Features of Familial Alzheimer's Disease[a]

M.N. ROSSOR,[b,c,e] A.M. KENNEDY,[b,c] AND R.S.J. FRACKOWIAK[d]

[b] *Dementia Research Group, The National Hospital for Neurology and Neurosurgery, 8-11 Queen Square, London WC1N 3BG United Kingdom*

[c] *Department of Neurology, St Mary's Hospital, Praed Street London W2 1NY United Kingdom*

[d] *MRC Cyclotron Unit, Royal Postgraduate Medical School, Hammersmith Hospital, 150 Ducane Road, London W12 ONN United Kingdom*

ABSTRACT: Subtle phenotypic differences between familial Alzheimer's disease (FAD) pedigrees can be identified which may reflect the genetic and allelic heterogeneity of the disease. Positron emission tomography (PET) of APP mutation and chromosome 14-linked FAD pedigree members reveals biparietal bitemporal hypometabolism. Scanning of asymptomatic at-risk individuals reveals a similar, but quantitatively less severe, pattern of hypometabolism.

INTRODUCTION

Since the early identification of familial Alzheimer's disease (FAD) pedigrees, the role of genetic factors has been increasingly recognized and the heterogeneity of FAD established with the identification of at least 5 separate genetic loci. Inheritance of the apolipoprotein E4 allele is a major determinant of risk for Alzheimer's disease and inheritance of the Apo E2 allele confers some protection. A number of late onset autosomal dominant pedigrees cannot be explained by the inheritance of the Apo E4 allele and may represent an additional locus. Very early onset autosomal dominant FAD with onset at around 40 years of age is associated with linkage to markers on chromosome 14.[1] By contrast, onset at around the age of 50 years is characteristic of FAD pedigrees associated with a variety of mutations in the amyloid precursor protein (APP) gene: three mutations at position APP 717 resulting in isoleucine, glycine or phenylalanine substitutions for valine, a double mutation at APP 670/671 and a mutation at APP 692 resulting in an alanine to glycine

[a] These studies were supported by the Medical Research Council, David and Fredrick Barclay Foundation, Wolfson Foundation, Wellington Foundation, and Alzheimer's Disease Society.

[e] *Send correspondence to:* Dr. Martin Rossor, The National Hospital for Neurology and Neurosurgery, 8-11 Queen Square, London WC1N 3BG United Kingdom; Tel: 44-171-837 3611 ext. 3854; Fax: 44-171-209-0182; E-Mail: K.homan@sm.ic.ac.uk.

substitution have all been described.[2–6] Although the APP mutation families have been important to our understanding of the pathophysiology of Alzheimer's disease, in clinical terms these account for less than 20 families worldwide. The Volga German pedigrees[7] are an additional early onset group which are unassociated with APP mutations and do not show linkage to chromosome 14. The overall clinical features of FAD with early memory deficits progressing to multiple domains of cognitive impairment and characteristic neurofibrillary tangles, senile plaques and vascular amyloid deposition at autopsy are all similar to sporadic Alzheimer's disease. However, there are subtle differences that have been observed between families[8] suggesting that the genetic and allelic heterogeneity of FAD may be reflected in subtle phenotypic differences. The study of FAD pedigrees thus offers an opportunity to relate specific molecular defects to the selective vulnerability of specific neural systems which might themselves result in characteristic clinical deficits. Moreover, the examination of at risk individuals in FAD families provides a unique opportunity to determine the earliest and presymptomatic clinical features. We have used this strategy to study at risk and affected individuals in APP mutation and chromosome 14-linked FAD pedigrees.

APP PEDIGREES

There are three known APP pedigrees in the UK: one APP 717val-gly (Family 19)[2,9] and two unrelated APP 717 val-ile pedigrees Family 23 and 172).[3,10] The mean age at onset of the APP 717 val-gly family is 52.1 years (SD ± 7.8).[9] The clinical features of affected individuals include prominent early myoclonus and seizures, depression and lack of insight into their cognitive deficit. Naming skills are relatively preserved. Magnetic resonance imaging (MRI) performed on a GE 1.5 Tesla Signa Unit in this family included studies of at risk presymptomatic individuals and indicated early hippocampal involvement (Fox *et al.* 1995, this volume). The mean age at onset of family 23 (APP 717 val-ile) was 53.7 years. Again, myoclonus was prominent but with variable insight and in the later stages of the disease, prominent increased tone emerged, initially of the Gegenhalten type but with late cogwheeling and impairment of gait. The two patients from this family who have come to autopsy have both shown widespread neocortical Lewy bodies.[11] Family 172 has a similar age at onset of 52.7 years, again with memory impairment and early myoclonus. MRI in two affected individuals showed hippocampal atrophy more prominent on the left than the right and one case showed scattered white matter lesions.

Positron Emission Tomography (PET) was performed on one member of family 23, two members of family 172 and two of family 19. One patient was mildly affected at the time of scanning (MMSE 24/30). All of the others had moderate to severe disease (MMSE 9-14/30) at the time of scanning. Cerebral glucose metabolism was measured following the intravenous administration of 18 [F] fluorodeoxyglucose (FDG).[12] Absolute quantification of the cerebral metabolic rate for glucose (CMR glu) was calculated using a regions of interest (ROI) technique. In addition, in order to demonstrate the relative changes in the distribution of regional CMR glu, scans were analyzed using statistical parametric mapping.[13,14] Differences in global CMR glu across all subjects and patients were normalized using an analysis of covariance. The adjusted regional CMR glu for each voxel in each group of patients was compared with that found in the control group using the t-statistic with a significance threshold of $p < 0.01$ (Scans were performed at the MRC Cyclotron Unit, Hammersmith Hospital using a Siemens/CTI ECAT 953B scanner). Bilateral parietotemporal hypometabolism was apparent in all APP mutation cases studied and additional frontal deficits were apparent in the more severely affected cases of APP 717 val-ile FAD but less apparent in the APP 717 val-gly FAD cases. We have also had the opportunity to perform PET scanning on two affected members (MMSE 15/30 and 26/30) of the Toronto 3 APP 717 val-ile family[15] by courtesy of Professor D. McLachlan and Dr. P. St. George-Hyslop.[16] These two cases also showed biparietal bitemporal hypometabolism.

CHROMOSOME 14 PEDIGREES

Seven pedigrees from a group which have shown chromosome 14 linkage[17] have been studied with PET. The mean age at onset of this group of 7 pedigrees was 41.9 years (SD ± 3.6) and all patients showed early myoclonus and seizures with increasing disease severity. Biparietal bitemporal hypometabolism were seen in all affected individuals, but there was an additional asymmetric frontal hypometabolism most marked on the left suggesting left hemisphere vulnerability in established disease. One additional family (F148) presented an unusual clinical phenotype. Three affected individuals have been studied who showed early memory deficit with dyscalculia, and relative preservation of naming, but in two patients with established disease a speech production deficit was noted.[12] The speech production deficit was characterized by impaired spontaneous speech output with stuttering, hesitancies and frequent phonemic paraphrases but with relatively intact word retrieval and comprehension skills. Speech production deficits were noted on formal tests of both reading and repetition suggesting that the deficit occurred at or after a common language output system with relative reliance neither on an auditory/phonological transcoding route nor on a semantic route of speech production. All three patients had a biparietal bitemporal pattern of hypometabolism more marked on the left than on the right and the two patients with demonstrable speech production deficits had additional inferior frontal hypometabolism. Two members of this family had periventricular white matter lesions on MRI.

AT RISK STUDIES

In order to study whether PET scanning can identify these metabolic changes at a presymptomatic stage we examined 24 asymptomatic at risk individuals from 12 early onset pedigrees with histologically confirmed disease (11 were from the 3 pedigrees with APP mutations—see above: 11 were from 5 of the families with chromosome 14 linkage; and 2 cases were from 2 additional pedigrees in which the genetic status could not be determined). We compared these with 19 affected FAD patients and 16 cognitively normal, asymptomatic, medication-free volunteers. The criterion for entry as an asymptomatic at risk individual was the absence of symptoms reported either by the individual or their spouse. The mean MMSE score for the at risk group was 29/30 with a range of 25–30. The one at risk individual who scored only 25/30 was considered to be asymptomatic by himself and the family and exclusion of this individual from the analysis did not alter the results nor their statistical significance. The mean MMSE of the affected FAD group was 16/30. The mean ages at scanning of the affected cases was 49.7y, range 39–65 years; of the at risk cases 44y, range 31-60 years, of the controls 50.1y, range 35–70. The at risk group were scanned on average 2.7 years before the mean age at onset for the respective pedigree whereas the affected FAD group were scanned on average 4.8 years after the mean age at onset for the family. The affected FAD cases had a mean global MRglu of 0.379 micromols/min/g (SD ± 0.060) which differed significantly from the value of 0.550 micromols/min/g (SD ± 0.080) found in the controls ($p < 0.0001$) (FIG. 1a). Similarly, the global MRglu in the at risk FAD subjects was 0.482 micromols/min/g (SD ± 0.06) which was also significantly less than that of the normal controls ($p < 0.011$). FIGURE 1b depicts the SPM map of the significant differences in regional CMRglu between all the affected FAD patients and controls. Significant areas of hypometabolism are seen in parietal, temporal and frontal cortices with relative sparing of the motor and subcortical structures. A similar parietotemporal pattern is also seen when the at risk FAD subjects are compared with the controls (FIG. 1c). The 7 cases with a global MRglu which lay more than one standard deviation below the mean of the normal controls were then selected and a further comparison of the regional CMRglu against that of the controls was made: the parietotemporal deficit in this group is even more pronounced and there is an additional minor dorsolateral prefrontal impairment (FIG. 1d).

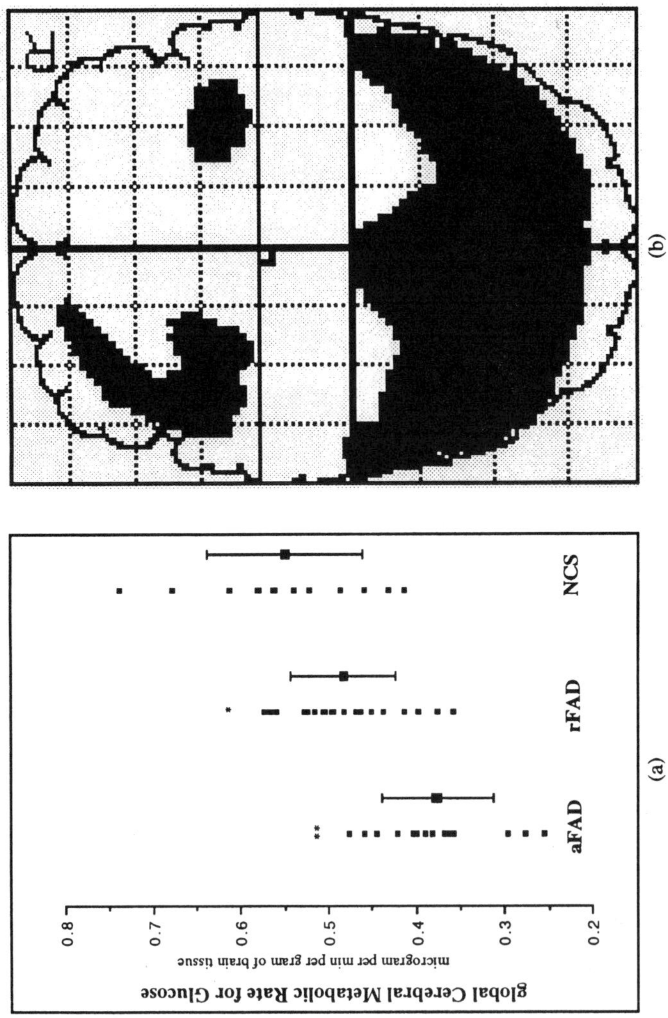

FIGURE 1. (Legend appears on p. 54.)

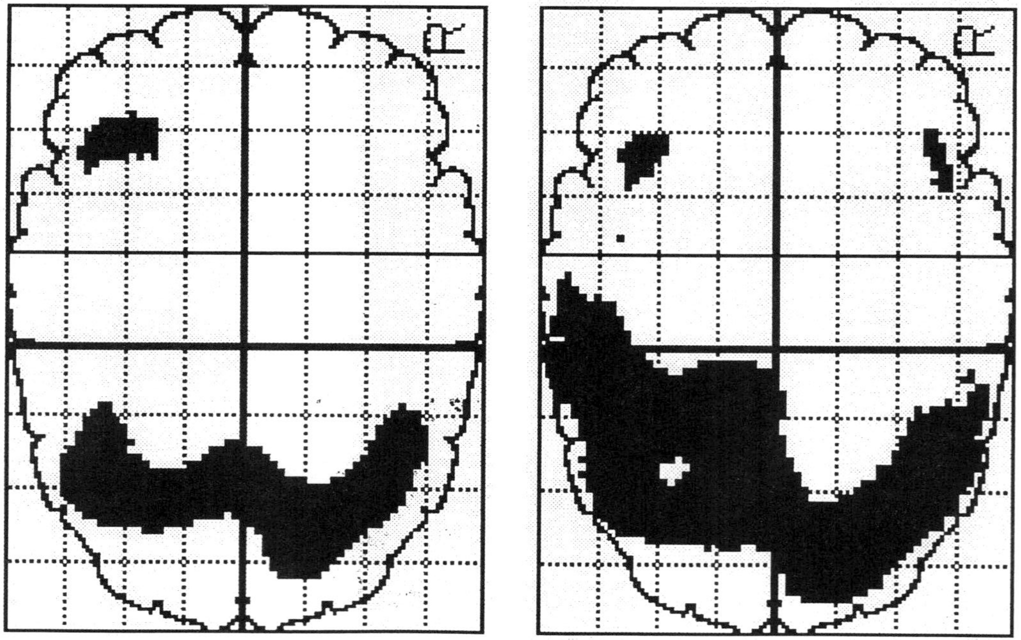

(c) (d)

FIGURE 1. Comparison of brain metabolism (global MRglu) between affected FAD patients, at risk members of FAD pedigrees and normal volunteers. Panels (b), (c) and (d) are the transverse SPM maps in which black areas indicated those pixels which are significantly hypometabolic ($p < 0.01$) when compared with controls: (a) affected FAD patients; (b) all at risk individuals; (d) the seven at risk individuals who had global MRglu are standard deviation blow the control group near. (Derived from Kennedy *et al.*[18])

CONCLUSION

FAD pedigrees provide an ideal opportunity to explore the relation between genetic and phenotypic heterogeneity. Although overall FAD cases are similar to those of sporadic Alzheimer's disease cases, subtle differences can be discerned between the different FAD pedigrees. The study of at risk individuals provides the ideal opportunity to assess the earliest changes in neuroimaging and clinical features. Such studies can further the exploration of brain behavior relationships. Moreover, identification of presymptomatic abnormalities will be important for early diagnosis particularly with the development of therapeutic strategies aimed at slowing progression of disease.

REFERENCES

1. SCHELLENBERG, G. D., T. BIRD, E. M. WILLSMAN, *et al.* 1992. Genetic linkage evidence for a familial Alzheimer's Disease locus on chromosome 14. Science **258:**668–671.
2. CHARTIER-HARLIN, M-C., F. CRAWFORD, H. HOULDEN, *et al.* 1991. Early onset Alzheimer's disease caused by mutations at codon 717 of the beta amyloid precursor gene. Nature **353:**844–846.
3. GOATE, A., M-C. CHARTIER-HARLIN, M. MULLAN, *et al.* 1991. Segregation of a mis-sense mutation in the amyloid precursor protein gene with familial Alzheimer's disease. Nature **349:**704–706.
4. HENDRIKS, L., C. VAN DUIJN, P. CRAS, *et al.* 1992. Presenile dementia and cerebral haemorrhage linked to a mutation at codon 692 of the beta amyloid precursor protein gene. Nature Genetics **1:**218–221.
5. MULLAN, M., F. CRAWFORD, K. AXELMAN, *et al.* 1992. A pathogenic mutation for probable Alzheimer's disease in the APP gene at the N-terminus of B-amyloid. Nature Genetics **1:**345–347.
6. MURRELL, J., M. FARLOW, B. GHETTI & M. BENSON. 1991. A mutation in the amyloid precursor protein associated with hereditary Alzheimer's disease. Science **253:**97–98.
7. BIRD, T. D., T. H. LAMPE, E. J. NEMENS, *et al.* 1988. Familial Alzheimer's disease in American descendants of the Volga Germans: Probable genetic founder effect. Ann Neurol. **23:**25–31.
8. BIRD, T. D., S. M. SUMI, E. J. NEMENS, *et al.* 1989. Phenotypic heterogeneity in familial Alzheimer's disease: A study of 24 kindreds. Ann Neurol. **25:**12–21.
9. KENNEDY, A. M., S. NEWMAN, A. MCCADDON, *et al.*. 1993. Familial Alzheimer's disease. A pedigree with a mis-sense mutation in the amyloid precursor protein gene (amyloid precursor protein 717 valine > glycine). Brain **116:** 309–324.
10. FIDANI, L., K. ROOKE, M-C. CHARTIER-HARLIN, *et al.* 1992. Screening for mutations in the open reading frame and promoter of the beta-amyloid precursor protein gene in familial Alzheimer's disease: Identification of a further family with APP 717 val > ile. Human Mol. Genet. **1:**165–168.

11. LANTOS, P. L. L., M. K. OVENSTONE, J. JOHNSTON, *et al.* 1994. Lewy bodies in the brain of two members of a family with the 717 (val to ile) mutation of the amyloid precursor protein gene. Neurosci. Lett. **172:**77–79.
12. KENNEDY, A. M., S. K. NEWMAN, R. S. J. FRACKOWIAK, *et al.* 1995. Chromosome 14 linked familial Alzheimer's disease: A clinico-pathological study of a single pedigree. Brain. **118:**185–205.
13. FRACKOWIAK, R. S. J. & K. J. FRISTON. 1994. Functional neuroanatomy of the human brain positron emission tomography—a new neuroanatomical technique. J. Anatomy **184:**211–225.
14. FRISTON, K., C. FRITH, P. LIDDLE & R. S. J. FRACKOWIAK. 1991. Comparing functional (PET) images; The assessment of significant change. J. Cereb Blood Flow Metab. **11:**690–699.
15. KARLINSKY, H., G. VAULA, J. HAINES, *et al.* 1992. Molecular and prospective phenotypic characterization of a pedigree with Familial Alzheimer's Disease and a mis-sense mutation in codon 717 of the beta amyloid precursor protein gene. Neurology **42:**1445–1453.
16. KENNEDY, A. M., M. N. ROSSOR, S. K. NEWMAN, *et al.* The neuroimaging features of individuals with APP 717 mutations. Submitted.
17. MULLAN, M., H. HOULDEN, M. WINDELSPECHT, *et al.* 1992. A locus for familial early onset Alzheimer's disease on the long arm of chromosome 14, proximal to the alpha 1 antichymotrypsin gene. Nature Genetics **2:**340–342.
18. KENNEDY, A. M., R. S. J. FRACKOWIAK, S. K. NEWMAN, *et al.* 1995. Deficits in cerebral glucose metabolism demonstrated by positron emission tomography in individuals at risk of familial Alzheimer's disease. Neurosci. Lett. **186**:17–20.

The Role of APP Processing and Trafficking Pathways in the Formation of Amyloid β-Protein[a]

D. J. SELKOE,[b] T. YAMAZAKI, M. CITRON, M. B. PODLISNY, E. H. KOO, D. B. TEPLOW, AND C. HAASS

Center for Neurologic Diseases, Brigham and Women's Hospital, Department of Neurology, Harvard Medical School, Boston, Massachusetts, USA

ABSTRACT: The amyloid β-protein (Aβ) is a proteolytic fragment of the β-amyloid precursor protein (βAPP). We previously reported the constitutive secretion of Aβ peptides from a variety of cells expressing βAPP under normal culture conditions. These endogenously produced Aβ peptides have heterogeneous N- and C-termini that vary as a function of βAPP missense mutations. Treatment of Aβ-secreting cells with agents that alter intravesicular pH showed that an acidic compartment is required for proper Aβ generation. One such compartment appears to be the endosome. Immunolabeling of cell-surface βAPP in living neurons and non-neuronal cells directly demonstrated the endocytosis of the protein and its rapid recycling (within 5–10 minutes) to the cell surface, as well as the trafficking of some βAPP to lysosomes. Expression of βAPP with various deletions of the cytoplasmic domain, including the NPTY motif, leads to decreased internalization and an associated decrease in the production of Aβ peptides that begin at the usual asp_1 start site. These and other data suggest that Aβ production begins with cleavage of βAPP by a still unknown protease(s) (β-secretase[s]) at the met-asp bond preceding the Aβ N-terminus and that this occurs in part in early endosomes. To characterize the substrate requirements of β-secretase, βAPP was mutagenized by placing stop codons within or at the end of the transmembrane domain or substituting other amino acids for the wild-type met and asp at the P_1 and P_1' positions. These experiments showed that proper β-secretase cleavage requires the precursor to be membrane-anchored and is highly sequence specific; most substitutions at met or asp substantially decrease Aβ production. Analogous mutagenesis experiments around the Aβ C-terminus revealed that the unknown protease(s) cleaving here ("γ-secretase[s]") does not show such specificity. Cells secreting Aβ may also be useful for examining the critical issue of the aggregation of Aβ into its neurotoxic polymeric form under physiological conditions. In this regard, we have found that βAPP-expressing CHO cells show aggregation of ≥ 10–20% of their secreted Aβ peptides into SDS-stable dimers, trimers and sometimes tetramers under normal culture conditions. The identity of these small multimers was confirmed by extensive immunochemical characterization and radiosequencing. They are present at

[a] This research was sponsored by NIH Grant AG 07911-07 and AG06173-10.

[b] *Send correspondence to:* Dr. Dennis J. Selkoe, Brigham and Women's Hospital, Center for Neurologic Diseases, 221 Longwood Avenue, Boston, MA 02115; Tel: (617) 732-6454; Fax: (617) 732-7787; Email: Selkoe@cnd.bwh.harvard.edu.

~ 100–500 pM levels in conditioned medium of CHO transfectants. Using this endogenous Aβ aggregating system, we have begun to examine variables that influence aggregation and compounds which may retard it. In conclusion, studies of the regulation of Aβ production and aggregation in cell culture can provide information under physiological conditions that can complement analyses of these processes *in vivo*.

INTRODUCTION

Progressive cerebral dysfunction in Alzheimer's disease and Down's syndrome is accompanied by the formation of innumerable extracellular amyloid deposits in the form of senile plaques and microvascular amyloid. The amyloid fibrils are composed of the 39–43 residue amyloid β-protein (Aβ), a fragment of the integral membrane polypeptide, β-amyloid precursor protein (βAPP). Evidence from several laboratories has shown that amorphous, largely nonfilamentous deposits of Aβ ("diffuse" or "preamyloid" plaques) precede the development of fibrillar amyloid, dystrophic neurites, neurofibrillary tangles, and other cytopathological changes in Down's syndrome and, by inference, in Alzheimer's disease. This finding suggests that β-amyloidosis, like certain other amyloidoses, does not occur secondary to local cellular pathology (*e.g.*, dystrophic neurites) but rather precedes it. The clearest evidence that the processing of βAPP into Aβ can actually cause Alzheimer's disease has come from the identification by several laboratories of missense mutations in the βAPP gene within and flanking the Aβ region in affected members of certain families having Alzheimer's disease or hereditary cerebral hemorrhage with amyloidosis of the Dutch type.

RESULTS AND DISCUSSION

The mechanism of proteolytic release of the Aβ fragment from βAPP is poorly understood. Because the normal secretion of the large extramembranous portion of βAPP (APP_s) from cells involves a proteolytic cleavage within Aβ, we searched for evidence of an alternate pathway of βAPP processing that leaves Aβ intact. In view of the presence of a consensus sequence (NPXY) in the cytoplasmic tail of βAPP that could mediate internalization of the protein from the cell surface and its targeting to endosomes/lysosomes, we looked specifically for evidence of endocytotic trafficking of βAPP. Incubation of an antibody to the extracellular region of βAPP with living human cells led to binding of the antibody to cell-surface βAPP and trafficking of the antigen-antibody complex to endosomes/lysosomes.[1] The resultant βAPP-immunoreactive pattern closely resembled that seen after incubating the same cells with rhodamine-tagged albumin, a marker for fluid-phase endocytosis.

Late endosomes/lysosomes purified from the cells by step-gradient centrifugation contained abundant full-length βAPP plus an array of low molecular weight, carboxyl terminal fragments ranging from ~ 10 to 22 kD, most of which are of a size and immunoreactivity suggesting that they contain the intact Aβ peptide.[1] Similar potentially amyloidogenic C-terminal fragments had been detected previously in whole lysates of βAPP-expressing cells.[2–4] These various results provide evidence that some βAPP molecules are normally reinternalized from the cell-surface and targeted to lysosomes. This second cellular pathway for βAPP processing (the first pathway described was the α-secretase pathway)[5,6] is capable of producing potentially amyloidogenic fragments. Recent studies by Koo and Squazzo have shown directly that at least a portion of secreted Aβ is generated during early endocytic trafficking of βAPP.[7] However, it is not yet clear to what extent the endocytic pathway, an alternative proteolytic cleavage (other than that by α-secretase) occurring within the secretory pathway,[8] or a combination of these two routes is responsible for Aβ formation in different cell types.

During the course of the aforementioned studies, we searched for evidence of the production and release of the Aβ peptide itself during normal cellular metabolism, based in part on the hypothesis that some Aβ deposits (*e.g.*, those in capillary walls and the subpial cortex) might arise from a circulating (plasma or CSF) source of the peptide. To this end, a series of antibodies to Aβ were used to screen the conditioned media of several cell types for the presence of soluble Aβ. These experiments demonstrated that Aβ is continuously produced as a soluble 4 kD peptide and is released into the media of normal cells.[9] In simultaneously conducted experiments, Aβ was detected in cerebrospinal fluid in addition to various cultured cell media[10,11] and in plasma.[10] The form in CSF has been purified and sequenced, confirming that it is authentic Aβ,[10] whereas the plasma form has clearly been detected immunochemically but not yet fully characterized at the structural level. Aβ peptides with heterogeneous N- and C-termini are now known to be released by all βAPP-expressing cells studied to date under normal culture conditions.[9–15] Aβ in culture supernatants is entirely soluble and generally present in high picomolar to low nanomolar concentrations.[10] Pulse-chase and biological toxin experiments have suggested that Aβ production follows full maturation of βAPP and involves an acidic compartment other than lysosomes, *e.g.*, early endosomes or the late Golgi.[12] The two proteolytic cleavages generating Aβ (*i.e.*, β-secretase and γ-secretase cleavages) may therefore occur in an acidic vesicle near the cell-surface, after which Aβ is rapidly released into the medium, with little or no Aβ detected intracellularly.[9,12]

The relevance of such *in vitro* Aβ production to the pathogenesis of Alzheimer's disease (AD) is demonstrated by the finding that a βAPP missense mutation causing a Swedish form of familial AD, when expressed in cultured cells, leads to a marked increase in Aβ production.[16] Moreover, primary fibroblasts cultured from the skin of living subjects with this Swedish mutation show hypersecretion of Aβ, and this can be seen at least a decade prior to onset of clinical symptoms.[17] The amyloidogenic mechanism of three other FAD-linked βAPP mutations, the mutations at βAPP_{717val}, have now been shown to involve increased production of the longer (42-residue), more amyloidogenic form of Aβ.[15] Also, the mutation at βAPP_{692Ala} that causes both AD and hemorrhagic amyloid angiopathy has been found to lead to increased Aβ secretion, particularly of N-terminally truncated forms.[14]

Because cell-surface βAPP molecules appear to provide one source of precursors for Aβ production, it is important to understand the precise pathway and kinetics of cell-surface βAPP trafficking. To this end, we have applied biochemical and morphological methods to βAPP-transfected Chinese hamster ovary cells. By binding a labeled monoclonal βAPP antibody to living cells, we demonstrated that some surface βAPP molecules were rapidly secreted into the medium with a tl/2 under ten minutes. At the same time, some surface βAPP was rapidly internalized. A cell line transfected with a deletion construct lacking the βAPP cytoplasmic tail showed enhanced secretion and a corresponding decrease in internalization, as expected. Using immunofluorescence and immunoelectron microscopy, βAPP was shown to be rapidly internalized via coated pits and vesicles, after which the molecules were transported to endosomes, prelysosomes and lysosomes. Using a modified immunodetection protocol, we directly demonstrated the rapid recycling (within 5–10 minutes) of endocytosed βAPP to the cell surface and the ultimate targeting of a portion of βAPP to lysosomes. Because endocytosis of cell-surface βAPP has been shown to be one route for the constitutive production of Aβ,[7] the recycling pathway for cell-surface βAPP delineated in these studies is a probable route for production of the critical Aβ fragment.

Because the β-secretase cleavage of βAPP to generate the Aβ N-terminus appears to occur prior to the γ-secretase cleavage that generates the C-terminus, it is important to characterize the substrate requirements of the β-secretase(s) and ultimately identify the responsible protease(s). To this end, we have mutagenized βAPP by placing stop codons within or at the end of the transmembrane domain or by substituting various amino acids for the wild-type methionine and aspartate at the P_1 and P_1' positions. These and related experiments have led to the conclusion that β-secretase-type cleavage requires a membrane-anchored βAPP substrate but can tolerate shifts in the distance of the hydrolyzed peptide bond from the membrane, in contrast to findings reported for α-secretase-type cleavage. The major β-secretase-type protease appears to have a minimum recognition region of val669 to ala673 (βAPP_{770}); most substitutions in this sequence strongly decrease or eliminate Aβ production when expressed in intact human cells. Only the Swedish familial AD mutation (lys670 asn/met671leu) strongly increases Aβ production. Moreover, in the Swedish mutant but not in wild-type βAPP, the entire cytoplasmic tail with its reinternalization signals can be deleted without affecting the β-secretase cleavage, consistent with the concept that Aβ-generating cleavage of this mutant form occurs in a different cellular compartment than that of wild-type molecules. Although our experiments demonstrate a high degree of sequence specificity for the β-secretase cleavage in intact neural and non-neural cells, analogous mutagenesis experiments around the Aβ C-terminus suggest that the unknown protease(s) cleaving here [*i.e.*, γ-secretase(s)] does not show much peptide bond specificity. Our results about the characteristics of these cleavages have implications for current intensive approaches to develop assays for, identify and purify enzymes with β-secretase or γ-secretase activity.

The cellular trafficking of βAPP is of interest not only because it should provide information about where within the cell the critical Aβ-generating cleavages occur, but also because it provides more general information about the mechanisms for trafficking of membrane-spanning proteins that are both inserted as holoproteins at the cell surface and serve as the precursors of various secreted derivatives. We have previously reported that βAPP undergoes highly polarized sorting in Maden-Darby canine kidney (MDCK) epithelial cells, a well characterized model system for studying protein trafficking in polarized cells. βAPP is trafficked principally to the basolateral surface of MDCK cells, and the major soluble derivative (APP_s) as well as Aβ are preferentially secreted from the basolateral surface.[18] In more recent studies that will be published shortly,[19] we have extended this work by conducting an analysis of the sorting signals in the βAPP cytoplasmic tail, using cDNA constructs containing a variety of deletions and substitutions.

Deletion of the last 32 amino acids (residues 664–695) of the βAPP_{695} cytoplasmic tail had no influence on either the constitutive ~90% level of basolateral sorting of surface βAPP, or the strong basolateral secretion of APP_s, Aβ, and p3. However, deleting the last 42 amino acids (residues 654–695) or changing tyrosine 653 to alanine altered the distribution of cell-surface βAPP so that ~40–50% of the molecules were inserted apically. In parallel, Aβ was now secreted from both surfaces. Surprisingly, this change in surface βAPP had no influence on the basolateral secretion of APP_s and p3. This result suggests that most βAPP molecules which give rise to APP_s in MDCK cells are cleaved intracellularly before reaching the surface. Consistent with this conclusion, we readily detected intracellular APP_s in carbonate extracts of isolated membrane vesicles. Moreover, ammonium chloride treatment resulted in the equal secretion of APP_s into both compartments, as occurs with other non-membranous, basolaterally secreted proteins, but it did not influence the polarity of cell-surface βAPP. These results demonstrate that in epithelial cells two independent mechanisms mediate the polarized trafficking of βAPP holoprotein and its major secreted derivative (APP_s) and that Aβ peptides are derived in part from βAPP holoprotein targeted to the cell surface by a signal that includes tyrosine 653.

The studies summarized above demonstrate the usefulness of cultured cells expressing normal or mutant βAPP molecules for deciphering both the proteolytic processing and the intercellular trafficking of βAPP. In addition to their utility in such studies, cells constitutively secreting Aβ might be useful for examining the critical issue of the aggregation of the soluble Aβ monomer into its potentially neurotoxic aggregated (polymeric) form under physiological conditions. Attempts to model the cytotoxicity of Aβ *in vitro* using synthetic peptides have shown that monomeric Aβ is relatively inert, whereas aggregated, polymeric Aβ reproducibly exerts a variety of neurotoxic effects. The processes that initiate the conversion of soluble Aβ to a polymeric, toxic state are of great interest, but previous studies of this conversion have employed high doses (10^{-3}–10^{-5} M) of synthetic Aβ peptides under nonphysiological conditions. We have found that Chinese hamster ovary cells expressing endogenous or transfected βAPP spontaneously form small amounts ($< 10^{-9}$M) of Aβ oligomers from their constitutively secreted Aβ monomers under normal culture conditions. The identity of these low oligomers (primarily dimers and trimers) was confirmed by selective and specific immunoprecipitation with a panel of Aβ antibodies, by electrophoretic comigration with synthetic Aβ oligomers and by amino acid sequencing. The oligomeric Aβ species comprised ~ 10–20% of the total immunoprecipitable Aβ in these cultures. A truncated Aβ species beginning at arg 5 was found to be enriched in the oligomers, suggesting that N-terminal heterogeneity can influence Aβ oligomerization in this system. Addition of Congo red (10 μm) during metabolic labeling of the cells led to increased monomeric and decreased oligomeric Aβ. Similarly, Aβ multimers have recently been detected in another cell type, but in much lower amounts. The ability to detect and quantitate low oligomers of secreted Aβ peptides in cell culture should now enable dynamic studies of the critical process of initial Aβ aggregation under physiological conditions, and could potentially be used to screen and characterize compounds that inhibit Aβ polymerization.

CONCLUSION

The studies summarized above as well as similar studies being conducted in numerous other laboratories demonstrate that the regulation of Aβ production, secretion and extracellular aggregation can be effectively studied under controlled conditions in a variety of cultured cells. Such work can provide information about βAPP processing and the economy of Aβ under physiological conditions that can complement analyses of these processes *in vivo*. Moreover, cells stably expressing wild-type or mutant βAPP provide an excellent screening system to identify potentially therapeutic compounds which decrease Aβ production or enhance its clearance. Positive compounds identified by such *in vitro* screening can then be tested in *in vivo* models of Alzheimer's disease.

REFERENCES

1. HAASS, C., E. H. KOO, A. MELLON, *et al.* 1992a Targeting of cell-surface β-amyloid precursor protein to lysosomes: alternative processing into amyloid-bearing fragments. Nature **357:**500–503.
2. COLE, G. M., T. V. HUYNH & R. T. SAITOH. 1989. Evidence for lysosomal processing of beta-amyloid precursor in cultured cells. Neurochem. Res. **14:** 933–939.
3. GOLDE, T. E., S. ESTUS, L. H. YOUNKIN, *et al.* 1992. Processing of the amyloid protein precursor to potentially amyloidogenic carboxyl-terminal derivatives. Science **255:**728–730.
4. CAPORASO, G. L., S. E. GANDY, J. D. BUXBAUM, *et al.* 1992. Chloroquine inhibits intracellular degradation but not secretion of Alzheimer β/A4 amyloid precursor protein. Proc. Natl. Acad. Sci. USA **89:**2252–2256.
5. ESCH, F. S., P. S. KEIM, E. C. BEATTIE, *et al.* 1990. Cleavage of amyloid β-peptide during constitutive processing of its precursor. Science **248:** 1122–1124.
6. SISODIA, S. S., E. H. KOO, K. BEYREUTHER, *et al.* 1990. Evidence that β-amyloid protein in Alzheimer's disease is not derived by normal processing. Science **248:**492–495.
7. KOO, E. H. & S. SQUAZZO. 1994. Evidence that production and release of amyloid β-protein involves the endocytic pathway. J. Biol. Chem. **269:** 17386–17389.
8. SEUBERT, P., T. OLTERSDORF, M. G. LEE, *et al.* 1993. Secretion of β-amyloid precursor protein cleaved at the amino-terminus of the β-amyloid peptide. Nature **361:**260–263.
9. HAASS, C., M. G. SCHLOSSMACHER, A. Y. HUNG, *et al.* 1992b. Amyloid β-peptide is produced by cultured cells during normal metabolism. Nature **359:** 322–325.
10. SEUBERT, P., C. VIGO-PELFREY, F. ESCH, *et al.* 1992. Isolation and quantitation of soluble Alzheimer's β-peptide from biological fluids. Nature **359**:325–327.
11. SHOJI, M., T. E. GOLDE, J. GHISO, *et al.* 1992. Production of the Alzheimer amyloid β protein by normal proteolytic processing. Science **258:**126–129.
12. HAASS, C., A. Y. HUNG, M. G. SCHLOSSMACHER, *et al.* 1993. β-Amyloid peptide and a 3-kDa fragment are derived by distinct cellular mechanisms. J. Biol. Chem. **268:**3021–3024.
13. BUSCIGLIO, J., D. H. GABUZDA, P. MATSUDAIRA, *et al.* 1993. Generation of β-amyloid in the secretory pathway in neuronal and nonneuronal cells. Proc. Natl. Acad. Sci. USA **90:**2092–2096.

14. HAASS, C., A. Y. HUNG, D. J. SELKOE, *et al.* 1994. Mutations associated with a locus for familial Alzheimer's disease result in alternative processing of amyloid β-protein precursor. J. Biol. Chem. **269:**17741–17748.
15. SUZUKI, N., T. T. CHEUNG, X.-D. CAI, *et al.* 1994. An increased percentage of long amyloid β-protein secreted by familial amyloid β-protein precursor (βAPP717) mutants. Science **264:**1336–1340.
16. CITRON, M., T. OLTERSDORF, C. HAASS, *et al.* 1992. Mutation of the β-amyloid precursor protein in familial Alzheimer's disease increases β-protein production. Nature **360:**672–674.
17. CITRON, M., C. VIGO-PELFREY, D. B. TEPLOW, *et al.* 1994. Excessive production of amyloid β-protein by peripheral cells of symptomatic and presymptomatic patients carrying the Swedish familial Alzheimer's disease mutation. Proc. Natl. Acad. Sci. USA. **91**:11993–11997.
18. HAASS, C., E. H. KOO, D. B. TEPLOW, *et al.* 1994. Polarized secretion of β-amyloid precursor protein and amyloid β-peptide in MDCK cells. Proc. Natl. Acad. Sci. USA **91:**1564–1568.
19. HAASS, C., E. H. KOO, A. CAPELL, *et al.* 1995. Polarized sorting of β-amyloid precursor protein and its proteolytic products in MDCK cells is regulated by two independent signals. J. Cell Biol. **128**:537–547.

Mice Homozygous for a Modified β-Amyloid Precursor Protein (βAPP) Gene Show Impaired Behavior and High Incidence of Agenesis of the Corpus Callosum[a]

U. MÜLLER,[b] N. CRISTINA,[b] Z.-W. LI,[b] D.P.Y. WOLFER,[c] H.-P. LIPP,[c] T. RÜLICKE,[d] S. BRANDNER,[e] A. AGUZZI,[e] AND C. WEISSMAN[b,f]

[b] *Institut für Molekularbiologie I, Universität Zürich, 8093 Zürich, Switzerland*
[c] *Anatomisches Institut, Universität Zürich, 8057 Zürich, Switzerland*
[d] *Biologisches Zentrallobor and* [e] *Institut für Neuropathologie, Universitätsspital Zürich, 8091 Zürich, Switzerland*

ABSTRACT: The amyloid precursor protein (βAPP) gene of the mouse was disrupted by homologous recombination; however, contrary to expectation, brain and other tissues still contained βAPP-specific RNA, albeit at a level 5-10 fold lower than wild-type and lacking the disrupted exon, which had been spliced out. The brain contained shortened βAPP-specific protein at a low level. Mutant mice were severely impaired in spatial learning and exploratory behavior and showed increased incidence of agenesis of the corpus callosum.

INTRODUCTION

Several possible functions have been proposed for βAPP (for reviews see refs. 1–3). Nonetheless, the physiological role of βAPP is not well understood and because it was of interest to determine whether βAPP was an essential protein, we undertook to disrupt the βAPP gene of the mouse by inserting into exon 2 a cassette containing a neomycin resistance gene and a putative transcription termination sequence. Contrary to expectation, brain and other tissues from mice homozygous for the insertion still contained βAPP-specific RNA, albeit at a level 5-10 fold lower than wild-type and lacking the disrupted

[a] This work was supported by the Kanton of Zürich, the Swiss National Foundation (NF 31-37497) and by a grant of EMBO to U.M. and of the Roche Research Foundation to N. Cristina.

[f] *Send correspondence to:* Dr. Charles Weissmann, Institut für Molekularbiologie I, Universität Zürich, 8093 Zürich, Switzerland; Tel: 411-63-2491; Fax: 411-371-7205; E-mail: Weissmann @aeolus.ethz.ch.

exon, which had been spliced out. Brain contained slightly shortened βAPP-specific protein at about 5% of wild-type level, and what appear to be βAPP-specific fragments at the 30% level. These mice showed a severely impaired behavioral phenotype and an incidence of agenesis of the corpus callosum substantially above that of wild-type littermates.

RESULTS

Generation of Mice Homozygous for Interrupted βAPP Genes

ES cells with one disrupted βAPP allele ($\beta APP^{\Delta/+}$) cells) were generated by homologous recombination with a targeting vector in which a cassette containing a neomycin resistance gene in the antisense orientation and the UMS sequence, described as a transcription termination sequence[4] had been inserted into exon 2. ES cells with an interrupted βAPP allele were injected into blastocysts and the blastocysts implanted into foster mothers. Chimeric males with a high frequency of germ line transmission were obtained. Mating of heterozygous offspring (with a 129/Sv(ev) × C57BL/6 background) yielded 19% $\beta APP^{\Delta/\Delta}$ homozygous pups. Southern analysis of tail DNA showed the pattern typical for the interrupted βAPP genes and the absence of wild-type genes in some progeny.[5]

Physical Characterization of βAPP^Δ/Δ Mice

$\beta APP^{\Delta/\Delta}$ mice were fertile and healthy. Seven of 9 $\beta APP^{\Delta/\Delta}$ brains examined lacked a corpus callosum, while only 1 of 8 wild-type and 1 of 8 heterozygous brains (all derived from 129/SV(ev) × C57BL/6 progeny) showed a similar defect (FIG. 1). The incidence of corpus callosum agenesis in 129/Sv(ev) parental mice was 14%. It would seem that the $\beta APP^{\Delta/\Delta}$ genotype causes agenesis or contributes to an underlying genetic predisposition for this trait.

Analysis of βAPP Gene Expression

Northern blot analysis of RNA from organs of βAPP mice hybridized with an exon 2 antisense probe revealed no 3.5-kb wild-type mRNA. Hybridization with an exon-17–containing antisense RNA probe revealed a transcript with a mobility marginally higher than that in RNA from wild-type organs, suggesting that the targeted exon had been skipped during splicing, giving rise to an mRNA shorter by 168 nucleotides than its wild-type counter-

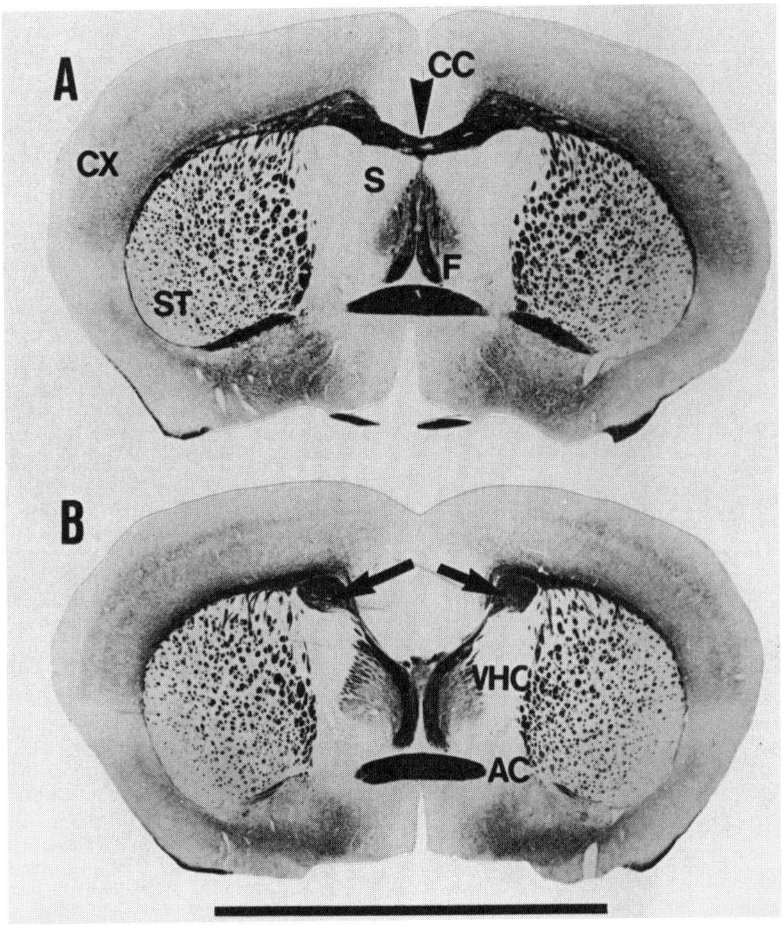

FIGURE 1. Brain sections through representative $\beta APP^{\Delta/\Delta}$ and wild brains. Histographies of coronal frozen sections (20 μm) at the level of the anterior commissure in a wild-type $\beta APP^{+/+}$ **(A)** and a mutant ($\beta APP^{\Delta/\Delta}$) mouse **(B)**. Myelin silver stain modified after Gallyas.[11] Note the presence of a fully developed corpus callosum *(large arrow-head)* in **A,** and the complete agenesis in the mutant animal **(B)**. Large arrows indicate longitudinal fiber tracts (Probst bundles) that are formed by commissural axons failing to cross the midline. *Scale bar:* 5 μm. **Abbreviations:** AC, anterior commissure; CC, corpus callosum; CX, neocortex; F, descending columns of the fornix; S, septum; ST, striatum; VHC, ventral hippocampal commissure. (From Müller *et al.*[5])

part. This interpretation was confirmed by RT-PCR analysis. Sequencing of the cloned, shortened PCR products revealed precise excision of exon 2 with retention of the reading frame.

Western analysis with the antibody 22C11, which is directed against an epitope overlapping the skipped exon (amino acids 66-81),[6] gave barely de-

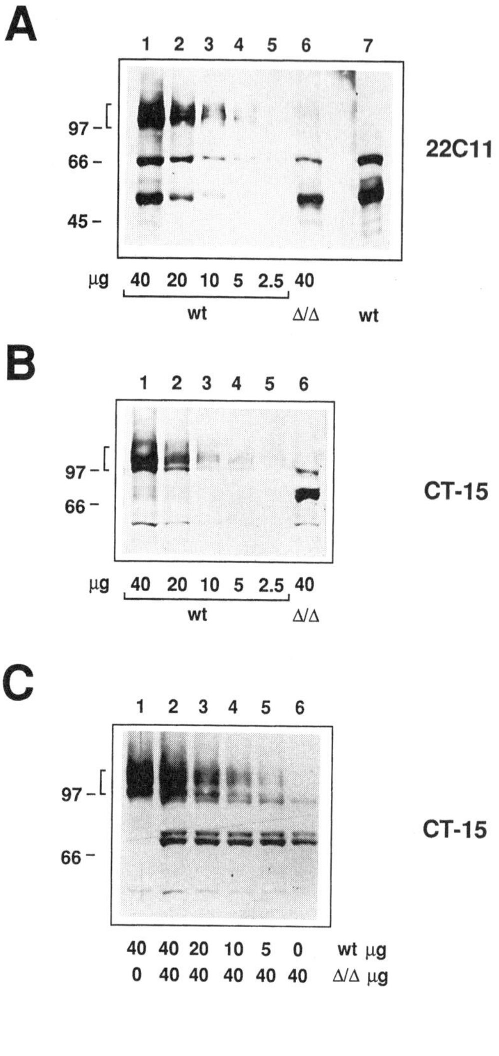

FIGURE 2. Western blot analysis of brain extracts from wild-type and $\beta APP^{\Delta/\Delta}$ mice. Western blots of wild-type (wt) and $\beta APP^{\Delta/\Delta}$ (Δ/Δ) mice were prepared as described in Müller *et al.*[5] **(A)** Different amounts of wild-type brain extracts (lanes 1–5) compared to 40 μg $\beta APP^{\Delta/\Delta}$ (lane 6). Incubation with the monoclonal antibody 22C11 (directed against an epitope overlapping exons 2 and 3[6] followed by anti-mouse HRP-linked antibody. Lane 7: first antibody omitted, detects unspecific reaction with 68- and 55-kD proteins. The brackets indicate βAPP-specific bands of 100-135 kD. After exposure, the blot was stripped and incubated with a monoclonal antibody against NCAM which revealed that wild-type lanes (1–5) were slightly underloaded compared to extract from the mutant, lane 6 (data not shown). **(B)** Different amounts of wild-type brain extracts (lanes 1–5) compared to 40 μg $\beta APP^{\Delta/\Delta}$ (lane 6). Incubation with the antiserum CT-15 directed against the 15 carboxy terminal amino acids of βAPP[7] followed by anti-rabbit HRP-linked antibody. **(C)** Extracts from wild-type (lane 1), $\beta APP^{\Delta/\Delta}$ (lane 6) and mixtures of the two in the proportions indicated (lanes 2–5). Incubation was with antiserum CT-15. All blots were developed with chemiluminescence reagents (ECL, Amersham). Biotinylated marker proteins (high molecular weight kit) were from Biorad. (From Müller *et al.*[5])

tectable APP bands in mutant mice (FIG. 2A). Analysis with the polyclonal antibody CT-15, which is directed against the 15 carboxy terminal amino acids of βAPP[7] and discriminates (S. Sisodia, personal communication) against the βAPP-like proteins APLP-1 and APLP-2, gave a set of bands of 100-135 kD for wild-type brain homogenates. In contrast, extracts from $\beta APP^{\Delta/\Delta}$ brains showed a weaker band of about 95 kD and two stronger bands of about 78 and 80 kD (FIG. 2B, lane 6), which were not detected in the wild-type pattern (lanes 1–5) and must therefore be APP-specific. A mixture of wild-type and mutant brain extracts (FIG. 2C, lanes 2–5) clearly shows that the 95-kD mutant band has a slightly higher mobility than the fastest wild type band, and may therefore correspond to APP with the expected deletion. The high molecular weight βAPP-specific band in the $\beta APP^{\Delta/\Delta}$ sample was about 5% of the sum of βAPP-specific bands in the wild-type extract (FIG. 2B, compare lane 6 to lanes 1–5), and the sum of all APP-specific immunoreactive material in the mutant sample was about 30%. From the Northern and the Western data we conclude that considerably reduced levels of βAPP lacking amino acids 20–75 accumulated in brain and, as judged by Northern analyses, also in other tissues of $\beta APP^{\Delta/\Delta}$-mice.

Behavioral Analysis

Open-field Behavior

$\beta APP^{\Delta/\Delta}$ mice showed distinctly lowered locomotor activity (data not shown). The total path length covered by $APP^{\Delta/\Delta}$ mice was about 70% of that measured for the control group. The difference disappeared only at the end of the second observation period due to the within-session habituation of the control group. The mutants did not show such habituation in either session. No significant differences were noticed for the speed of locomotion (16.8 ± 0.3 cm/sec for the entire sample).

Morris Swimming Navigation Test

During the three days of acquisition, all mice showed significant learning as evidenced by increasingly shorter pathways (ANOVA $p < 0.001$; FIG. 3A), but $\beta APP^{\Delta/\Delta}$ mice were, on average, significantly inferior in this regard ($p < 0.01$). On days 2 and 3 of acquisition, mutant mice showed an 84% and 188% total swim path, respectively, than the controls, despite a single extremely poorly behaving wild-type mouse. Detailed analysis of the swimming paths (for an example see FIG. 3B) showed that during the second and third day of acquisition about half the mutant mice found the platform occasionally, but such successful trials alternated with trials in which the mice swam seemingly aimlessly. But even $\beta APP^{\Delta/\Delta}$ mice which had successfully acquired swimming navigation seemed mildly impaired inasmuch as only 3 of them performed above the median level of the entire control sample.

During the first trial after platform relocation, controls crossed the former platform position significantly more often ($p < 0.05$) than the $\beta APP^{\Delta/\Delta}$ mice, while the number of crossings over symmetrical positions (annuli) did not show group differences (data not shown). Similarly, the proportion of the swim path length in the old target quadrant was much higher for controls

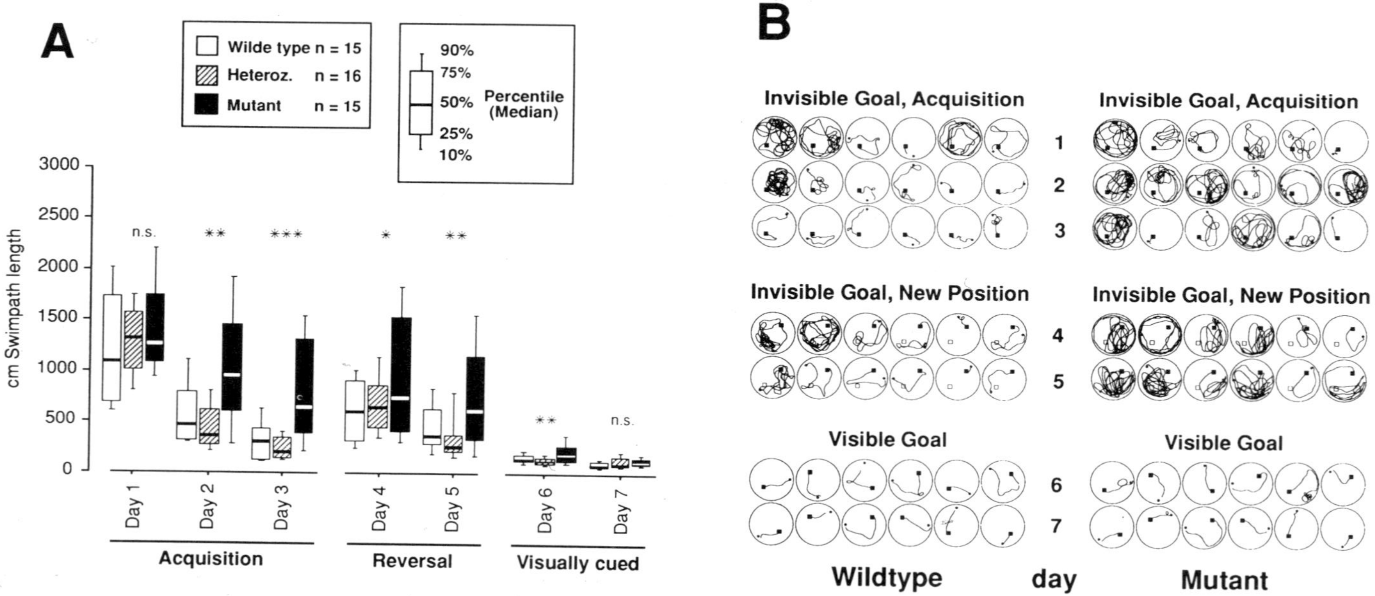

FIGURE 3. Behavior of wild-type, heterozygous and $\beta APP^{\Delta/\Delta}$ litter mates in the Morris water maze. The assay conditions are described in Müller *et al.*[5] n.s., not significant; *, $p < 0.05$; **, $p < 0.01$, ***, $p < 0.001$; two-tailed t-test (wild-type and heterozygous mice versus βAPP mutants). **(A)** "Box-and-whisker" plot of the average swim path length during five days of swimming navigation learning ("acquisition": invisible platform, standard position: "reversal": invisible platform, new position) and two days of visually guided water escape behavior ("visually cued": visible platform, variable position). **(B)** Swim paths of a representative βAPP mutant and of a wild-type mouse during swimming navigation learning ("invisible goal, acquisition" and "invisible goal, new position") and water escape learning using a flagged platform ("visible goal"). The wild-type mouse shows well-directed learning during the acquisition phase in the Morris test, intensive search during the probe trial and smooth learning of the new platform position, (with short visits of the old platform position). With the escape platform flagged, the swim paths appear well-directed and short. In contrast, the βAPP mutant shows erratic performance during acquisition and reversal learning. Visually guided escape behavior is performed smoothly. (Modified from Müller *et al.*[5])

(45%) than for $\beta APP^{\Delta/\Delta}$ mice (29%) during the first two trials of reversal learning ($p < 0.01$). During the first 2-3 trials with the relocated platform, controls showed the expected prolonged swim paths and then learned to find the new position, while most mutants had swim paths about twice as long as those of the controls throughout.

Visually Cued Water Escape Learning

This paradigm tests whether poor swimming navigation is due to visual problems rather than learning per se. All mice showed quick orientation towards the visually marked platform location (for examples, see FIG. 3B).

DISCUSSION

So far, total ablation of βAPP in the mouse has not been reported; in Drosophila, deletion of an βAPP-like gene results in a viable fly exhibiting subtle behavioral deficits which can be complemented by the human βAPP gene.[8]

We were not successful in our original goal of completely abolishing expression of βAPP in the mouse because the disrupted exon was skipped during splicing of the readthrough transcript. As a result a reduced amount of βAPP lacking amino acids 20 to 75 was still formed.

Behavioral analyses revealed remarkably strong deficits in the mutant animals. In comparison to a much larger sample of about 300 mice of comparable genetic background (129/Sv(ev) × C57BL/6) carrying a variety of homozygous gene disruptions, the $\beta APP^{\Delta/\Delta}$ mice stood out as the only group showing clearly abnormal acquisition of swimming navigation. We ascribe the behavioral impairments to the mutation in the βAPP gene (or in a gene very closely linked to it) rather than to an incidental mutation, because the mice tested were littermates derived from heterozygous crosses and selected for their βAPP genotype.

The results of swimming navigation indicate that the mutation impaired cognitive processes in about half of the $\beta APP^{\Delta/\Delta}$ mice, while others seemed less affected. The cognitive nature of this deficit is perhaps characterized best as unstable and retarded acquisition of complex tasks rather than a principal learning inability. The results of the open-field test showing strongly reduced locomotor activity as well as reduced exploratory behavior suggest that the mutation influenced motivational systems responsible for behavioral control in stress situations.

Several findings suggested that βAPP may have a role in developmental or synaptic plasticity, notably in neurite outgrowth (reviewed in Small *et al.*,

1994[12]) and regulation of intraneuronal Ca^{2+}-level.[9] Since the $\beta APP^{\Delta/\Delta}$ protein has a deletion in the extracellular domain cell adhesion or axon outgrowth could be impaired.

Complete agenesis of the corpus callosum was observed in about 80% of the $\beta APP^{\Delta/\Delta}$ and 12% of the wild-type and heterozygous littermates. In 129/Sv(ev) wild-type mice of pure genetic background the incidence of corpus callosum agenesis was 14%. The trait, which does not occur in C57BL/6 mice, is attributed to recessive, multifactorial inheritance and is due to retarded outgrowth of neurites during embryonal development (reviewed by Lipp and Wahlsten[10]). The defect is not expected to occur in F1 offspring of 129/Sv(ev) × C57BL/6 but may reappear in offspring of F1 × F1 and of subsequent crosses and thus account for our findings. It would seem that the very high incidence of total agenesis in $\beta APP^{\Delta/\Delta}$ mice is due to the APP mutation; whether or not expression of this phenotype requires the 129 genetic background remains to be determined.

In summary, the analysis of the behavioral deficits suggests that several brain systems were affected by the mutation. Because there was no evidence for motor dyscoordination or sensory disability, one may infer that the main effects were mediated by reduced functionality of the forebrain. The wide variance in the performance of the mutant mice may be ascribed to developmental plasticity. Agenesis of the corpus callosum comes about when the outgrowth of the relevant neurites, which must occur within a narrow temporal window during development, is retarded.[10] Incomplete penetrance of the agenesis trait could come about if the timing of outgrowth in the $APP^{\Delta/\Delta}$ mutant varies around the critical value.

It is not clear whether the phenotype we have observed is due to impaired functionality of the shortened βAPP protein, to the low level of a functional albeit shortened βAPP protein, to abnormal processing of the mutant protein or to a combination of the above. In any event, it is clear that the βAPP mutation we describe is recessive, because no difference in behavior and anatomy between heterozygote and wild-type animals were noted.

ACKNOWLEDGMENTS

We thank K. Beyreuther (Heidelberg), S. Sisodia (Baltimore) and P. Frey (Sandoz, Basel) for their kind gifts of antibodies, Y. Sakaki (Tokyo) for βAPP cDNAs, G. Huber and M. Gschwind (Hoffmann-LaRoche, Basel) for a βAPP genomic clone and Mrs. M. Stagljar for carrying out behavioral tests.

REFERENCES

1. MULLAN, M. & F. CRAWFORD. 1993. Genetic and molecular advances in Alzheimer's disease. Trends Neurosci. **16:**398–403.

2. SELKOE, D. J. 1993 Physiological production of the β-amyloid protein and the mechanism of Alzheimer's disease. Trends Neurosci. **16:**403–409.
3. MATTSON, M. P., S. W. BARGER, B. CHENG, I. LIEBERBURG, S. V. SMITH & R. E. RYDEL. 1993. Beta-amyloid precursor protein metabolites and loss of neuronal CA2+ homeostasis in Alzheimer's disease. Trends Neurosci. **16:** 409–414.
4. HEARD, J. M., P. HERBOMEL, M. O. OTT, R. A. MOTTURA, M. WEISS & M. YANIV. 1987. Determinants of rat albumin promoter tissue specificity analyzed by an improved transient expression system. Mol. Cell Biol. **7:**2425–2434.
5. MÜLLER, U., N. CRISTINA, Z-W. LI, *et al.* 1994. Behavioral and anatomical deficits in mice homozygous for a modified β-amyloid precursor protein gene. Cell **79:**755–765.
6. HILBICH, C., U. MONNING, C. GRUND, C. L. MASTERS & K. BEYREUTHER. 1993. Amyloid-like properties of peptides flanking the epitope of amyloid precursor protein-specific monoclonal antibody 22C11. J. Biol. Chem. **268:** 26571–26577.
7. SISODIA, S. S., E. H. KOO, P. N. HOFFMAN, G. PERRY & D. L. PRICE. 1993. Identification and transport of full-length amyloid precursor proteins in rat peripheral nervous system. J. Neurosci. **13:**3136–3142.
8. LUO, L., T. TULLY & K. WHITE. 1992. Human amyloid precursor protein ameliorates behavioral deficit of flies deleted for APPI gene. Neuron **9:** 595–605.
9. MATTSON, M. P., B. CHENG, A. R. CULWELL, F. S. ESCH, I. LIEBERBURG & R. E. RYDEL. 1993. Evidence for excitoprotective and intraneuronal calcium-regulating roles for secreted forms of the beta-amyloid precursor protein. Neuron **10:**243–254.
10. LIPP, H-P. & D. WAHLSTEN. 1992. Absence of the corpus callosum. *In* Genetically Defined Animal Models of Neurobehavioral Dysfunction. P. Driscoll, Ed. : 217–252. Boston/Basel: Birkhäuser.
11. GALLYAS, F. 1979. Silver staining of myelin by means of physical development. Neurol. Res. **1:**203–209.
12. SMALL, D. H., V. NURCOMBE, G. REED, H. CLARRIS, R. MOIR, K. BEYREUTHER & C. L. MASTERS. 1994. A heparin-binding domain in the amyloid protein precursor of Alzheimer's disease is involved in the regulation of neurite outgrowth. J. Neurosci. **14:**2117–2127.

Regulation of APP Expression, Biogenesis and Metabolism by Extracellular Matrix and Cytokines[a]

K. BEYREUTHER,[b,d] G. MULTHAUP,[b] U. MÖNNING,[b] R. SANDBRINK,[b] D. BEHER,[b] L. HESSE,[b] D. H. SMALL,[c] AND C. L.MASTERS[c]

[b] *ZMBH, Center for Molecular Biology, University of Heidelberg, D-69120 Heidelberg, Germany*

[c] *Department of Pathology, The University of Melbourne, and Mental Health Research Institute, Parkville, Victoria 3052, Australia*

ABSTRACT: We have identified and characterized the ligand binding properties of the Alzheimer's disease (AD) βA4 amyloid protein precursor (APP), mapped the APP ligand binding sites and analyzed the regulation of APP expression, biogenesis and metabolism by components of the extracellular matrix (ECM) and cytokines.

We find that the sequence of APP includes a binding site for copper(II) within exon IV,[1] for zinc(II) within exon V,[2] a high-affinity binding site for the heparin glycosaminoglycan side chains of proteoglycans (HSPG) within exons IX and X,[3,4] a homophilic binding site within exon XII and XIII[2] and a collagen and laminin site within exons XII and XIII.[5] Homophilic binding of APP to APP and binding of APP to heparin are modulated by zinc(II).[1,3] A low-affinity binding site for proteoglycans which mediates the neurite outgrowth promoting effects of APP has been mapped within exon III.[6]

By examining APP expression in different immuno-competent cells and cells of the nervous system we found that APP expression, biogenesis and metabolism is also regulated by components of the extracellular matrix (ECM) and by cytokines which are produced in response to tissue damage.[7–10]

Our axotomy studies revealed that APP participates in the microglial im-

[a] This work was supported by grants from the Deutsche Forschungsgemeinschaft, the Bundesministerim für Forschung und Technologie; the Landesforschungsschwerpunkt Baden-Wrttemberg, the Metropolitan-Life Foundation, from the Fonds der Chemischen Industrie, the National Health and Medical Research Council of Australia, and from the Victorian Health Promotion Foundation.

[d] *Send correspondence to:* Dr. Konrad Beyreuther, ZMBH, Zentrum für Molekulare Biologie, Universität Heidelberg, Im Neuenheimer Feld 282, D-69120 Heidelberg, Germany; Tel: +49-6221-566845; Fax: +49-62211-565891; Email: bsa784@sun0.urz.uni-heidelberg.de.

mune response of the brain.[8] Interaction of microglia with fibronectin upregulates APP secretion and decreases the level of cellular amyloidogenic APP.[9,10] Laminin and collagen produce an increase in amyloidodogenic APP and APP fragments in the cells and a decrease in APP secretion. The immunosuppressive cytokine TGFβ induces also this accumulation of cellular βA4-containing APP.[9]

These studies show first that APP is a metal binding protein with binding characteristics of cell-adhesion and substrate-adhesion molecules, and second that APP expression, biogenesis and metabolism is strongly influenced by tissue remodeling and repair processes in the central nervous system and peripheral organs. Because APP expression in neurons is characterized by the absence of splice isoforms lacking APP exon 15 and thus distinct from that in peripheral and non-neuronal cells which produce high levels of APP isoforms without exon 15 (L-APP),[11–14] and because βA4 amyloid pathology in Alzheimer's disease is restricted to the CNS, we propose this pathology to be induced and propagated by a dysregulation of neuronal APP biogenesis and metabolism in response to injury of central nerve terminals. Once βA4 protein aggregation is started by the burst of radicals released upon this injury[15,16] βA4 protein aggregation appears to spread along the nerve projections. Whether this spread is also radical driven and whether copper(II) or zinc(II) bound to and transported by transcytoting APP molecules[17] are involved in radical production and spread of βA4 deposition remains to be elucidated.

REFERENCES

1. HESSE, L., D. BEHER, C. L. MASTERS & G. MULTHAUP. 1994. The βA4 amyloid precursor protein binding to copper. FEBS Lett. **349:**109–116.
2. BUSH, A. I. G. MULTHAUP, R. D. MOIR, T. G. WILLIAMSON, D. H. SMALL, B. RUMBLE, P. POLLWEIN, K. BEYREUTHER & C. L. MASTERS. 1993. A novel zinc(II) binding site modulates the function of the βA4 amyloid protein precursor of Alzheimer's disease. J. Biol. Chem. **268:**16109–16112.
3. MULTHAUP, G., A. I. BUSH, P. POLLWEIN & C. L. MASTERS. 1994. Interaction between the zinc(II) and the heparin binding site of the Alzheimer's disease βA4 amyloid precursor protein (APP). FEBS Lett. **355:**151–154.
4. MULTHAUP, G. 1994. Identification and regulation of the high affinity binding site of the Alzheimer's disease amyloid protein precursor (APP) to glycosaminoglycans. Biochemie **76:**304–311.
5. MULTHAUP, G., A. I. BUSH, P. POLLWEIN, C. L. MASTERS & K. BEYREUTHER. 1992. Specific binding of the Alzheimer βA4 precursor protein to collagen, laminin and heparin. J. Protein Chem. **11:**398–399.
6. SMALL, D. H., V. NURCOMBE, G. REED, H. CLARRIS, R. MOIR, K. BEYREUTHER & C. L. MASTERS. 1994. A heparin-binding domain in the amyloid protein precursor of Alzheimer's disease is involved in the regulation of neurite outgrowth. J. Neurosci. **14:**2117–2127.

7. MÖNNING, U., G. KÖNIG, R. B. BANATI, H. MECHLER, C. CZECH, J. GEHRMANN, V. SCHREITER-GASSER, C. L. MASTERS & K. BEYREUTHER. 1992. Alzheimer βA4-amyloid protein precursor in immunocompetent cells. J. Biol. Chem. **267**:23950–23956.
8. BANATI, R. B., J. GEHRMANN, C. CZECH, V. MÖNNING, L. L. JONES, G. KÖNIG, K. BEYREUTHER & G. W. KREUTZBERG. 1993. Early and rapid de novo synthesis of Alzheimer βA4-amyloid precursor protein (APP) in activated microglia. Glia **9**:199–210.
9. MÖNNING, U., R. SANDBRINK, R. B. BANATI, C. L. MASTERS & K. BEYREUTHER. 1994. Transforming growth factor β mediates increase of mature transmembrane amyloid precursor protein in microglial cells. FEBS Lett. **342**: 267–272.
10. MÖNNING, U., R. SANDBRINK, A. WEIDEMANN, R. B. BANATI, C. L. MASTERS & K. BEYREUTHER. 1995. Extracellular matrix influences the biogenesis of amyloid precursor protein in microglial cells. J. Biol. Chem. **270**:7104–7110.
11. KÖNIG, G., U. MÖNNING, R. PRIOR, R. BANATI, V. SCHREITER-GASSER, J. BAUER, C. L. MASTERS & K. BEYREUTHER. 1992. Identification of a novel, leukocyte-derived isoform of the Alzheimer amyloid precursor protein. J. Biol. Chem. **267**:10804–10809.
12. SANDBRINK, R., C. L. MASTERS & K. BEYREUTHER. 1994. Amyloid protein precursor mRNA isoforms without exon 15 (L-APP mRNA's) are ubiquitously expressed in rat tissues including brain but not in neurons. J. Biol. Chem. **269**: 1510–15
13. SANDBRINK, R., C. L. MASTERS & K. BEYREUTHER. 1994. Similar alternative splicing of non-homologous domain in βA4-amyloid protein precursor-like proteins. J. Biol. Chem. **269**:14227–14234.
14. SANDBRINK, R., C. L. MASTERS & K. BEYREUTHER. 1994. APP gene family: Unique age-associated changes in splicing of Alzheimer's βA4-amyloid protein precursor. Neurobiol. Dis. **1**:13–24.
15. DYRKS, T., E. DYRKS, T. HARTMANN, C. L. MASTERS & K. BEYREUTHER. 1992. Amyloidogenicity of βA4 and βA4-bearing APP fragments by metal catalysed oxidation. J. Biol. Chem. **267**:18210–18217.
16. DYRKS, T., E. DYRKS, C. L. MASTERS & K. BEYREUTHER. 1993. Amyloidogenicity of rodent and human βA4 sequences. FEBS Lett. **324**:231–236.
17. SIMONS, M., E. IKONEN, P. J. TIENARIA, A. CID-ARREGUI, V. MÖNNING, K. BEYRENTHER & C. G. DOTTI. 1995. Intracellular routing of human amyloid protein precursor: Axonal delivery followed by transport to the dendrites. J. Neurosci. Res. **41**:121–128.

Studies on the Metabolism and Biological Function of APLP2[a]

S. S. SISODIA,[b,f,i] G. THINAKARAN,[b,f] H. H. SLUNT,[f] C. A. KITT,[b,f] C. S. VON KOCH,[c,f] R. R. REED,[c,d,g] H. ZHENG,[h] AND D. L. PRICE[b,c,e,f]

[b] *Departments of Pathology,[c] Neuroscience,[d] Molecular Biology & Genetics and [e]Neurology and the [f]Neuropathology Laboratory, [g]Howard Hughes Medical Institute, The Johns Hopkins University School of Medicine, Baltimore, Maryland, USA*

[h] *Merck Research Laboratories, Rahway, New Jersey, USA*

ABSTRACT: Amyloid precursor proteins (APP) are a member of a larger family of proteins that include the amyloid precursor-like proteins (APLP) APLP1 and APLP2. We have examined the expression and metabolism of APLP2 and document that APLP2 is expressed at high levels in the nervous system and in peripheral tissues. Furthermore, several APLP2 isoforms encoded by alternatively spliced transcripts are posttranslationally modified by a chondroitin sulfate glycosaminoglycan (CSGAG) chain. Furthermore, CSGAG modification is regulated by the insertion of sequences encoded by an alternatively spliced exon. Notably, expression of the CSGAG form of APLP2 appears restricted to embryonic neurons and mature neuronal populations that undergo regeneration, such as olfactory sensory neurons. Thus, differences in posttranslational modifications between the APLP2 isoforms and APP are likely to underlie differences in the regulation and function of these homologues. Our present efforts are directed towards using gene targeting strategies to disrupt the expression of the mouse *APP/APLP2* genes to define the normative roles of the encoded molecules in development, plasticity, regeneration, and repair.

INTRODUCTION

A neuropathological hallmark of Alzheimer's disease (AD) is the presence of amyloid deposits in senile plaques[1,2] and in the walls of cerebral blood vessels.[3] The principal component of amyloid deposits is the 39-43 amino acid Aβ[1,3] derived from much larger integral membrane glycoproteins,

[a] This work was supported by grants from the U.S. Public Health Service (NIH NS 20471, AG 05146, NS 097179) and the American Health Assistance Foundation, the Alzheimer's Association, and the Adler Foundation.

[i] *Send correspondence to:* Dr. Sangram S. Sisodia, The Johns Hopkins University School of Medicine, Neuropathology Laboratory, 558 Ross Research Building, 720 Rutland Avenue, Baltimore, Maryland 21205-2196; Tel: (410)550-5173; Fax: (410)955-9777.

termed APP.[4] APP is a member of a family of homologous APLP, including APLP1[5] and APLP2.[6–8]

In the present discussion, we will focus on expression, posttranslational processing, and potential biological functions of APLP2 in the mouse central nervous system (CNS).

DISTRIBUTION OF APLP2 AND APP IN THE MOUSE CNS

At present, APLP2 is the best characterized member of the APLPs. Like APP, APLP2 is encoded by several mRNA derived by alternative splicing.[7–9] Both APP and APLP2 are expressed in brain and peripheral tissues,[7,8] and *in situ* hybridization studies disclose similarities in the distributions of APP and APLP2 mRNA in mouse[8] and humans[7] CNS. Finally, APLP2 matures through the same unusual secretory/cleavage pathway as APP.[8] Moreover, APLP2-751 is an efficient substrate for modification by the addition of a CSGAG chain.[10]

Because of the extensive homologies between APP and APLP2, many of the widely utilized antibodies raised against APP cross-react with APLP2,[8] potentially confounding the interpretations of a large battery of immunochemical and biochemical studies that attempted to characterize the distributions of APP immunoreactivity (IR) in brain. To overcome this problem, we generated antibodies that uniquely recognize the C-terminus of APP or APLP2. The specificities of these antibodies were documented by immunoprecipitation and Western blot analysis of extracts prepared from mouse cells transiently transfected with cDNA encoding APP, APLP1, and APLP2. Consistent with previous *in situ* hybridization studies,[8] our initial investigations with APP- and APLP2-specific antibodies failed to demonstrate significant differential distributions of APP and APLP2 in a variety of regions, including cerebral cortex, hippocampus, and thalamus. However, we document that the cellular and anatomical distributions of APP and APLP2 are strikingly different in the olfactory system. In the olfactory epithelium, both APP- and APLP2-IR are present in olfactory sensory neurons, whereas support cells show intense APP-IR but stain only lightly with APLP2 antibodies. On the other hand, APLP2-specific antibodies specifically recognize olfactory cilia as well as sensory axons and their terminal fields in glomeruli of the olfactory bulb—structures that are not stained by APP-specific antiserum. In contrast to the striking localization of APLP2-IR in sensory axons and glomeruli in the olfactory bulb, APP-IR is not detected in these structures. However, the cytoplasm of periglomerular granule and mitral cells, the principal targets of olfactory inputs, are uniformly stained for APP. Because olfactory sensory neurons are the only regenerating neuronal population in the adult mammalian CNS, our observation that APLP2 is uniquely localized in sensory axons and termini in the olfactory bulb suggests that the expression of this APP homologue may play important roles in axon outgrowth and/or establishment of synaptic connection of olfactory sensory neurons with their targets in the CNS.

On the basis of earlier studies documenting that the primary APLP2 transcript undergoes alternate splicing,[6–9] we carried out reverse transcriptase-polymerase chain reaction (RT-PCR) analysis of mRNA prepared from olfactory epithelium. The alternatively spliced APLP2 transcripts encode APLP2 polypeptides of 694, 706, 751, or 763 amino acids. The 751 and 763 isoforms contain a domain homologous to the Kunitz protease inhibitors, whereas the 694 and 706 forms lack this region. Moreover, the 706 and 763 forms contain an additional 12 amino acid sequence encoded by an alternatively spliced exon. Notably, this sequence is inserted immediately upstream of the chondroitin sulfate addition site in APLP2-751. The sequence of the insert and flanking residues is ENE<u>DTQPELYHPMKK</u>GSGMAEQ. To assess whether the insert influences the CSGAG modification of APLP2, we expressed cDNA encoding APLP2-751 and APLP2-763 isoforms in COS-1 cells. Immunoprecipitation analysis of detergent lysates prepared from transfected cells reveals that APLP2-751 is synthesized as an ~120-kD polypeptide and matures into a heterogeneous set of polypeptides of ~130–200 kD, the result of the addition of variable-length CSGAG to the core protein (Thinakaran and Sisodia, 1994). In contrast, APLP2-763 is synthesized as a ~120 kD molecule and matures into a homogeneous polypeptide of ~130 kD. Thus, APLP2-763, unlike APLP2-751, is not a substrate for modification by chondroitin sulfate glycosaminoglycan chains.

To determine whether APLP2 mRNA expressed in olfactory epithelial cells contains the alternatively spliced insert encoding the 12 amino acid insert, termed *Ch*ondroitin sulfate *I*nhibitory *P*eptide [ChIP], we performed RT-PCR analysis with primers that flank sequences encoding ChIP. In striking contrast to brain RNA, in which ~90% of APLP2 mRNA contains the sequence that encodes ChIP, we demonstrate that ~90% of APLP2 mRNA in olfactory epithelium *lacks* the ChIP sequence. Furthermore, it appears that APLP2-694 (an isoform that lacks the Kunitz domain) represents ~80% of the APLP2 isoforms expressed in olfactory epithelium.

DISCUSSION

Although speculative, we suggest that the unique distributions of APLP2 to cilia, sensory afferents, and terminals may be linked to its generalized role in process formation or pathfinding and connectivity. Olfactory sensory neurons are the only neuronal population in the adult mammalian CNS that are continuously renewed from a population of mitotic precursors, with replacement of neurons in the olfactory epithelium occurring every 4–6 weeks.[11] Hence, axonal outgrowth from newly born sensory neurons towards the olfactory bulb is a process that requires molecular clues involved in guidance, adhesion, and synaptogenesis. Because APLP2 is a core protein for CSGAG, we suggest that these molecules are intimately involved in the process of axogenesis or pathfinding. It will be important to assess the contribution of APP/APLP2 to neuronal regeneration by examination of the expression of each molecule following experimental lesions that disconnect sensory projections from target fields in the olfactory bulb.

Our parallel studies of APP/APLP2 expression during mouse development support the view that APLP2 may be involved in axonal growth and establishment of synapse connectivity (S Fisher and SS Sisodia, unpublished observations). We document that although APLP2 is expressed at lower levels in neuroblasts (days 9.5–11.5 postcoitum), it is expressed at high levels in ventral horn motor neurons and dorsal root ganglia sensory neurons at a time when these populations of cells are reaching their targets and establishing synapses (days 12–13 postcoitum). Furthermore, our RT-PCR studies of RNA prepared during these stages in development reveal that >90% of APLP2 mRNA expressed in early mouse development lacks the ChIP (see above). Taken together with our present data, we hypothesize that CSGAG-modified forms of APLP2 in neuronal cells play an important role in axonal pathfinding/guidance or synaptic formation.

To begin to define the functional roles of APP/APLP2 in the CNS and, more specifically, in the olfactory system, it will be critical to generate mice that contain mutated alleles of *APP/APLP2* generated by gene targeting strategies. These efforts, presently ongoing in several laboratories, will provide a model system to examine the roles of APP/APLP in development, plasticity, regeneration, and repair of the nervous system.

REFERENCES

1. MASTERS, C. L., G. MULTHAUP, G. SIMMS, J. POTTGIESSER, R. N. MARTINS & K. BEYREUTHER. 1985. Neuronal origin of a cerebral amyloid: Neurofibrillary tangles of Alzheimer's disease contain the same protein as the amyloid of plaque cores and blood vessels. EMBO J. **4**:2757–2763.
2. MASTERS, C. L., G. SIMMS, N. A. WEINMAN, G. MULTHAUP, B. L. MCDONALD & K. BEYREUTHER. 1985. Amyloid plaque core protein in Alzheimer disease and Down syndrome. Proc. Natl. Acad. Sci. USA **82**:4245–4249.
3. GLENNER, G. G. & C. W. WONG. 1984. Alzheimer's disease: Initial report of the purification and characterization of a novel cerebrovascular amyloid protein. Biochem. Biophys. Res. Commun. **120**:885–890.
4. KANG, J., H-G. LEMAIRE, A. UNTERBECK, J. M. SALBAUM, C. L. MASTERS K. H. GRZESCHIK, G. MULTHAUP, K. BEYREUTHER & B. MÜLLER-HILL. 1987.

The precursor of Alzheimer's disease amyloid A4 protein resembles a cell-surface receptor. Nature **325:**733–736.

5. WASCO, W., K. BUPP, M. MAGENDANTZ, J. F. GUSELLA, R. E. TANZI & F. SOLOMON. 1992. Identification of a mouse brain cDNA that encodes a protein related to the Alzheimer disease-associated amyloid-beta-protein precursor. Proc. Natl. Acad. Sci. USA **89:**10758–10762.
6. SPRECHER, C. A., F. J. GRANT, G. GRIMM, P. J. O'HARA, F. NORRIS, K. NORRIS & D. C. FOSTER. 1993. Molecular cloning of the cDNA for a human amyloid precursor protein homolog: Evidence for a multigene family. Biochemistry **32:**4481–4486.
7. WASCO, W., S. GURUBHAGAVATULA, M. D. PARADIS, D. M. ROMANO, S. S. SISODIA, B. T. HYMAN, R. L. NEVE & R. E. TANZI. 1993. Isolation and characterization of *APLP2* encoding a homologue of the Alzheimer's associated amyloid β protein precursor. Nature Genetics **5:**95–99.
8. SLUNT, H. H., G. THINAKARAN, C. VON KOCH, A. C. Y. LO, R. E. & S. S. SISODIA. 1994. Expression of a ubiquitous, cross-reactive homologue of the mouse β-amyloid precursor protein (APP). J. Biol. Chem. **269:**2637–2644.
9. SANDBRINK, R., C. L. MASTERS & K. BEYREUTHER. 1994. Similar alternative splicing of a non-homologous domain in βA4-amyloid protein precursor-like proteins. J. Biol. Chem. **269:**14227–14234.
10. THINAKARAN, G. & S. S. SISODIA. 1994. Amyloid precursor-like protein 2 (APLP2) is modified by the addition of chondroitin sulfate glycosaminoglycan at a single site. J. Biol. Chem. **269:**22099–22104.
11. GRAZIADEI, P. P. C. & G. A. MONTI GRAZIADEI. 1979. Neurogenesis and neuron regeneration in the olfactory system of mammals. I. Morphological aspects of differentation and structural organization of the olfactory sensory neurons. J. Neurocytol. **8:**1–18.

Neurotrophic and Neuroprotective Effects of hAPP in Transgenic Mice[a]

L. MUCKE,[b] C. R. ABRAHAM,[c] AND E. MASLIAH[d]

[b] *Department of Neuropharmacology, The Scripps Research Institute, La Jolla, California 92037 USA*

[c] *Departments of Medicine and Biochemistry, Boston University School of Medicine, Boston, Massachusetts, USA*

[d] *Departments of Neurosciences and Pathology, University of California San Diego, La Jolla, California, USA*

ABSTRACT: To better understand the role the human amyloid precursor protein (hAPP) plays in Alzheimer's disease (AD), it is essential to define its primary function(s). Here we expressed different hAPPs in neurons of transgenic (tg) mice to characterize their effects on the intact central nervous system (CNS). Immunolabeled brain sections of tg and non-tg mice were compared quantitatively by microdensitometry and computer-aided analysis of laser scanning confocal digitized images. Compared with non-tg mice, tg mice overexpressing hAPPs showed an increase in the number of synaptophysin immunoreactive presynaptic terminals as well as in the expression of the growth-associated marker GAP-43. While non-tg controls and tg mice expressing hAPP751 at moderate levels displayed a normal pattern of reinnervation of the dentate gyrus following perforant pathway transection, tg mice expressing hAPP695 at severalfold higher levels showed an accentuation of the synaptic loss and no sprouting reaction. In addition, expression of hAPP751 at moderate levels effectively protected neurons against excitotoxic injury induced either acutely by systemic injection of kainic acid or chronically by transgene-driven glial production of the soluble HIV-1 protein gp120. Neuronal expression of hAPP695 at higher levels provided less excitoprotection. Our findings are consistent with the postulate that APP plays a role in the formation/maintenance of synapses and that processes which affect this function could contribute to the synaptic pathology seen in AD. Our study also revealed that hAPPs can exert important excitoprotective functions *in vivo* and that the efficiency of this protection may depend on the hAPP isoform expressed as well as on the level of neuronal hAPP expression. Neuronal overexpression of hAPP beyond a certain level may have detrimental effects on the CNS, particularly in the context of secondary neural injuries.

[a] This research was sponsored by NIH grants AG-04342 (LM), AG-11385 (LM), MH-47680 (LM), AG-00001 (CRA), AG-09905 (CRA), AG-05131 (EM) and AG-10689 (EM), by the Alzheimer's Association [FSA-91-010 (LM), IIRG-89-125 (CRA)] and by the Department of Health Services, State of California [92-1539 (EM)].

[e] *Send correspondence to:* Dr. Lennart Mucke, Department of Neuropharmacology, The Scripps Research Institute (IMM6), 10666 North Torrey Pines Road, La Jolla, CA 92037; Tel: (617) 554-8372; Fax: (617) 554-9981; E-mail: lmucke@scripps.edu.

INTRODUCTION

The human amyloid precursor protein (hAPP) is involved in Alzheimer's disease (AD) since hAPP derivatives accumulate abnormally in brains of AD victims and APP point mutations are associated with familial AD.[1] Within the central nervous system (CNS), APP is found primarily in neurons with a preferential localization at central and peripheral synaptic sites.[2]

To better understand the role APP plays in AD, it is essential to define its primary function(s). Its strong conservation across species and prominent expression in brain and other organs clearly suggest that *in vivo* APP normally fulfills important beneficial functions. However, a number of *in vitro* studies indicate that APP derivatives can have either neurotrophic/neuroprotective or neurotoxic effects.[3] The current investigation aimed to characterize the effects of hAPPs on the intact CNS by expressing them in neurons of transgenic (tg) mice.

METHODS

Care of tg and non-tg B6xSJL mice (2-24 months of age) was in accordance with institutional guidelines. For detailed descriptions of the establishment and characterization of NSE-hAPP (hAPP) and GFAP-gp120 (gp120) tg mice see Mucke *et al.*[4] and Toggas *et al.*,[5] respectively. Perforant pathway transections and subsequent analyses were carried out as described.[6] Kainic acid (25 mg/kg) diluted in normal saline (NS) was injected intraperitoneally (i.p.): controls were sham-injected with NS. Immuno-histopathological evaluations were carried out essentially as described.[4,5] For quantitative comparisons, 3-10 mice were analyzed per group. Hemibrains were assigned code numbers to ensure objective assessment and codes were broken only after the analysis was complete. For each mouse and immunostain, three serial sections of corresponding brain regions were analyzed. For the assessment of neuronal changes, sections were examined using the Bio-Rad MRC-600 laser scanning confocal microscope mounted on an Axiovert Zeiss microscope. Digitized images (4/section/case), 0.5 μm in thickness, were transferred to a Macintosh IIci, running the public domain program of Wayne Rasband (Image 1.23). Microdensitometric measurements were carried out using the Quantimet 570C. Statistical analyses of the results were conducted using the STAT VIEW II (Abacus Concepts, Berkeley, CA) software package.

RESULTS

Fusion gene constructs expressed in tg animals allow for the prolonged, reproducible delivery of selected proteins (devoid of any contaminants) to specific areas/cells within the intact CNS.[7] To elucidate the effects hAPPs have on the CNS *in vivo,* cDNAs encoding full-length hAPP695 or hAPP751 were placed downstream of the neuron-specific enolase (NSE) promoter; the resulting fusion genes were used to generate transgenic mice. In multiple lines of hAPP tg mice, neuronal overexpression of hAPP resulted in an increase in the number of synaptophysin immunoreactive (SYN-IR) presynaptic terminals and in the expression of the growth-associated marker GAP-43 (see ref. 4 for details). In lines expressing hAPP751 or hAPP695 at moderate levels, synaptotrophic effects were more prominent in homozygous than in heterozygous tg mice. None of the tg lines (age of mice analyzed: 2–24 months) showed any evidence for amyloid deposits or neurodegeneration. However, hAPP695 line 19 which expressed hAPP at severalfold higher levels than the other lines showed less increase in SYN-IR presynaptic terminals per amount of hAPP expressed than the lower expressor lines and a decrease in synaptotrophic effects in homozygous compared with heterozygous tg offspring.

To further assess the neurotrophic/neuroprotective potential of hAPP *in vivo,* the neuropathological consequences of distinct neural lesions were compared quantitatively across different lines of hAPP tg mice and non-tg controls. Specifically, mice were challenged with three well-characterized experimental paradigms: i) perforant pathway transection, ii) systemic injection of kainic acid, or iii) transgene-driven expression of HIV-1 gp120.

i) Following perforant pathway transection, the patterns of denervation and reinnervation in the dentate gyri of hAPP tg and non-tg mice were analyzed at 3 and 30 days, respectively (see ref. 6 for details). At 3 days post lesion, all mice showed a significant decrease in the number of SYN-IR presynaptic terminals in the outer molecular layer (OML). At 30 days post lesion, non-tg mice and tg mice expressing hAPP751 at moderate levels displayed a normal pattern of reinnervation of the dentate gyrus, whereas tg mice expressing hAPP695 at severalfold higher levels showed an accentuation of the synaptic loss and no sprouting reaction in the OML.

ii) Excessive stimulation of excitatory amino acid receptors and consequent derangements of the neuronal calcium homeostasis appear to constitute a final common pathway in the etiology of diverse neurodegenerative disorders.[8] To directly assess the excitoprotective potential of hAPPs *in vivo,* tg and non-tg mice were analyzed 4 days following the intraperitoneal injection of kainic acid. In non-tg mice, this treatment induced typical excitotoxic brain lesions, particularly in the hippocampus and cerebral cortex. Compared with these controls, tg mice overexpressing hAPP751 at moderate levels showed significantly less neuronal loss and synapto-dendritic degeneration. Tg mice expressing hAPP695 at higher levels showed less neuroprotection.

iii) Glial production of the soluble HIV-1 envelope glycoprotein gp120 in brains of tg mice induces extensive neurodegeneration.[5] Notably, this process is chronically progressive and appears to involve an NMDA receptor-mediated destabilization of the neuronal calcium homeostasis (Toggas, Masliah and Mucke, unpublished data, see ref. 9 for review of other pertinent studies). Hence, gp120 tg mice may provide a suitable *in vivo* model of chronic excitotoxicity-related neurodegeneration. To determine the effects of hAPPs on gp120-induced brain damage, singly tg hAPP heterozygotes were crossed with singly tg gp120 heterozygotes. Gp120/APP bigenic offspring overexpressing hAPP at moderate levels showed significantly less neuronal loss, synapto-dendritic damage and gliosis than gp120 singly tg littermate controls, whereas severalfold higher levels of hAPP695 expression failed to prevent gp120-induced brain damage (10).

DISCUSSION

In multiple independent lines of tg mice, neuronal overexpression of hAPP increased the number of SYN-IR presynaptic terminals in the CNS. It is possible that overexpression of hAPP in these mice increased synapse formation or decreased the synapse drop out which usually follows the period of maximal synapse formation during development. In either case, our results are consistent with the recent observation that infusion of an APP 17mer peptide into the lateral ventricles induces synaptotrophic effects in the neocortex of aged rats[11] as well as with previous studies which identified APP in growing neurites of immature rat brain[12] and demonstrated that APP reaches its highest level of expression in the CNS at the time of brain maturation and completion of synaptic connections.[13] Taken together, these observations indicate that APPs may play a key role in synaptic plasticity and/or homeostasis *in vivo.*

In vitro studies indicate that secretable forms of hAPP can protect neurons against excessive stimulation of excitatory amino acid receptors and consequent derangements of the neuronal calcium homeostasis.[14] Our demonstration of neuroprotective hAPP effects in kainate challenged hAPP singly tg and unmanipulated gp120/hAPP bigenic mice indicates that these molecules may have similar effects *in vivo.* A recent study which demonstrated that intracerebroventricular infusion of APP peptides can protect the CNS against ischemic damage[15] is also consistent with such an excitoprotective function of APP. It is interesting that $[Ca^{2+}]_i$-imaging of cultured neurons revealed the APP-mediated reduction in neuronal $[Ca^{2+}]_i$ to be especially marked in dendrites,[14] since these neuronal structures showed particularly prominent damage in kainate treated non-tg and unmanipulated gp120 tg mice.

While it is tempting to speculate that kainate- and gp120-induced neurotoxic effects were prevented in hAPP tg mice by hAPP-mediated stabilization of the intraneuronal calcium homeostasis, alternative mechanisms also deserve consideration. Both hAPP695 and hAPP751 contain a metalloprotease inhibitor domain[16] and protease inhibitors could mediate a variety of important biological effects in the CNS.[17,18] It is interesting in this context that hAPP751 appeared to be somewhat more neuroprotective than hAPP695, both *in vitro*[10] and *in vivo* (see above). It is conceivable that this difference relates to an activity of the Kunitz protease inhibitor–like domain which is present in hAPP751 and absent from hAPP695: the secreted form of hAPP751 appears to be identical with protease nexin II (PNII).[19,20] However, *in vivo* comparisons between hAPP695 and hAPP751 are complicated by the fact that most of the endogenous APP expressed in rodent brain is of the APP695 isoform. This implies that similar absolute levels of hAPP695 and hAPP751 expression make very different relative contributions to the pre-existing pools of the corresponding endogenous murine APP isoforms. Experiments involving multiple lines of tg mice are currently in progress to determine more conclusively to what extent the identified biological hAPP effects are influenced by i) hAPP expression levels, ii) differences between hAPP695 versus hAPP751, and iii) differences between non-mutated versus mutated forms of hAPP.

Evidence is accumulating that the neurodegeneration of AD may involve complex interactions among different pathogenetic pathways. It is interesting in this context that increasing neuronal hAPP expression to the level found in hAPP695 line 19 decreased synaptotrophic effects in unmanipulated tg mice,[4] inhibited neuroregeneration following perforant pathway transection,[6] and failed to provide neuroprotection in gp120/hAPP bigenic mice. These findings suggest that above a certain level, overexpression of hAPP may have detrimental effects on the CNS, particularly in the context of secondary neural injuries.

In conclusion, the above findings indicate that APP plays a role in the formation/maintenance of synapses *in vivo* and that its synaptotrophic effects may be critically dependent on the expression levels of different APP isoforms. They are consistent with the postulate that processes which alter the expression, metabolism or function of APPs could contribute to the synaptic pathology seen in AD and, hence, form potential targets for therapeutic interventions. Our study also revealed that hAPP can exert excitoprotective functions *in vivo* and that the efficiency of this protection may depend on the hAPP isoform expressed as well as on the level of neuronal hAPP expression.

ACKNOWLEDGMENTS

We thank C. Westland, M. Ruppe and M. Mallory for excellent technical assistance.

REFERENCES

1. CLARK, R. F. & A. M. GOATE. 1993. Molecular genetics of Alzheimer's disease. Arch. Neurol. **50:**1164–1172.
2. SCHUBERT, W., R. PRIOR, A. WEIDEMANN, H. DIRCKSEN, G. MULTHAUP, C. L. MASTERS & K. BEYREUTHER. 1991. Localization of Alzheimer βA4 amyloid precursor protein at central and peripheral synaptic sites. Brain Res. **563:** 184–194.
3. MATTSON, M. P., S. W. BARGER, B. CHENG, I. LIEBERBURG, V. L. SMITH-SWINTOSKY & R. E. RYDEL. 1993. β-Amyloid precursor protein metabolites and loss of neuronal Ca^{2+} homeostasis in Alzheimer's disease. TINS **16:** 409–414.
4. MUCKE, L., E. MASLIAH, W. B. JOHNSON, M. D. RUPPE, M. ALFORD, E. M. ROCKENSTEIN, S. FORSS-PETTER, M. PIETROPAOLO, M. MALLORY & C. R. ABRAHAM. 1994. Synaptotrophic effects of human amyloid β protein precursor in the cortex of transgenic mice. Brain Res. **666:**151–167.
5. TOGGAS, S. M., E. MASLIAH, E. M. ROCKENSTEIN, G. F. RALL, C. R. ABRAHAM & A. MUCKE. 1994. Central nervous system damage produced by expression of the HIV-1 coat protein gp120 in transgenic mice. Nature **367:**188–193.
6. MASLIAH, E., M. MALLORY, M. ALFORD, N. GE & L. MUCKE. 1995. Abnormal synaptic regeneration in hAPP695 transgenic and ApoE knockout mice. *In* Research Advances in Alzheimer's Disease and Related Disorders. K. Iqbal, J. Mortimer, B. Winblad & H. Wisniewski, Eds. Chapter 44: 405–414. John Wiley & Sons, Ltd.
7. TOGGAS, S. M. & L. MUCKE. 1994. Transgenic models in the study of AIDS dementia complex. Curr. Top. Microbiol. Immunol. In press.
8. LIPTON, S. A. & P. A. ROSENBERG. 1994. Mechanisms of disease: Excitatory amino acids as a final common pathway for neurologic disorders. N. Engl. J. Med. **330:**613-622.
9. LIPTON, S. A. 1994. HIV-related neuronal injury: Potential therapeutic intervention with calcium channel antagonists and NMDA antagonists. Mol. Neurobiol. **8:**181–196.
10. MUCKE, L., C. R. ABRAHAM, M. D. RUPPE, E. M. ROCKENSTEIN, S. T. TOGGAS, M. ALFORD & E. MASLIAH. 1995. Protection against HIV—1 gp120-induced brain damage by neuraonal overexpression of amyloid protein precursor (APP). J. Exp. Med. **181**:1551–1556.
11. ROCH, J-M., E. MASLIAH, A-C. ROCH-LEVECQ, M. P. SUNDSMO, D. A. C. OTERO, I. VEINBERGS & T. SAITOH. 1994. Increase of synaptic density and memory retention by a peptide representing the trophic domain of the amyloid β/A4 protein precursor. Proc. Natl. Acad. Sci. USA **91:**7450–7454.
12. MASLIAH, E., M. MALLORY, N. GE & T. SAITOH. 1992. Amyloid precursor protein is localized in growing neurites of neonatal rat brain. Brain Res. **593:** 323–328.
13. LÖFFLER, J. & G. HUBER. 1992. β-amyloid precursor protein isoforms in various rat brain regions and during brain development. J. Neurochem. **59:** 1316–1324.
14. MATTSON, M. P., B. CHENG, A. R. CULWELL, F. S. ESCH, I. LIEBERBURG &

R. E. Rydel. 1993. Evidence for excitoprotective and intraneuronal calcium-regulating roles for secreted forms of the β-amyloid precursor protein. Neuron **10**:243–254.

15. Smith-Swintosky, V. L., L. C. Pettigrew, S. D. Craddock, A. R. Culwell, R. E. Rydel & M. P. Mattson. 1994. Secreted forms of β-amyloid precursor protein protect against schemic brain injury. J. Neurochem. **63**: 781–784.
16. Miyazaki, K., M. Hasegawa, K. Funahashi & M. Umeda. 1993. A metalloproteinase inhibitor domain in Alzheimer amyloid protein precursor. Nature **362**:839–842.
17. Abraham, C. R., K. Kanemaru & L. Mucke. 1993. Expression of cathepsin G-like and α1-antichymotrypsin-like proteins in reactive astrocytes. Brain Res. **621**:222–232.
18. Akopyan, T. 1991. Protein inhibitors of proteinases from brain. Neurochem. Res. **16**:513–517.
19. Oltersdorf, T., L. C. Fritz, D. B. Schenk, I. Lieberburg, K. L. Johnson Wood, E. C. Beattie, P. J. Ward, R. W. Blacher, H. F. Dovey & S. Sinha. 1989. The secreted form of the Alzheimer's amyloid precursor protein with the Kunitz domain is protease nexin-II. Nature **341**:144–147.
20. Van Nostrand, W. E., S. L. Wagner, M. Suzuki, B. H. Choi, J. S. Farrow, J. W. Geddes, C. W. Cotman & D. D. Cunningham. 1989. Protease nexin-II, a potent antichymotrypsin, shows identity to amyloid beta-protein precursor. Nature **341**:546–549.

Amyloid Fibril Toxicity in Alzheimer's Disease and Diabetes[a]

A. LORENZO AND B. A. YANKNER[b]

Department of Neurology, Harvard Medical School and The Children's Hospital, Boston, Massachusetts 02115 USA

ABSTRACT: Several lines of evidence suggest that amyloid deposition in the brain contributes to neuronal degeneration in Alzheimer's disease (AD). In the AD brain, diffuse plaques composed mostly of amorphous β amyloid (Am-βA) are inert, whereas compact plaques composed of β amyloid fibrils (Fib-βA) are associated with neurodegenerative changes. The effects of these two types of amyloid were tested on primary rat hippocampal neurons. Fib-βA induced the formation of dystrophic neurites and caused neuronal cell death, whereas Am-βA was not toxic. In addition, Fib-βA caused synapse loss in the remaining viable neurons, whereas Am-βA did not significantly affect synapse number. We also examined the effects of amylin, the primary constituent of the amyloid fibrils that form in the pancreas in adult-onset diabetes. Amylin was toxic to rat and human insulin-producing islet cells in the concentration range of fibril formation. The relative toxic potencies of amylin peptides of different species correlated with their fibril-forming capacity. Soluble amylin was not toxic. The amyloid fibril-binding dye Congo red inhibited the toxicity of both βA and amylin. Congo red afforded protection against toxicity by a dual mechanism. When present during the phase of fibril polymerization, Congo red could inhibit fibril formation from some peptides. When added to preformed fibrils, Congo red bound to fibrils rendering them nontoxic. These results suggest that fibril formation is necessary for both βA and amylin toxicity. Congo red appears to be a general inhibitor of amyloid fibril toxicity and may therefore be a useful prototype for drugs targeted to the amyloid pathology of AD and adult-onset diabetes.

INTRODUCTION

The formation and accumulation of amyloid fibrils is a pathological feature of a variety of degenerative disorders associated with aging. Amyloid fibrils are formed from a large number of different proteins and peptides that share the property of adopting a crossed β-sheet conformation under particular

[a] Supported by NIH Grants AG09220 and NS30352 and grants from the Alzheimer's Association, and Sandoz Pharma. Ltd.

[b] *Send correspondence to:* Dr. B. A. Yankner, Department of Neurology, Harvard Medical School and The Children's Hospital, Enders 260, 300 Longwood Avenue, Boston, MA 02115; Tel: (617) 355-7220; Fax: (617) 738-1542.

conditions. Amyloid fibrils rarely form in young adults, suggesting that fibrillogenesis depends on biophysical changes in proteins which occur during the aging process and are exaggerated in disease. Although amyloid deposition has been well-characterized at the biophysical level, the cellular consequences of this process are poorly understood. A model of amyloidosis and its biological consequences is proposed here based on studies of the effects of β amyloid (βA) in Alzheimer's disease (AD) and amylin in diabetes.

Considerable evidence now suggests that βA is neurotoxic[1–4] and that peptide aggregation is necessary for βA to achieve a neurotoxic state.[4–8] In this report, we review our recent studies which suggest that aggregation of βA is not sufficient for neurotoxicity, but rather that amyloid fibril formation is required.[9] We will also review recent studies which suggest that the amyloid fibrils that form in the pancreas in adult-onset diabetes are toxic to insulin-producing islet cells. These findings give rise to a unified hypothesis for the pathogenesis of amyloidoses which is based on the common cytopathic properties of the amyloid fibril.

METHODS

Establishment of primary cultures of fetal rat hippocampal neurons and adult rat pancreatic islet cells and protocols for viability assays and the use of peptides have been described.[9,10]

RESULTS AND DISCUSSION

In the AD brain, βA deposition occurs in diffuse plaques that are inert and in mature compact plaques that are associated with neurodegeneration. An important question is whether the neurodegenerative changes surrounding the mature plaque is due to a change which has occurred in the physical state of βA or whether the change in βA is secondary to neurodegeneration. To address this issue, we created *in vitro* analogs of the βA in diffuse and compact plaques using the 1-42 βA peptide, the predominant form of βA in senile plaques. Rapid aggregation of 1-42 βA in a salt-containing buffer at pH 7.3 produced amorphous aggregated βA (Am-βA) which was not birefringent after Congo red staining, similar to the diffuse βA in immature plaques (FIG. 1A, C). Slow aggregation of 1-42 βA produced 6-9 nm βA fibrils (Fib-βA) that were strongly Congo red-birefringent, similar to the βA in compact plaques (FIG. 1B, D).

The neurotoxicity of the two βA preparations was assessed in primary rat hippocampal cultures. Exposure to Fib-βA resulted in profound neurodegenerative changes including the formation of dystrophic neurites, shrinkage of

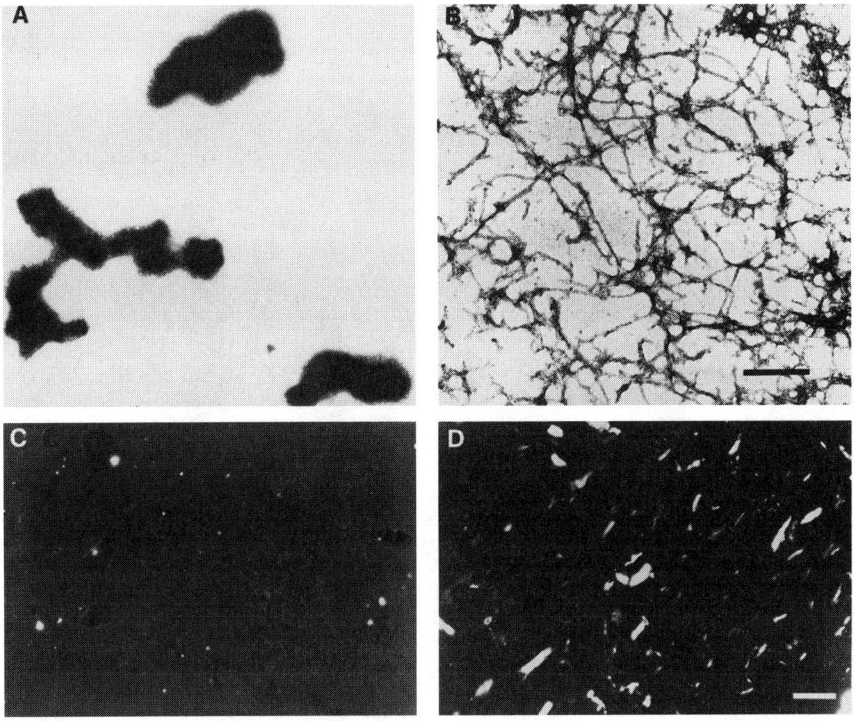

FIGURE 1. Preparations of amorphous and fibrillar βA 1-42. βA preparations were analyzed by transmission electron microscopy **(A, B)** and Congo red staining followed by illumination with polarized light **(C, D)**. Rapid aggregation of the 1-42 βA peptide in PBS produces amorphous aggregates which are not Congo red birefringent **(A, C)**. Incubation of 1-42 βA for 3 days in water produces 6-9 nm βA fibrils which are intensely Congo red birefringent **(B, D)**. *Scale bars*: B, 200 nm; **C**, 50 ím. (Reproduced with permission from Yankner.[4] Copyright National Academy of Sciences.)

the neuronal soma and neuronal death (FIG. 2C). In contrast, exposure to Am-βA had no significant effect on neuronal morphology or viability (FIG. 2B). These observations were confirmed by quantitative assessment of neuronal viability, which showed 47 ± 3% neuronal cell death in cultures treated with Fib-βA ($p < 0.001$), and no statistically significant effect on cell viability in cultures treated with Am-βA.

Since synapse loss is an early pathological indicator of the severity of cognitive impairment in AD, we examined the effect of βA on synapse number. Immunocytochemical staining for synaptophysin showed punctate clustering of synaptophysin indicative of presynaptic terminals along the dendrites of hippocampal neurons maintained for 8 days in culture. Treatment with Fib-βA resulted in a significant decrease in synaptic density along the dendrites of

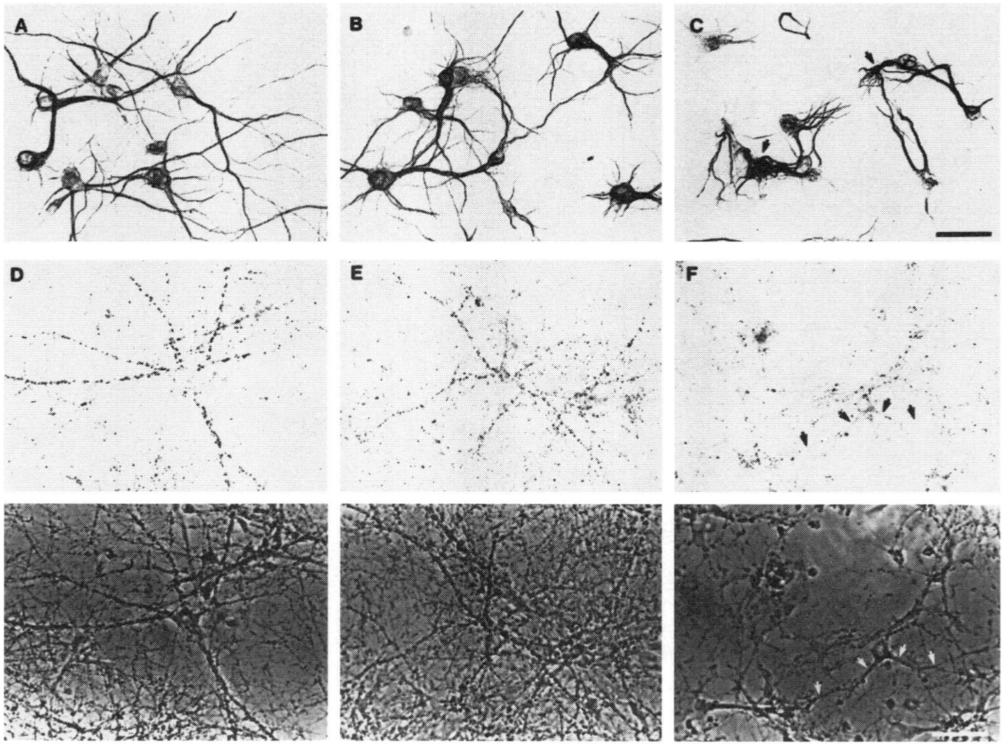

FIGURE 2. Fibrillar βA mediates neurodegeneration and synapse loss. Hippocampal cultures were immunocytochemically stained for MAP-2 (**A, B, C**) or synaptophysin (**D, E, F:** *upper panel*—bright field; *lower panel*—phase contrast). (**A, D**) Control. (**B, E**) Am-βA. (**C, F**) Fib-βA. Note that Fib-βA causes neurite retraction and dystrophic changes (**C,** *arrows*) and a marked decrease in the density of presynaptic terminals along dendrites (**F,** *arrows*). Neurons treated with Am-βA (**B, E**) are indistinguishable from controls (**A, D**). Hippocampal cultures were incubated with 20 íM of βA peptides for 3 days. Scale bars, 40 ím. (Reproduced with permission from Yankner.[4] Copyright National Academy of Sciences.)

remaining viable neurons (FIG. 2D, F). This was confirmed by quantititative assessment of synapse number which showed that Fib-βA induced the loss of 60 $\pm$ 6% of the synapses ($p < 0.001$). In contrast, treatment with Am-βA did not have a statistically significant effect on synapse number (FIG. 2D, E). These results suggest that βA-induced neurotoxicity and synapse loss are mediated by the amyloid fibril. Soluble βA and amorphous aggregated βA are not toxic.

In adult-onset diabetes, amyloid fibrils form in the pancreatic islets of Langerhans. The major constituent of islet amyloid is the 37 amino acid peptide amylin.[11] It is of interest to note that diabetic amyloid was first described in 1900, seven years before Alzheimer's landmark report of amyloid in the brain of a woman suffering from a dementing illness. Since that time, the role of amylin in the pathogenesis of diabetes has been controversial, with arguments both pro and con quite similar to those regarding the role of βA in AD. Although amylin and βA do not share significant amino acid homology, they share the property of amyloid fibril formation. This fact prompted us to determine if a pathogenic situation analogous to βA neurotoxicity occurs in adult-onset diabetes.

Using a cell culture system for the maintenance of adult rat and human pancreatic islet cells, we examined the effect of human amylin.[10] Human amylin caused profound degeneration and death of greater than 90% of both rat and human insulin-producing islet cells. We then examined the concentration dependence of amylin fibril formation by electron microscopy and Congo red staining. The concentration dependence of islet cell toxicity paralleled the concentration dependence of fibril formation.

The toxicity of amylins from different species was then assessed, since these different amylin peptides are known to have different fibril-forming capacities. Human and cat amylin form pancreatic amyloid deposits and develop maturity-onset diabetic syndromes, whereas rat amylin does not form amyloid deposits and rats do not develop this syndrome. We found that cat amylin, like human amylin, caused significant islet cell toxicity and the concentration dependence paralleled that of fibril formation. Rat amylin was not toxic and did not form fibrils at any concentration tested. Thus, there is a strong correlation between amylin fibril formation and islet cell toxicity.

To determine if soluble amylin contributes to islet cell toxicity, we used a "coverslip sandwich" in culture to create a situation in which some islet cells were accessible only to soluble peptide, whereas others were accessible to amylin aggregates. In this culture, only the cells in contact with fibrillar aggregates degenerated. Cells exposed to soluble amylin alone remained viable. These results suggest that fibril formation is required for amylin-induced islet cell toxicity.

If amyloid fibril formation is sufficient to induce βA and amylin toxicity, then inhibition of fibril formation or fibril binding to the cell surface would be expected to inhibit the toxicity of both peptides. We examined the effects of the amyloid fibril-specific dye Congo red on the neurotoxicity of βA and the islet cell toxicity of amylin.[9] When Congo red was added to soluble 1-40 βA, it almost completely inhibited fibril formation at a molar ratio 0.35 : 1. In addition, Congo red inhibited the neurotoxicity of preformed βA fibrils by binding to fibrils without causing fibril dissolution. Thus, Congo red inhibits the neurotoxicity of 1-40 βA by a dual mechanism. For the 1-42 βA peptide, however, Congo red inhibited neurotoxicity by binding to the fibril but did not inhibit fibril formation in the concentration range tested.

We also examined the effects of Congo red on the islet cell toxicity of amylin. Congo red completely inhibited the islet cell toxicity of human amylin. The formation of amylin fibrils did not appear to be significantly inhibited by Congo red. Rather, inhibition of toxicity was due to the binding of Congo red to preformed amylin fibrils.

These experiments demonstrate that βA neurotoxicity and amylin islet cell toxicity are both mediated by amyloid fibrils. In both cases, the soluble peptide is not toxic and in the case of βA, the amorphous aggregated peptide is not toxic. The idea that amyloid fibrils are generally cytopathic is further supported by the observation that βA is toxic to both neurons and islet cells, and that amylin is toxic not only to islet cells but also to neurons, endothelial cells, COS cells and PC12 cells.[10,12] However, we have shown that both βA and amylin toxicity exhibit cell-type specificity. For example, βA is not toxic to astrocytes,[3] and amylin has low toxic potency for fibroblasts and H4 hematoma cells.[10] The basis for the resistance of some cells to amyloid toxicity is unclear, but may involve the ability of these cell types to phagocytose and degrade amyloid fibrils.

The general cytopathic effect of amyloid fibrils suggests that the conformation of the fibril is responsible for toxicity. It is important to note that this paradigm does not exclude the possibility that fibrils can exert their toxic effects by interacting with cell surface receptors, since many receptors recognize conformation-related motifs. The causes of fibril formation and their relationship to the normal biology of the aging process have now become key issues in understanding the etiology of AD and other amyloidoses. A common therapeutic approach to amyloid-related diseases may be possible, as suggested by the ability of Congo red to prevent the toxicity of both βA and amylin.

REFERENCES

1. Roher, A. E., M. J. Ball, S. V. Bhave & A. R. Wakade. 1991. β-amyloid from Alzheimer's disease brain inhibits sprouting and survival of sympathetic neurons. Biochem. Biophys. Res. Commun. **174:**572–579.

2. YANKNER, B. A., L. R. DAWES, S. FISHER, L. VILLA-KOMAROFF, M. L. OSTER-GRANITE & R. L. NEVE. 1989. Neurotoxicity of a fragment of the amyloid precursor associated with Alzheimer's disease. Science **245**:417–420.
3. YANKNER, B. A., L. K. DUFFY & D. A. KIRSCHNER. 1990. Neurotrophic and neurotoxic effects of amyloid β protein: Reversal by tachykinin neuropeptides. Science **250**:279–282.
4. YANKNER, B. A. 1992. Commentary and perspective on studies of beta amyloid neurotoxicity. Neurobiol. Aging **13**:615–616.5.
5. BUSCIGLIO, J., A. LORENZO & B. A. YANKNER. 1992. Methodological variables in the assessment of β amyloid neurotoxicity. Neurobiol. Aging **13**:609–612.
6. MATTSON, M. P., K. J. TOMASELLI & R. E. RYDEL. 1993. Calcium-destabilizing and neurodegenerative effects of aggregated β-amyloid peptide are attenuated by basic FGF. Brain Res. **621**:35–49.
7. PIKE, C. J., D. BURDICK, A. J. WALENCEWICZ, C. G. GLABE & C. W. COTMAN. 1993. Neurodegenertion induced by β-amyloid peptides *in vitro*: The role of peptide assembly state. J. Neurosci. **13**:1676–1687.
8. PIKE, C. J., A. J. WALENCEWICZ, C. G. GLABE & C. W. COTMAN. 1991. *In vitro* aging of β-amyloid protein causes peptide aggregation and neurotoxicity. Brain Res. **563**:311–314.
9. LORENZO, A. & B. A. YANKNER. 1994. Beta amyloid neurotoxicity requires fibril formation and is inhibited by Congo red. Proc. Natl. Acad. Sci. USA **91**: 12243–12247.
10. LORENZO, A., B. RAZZABONI, C. WEIR & B. A. YANKNER. 1994. Pancreatic islet cell toxicity of amylin associated with type-2 diabetes mellitus. Nature **368**: 756–760.
11. JOHNSON, K. H., T. D. O'BRIEN, C. BETSHOLTZ & P. WESTERMARK. 1989. Islet amyloid, islet amyloid polypeptide, and diabetes mellitus. New Engl. J. Med. **321**:513–518.
12. MAY, P. C., L. N. BOGGS & K. S. FUSON. 1993. Neurotoxicity of human amylin in rat primary hippocampal cultures: Similarity to Alzheimer's disease amyloid-beta neurotoxicity. J. Neurochem. **61**:2330–2333.

Structure, Microtubule Interactions, and Phosphorylation of Tau Protein[a]

E.-M. MANDELKOW,[b] O. SCHWEERS, G. DREWES, J. BIERNAT, N. GUSTKE, B. TRINCZEK, AND E. MANDELKOW

Max-Planck-Unit for Structural Molecular Biology, Hamburg, Germany

ABSTRACT: This paper summarizes recent structural and functional studies on tau protein, its interactions with microtubules, its self-assembly into paired helical filaments (PHF)-like fibers, and its modification by phosphorylation. The structure of tau in solution resembles that of a random coil. Both tau and Alzheimer PHFs have very little secondary structure, making it improbable that the assembly of tau into PHFs is based on interacting beta sheets. Tau's binding to microtubules can be described by a "jaws" effect. The domain containing the repeats binds very weakly, while the flanking regions (jaws) bind strongly, even without the repeats. However, only the combination of flanking regions and repeats makes binding productive in terms of microtubule nucleation and assembly. Although the majority of Alzheimer-like phosphorylation sites are outside the repeats they have only a weak influence on binding, whereas the phosphorylation at Ser262 inside the repeats inhibits binding and makes microtubules dynamically unstable. This site can be phosphorylated by kinases present in brain tissue, and it is uniquely phosphorylated in Alzheimer brain.

INTRODUCTION

Tau (FIG. 1) is a largely axonal protein that normally functions as a microtubule-associated protein (MAP), presumably by stabilizing the microtubules which serve as tracks for cytoplasmic transport. Tau is also the main component of the paired helical filaments of Alzheimer's disease (AD). It is therefore pertinent to ask in what sense is AD tau different from normal tau? The differences include aggregation, phosphorylation, proteolytic degradation, loss of microtubule binding and assembly competence, glycation, and ubiquitination (see review by Kosik and Greenberg[1]). This illustrates the complexity of tau's degeneration in AD, but since it is a faithful marker of the progression of the disease[2] the analysis of the modifications may eventually lead to the

[a] This work was supported by grants from the Deutsche Forschungsgemeinschaft and the Federal Ministry of Science and Technology.

[b] *Send correspondence to:* Dr. E.-M. Mandelkow, Max-Planck-Unit for Structural Molecular Biology, c/o DESY, Notkestrasse 85, D-22603 Hamburg, Germany; Tel: (+49)40-89982810; Fax: (+49)40-891314; E-mail: mand@mpasmb.desy.de.

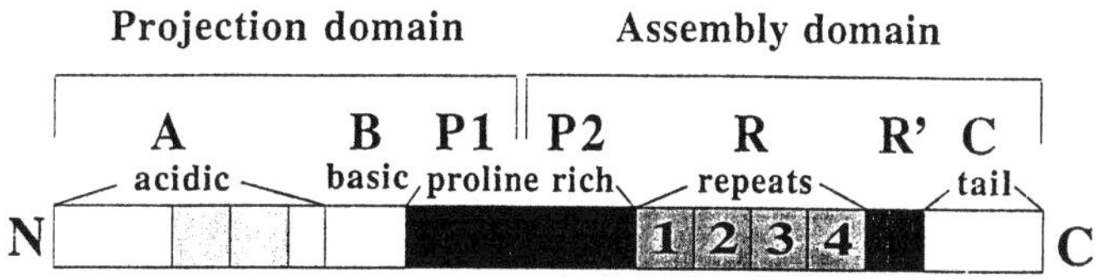

FIGURE 1. Bar diagram and domain structure of tau. For details see text. The residues are numbered according to the longest isoform in human neurons.[46]

factors that cause the disease. In this paper we summarize some recent studies on tau in four areas. First we describe some structural features of tau and PHFs. Next we deal with the interaction between tau and microtubules. This is followed by a consideration of how phosphorylation regulates tau-microtubule interactions, and what kinases might be involved.

RESULTS AND DISCUSSION

Structural Features of Tau and Paired Helical Filaments

Tau Protein

Electron microsocopy of tau shows an elongated molecule, with lengths around 35 nm.[3] However, X-ray solution scattering shows that tau does not have a well-defined shape, unlike most other soluble proteins. There is no defined radius of gyration, and the molecule is best described as a Gaussian or random polymer, with a persistence length of about 2 nm.[4] In other words, tau in solution resembles a denatured protein. This agrees well with the observation that tau can be treated with denaturants (heat, dilute acids), and yet it remains soluble and does not lose its biological activity of microtubule binding and stabilization. In keeping with the "denatured" appearance, spectroscopic experiments by CD do not yield any evidence for regular secondary structure either (such as alpha helix or beta sheet). These observations matter in the context of current paradigms for the pathological aggregation of proteins: Prion protein or the amyloid beta peptide are thought to aggregate by interactions of strands of beta sheet.[5–7] Tau appears to be different in this respect.

Alzheimer Paired Helical Filaments

PHFs isolated from Alzheimer brain tissue have been investigated mainly by electron microscopy and image reconstruction which yields the well known features of an apparent pair of strands, wound around one another, with cross-over repeats around 75-80 nm and widths varying between 5 and 22 nm.[8] However, the question of the subunit structure and arrangement has yet to be answered. X-ray fiber diffraction shows weak maxima in the range of 10 nm (perhaps indicating an average diameter of 15 nm) and 3 nm (possibly arising from some axial periodicity). More significant is the absence of reflections typical of secondary structure, confirming that Alzheimer PHFs, like tau itself, do not show obvious alpha-helix or beta-sheet contributions. A similar result is obtained by spectroscopic methods such as FTIR.[4]

Dimerization and Aggregation of Tau in Vitro

In order to understand PHF assembly and to find ways to prevent this from happening in the brain it would be important to reassemble them *in vitro*. Most proteins aggregate in some conditions (*e.g.* upon denaturation), but in order to be meaningful for AD research the aggregates need to show the characteristics of PHFs (crossover repeats, width); secondly, the aggregation *in vitro* should reproducibly yield a reasonably homogeneous population of PHF-like fibers. These criteria have been met by allowing bacterially expressed constructs of tau to aggregate which contain essentially the repeat domain.[3] Full-length tau does not assemble into bona fide PHF-like fibers *in vitro,* although fibers that are possibly related have been observed in some conditions.[9,47] In this respect, tau research lags behind amyloid research where large quantities of amyloid fibers can be reassembled from the amyloid peptide *in vitro.*

The assembly of PHF-like fibers *in vitro* works best under two conditions: i) the starting material is roughly equivalent to the repeat domain of tau, as mentioned above, ii) the monomers are cross-linked into dimers. The production of stable dimers of tau can be achieved either by oxidation (using cysteine 322 in repeat 3, ref. 51), or by chemical cross-linkers that link cysteines to cysteines or cysteines to lysines.[3] Analysis by EM and decoration with specific antibodies shows that such dimers are antiparallel and roughly in register. On the basis of these observations one can postulate an assembly mechanism where dimerization precedes polymerization (FIG. 2). This scheme is consistent with reports that the repeat domain forms the core of the Alzheimer PHFs.[10,11] In this context it is notable that the *in vitro* assembly of PHF-like fibers from the repeat domain does not require phosphorylation. Thus the relationship between assembly of Alzheimer PHFs and hyperphosphorylation[12] remains to be explained.

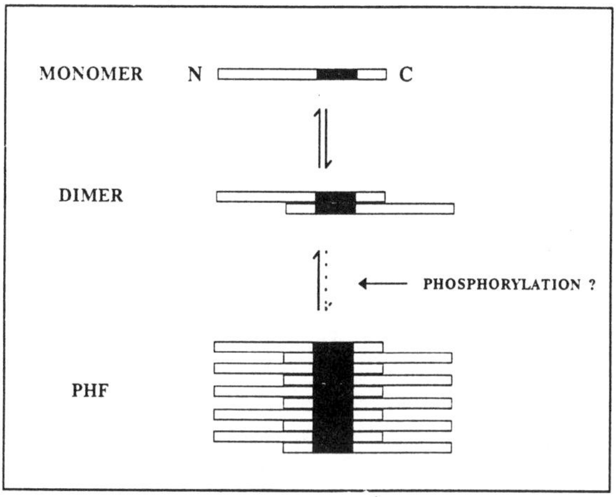

FIGURE 2. Hypothetical assembly scheme of tau monomers, first into antiparallel dimers, then into paired helical filaments.

Interaction of Tau with Microtubules

Domains of Tau Interacting with Microtubules

Tau stabilizes neuronal microtubules and promotes neurite outgrowth.[13,14] We wanted to test which domains are responsible for the binding to microtubules and for promoting their assembly. A protein domain is usually defined as a folding unit, but—in the absence of a high resolution structure—the best alternative to define tau's domains is on the basis of sequence, as shown in FIGURE 1. We distinguish acidic, basic, proline-rich, repeat, or tail domains. The microtubule interacting domains reside somewhere in the C-terminal half of the molecule, the N-terminal half probably serves as spacer between microtubules, possibly as anchor for other proteins, or in some other tissue-specific function. Of particular importance are the 3 or 4 repeats, R1-R4, and the flanking regions P and R′ (note that R′ bears some sequence relationship to the other repeats and thus is termed a "fifth" repeat).

It is usually thought that the repeats interact with microtubules, and indeed the repeat peptides or tau or related MAPs bind to microtubules.[15,16] However, while full-length tau binds with dissociation constants in the micromolar range, the peptides bind a thousand times more weakly, begging the question of what determines the high affinity of tau. This question has been addressed by several authors using constructs of tau truncated to various extents.[17–19] We used a somewhat different approach, exchanging domains or areas of tau in a cassette-like fashion, and analyzing the tau-microtubule interactions in several ways (binding stoichiometry and binding constant, nucleation capacity, induction of microtubule bundles, effect on dynamic instability.[20] Some key results can be summarized as follows: The maximum binding stoichiometry (about 1 tau per 2 tubulin dimers) is surprisingly reproducible for different constructs, whereas the dissociation constant varies from strong binding (micromolar Kd) to very weak (millimolar Kd). The number and choice of repeats is of little effect. The repeats alone bind only very weakly, but binding and nucleation capacity improves dramatically when the repeats are combined with either or both of the flanking regions (domains P or R′). Perhaps the most unexpected result is that repeat-less tau binds almost as tightly as full-length tau; in other words, the repeats are not necessary for the high affinity. However, such repeat-less tau constructs are unproductive in terms of microtubule assembly. This means that we have to view tau's binding to microtubules and its effect on assembly as two separate features. In FIGURE 3 we

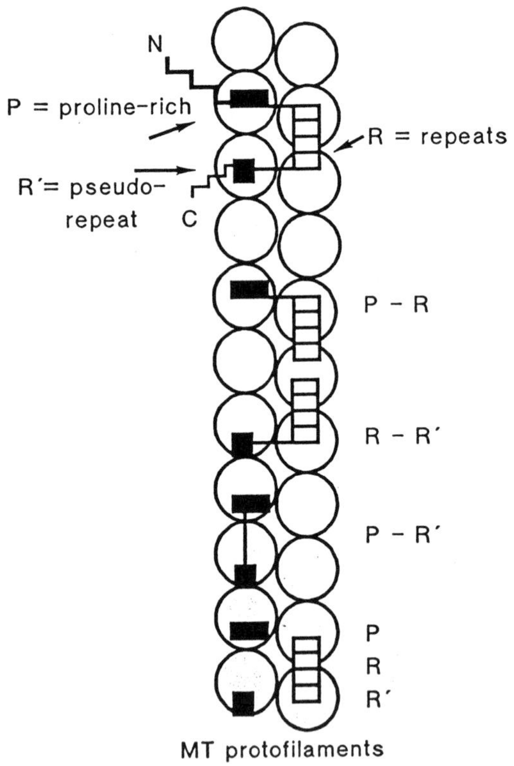

FIGURE 3. Interactions of tau domains with microtubules. Jaw model of tau-microtubule interactions. The rows of spheres represent two protofilaments of a microtubule. It is assumed that tau domains may bind to one or more protofilaments, but microtubules are stabilized only when protofilaments are cross-connected. Tau is shown with the repeat domain (four white boxes), the flanking regions (black), and the N- and C-terminal tails. Several binding modes are shown from top to bottom: (a) full length tau, containing both flanks and the repeats, P-R-R′; (b) a construct containing the left flank and the repeats, P-R; (c) a construct containing the right flank and the repeats, R-R′. In these three cases the combination of the repeats and at least one of the flanking regions causes tight binding and is productive in terms of microtubule assembly. (d) A construct with only the two flanks but without repeats, P-R′: This construct binds tightly but does not cause assembly. (e) Constructs P, R, or R′ alone bind weakly and are inefficient in stabilizing microtubules.

envisage that certain domains of tau (the flanking regions, black, compare FIG. 1) may bind tightly to a single protofilament, but without cross-linking adjacent protofilaments and thus stabilizing their interaction. The cross-linking of protofilaments is achieved by combining the flanking regions with the repeat domain. Thus the flanking regions can be likened to "anchoring" or "targeting" domains, while the repeats act as "actuators" of microtubule assembly.

Effect of Phosphorylation on Tau-Microtubule Interactions

Native tau is phosphorylated (about 2 Pi per mole, distributed over several sites,[21,22,48] and it is thought that this somehow regulates tau's affinity for microtubules.[23] AD tau is more highly phosphorylated (about 6-8 Pi per mole), and this is considered to be responsible for the observation that AD tau is a poor microtubule stabilizer.[24] As described below, several of the "abnormal" phosphorylation sites are of the Ser-Pro type in the flanking regions, and it was therefore natural to anticipate that phosphorylation in the flanking regions would strongly affect tau's binding to microtubules. However, this is not observed: Tau can be phosphorylated very efficiently with MAP kinase which phosphorylates most Ser-Pro or Thr-Pro motifs (to about 15 Pi per molecule). This has only a moderate effect on tau-microtubule binding (about 20% reduction in stoichiometry). By contrast, the phosphorylation at the single site Ser262 (in the first repeat) has a very strong effect and virtually abolishes tau-microtubule binding; similarly, it abrogates tau's stabilizing influence on microtubules and makes them dynamically unstable. The kinases affecting Ser262 are described below. The significance of this finding lies in the fact that Ser262 is a unique site found in tau from AD brains[25] (results confirmed in our lab[49]). Phosphorylation at this site could therefore explain why AD tau fails to stabilize microtubules.

These results pose a curious discrepancy (FIG. 4): The flanking regions are important for the tight binding of tau to microtubules—yet their phosphorylation (*e.g.*, by MAP kinase) makes little difference to the binding. The repeat domain binds only weakly to microtubules—yet its phosphorylation has a very strong influence on binding and dynamic instability. It is difficult to explain this phenomenon at present, but presumably it is related to the three-dimensional folding of tau and tubulin, both of which are unknown.

Phosphorylation of Tau

Conceptionally we can distinguish several classes of kinases or kinase activities that can phosphorylate tau *in vitro* (FIG. 5):

i) Tau is a target of several proline-directed kinases such as MAP kinase,[26] GSK-3,[27–29] and several kinases related to the cell cycle kinase cdc2, e.g. cdk2, cdk5, PDPK, or nclk.[30–33] These kinases phosphorylate some sites that

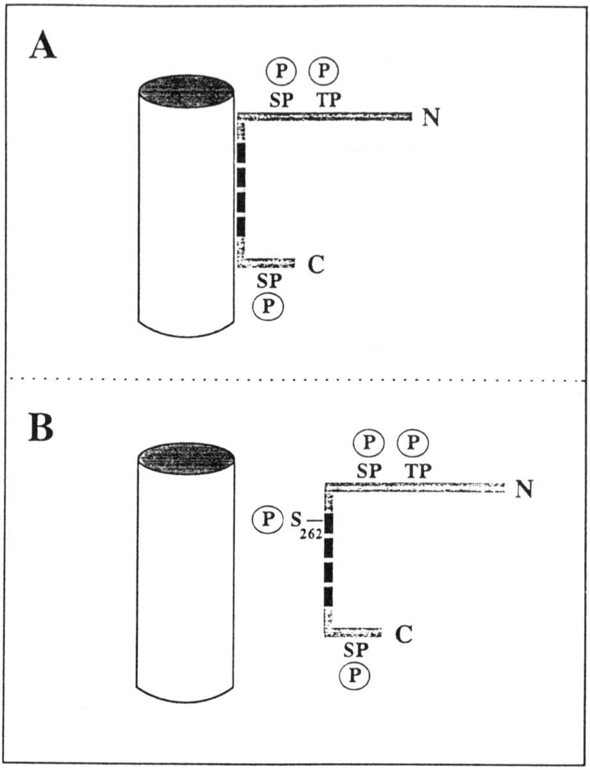

FIGURE 4. Diagram describing the effect of tau's phosphorylation on binding to microtubules. The repeats are represented by black bars. **(A)**: Although the regions flanking the repeats are important for tight binding, their phosphorylation (mostly Ser/Thr-Pro motifs) has little effect. **(B)**: By contrast the phosphorylation within the repeats at Ser262 strongly reduces the affinity between tau and microtubules.

are also found in AD tau. Several antibodies that are of diagnostic value for distinguishing normal and AD tau are directed against epitopes containing Ser/Thr-Pro motifs, either in a phosphorylated or non-phosphorylated form.[34–36,50] When preparing tau from adult rat or pig brains according to standard procedures the Ser/Thr-Pro motifs are normally not found to be phosphorylated, whereas they are partly phosphorylated in AD tissue,[25] in fetal tissue,[37,38] or when tau is rapidly extracted from biopsy tissue.[39] All of the sites induced by the proline directed kinases can be dephosphorylated by phosphatases PP-2A or PP-2B (calcineurin).[40] Note that these sites lie outside the repeat domain and have only a weak influence on the tau-microtubule interaction.

ii) Several second-messenger dependent kinases can phosphorylate tau at

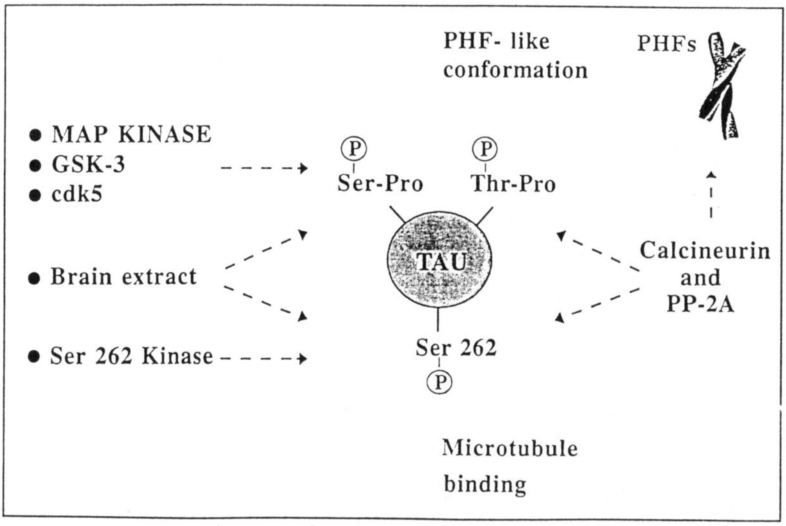

FIGURE 5. Summary of tau kinases and phosphatases. Brain extract *(left, middle)* contains proline-directed kinases *(left, top)* which confer an "Alzheimer-like" immunoreactivity on tau, and non-proline directed kinases, including MARK which phosphorylates Ser262 and thus suppresses tau's binding to microtubules *(left, bottom)*. The phosphorylation sites can be cleared by phosphatases PP-2A and calcineurin; the same phosphatases can dephosphorylate Alzheimer PHFs.

non-Ser-Pro motifs, and this includes some sites inside the repeat domain. Examples are CaM kinase, PKA, PKC, or casein kinase II.[41–44] These kinases, too, have only a weak influence on tau-microtubule interaction, and the sites can be dephosphorylated by the same phosphatases as above (PP-2A, PP-2B).

iii) Kinases which affect tau's binding to microtubules include those that phosphorylate Ser262.[44] We recently prepared a kinase from brain which phosphorylates Ser262 and equivalent serines in the other repeats (in the KXGS motifs), termed MARK because it acts as a MAP/microtubule affinity regulating kinase. FIGURE 6 shows an example of how the average length of microtubules is dramatically reduced when tau becomes phosphorylated by MARK.[45] This effect is caused by the detachment of tau from microtubules which leads to a high dynamic instability. In fact when tau is phosphorylated by MARK, microtubules behave as if there was no tau present in the solution. An effect of this magnitude has not been observed with other kinases tested. The kinase therefore qualifies as a potential culprit for causing tau's dissociation from microtubules and their subsequent destruction in pathological conditions. Whether MARK has an effect on tau's aggregation as well is currently under study.

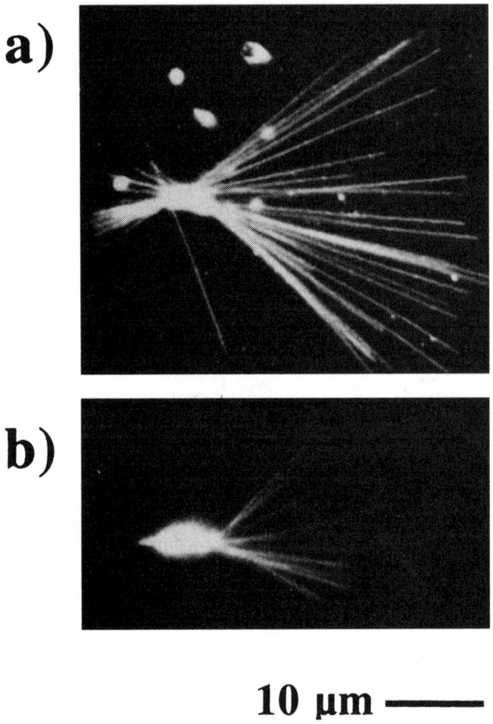

FIGURE 6. Effect of tau's phosphorylation by the microtubule affinity regulating kinase on the dynamics of microtubules. **a)** Microtubules were nucleated on sea urchin sperm axonemes in the presence of tau and the kinase, but without ATP. In this case the microtubules grow continuously for 20 min. **b)** When ATP is added, tau can be phosphorylated, leading to a destabilization and shortening of microtubules. Bar = 10 μm.

ACKNOWLEDGMENTS

Brain tissue was generously provided by the Bryan Alzheimer Disease Research Center (Duke University Medical Center, Durham, NC), the Brain Tissue Resource Center (McLean Hospital/Harvard Medical School, Belmont, MA), and the Alzheimer Research Center (University of Rochester Medical School, Rochester, NY). Phosphopeptide sequencing was done in collaboration with Dr. H. E. Meyer, University of Bochum.

REFERENCES

1. Kosik, K. S. & S. M. Greenberg. *In* Alzheimer Disease. R. Terry, R. Katzman & K. Bick, Eds.: 335–344 New York: Raven Press.
2. Braak, H. & E. Braak. 1994. *In* Neurodegenerative Diseases. D. Calne *et al.*, Eds.: 585–613. Philadelphia: W.B. Saunders Co..
3. Wille, H., G. Drewes, J. Biernat, E.-M. Mandelkow & E. Mandelkow. 1992. J. Cell Biol. **118:**573–584.

4. Schweers, O., E. Schönbrunn-Hanebeck, A. Marx, E. Mandelkow. 1994. J. Biol. Chem. **269:**24290–24297.
5. Gasset, M., M. A. Baldwin, R. J. Fletterick & S. B. Prusiner. 1993. Proc. Natl. Acad. Sci. USA **90:**1–5.
6. Hilbich, C., B. Kisters-Woike, J. Reed, C. Masters & K. Beyreuther. 1992. J. Mol. Biol. **228:**460–473.
7. Inouye, H., P. E. Fraser & D. A. Kirschner. 1993. Biophys. J. **64:**502–519.
8. Crowther, R. A. & C. M. Wischik. 1985. EMBO J. **4:**3661–3665.
9. Crowther, R. A., O. F. Olesen, M. J. Smith, R. Jakes & M. Goedert. 1994. FEBS Letters **337:**135–138.
10. Kondo, J., T. Honda, H. Mori, Y. Hamada, R. Miura, M. Ogawara & Y. Ihara. 1988. Neuron **1:**827–834.
11. Novak, M., J. Kabat & C. M. Wischik. 1993. EMBO J. **12:**365–370.
12. Grundke-Iqbal, I., K. Iqbal, Y. Tung, M. Quinlan, H. Wisniewski & L. Binder. 1986. Proc. Natl. Acad. Sci. USA **83:**4913–4917.
13. Drubin, D. & M. Kirschner. 1986. J. Cell Biol. **103:**2739–2746.
14. Knops, J., K. Kosik, G. Lee, J. Pardee, L. Cohengould & L. McConlogue. 1991. J. Cell Biol. **114:**725–733.
15. Ennulat, D. J., R. K. H. Liem, G. A. Hashim & M. L. Shelanski. 1989. J. Biol. Chem. **264:**5327–5330.
16. Lee, G., R. L. Neve & K. S. Kosik. 1989. Neuron **2:**1615–1624.
17. Butner, K. A. & M. W. Kirschner. 1991. J. Cell Biol. **115:**717–730.
18. Brandt, R. & G. Lee. 1993. J. Biol. Chem. **268:**3414–3419.
19. Goode, B. L. & S. C. Feinstein. 1994. J. Cell Biol. **124:**769–782.
20. Gustke, N., B. Trinczek, J. Biernat, E.-M. Mandelkow & E. Mandelkow. 1994. Biochemistry **33:**9511–9522.
21. Ksiezak-Reding, H., W. K. Liu & S. H. Yen. 1992. Brain Res. **597:**209–219.
22. Watanabe, A., M. Hasegawa, M. Suzuki, K. Takio, M. Morishima-Kawashima, K. Titani, T. Arai, K. S. Kosik & Y. Ihara. 1993. J. Biol Chem. **268:**25712–25717.
23. Drechsel, D. N., A. A. Hyman, M. H. Cobb & M. W. Kirschner. 1992. Mol. Biol. Cell **3:**1141–1154.
24. Yoshida, H. & T. Ihara. 1993. J. Neurochem. **61:**1183–1186.
25. Hasegawa, M., M. Morishima-Kawashima, K. Takio, M. Suzuki, K. Titani & Y. Ihara. 1992. J. Biol. Chem. **26:**17047–17054.
26. Drewes, G., B. Lichtenberg-Kraag, F. Döring, E.-M. Mandelkow, J. Biernat, J. Goris, M. Doree & E. Mandelkow. 1992. EMBO J. **11:** 2131–2138.
27. Mandelkow, E.-M., G. Drewes, J. Biernat, N. Gustke, J. Van Lint, J. R. Vandenheede & E. Mandelkow. 1992. FEBS Lett. **314:**315–321.
28. Hanger, D., K. Hughes, J. Woodgett, J. Brion & B. Anderton. 1992. Neurosci. Lett. **147:**58–62.
29. Ishiguro, K., A. Shiratsuchi, S. Sato, A. Omori, M. Arioka, S. Kobayashi, T. Uchida & K. Imahori. 1993. FEBS Letters **325:**167–172.
30. Baumann, K., E.-M. Mandelkow, J. Biernat, H. Piwnica-Worms & E. Mandelkow. 1993. FEBS Lett. **336:**417–424.

31. Vulliet, R., S. Halloran, R. Braun, A. Smith & G. Lee. 1992. J. Biol. Chem. **267:**22570–22574.
32. Paudel, H., J. Lew, Z. Ali & J. Wang. 1993. J. Biol. Chem. **268:** 23512–23518.
33. Kobayashi, S., K. Ishiguro, A. Omori, M. Takamatsu, M. Arioka, K. Imahori & T. Uchida. 1993. FEBS Letters **335:**171–175.
34. Lee, V. M. Y., B. J. Balin, L. Otvos & J. Q. Trojanowski. 1991. Science **251:**675–678.
35. Biernat, J., E.-M. Mandelkow, C. Schröter, B. Lichtenberg-Kraag, B. Steiner, B. Berling, H. E. Meyer, M. Mercken, A. Vandermeeren, M. Goedert & E. Mandelkow. 1992. EMBO J. **11:**1593–1597.
36. Lichtenberg-Kraag, B., E.-M. Mandelkow, J. Biernat, B. Steiner, C. Schroter, N. Gustke, H. E. Meyer & E. Mandelkow. 1992. Proc. Natl. Acad. Sci. USA **89:**5384–5388.
37. Brion, J. P., C. Smith, A. M. Couck, J. M. Gallo & B. H. Anderton. 1993. J. Neurochem. **61:**2071–2080.
38. Bramblett, G. T., M. Goedert, R. Jakes, S. E. Merrick, J. Q. Trojanowski & V. M. Y. Lee. 1993. Neuron **10:**1089–1099.
39. Matsuo, E. S., R. W. Shin, M. L. Billingsley, A. Vandevoorde, M. Oconnor, J. Q. Trojanowski & V. M. Y. Lee. 1994. Neuron **13:**989–1002.
40. Drewes, G., E.-M. Mandelkow, K. Baumann, J. Goris, W. Merlevede & E. Mandelkow. 1993. FEBS Lett. **336:**425–432.
41. Steiner, B., E.-M. Mandelkow, J. Biernat, N. Gustke, H. E. Meyer, B. Schmidt, G. Mieskes, H. D. Söling, D. Drechsel, M. W. Kirschner, M. Goedert & E. Mandelkow. 1990. EMBO J. **9:**3539–3544.
42. Correas, I., J. Diaznido & J. Avila. 1992. J. Biol. Chem. **267:** 15721–15728.
43. Scott, C., R. Spreen, J. Herman, F. Chow, M. Davison, J. Young & C. Caputo. 1993. J. Biol. Chem. **268:**1166–1173.
44. Biernat, J., N. Gustke, G. Drewes, E.-M. Mandelkow & E. Mandelkow. 1993. Neuron **11:**153–163.
45. Drewes, G., B. Trinczek, S. Illenberger, J. Biernat, G. Schmitt-Ulms, H. E. Meyer, E.-M. Mandelkow & E. Mandelkow. 1995. J. Biol. Chem. **270:**7679–7688.
46. Goedert, M., M. Spillantini, R. Jakes, D. Rutherford & R. A. Crowther. 1989. Neuron **3:**519–526.
47. Degarcini, E. M., J. L. Carrascosa, I. Correas, A. Nieto & J. Avila. 1988. FEBS LETT. **236:**150–154.
48. Köpke, E., Y. Tung, S. Shaikh, A. Alonso, K. Iqbal & I. Grundke-Iqbal. 1993. J. Biol. Chem. **268:**24374–24384.
49. Gross, U., E.-M. Mandelkow, J. Biernat, A. Marx, A. Thiemann, H. E. Meyer, J. Metzger & E. Mandelkow. 1994. Mol. Biol. Cell **5:**289a.
50. Ishiguro, K., A. Omori, K. Sato, K. Tomizawa, K. Imahori & T. Uchida. 1991. Neurosci. Lett. **128:**195–198.
51. Schweers, O., E.-M. Mandelkow, J. Biernat & E. Mandelkow. 1995. Proc. Natl. Acad. Sci. USA **92:**8463–8467.

Regulation of Tau Phosphorylation in Alzheimer's Disease

V. M.-Y. LEE[a]

Department of Pathology and Laboratory Medicine, University of Pennsylvania School of Medicine, Philadelphia, Pennsylvania 19104 USA

ABSTRACT: Alzheimer's disease (AD) is a heterogeneous dementing disorder of the elderly that is characterized by progressive cognitive impairments and the accumulation of abundant amyloid or senile plaques (SPs) and neurofibrillary tangles (NFTs) as well as the massive loss of neurons in the AD brain. Indeed, a secure diagnosis of AD in patients with a chronic progressive dementia requires evidence of numerous SPs and NFTs in the postmortem brain. Although the deposition of fibrillar amyloid or Aβ-peptides in extracellular plaques and the accumulation of tau-rich intraneuronal NFTs are not restricted exclusively to AD, there is a close correlation between the burden of tau-rich neurofibrillary lesions in selected telencephalic regions of the brain and the dementia in AD. Since the formation of neurofibrillary lesions from hyperphosphorylated tau proteins may compromise the function and viability of neurons in the AD brain, this review summarizes recent insights into mechanisms that regulate the phosphorylation state of tau in AD.

Paired helical filaments (PHFs) are composed of each of the 6 CNS tau isoforms, and PHFs form the principal fibrous components of neurofibrillary tangles (NFTs), major neuropathological lesions found in Alzheimer's Disease (AD).[1–6] The cloning and sequencing of the tau gene demonstrated that a total of 6 different tau isoforms can be generated from a single tau gene by alternative splicing.[7,8] The expression of tau is developmentally regulated, since all 6 tau isoforms are found in the adult human brain and only the smallest isoform known as fetal tau is present in the fetal brain.[4,7,9] Although all 6 normal human tau isoforms are phosphoproteins, PHF-tau from the AD brain has been reported to be "hyperphosphorylated" or "abnormally" phosphorylated on sites that are not found in normal adult human tau.[1,4,10] These and other studies led to the hypothesis that excessive or abnormal phosphorylation of tau is a key event in the transformation of normal adult tau into PHF-tau.

Recent studies using direct chemical analysis have identified phosphorylation sites on both normal human tau, PHF-tau, fetal rat tau as well as adult

[a] *Send correspondence to:* Dr. Virginia M.-Y. Lee, Department of Pathology and Laboratory Medicine, University of Pennsylvania School of Medicine, Philadelphia, PA 19104-4283; Tel: (215)662-6427; Fax: (215)349-5909.

rat tau.[10,11] These studies showed that the sites Thr^{181}, Ser^{199}, Ser^{202}, Thr^{231}, Ser^{235}, Ser^{396} and Ser^{404} are phosphorylated in PHF-tau and fetal tau but not in normal adult tau. Recently, we demonstrated that many of these same sites are phosphorylated in the fetal human brain but around birth, they are preferentially dephosphorylated and they appear to remain in that state even into the later decades of life.[4,9,12] However, it is important to recognize that postmortem human brain tissues were used in previous studies of human tau conducted in our laboratory and by other investigators. Hence, it is unclear whether or not the lack of phosphorylation at these sites in adult human tau is due to the more rapid dephosphorylation of adult versus fetal tau by phosphatases during similar postmortem intervals or if this reflects a developmental down-regulation of kinases and/or upregulation of phosphatases in the fetal versus the adult CNS. Several kinases including MAP kinase,[13,14] GSK3,[15,16] CDK5,[17–19] CDC2 kinase[20] as well as a proline-directed kinase[21] have been shown to phosphorylate tau *in vitro*. Other studies have demonstrated that the protein phosphatase 2A[14] and 2B[22,23] are the principal phosphatases that dephosphorylate tau *in vitro*. Since phosphorylation at specific sites negatively regulate the binding of tau to microtubules and hyperphosphorylated PHF-tau is incapable of binding to microtubules unless it is first dephosphorylated,[9] an understanding of the regulation of the dynamic phosphorylation/dephosphorylation of tau will provide important insights into the pathogenesis of NFTs, and also shed light into the molecular mechanism leading to the collapse of the neuronal cytoskeleton and the demise of the neuron. An essential first step in the understanding of these processes is to determine the exact state and sites of phosphorylation of tau in fresh adult human neurons.

A recent study demonstrated that the treatment of human biopsy brain slices with the phosphatase inhibitor okadaic acid resulted in tau that comigrates with PHF-tau, and this tau preparation is presumably more phosphorylated than normal adult human tau isolated from postmortem brain tissue.[23] Since okadaic acid is capable of increasing phosphorylation by the inhibition of phosphatases as well as by the stimulation of kinases, it is unclear whether or not normal fresh human biopsy brain tissue processed with no postsurgical delay would be more phosphorylated than normal adult human tau isolated from the postmortem brain. Previous studies on another brain-specific, microtubule-associated protein (*i.e.,* MAP2) showed that it is exquisitely sensitive to dephosphorylation during postmortem delay.[24] These studies prompted us to reconsider the possibility that adult human tau found *in situ* may be more phosphorylated than we and others have previously recognized. If this is the case, we have speculated that adult CNS tau like fetal tau could be phosphorylated on sites that are similar to those found on PHF-tau. If we demonstrate that normal adult human tau proteins are phosphorylated at sites similar to those previously identified for PHF-tau, this would have significant implications for current hypothesis concerning the pathogenesis of PHFs since the abnormal phosphorylation of tau proteins by improperly regulated kinases is considered to be the leading hypothesis to explain the conversion of normal tau into PHF-tau. However, if freshly isolated brain tau is phosphorylated at sites similar to fetal tau and PHF-tau, then we would shift our attention to specific phosphatase(s) that are down-regulated or inactivated in NFT bearing neurons since the inability to remove phosphate groups from normal sites of phosphorylation in tau may explain the dramatic differences in the phosphorylation state of PHF-tau compared to human CNS tau obtained from the postmortem brain.

Studies were conducted to characterize the phosphorylation state of tau freshly isolated from human biopsy samples.[25] Using a panel of well-defined, phosphorylation-dependent antibodies that recognized a number of phosphorylation dependent epitopes in PHF-tau in Western blot analyses, we demonstrated that biopsy-derived tau with no postsurgical delay is phosphorylated at almost all of the sites found in PHF-tau. Several criteria were used to verify the observation that biopsy-derived human adult brain tau is more phosphorylated than autopsy-derived human brain tau and that these tau proteins share many phosphorylation sites with PHF-tau. First, we showed that biopsy-derived brain tau nearly comigrated with PHF-tau but not autopsy-derived human adult brain tau. Second, biopsy-derived brain tau, like PHF-tau, was resolved as three major protein bands with Mr of about 55-68 kD, whereas autopsy-derived tau protein migrated as six protein bands on SDS-PAGE gels with a Mr of about 50-62 kD. Third, upon dephosphorylation, biopsy-derived adult human tau proteins showed as increase in electrophoretic mobility such that they migrated in a manner indistinguishable from that of autopsy-derived brain tau. Finally, we showed that a number of well-characterized phosphorylation site-specific antibodies[1,9,26–35] bind to biopsy-derived brain tau just like they bind to PHF-tau and autopsy-derived human fetal tau but not autopsy-derived human adult tau. Significantly, we also provided an explanation of why biopsy-derived tau is much more phosphorylated than autopsy-derived brain tau. We showed that tau in biopsy samples is rapidly dephosphorylated during postsurgical delay suggesting that endogenous phosphatases are either constitutively active or that there is a net increase in the activity of phosphatases relative to kinases in brain tissue during postmortem or postsurgical delay. Other studies also showed that both protein phosphatase 2A (PP2A) and protein phosphatase 2B (PP2B) in biopsy brain samples are responsible for the dephosphorylation of tau during postsurgical delay.

Similar observations were made using tau isolated from rat and monkey brains.[36,37] We showed that all the phosphorylation sites that have been identified in biopsy-human tau are also phosphorylated in fetal and adult rat brain tau.[37] Our studies using fetal and adult rat tau also provide an explanation of why autopsy-human fetal tau but not autopsy-human adult tau is phosphorylated at most of the sites found in PHF-tau. We showed that during development from embryonic day 18 (E18), fetal tau was highly phosphorylated until postnatal day 11 (P11). Thereafter, the levels of fetal tau diminished as did its phosphorylation state concomitant with the appearance of the adult tau isoforms, suggesting that kinases were down-regulated as the animal matures. Other evidence also suggests that phosphatases were activated after P12. This is based on the observation that PP2A and PP2B activities in the adult rat brain extracts dephosphorylate tau efficiently whereas phosphatases in extracts of the P6 rat brain did not have the same effects. Further, during postmortem delay, fetal rat tau is dephosphorylated much more slowly than adult rat tau confirming that the phosphatases involved in dephosphorylating tau may not be as active in immature animals as they are in adult animals. These studies provide evidence to substantiate that adult tau is phosphorylated at similar sites as fetal tau but that fetal tau remains more phosphorylated because it is more phosphorylated *in situ* and is dephosphorylated less efficiently during postmortem or postsurgical delay.

Our observation that biopsy-derived adult brain tau proteins are phosphorylated at the same sites as PHF-tau and that these sites can be dephosphorylated easily and efficiently by endogenous phosphatases will require a reassessment of current hypotheses about the pathogenesis of PHF-rich neurofibrillary lesions such as NFTs and dystrophic neurites in the AD brain. For example, our data imply that the activities of these phosphatases may be down-regulated or defective in those neurons in the AD brain that accumulate NFTs or develop PHF-tau in their dystrophic processes. Indeed, since poorly phosphorylated, autopsy-derived tau and highly phosphorylated PHF-tau can be isolated from the AD brain, this suggests that phosphatases may be defective only in neurons that are affected by neurofibrillary pathology. Thus, defective dephosphorylation rather than aberrant phosphorylation cascades may be chiefly responsible for the pathogenesis of neurofibrillary lesions. However, recent studies suggest that the signal transduction pathways involving kinases and phosphatases are intimately interrelated. The down-regulation of phosphatases such as PP2A may directly increase the phosphorylation of tau by blocking dephosphorylation. Additionally, the phosphorylation of tau can increase indirectly since the inactivation of PP2A could lead to the activation of kinases such as MAP kinase that phosphorylate tau. Thus, the data presented here emphasize the possible involvement of phosphatases in tipping the balance toward an increase in the phosphorylation of tau. Future studies should be directed toward determining whether the inactivation of phosphatases in neurons vulnerable to neurofibrillary lesions in the AD brain is responsible for the hyperphosphorylation of tau leading to the formation of PHFs.

REFERENCES

1. Lee, V. M.-Y., B. J. Balin, L. Otvos, Jr. & J. Q. Trojanowski. 1991. A68: A major subunit of paired helical filaments and derivatized forms of normal tau. Science **251**:675–678.

2. KOSIK, K. S. 1992. Alzheimer's disease: A cell biological perspective. Science **256:**780–783.
3. LEE, V. M.-Y. & J. Q. TROJANOWSKI. 1992. The disordered neuronal cytoskeleton in Alzheimer's disease. Curr. Opin. Neurobiol. **2:**653–656.
4. GOEDERT, M., R. JAKES, R. A. CROWTHER, J. SIX, U. LUBKE, M. VANDERMEEREN, P. CRAS, J. Q. TROJANOWSKI & J. M.-Y. LEE. 1993. The abnormal phosphorylation of tau proteins at Ser-202 in Alzheimer disease recapitulates phosphorylation during development. Proc. Natl. Acad. Sci. USA **90:** 5066–5070.
5. TROJANOWSKI, J. Q., M. L. SCHMIDT, R.-W. SHIN, G. T. BRAMBLETT, D. RAO & V. M.-Y. LEE. 1993. Altered tau and neurofilament proteins in neurodegenerative diseases: Diagnostic implications for Alzheimer's disease and Lewy body dementias. Brain Pathol. **3:**45–54.
6. TROJANOWSKI, J. Q., M. L. SCHMIDT, R.-W. SHIN, G. T. BRAMBLETT, M. GOEDERT & V. M.-Y. LEE. 1993. PHFτ (A68): From pathological marker to potential mediator of neuronal dysfunction and degeneration in Alzheimer' disease. Clin. Neurosci. **1:**184–191.
7. GOEDERT, M., M. G. SPILLANTINI, R. JAKES, D. RUTHERFORD & R. A. CROWTHER. 1989. Multiple isoforms of human microtubule-associated protein tau: Sequences and localization in neurofibrillary tangles of Alzheimer's disease. Neuron **3:**519–526.
8. HIMMLER, A. 1989. Structure of the bovine tau gene: Alternatively spliced transcripts generate a protein family. Mol. Cell Biol. **9:**1389–1395.
9. BRAMBLETT, G. T., M. GOEDERT, R. JAKES, S. E. MERRICK, J. Q. TROJANOWSKI & V. M.-Y. LEE. 1993. Abnormal tau phosphorylation at Ser396 in Alzheimer's disease recapitulates development and contributes to reduced microtubule binding. Neuron **10:**1089–1099.
10. HASAGEWA, M., M. MORISHIMA-KAWASHIMA, K. TAKIO, M. SUZUKI, K. LITANI & Y. IHARA. 1992. Protein sequence and mass spectrometric analyses of tau in the Alzheimer's disease brain. J. Biol. Chem. **267:**17047–17054.
11. WATANABE, A., M. HASEGAWA, M. SUZUKI, K. TAKIO, M. MORISHIMA-KAWASHIMA, K. TITANI, T. ARAI, K. S. KOSIK & Y. IHARA. 1993. In vivo phosphorylation sites in fetal and adult rat tau. J. Biol. Chem. **268:**25712–25717.
12. KANEMARU, K., K. TAKIO, R. MIURA, K. TITANI & Y. IHARA. 1992. Fetal-type phosphorylation of the tau in paired helical filaments. J. Neurochem. **58:** 1667–1675.
13. DREWES, G., B. LICHTENBERG-KRAGG, F. DORING, E. M. MANDELKOW, J. BIENART, M. DOREE & E. MANDELKOW. 1992. Mitogen activated protein (MAP) kinase transform tau protein into an Alzheimer-like state. EMBO J. **11:** 2131–2138.
14. GOEDERT, M., E. S. COHEN, R. JAKES & P. COHEN. 1992. p42 MAP kinase phosphorylation sites in microtubule-associated protein tau are dephosphorylated by protein phosphatase $2A_1$. FEBS Lett **312:**95–99.
15. HANGER, D. P., K. HUGHES, J. R. WOODGETT, J. P. BRION & B. H. ANDERTON. 1992. Glycogen synthase kinase-3 induces Alzheimer's disease-like phosphorylation of tau: Generation of paired helical filament epitopes and neuronal localization of the kinase. Neurosci. Lett. **147:**58–62.

16. MANDELKOW, E. M., G. DREWES, J. BIERNAT, N. GUSTKE, J. VAN LINT, J. R. VANDENHEEDE & E. MANDELKOW. 1992. Glycogen synthase kinase 3 and the Alzheimer's disease-like state of microtubule-associated protein tau. FEBS Lett **314**:315–321.
17. BAUMAN, K., E.-M. MANDELKOW, J. BIERNAT, H. PIWNICA-WORMS & E. MANDELKOW. 1993. Abnormal Alzheimer's like phosphorylation of tau protein by cyclin-dependent cdk2 and cdk5. Febs Letts. **336**:417–424.
18. KOBAYASHI, S., K. ISHIGURO, A. OMORI, M. TAKAMATSU, M. ARIOKA, K. IMAHORI & T. UCHIDA. 1993. A cdc-related kinase PSSALRE/cdk5 is homologous with the 30 kDa subunit of tau protein kinase II, a proline-directed protein kinase associated with microtubule. FEBS Lett. **335**:171–175.
19. PAUDEL, H. K., J. LEW, A. ZENOBIA & J. H. WANG. 1993. Brain-proline-directed protein kinase phosphorylates tau on sites that are abnormally phosphorylated in tau associated with Alzheimer's paired helical filaments. J. Biol. Chem. **268**:23512–23518.
20. LEDESMA, M. D., L. CORREAS, J. AVILA & J. DIAZ-NIDO. 1992. Implication of brain cdc2 and MAP kinases in the phosphorylation of tau protein in Alzheimer's disease. FEBS Letts. **308**:218–224.
21. VULLIET, R., S. M. HALLORAN, R. K. BRAUN, A. J. SMITH & G. LEE. 1992. Proline-directed phosphorylation of human tau protein. J. Biol. Chem. **267**: 22570–22574.
22. DREWES, G., E.-M. MANDELKOW, K. BAUMAN, J. GORIS, W. MERLEVEDE & E. MANDELKOW. 1993. Dephosphorylation of tau protein and Alzheimer paired helical filaments by calcineurin and phosphatase 2A. FEBS Lett. **336**:425–432.
23. HARRIS, K. A., G. A. OYLER, G. M. DOOLITTLE, I. VINCENT, R. A. W. LEHMAN, R. L. KINCAID & M. L. BILLINGSLEY. 1993. Okadaic acid induces hyperphosphorylated forms of tau protein in human brain slices. Ann. Neurol. **33**: 77–87.
24. TSUYAMA, S., Y. TERAYAMA & S. MATSUYAMA. 1987. Numerous phosphates of microtubule-associated protein 2 in living rat brain. J. Biol. Chem. **262**: 10886–10892.
25. MATSUO, E. S., R.-W. SHIN, M. L. BILLINSLEY, A. VAN DE VOORDE, M. O'CONNOR, J. Q. TROJANOWSKI & V. M.-Y. LEE. 1994. Biopsy-derived adult human brain tau is phosphorylated at many of the same sites as Alzheimer's disease paired helical filament tau. Neuron **13**:989–1002.
26. BINDER, L. I., A. FRANKFURTER & L. I. REBHUN. 1985. The distribution of tau in the mammalian central nervous system. J. Cell Biol. **101**:1371–1378.
27. WOLOZIN, B. L., A. PRUCHNICKI, D. W. DICKINSON & P. DAVIES. 1986. A neuronal antigen in the brains of Alzheimer patients. Science **232**:648–650.
28. KOSIK, K. S., L. D. ORECCHIO, L. BINDER, J. Q. TROJANOWSKI, V. M.-Y. LEE & G. LEE. 1988. Epitopes that span the tau molecule are shared with paired helical filaments. Neuron **1**:817–825.
29. TROJANOWSKI, J. Q., T. SCHUCK, M. L. SCHMIDT & V. M.-Y. LEE. 1989. Distribution of tau proteins in the normal human central and peripheral nervous system. J. Histochem. Cytochem. **37**:209–215.
30. GREENBERG, S. G. & P. DAVIES. 1990. A preparation of Alzheimer paired

helical filaments that display distinct tau proteins by polyacrylamide gel electrophoresis. Proc. Natl. Acad. Sci. USA **87**:5827–5831.
31. MERCKEN, M., M. VANDERMEEREN, V. LUBKE, J. BOONS, A. VAN DE VOORDE, J.-J. MARTIN & J. GHEUEN. 1992. Monoclonal antibodies with selective specificity for Alzheimer tau are directed against phosphatase-sensitive epitope. Acta Neuropathol. **84**:265–272.
32. GOEDERT, M., M. G. SPILLANTINI & R. JAKES. 1991. Localization of the Alz-50 epitope in recombinant human microtubule-associated protein tau. Neurosci. Lett. **126**:149–154.
33. GOEDERT, M. 1993. Tau protein and the neurofibrillary pathology of Alzheimer's disease. Trends Neurosci. **16**:460–465.
34. GOEDERT, M., R. JAKES, R. A. CROWTHER, P. COHEN, E. VANMECHELEN, M. VANDERMEEREN & P. CRAS. 1994. Epitope mapping of monoclonal antibodies to the paired helical filaments of Alzheimer's disease: Identification of phosphorylation sites in tau. Biochem. J. **301**:871–877.
35. OTVOS, I., JR., L. FEINER, E. LANG, G. I. SZENDREI, M. GOEDERT & V. M.-Y. LEE. 1994. Phosphorylated serine residues 396 and 404 in paired helical filaments define the epitope for the monoclonal antibody PHF1. J. Neurosci. Res. **39**:669–673.
36. GARVER, T. D., K. A. HARRIS, R. A. W. LEHMAN, V. M.-Y. LEE, J. Q. TROJANOWSKI & M. L. BILLINSLEY. 1994. τ Phosphorylation in human, primate, and rat brain: Evidence that a pool of τ is highly phosphorylated in vivo and is rapidly dephosphorylated in vitro. J. Neurochem. **63**:212–222.
37. MAWAL-DEWAN, M., J. HENLEY, A. VAN DE VOORDE, J. Q. TROJANOWSKI & V. M.-Y. LEE. 1994. The phosphorylation state of tau in the developing rat brain is regulated by phosphoprotein phosphatases. J. Biol. Chem. **49**:30981–30987.

Cellular Signaling Pathways and Cytoskeletal Organization

K. S. KOSIK,[a,c] W. Q. QIU,[a] AND S. GREENBERG[b]

[a] *Harvard Medical School and Center for Neurologic Diseases, Department of Neurology, Brigham and Women's Hospital, Boston, Massachusetts 02115 USA*

[b] *Department of Neurology, Massachusetts General Hospital, Boston, Massachusetts 02114 USA*

ABSTRACT: We describe a possible mechanism by which the amyloid precursor protein (APP) may be linked to the intraneuronal pathology of Alzheimer's disease (AD). Extracellular proteolytic products of APP may directly or indirectly transduce signals to cells under normal as well as pathological conditions. This activity appears to reside in the cysteine-rich amino terminus of APP. How APP or its metabolic products may affect the phosphorylation of tau is considered.

INTRODUCTION

Although accumulating evidence links the amyloid precursor protein (APP) to the intraneuronal pathology of Alzheimer's disease, the mechanism mediating this link is obscure. Owing to several structural features of APP and because some of the major proteolytic products of APP, including the Aβ fragment, are extracellular, it is reasonable to consider how these molecules may directly or indirectly transduce signals to cells under both normal and pathological conditions. Although some connections remain vague, current understanding allows one to trace a signal transduction path from the cell surface to components transcriptional regulation in the nucleus.[1] More problematic is signal transduction to the cytoskeleton. Among the few suggested intermediates are the Rho and Rac GTP-binding proteins that have been implicated in membrane ruffling[2,3] and the presence of SH3 domains in several cytoskeletal proteins including spectrin, myosin 1, and an actin-binding protein from yeast.[4–6]

THE APP HOLOPROTEIN

The extra-membranous portion of APP contains an amino terminal 17 residue signal sequence, followed by a cysteine-rich domain, and an anionic

[c] *Send correspondence to:* Dr. K. S. Kosik, Harvard Medical School and Center for Neurologic Diseases, Department of Neurology, Brigham and Women's Hospital, Boston, MA 02115.

domain of approximately 100 residues (reviewed in ref. 7). This ectodomain undergoes N- and O-glycosylation, tyrosine sulfation, and phosphorylation.[8] APP undergoes alternative splicing that results in several isoforms. These splice variants include exons in the ectodomain, one with homology to the Kunitz family of serine protease inhibitors, another with homology to the OX-2 surface antigen in leukocytes. A shorter 695-residue isoform is predominately neuronal; whereas a longer isoform with these exons tends to be expressed in non-neural tissue including glia.[9] Another exon, exon 15, is alternatively spliced in white cells where it may have a role in cell adhesion.[10] The holoprotein is internalized via clathrin-coated vesicles[11,12] probably utilizing the NPXY consensus sequence in the cytoplasmic tail. APP traffics to the endosomal-lysosomal pathway where a series of carboxy terminal fragments are generated.[13]

STIMULATION OF MAPK BY SECRETED APP

APP generates specific fragments with distinct fates in discrete cellular compartments. The most abundant secreted fragment, called APP_S, is a 100-110 kD polypeptide cleaved between residues Lys16-Leu17 of the Aβ peptide, which precludes the formation of Aβ.[14] Possible functions of APP_S are a serine protease inhibitor identified first as protease nexin II, and later as factor XI_a-inhibitor (reviewed in ref. 15). APP_S also has growth promoting properties,[16] which may follow from its ability to activate the mitogen-activated kinase in a ras-dependent manner,[17] which will be discussed extensively below. In the ECM, APP_S can bind to heparin,[18] laminin,[19] and a basement membrane HSPG,[20] where it may mediate adhesiveness.

The secreted form of APP, designated APP_S, activates MAPK in PC12 cells.[17] The MAPK (also known as ERKs) family of kinases has been reviewed by Cobb.[28] APP_S was purified from medium conditioned by CHO cells stably transfected with either 695, 751, or 770 amino acid forms of APP. Purification entailed a column to which two monoclonal antibodies against distinct sites in APP were bound. A single band was eluted from the column with high salt. PC12 cells treated with purified APP_{751} increased the activity of MAPK/ERK-1 in PC12 lysates, as measured in immunocomplexes using myelin basic protein as a substrate. Activity increased from 17- to 25-fold after 10 minutes. Stimulation declined after longer exposures with a 50% reduction in activity following one hour of treatment. Increased activation was observed with as little as 20 picomoles of APP_{751}. APP_{695} appears to have an identical effect on MAPK activity. As an independent assay for MAPK activation, antibodies against phosphotyrosine were used to blot lysates after stimulation. Increased tyrosine phosphorylation of proteins at M_r 44 and 42,000 that comigrate with ERK-1 and ERK-2 was readily detected. Synthetic Aβ peptide had no effect on MAPK activity.

Other growth factors that activate MAPK, such as nerve growth factor and basic fibroblast growth factor, do so via activation of ras.[21–24] The requirement for ras, following APP$_S$ stimulation was assayed in the PC12-derived cell line GSrasDN6. In this line, dexamethasone induces a dominant inhibitory form of ras.[21] Induction of inhibitory ras prevented APP$_S$-stimulated MAPK activation as determined by phosphorylation of the MBP substrate and tyrosine phosphorylation.

One possibility for artifact is that a small molecule bound to APP$_S$ is responsible for the observed activity. For instance TGFβ, which can activate MAPK, was shown to bind to soluble derivatives of APP.[25] To eliminate this possibility we also utilized a bacterial fusion protein to stimulate activity. This work is described in the next section.

SITE OF ACTIVITY WITHIN APP

The site within APP where the stimulatory activity resides corresponds best with the cysteine-rich region at the extreme amino terminus. These data are best presented in FIGURE 1 below, which indicates the amino acid sequence of APP with the cysteines in bold. The sites where certain restriction enzymes would cut the corresponding nucleotide sequence are superimposed on the amino acid sequence.

The first evidence for assigning activity to the amino terminus comes from the C7 construct expressed in CHO cells. We used eukaryotically expressed APP obtained from cells originally made by Sam Sisodia at Johns Hopkins.[26] A segment between the restriction sites XhoI and BglII from the middle of APP was deleted leaving 350 residues on the amino side of the cut and approximately 100 residues from the carboxy terminus. This fragment retained activity. APP bacterial fusion fragments were also effective in stimulating MAPK. A glutathione S-transferase (GST) fusion protein was cleaved with thrombin, and the cleavage product purified using a GST Purification Module (Pharmacia). We have also taken advantage of the heparin-binding properties of APP to obtain a high degree of purification. The fragments tested for activity were the following expressed constructs: Kpi to Bgl II, which is nearly identical to full length APP; Kpi to Xho I, which represents the amino half of APP; and Xho I to Bgl II, which represents the carboxy half of the molecule. The full length and amino constructs were active. Control bacterial lysates and GST alone did not show any stimulatory effect. Because thrombin used in the cleavage can also activate MAPK, we also showed that the non-cleaved Kpi to Xho I fragment retained activity when still fused to

Kpi
MLPGLALLLLAAWTARALEVPTDGNAGLLAEPQIAMF**C**GRLNMHMNVQNGK
AccI
WDSDPSGTKT**C**IDTKEGILQY**C**QEVYPELQITNVVEANQPVTIQNW**C**KRGR
KQ**C**KTHPHFVIPYR**C**LVGEFVSDALLVPDK**C**KFLHQERMDV**C**ETHLHWHTV
AKET**C**SEKSTNLHDYGMLLP**C**GIDKFRGVEFV**CC**PLAEESDNVDSADAEED
DSDVWWGGADTDYADGSEDKVVEVAEEEEVAEVEEEEADDDEDDEDGDEVE
EEAEEPYEEATERTTSIATTTTTTTTESVEEVVREV**C**SEQAETGP**C**RAMISR
WYFDVTEGK**C**APFFYGG**C**GGNRNNFDTEEY**C**MAV**C**GSAIPTTAASTPDAVD
XhoI
KYLETPGDENEHAHFQKAKERLEAKHRERMSQVMREWEEAERQAKNLPKAD
KKAVIQHFQEKVESLEQEAANERQQLVETHMARVEAMLNDRRRLALENYIT
ALQAVPPRPRHVFNMLKKYVRAEQKDRQHTLKHFEHVRMVDPKKAAQIRSQ
VMTHLRVIYERMNQSLSLLYNVPAVAEEIQDEVDELLQKEQNYSDDVLANM
ISEPRISYGNDALMPSLTETKTTVELLPVNGEFSLDDLQPWHSFGADSVPA
BglII
NTENEVEPVDARPAADRGLTTRPGSGLTNIKTEEISEVKMDAEFRHDSGYE
VHHQKLVFFAEDVGSNKGAIIGLMVGGVVIATVIVITLVMLKKKQYTSIHH
GVVEVDAAVTPEERHLSKMQQNGYENPTYKFFEQMQN

FIGURE 1. Human mRNA for amyloid precursor protein (APP) ACCESSION X06989 Y00297 08-MAY-1991.

GST. The carboxy fragment showed only modest stimulation when added at 100 times the concentration.

APP_S from CHO cells was treated with DTT/iodoacetamide to prevent the formation of disulfide linkages by modifying the cysteine to S-carboxymethylcysteine. This modification prevented activation of MAPK/ERK. Our interpretation of this result is that the active site has specific conformational requirements involving disulfide linkages at the amino terminus.

LINKING APP AND HYPERPHOSPHORYLATION OF TAU

How might APP or its metabolic products alter the phosphorylation of tau? To answer this question PC12 cells were preincubated with [^{32}P] orthophosphate to saturation followed by the addition of APP_S. Beginning 15 minutes after the addition of APP_S, tau was immunoprecipitated from the PC12 cell lysates. Quantitation of phosphate incorporation by Phosphor imaging revealed a doubling in the radioactive phosphate content of tau following stimulation. Medium from untransfected cells, Aβ peptide, and stimulation in GSrasDN6 cells all showed no enhanced tau phosphorylation.

In PC12 cells, as in other cell cultures, the majority of the tau is present in a fetal-like highly phosphorylated state. This baseline degree of phosphorylation may explain the 17- to 25-fold activation of MAPK/ERK, and only a concomitant doubling of the tau phosphorylation. Secondly, because there is similarity between the fetal phosphorylation of tau and its phosphorylation state in Alzheimer's disease, it is important to determine whether the enhanced phosphorylation of tau following APP stimulation bears any relationship to the Alzheimer state. To address this question we repeated the experiments in hippocampal cultures. These cultures have many pyramidal cells, one of the principal target cells in the disease process. We found that reactivity with AT8,[27] a monoclonal antibody that reacts with PHF-tau, but not with normal adult tau, was enhanced after stimulation of these cells. However, we cannot, as yet, tie the enhanced phosphorylation of tau seen here to the form of tau found in PHF.

REFERENCES

1. SASSONE-CORSI, P. 1994. Goals for signal transduction pathways: Linking up with transcriptional regulation. EMBO J. **13**:4717–4728.
2. RIDLEY, A. J. & A. HALL. 1992. The small GTP-binding protein rho regulates the assembly of local adhesions and stress fibers in response to growth factors. Cell **70**:389–399.
3. RIDLEY, A. J., H. F. PATERSON, C. L. JOHNSTON, D. DIEKMANN & A. HALL. 1992. The small GTP-binding protein rac regulates growth factor-induced membrane ruffling. Cell **70**:401–410.
4. JUNG, G., E. D. KORN & J. A. HAMMER III. 1987. The heavy chain of Acanthamoeba myosin IB is a fusion of myosin like and non-myosin like sequences. Proc. Natl. Acad. Sci. USA **84**:6720–6724.
5. DRUBIN, D. G., J. MULHOLLAND, Z. ZHU & D. BOTSTEIN. 1990. Homology of a yeast actin-binding protein to signal transduction proteins and myosin-1. Nature **343**:288–290.
6. RODWAY, A. R. F., M. J. E. STERNBERG & D. L. BENTLEY. 1989. Similarity in membrane proteins. Nature **342**:642.
7. KOSIK, K. S. 1992. Alzheimer's disease from a cell biological perspective. Science **256**:780–783.
8. HUNG, A. Y. & D. J. SELKOE. 1994. Selective ectodomain phosphorylation and regulated cleavage of β-amyloid precursor protein. EMBO J. **13**:534–542.
9. WERTKIN, A. M., R. S. TURNER, S. J. PLEASURE, T. E. GOLDE, S. G. YOUNKIN, J. Q. TROJANOWSKI & V. M. LEE. 1993. Human neurons derived from a teratocarcinoma cell line expresses solely the 695-amino acid amyloid precursor protein and produce intracellular β-amyloid of A4 peptides. Proc. Natl. Acad. Sci. USA **90**:9513–9517.
10. KOENIG, G., U. MONNING, C. CZECH, R. PRIOR, R. BANATI, U. SCHREITER-GASSER, J. BAUER, C. L. MASTERS & K. BEYREUTHER. 1992. Identification and differential expression of a novel alternative splice isoform of the βA4 amyloid

precursor protein (APP) mRNA in leukocytes and brain microglial cells. J. Biol. Chem. **267**:10804–10809.

11. Ferreira, A., A. Caceres & K. S. Kosik. 1993. Intraneuronal compartments of the amyloid precursor protein. J. Neurosci. **13**:3112–3123.
12. Nordstedt, C., G. L. Caporaso, J. Thyberg, S. E. Gandy & P. Greengard. 1993. Identification of the Alzheimer β/A4 amyloid precursor protein in clathrin-coated vesicles purified from PC12 cells. J. Biol. Chem. **268**:608–612.
13. Estus, S., T. E. Golde, T. Kunishita, D. Blades, D. Lowery, M. Eisen, M. Usiak, X. Qu, T. Tabira, B. D. Greenberg & S. G. Younkin. 1992. Potentially amyloidogenic carboxyl-terminal derivatives of the amyloid protein precursor. Science **255**:726–728.
14. Esch, F. S., P. S. Kiem, E. C. Beattie, R. W. Blacher, A. R. Culwell, T. Oltersdorf, D. McClure & P. J. Ward. 1990. Cleavage of amyloid β-peptide during constitutive processing of its precursor. Science **248**: 1122–1124.
15. Selkoe, D. J. 1994. Normal and abnormal biology of the β-amyloid precursor protein. Ann. Rev. Neurosci. **17**:489–517.
16. Ninomiya, H., J. Roch, M. P. Sundsmo, D. A. C. Otero & T. Saitoh. 1993. Amino acid sequence RERMS represents the active domain of amyloid β/A4 protein precursor that promotes fibroblast growth. J. Cell Biol. **121**: 879–886.
17. Greenberg, S. M., E. H. Koo, D. J. Selkoe, W. Q. Qiu & K. S. Kosik. 1994. Stimulation of MAP kinase and tau phosphorylation by secreted β-amyloid precursor protein. Proc. Natl. Acad. Sci. USA. **91**:7104–7108.
18. Schubert, D., M. LaCorbiere, T. Saitoh & G. Cole. 1989. Characterization of an amyloid β precursor protein that binds heparin and contains tyrosine sulfate. Proc. Natl. Acad. Sci. USA **86**:2066–2069.
19. Kibbey, M. C., M. Jucker, B. S. Weeks, R. L. Neve, W. E. Van Nostrand & H. K. Klienman. 1993. β-Amyloid precursor protein binds to the neurite-promoting IKVAV site of laminin. Proc. Natl. Acad. Sci. **90**:10150–10153.
20. Narindrasorasak, S., D. Lowery, P. Gonzalez-DeWhitt, R. A. Poorman, B. Greenberg & R. Kisilevsky. 1991. High affinity interactions between the Alzheimer's β-amyloid precursor proteins and the basement membrane form of heparan sulphate proteoglycan. J. Biol. Chem. **266**: 12878–12883.
21. Thomas, S. M., M. DeMarco, G. D'Arcangelo, S. Halegoua & J. S. Brugge. 1992. Ras is essential for nerve growth factor- and phorbol ester-induced tyrosine phosphorylation of MAP kinases. Cell **68**:1031–1040.
22. Wood, K. W., C. Sarnecki, T. M. Roberts & J. Blenis. 1992. Ras mediates nerve growth factor receptor modulation of three signal-transducing protein kinases: MAP kinase, Raf-1, and RSK. Cell **68**:1041–1050.
23. Kremer, N. E., G. D'Arcangelo, S. M. Thomas, M. DeMarco, J. S. Brugge & S. Halegoua. 1991. Signal transduction by nerve growth factor and fibroblast growth factor in PC12 cells requires a sequence of Src and Ras actions. J. Cell Biol. **115**:809–819.
24. Qui, M. S. & S. H. Green. 1992. PC12 cell neuronal differentiation is associ-

ated with prolonged p21ras activity and consequent prolonged ERK activity. Neuron **9:**705–717.

25. Bodmer, S., M. B. Poldlisny, D. J. Selkoe, I. Heid & A. Fontana. 1990. Transforming growth factor-beta bound to soluble derivatives of the beta amyloid precursor protein of Alzheimer's disease. Biochem. Biophys. Res. Commun. **171:**890–897.
26. Wang, R., J. F. Meschia, R. J. Cotter & S. S. Sisodia. 1991. Secretion of the beta/A4 amyloid precursor protein. Identification of a cleavage site in cultured mammalian cells. J. Biol. Chem. **266:**16960–16964.
27. Biernat, J., E.-M. Mandelkow, C. Schroter, *et al.* 1992. The switch of tau protein to an Alzheimer-like state includes the phosphorylation of two serine-proline motifs upstream of the microtubule binding region. EMBO J. **11:** 1593–1597.
28. Cobb, M. H., T. G. Boulton & D. J. Robbins. 1991. Extracellular signal-regulated kinases: ERKs in progress. Cell Regul. **2:**965–978.

Tau Protein and the Neurofibrillary Pathology of Alzheimer's Disease

M. GOEDERT

MRC Laboratory of Molecular Biology, Cambridge CB2 2QH, United Kingdom

ABSTRACT: Abundant neurofibrillary tangles, neuropil threads and senile plaque neurites constitute the neurofibrillary pathology of Alzheimer's disease. They form in the nerve cells that undergo degeneration in the disease, in which their regional distribution correlates with the degree of dementia. Each lesion contains the paired helical filament (PHF) as its major component. PHFs are composed of the microtubule-associated protein tau in a hyperphosphorylated state. PHF-tau is hyperphosphorylated on all six adult brain isoforms. As a consequence, tau is unable to bind to microtubules and is believed to self-assemble into the PHF. Several candidate protein kinases and protein phosphatases have been identified through *in vitro* experiments.

INTRODUCTION

The intellectual decline that is characteristic of Alzheimer's disease is accompanied by the accumulation in the brain of insoluble fibrous material, both extracellularly in the form of neuritic plaques and intracellularly in the form of neurofibrillary lesions. The formation of the latter is believed to produce the symptoms of the disease, which result most probably from the degeneration of nerve cells in cerebral cortex, hippocampal formation and some subcortical nuclei, with ensuing neuronal loss and reduction in synapse number. Neurofibrillary lesions are found in cell bodies and apical dendrites as neurofibrillary tangles, in distal dendrites as neuropil threads and in the abnormal neurites that are associated with some amyloid plaques (neuritic plaques). Ultrastructurally, all three lesions contain abnormal paired helical filaments (PHFs) as their major fibrous component and straight filaments (SFs) as a minor component. Neurofibrillary lesions develop in the vast majority of nerve cells that undergo degeneration in cerebral cortex and hippocampal formation.

Over the past several years significant progress has been made in unraveling the molecular composition of PHFs and in identifying the mechanisms that

Send correspondence to: Dr. Michel Goedert, MRC Laboratory of Molecular Biology, Hills Road, Cambridge CB2 2QH, U.K.; Tel: 0223-402036; Fax: 0223-412282.

may lead to their assembly. Current evidence indicates that they are made entirely of microtubule-associated protein tau in a hyperphosphorylated state.

TAU PROTEIN

Tau is an abundant protein in both central and peripheral nervous systems. In brain it is found predominantly in nerve cells, where it is concentrated in axons. This contrasts with the neurofibrillary lesions in Alzheimer's disease, which are found in nerve cell bodies, axons and dendrites. Multiple tau isoforms are produced from a single gene through alternative mRNA splicing. In adult human brain six tau isoforms are found, which range from 352 to 441 amino acids and differ from each other by the presence or absence of three inserts (FIG. 1).[1] The most striking feature of the tau sequence is the presence of three or four tandem repeats of 31 or 32 amino acids located in the carboxy-terminal half which constitute microtubule-binding domains. Besides being distinguished by the presence of three or four tandem repeats, some tau isoforms contain 29 or 58 amino acid inserts located near the amino-terminus. An isoform with a large additional insert in the amino-terminal half has been described in the peripheral nervous system.[2,3] Tau is subject to developmentally regulated alternative mRNA splicing, in that in immature brain only the transcript encoding the shortest isoform with three repeats is expressed. The developmental shift of tau bands from a simple fetal pattern to a more complex adult pattern thus involves the transition from the expres-

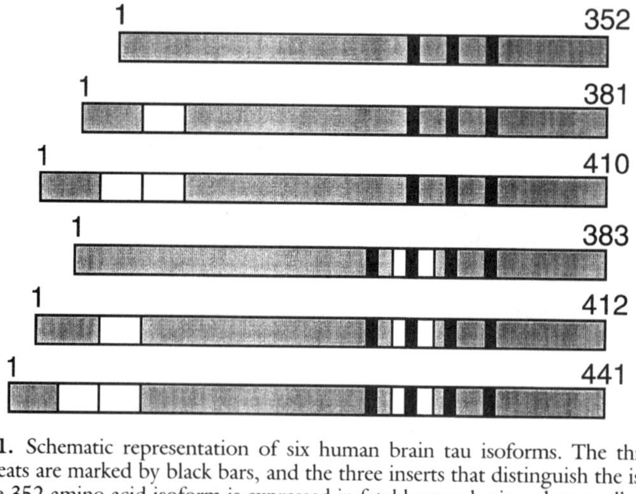

FIGURE 1. Schematic representation of six human brain tau isoforms. The three or four tandem repeats are marked by black bars, and the three inserts that distinguish the isoforms are shaded. The 352 amino acid isoform is expressed in fetal human brain, whereas all six isoforms are expressed in adult human brain.

sion of the isoform with three repeats and no inserts to the expression of all six isoforms.[1]

Tau is a phosphoprotein[4] and phosphorylation is also developmentally regulated in that tau from immature brain is phosphorylated at more sites than tau from adult brain. One way to identify amino acids that are phosphorylated in tau from developing brain, adult brain, and Alzheimer's disease brain is to use antibodies that recognize phosphorylated residues within a specific amino acid sequence. On immunoblots these antibodies do not recognize recombinant tau, but do so after phosphorylation with a brain extract or with certain purified protein kinases.[5] Recombinant human tau proteins mutated at specific amino acid residues can then be used to identify the epitopes of these antibodies. We have investigated monoclonal antibodies AT8, AT180, AT270, PHF1 and 12E8 in this manner.[6–10] AT8 fails to recognize phosphorylated recombinant tau mutated at serine 202 or threonine 205 (in the numbering of the longest human brain tau isoform) (FIG. 2). AT270 does not recognize phosphorylated recombinant tau mutated at threonine 181, whereas AT180 fails to recognize phosphorylated recombinant tau mutated at threonine 231 or serine 235. PHF1 recognizes tau phosphorylated at serines 396 and 404, while 12E8 recognizes tau phosphorylated at serines 262 and/or 356. With the exception of serines 262 and 356 these sites are all serine/threonine-prolines. The longest human brain tau isoform contains 17 such sites.[1] They show an intriguing distribution in that they are distributed throughout the sequence, with the notable exception of the microtubule-binding tandem repeat region. In particular, a cluster of serine/threonine-prolines is present in a proline-rich region of the protein located upstream of the repeats and a smaller cluster in the carboxy-terminal region. By contrast, serines 262 and 356 are located in the microtubule-binding region within the sequences KIGST (serine 262) and KIGSL (serine 356).

AT8, AT180, AT270, PHF1 and 12E8 recognize a significant fraction of tau from developing brain and a minor fraction of tau from adult brain, indicating that tau is normally phosphorylated at both serine/threonine-prolines and at sites that are not serine/threonine-prolines.[6–10] Additional phosphorylated residues have been identified by mass spectrometry and peptide sequencing.[11]

The epitopes of AT8, AT270 and PHF1 can be produced by phosphorylating recombinant tau with mitogen-activated protein (MAP) kinase or glycogen synthase kinase-3 (GSK3).[7,8] The epitope of AT180 is produced by phosphorylating recombinant tau with a mixture of MAP kinase and GSK3.[7] Similarly, the epitopes of these antibodies are produced upon phosphorylation of recombinant tau with cyclin-dependent protein kinase 5 (cdk5). The 12E8 epitope is produced by phosphorylating recombinant tau with cyclic AMP-dependent protein kinase (PKA) and Ca^{2+}/calmodulin-dependent protein kinase II (CaM kinase).[13] The phosphorylation state of a protein is the

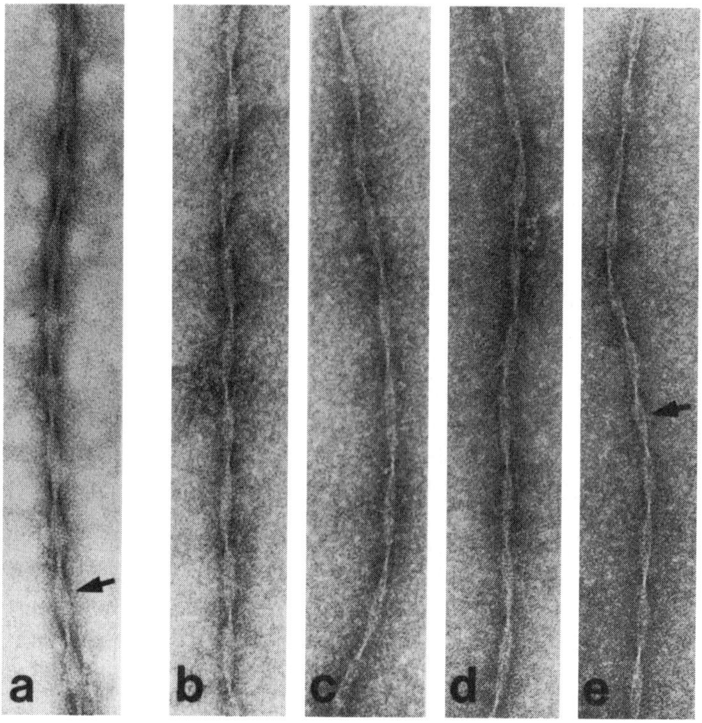

FIGURE 2. PHF from the brain of an Alzheimer's disease patient and filaments assembled in vitro from a non-phosphorylated 99 amino acid fragment of tau encompassing three repeats. **(a)** Pronase-treated Alzheimer's disease PHF. **(b–e)** Filaments resembling PHFs assembled from expressed tau protein. The arrows indicate regions of filament where the characteristic pattern of four white stain-excluding lines parallel to the filament axis can be seen.

result of a balance between protein kinase and protein phosphatase activities. Of the major phosphatase activities present in a brain extract, tau phosphorylated by MAP kinase is dephosphorylated predominantly by the trimeric form of protein phosphatase 2A.[14]

TAU PROTEIN AND THE PAIRED HELICAL FILAMENT

Protein chemistry and molecular cloning directly established that tau forms an integral component of the PHF,[15,16] putting previous immunohistochemical studies on a solid basis. These results were obtained using PHFs extracted from tangle fragments. Because of the insolubility of these filaments it was not possible to conclude that tau is the only component of the PHF. This situation has changed with the development of extraction techniques for dis-

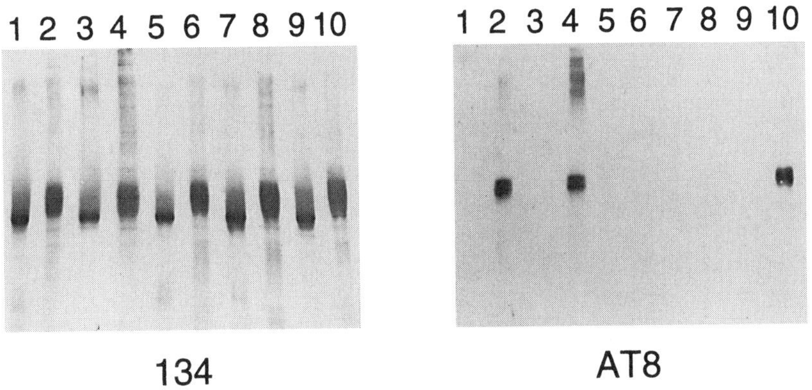

FIGURE 3. Mapping of the epitope of the phosphorylation-dependent anti-tau antibody AT8 to phosphoserine 202 and phosphothreonine 205 in tau (in the numbering of the longest adult human brain tau isoform). Wild-type and recombinant tau proteins (expressed from clone htau24) were phosphorylated with a rat brain extract in vitro. Immunoblots were stained with anti-tau serum 134 and monoclonal antibody AT8. *Lane 1,* htau24; *lane 2,* htau24 plus brain extract; *lane 3,* S199A htau24; *lane 4,* S199A htau24 plus brain extract; *lane 5,* S202A htau24; *lane 6,* S202A htau24 plus brain extract; *lane 7,* T205A htau24; *lane 8,* T205A htau24 plus brain extract; *lane 9,* T212A htau24; *lane 10,* T212A htau24 plus brain extract.

persed PHFs.[17] Purification of these filaments to apparent homogeneity has provided strong evidence that tau forms the PHF.[18] In addition, it has been shown that paired helical-like filaments can be assembled *in vitro* from bacterially expressed non-phosphorylated three-repeat or four-repeat fragments of tau (FIG. 3).[19,20]

Unlike tangle fragment PHFs, a large proportion of dispersed PHFs contains intact tau. This indicates that dispersed filaments constitute an earlier stage of PHFs than the bulk of filaments isolated from tangle fragments. A further difference between tangle fragment and dispersed PHFs is that the majority of tangle fragment PHFs is ubiquitinated.[21] The natural history of the PHF after assembly thus leads from an initial guanidine-soluble intracellular form to a very insoluble, proteolysed, ubiquitinated and extracellular form, with a multitude of intermediates.

Tau protein extracted from PHFs runs as three major bands of 60, 64 and 68 kD, with a variable amount of background smear.[17,18] These PHF-tau bands are retarded on gels relative to normal or recombinant tau. They contain intact tau, as they stain with antibodies directed against the amino- and carboxy-termini of tau. After treatment with alkaline phosphatase at high temperature the three PHF-tau bands run as six bands which align with the recombinant tau isoforms.[12] Thus, PHFs contain all six adult brain tau isoforms in a hyperphosphorylated state. The phosphorylation-dependent anti-tau antibodies AT8, AT180, AT270 and PHF1 stain neurofibrillary lesions and PHFs in a phosphorylation-dependent manner. PHF-tau has a greatly reduced ability to bind to microtubules.[22,23] This loss of function results entirely from hyperphosphorylation, since dephosphorylated PHF-tau binds as well to microtubules as does normal tau. The reduced binding of PHF-tau to microtubules may destabilise microtubules in Alzheimer's disease, resulting in the impairment of vital cellular processes, such as rapid axonal transport, and leading to the degeneration of affected nerve cells.

Hyperphosphorylation means that PHF-tau is phosphorylated at more sites than tau from adult brain and that, for a given site, a higher than normal percentage of tau molecules is phosphorylated. Sites identified through the use of phosphorylation-dependent anti-tau antibodies include threonines 181, 205 and 231 and serines 202, 235, 262, 396 and 404.[6–10] Additional sites have been identified by mass spectrometry and peptide sequencing.[24,25] To date, about half of the known sites are serine/threonine-prolines and half are serine/threonines not followed by a proline (FIG. 4). As discussed above some (but not all) of these sites are also partially phosphorylated in tau from developing and adult brain. MAP kinase, GSK3 and cdk5 phosphorylate some of the hyperphosphorylated sites of tau *in vitro*. However, it is unclear whether these kinases cause the hyperphosphorylation of tau in Alzheimer's disease.

Whatever the mechanisms involved, it appears unlikely that tau is the only protein that becomes hyperphosphorylated. In affected nerve cells PHFs probably represent only the most visible manifestation of the deregulation of phosphorylation/dephosphorylation mechanisms. In dividing cells such a deregulation is likely to result in cell transformation, since constitutive activation of proline-directed protein kinases occurs in response to a variety of oncogenes. Since brain cells cannot undergo cell division, the consequence of sustained activation of proline-directed protein kinases would not be cell transformation. Instead brain function might be disrupted by the hyperphosphorylation of proteins that are not normally phosphorylated to such an extent.

CONCLUSION

The degree of phosphorylation is the only known difference between PHF-tau and normal adult tau, with all but one of the hyperphosphorylated sites located outside the tandem repeat region. As a consequence tau is prevented from binding to microtubules and is believed to self-assemble through the tandem repeat region into the PHF. This view assumes a causal link between hyperphosphorylation of tau and PHF assembly. However, to date, there is no direct evidence to show either that hyperphosphorylation precedes PHF assembly or that it leads to PHF assembly. Several candidate protein kinases

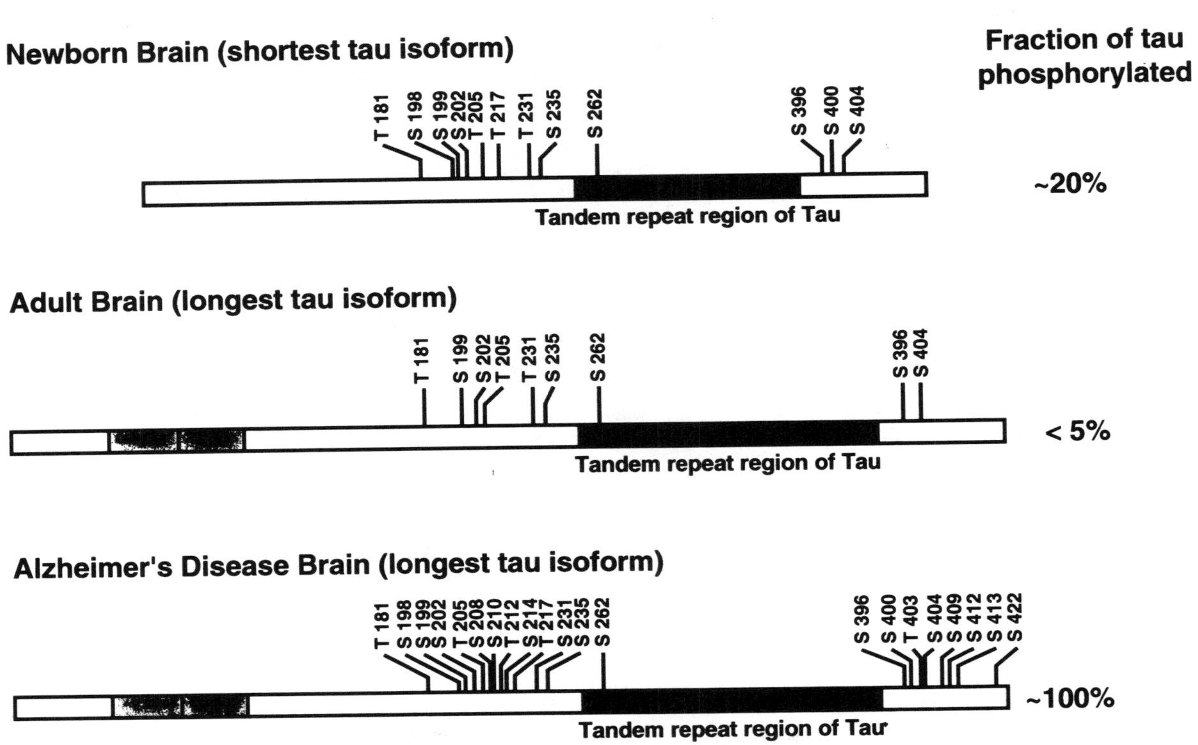

FIGURE 4. Known phosphorylation sites in fetal tau, adult tau and PHF-tau. All phosphorylated residues are serines (S) or threonines (T), many of which are serine/threonine-prolines. With the exception of S262 all sites are located outside of the microtubule-binding repeat region. Some of the sites are shared between fetal brain tau, adult brain tau and PHF-tau. However, for these sites, a larger fraction of tau is phosphorylated in PHF-tau than in fetal brain tau, with only a small fraction phosphorylated in adult brain tau. Approximate percentages are indicated. Current evidence thus indicates that PHF-tau is both hyperphosphorylated and abnormally phosphorylated.

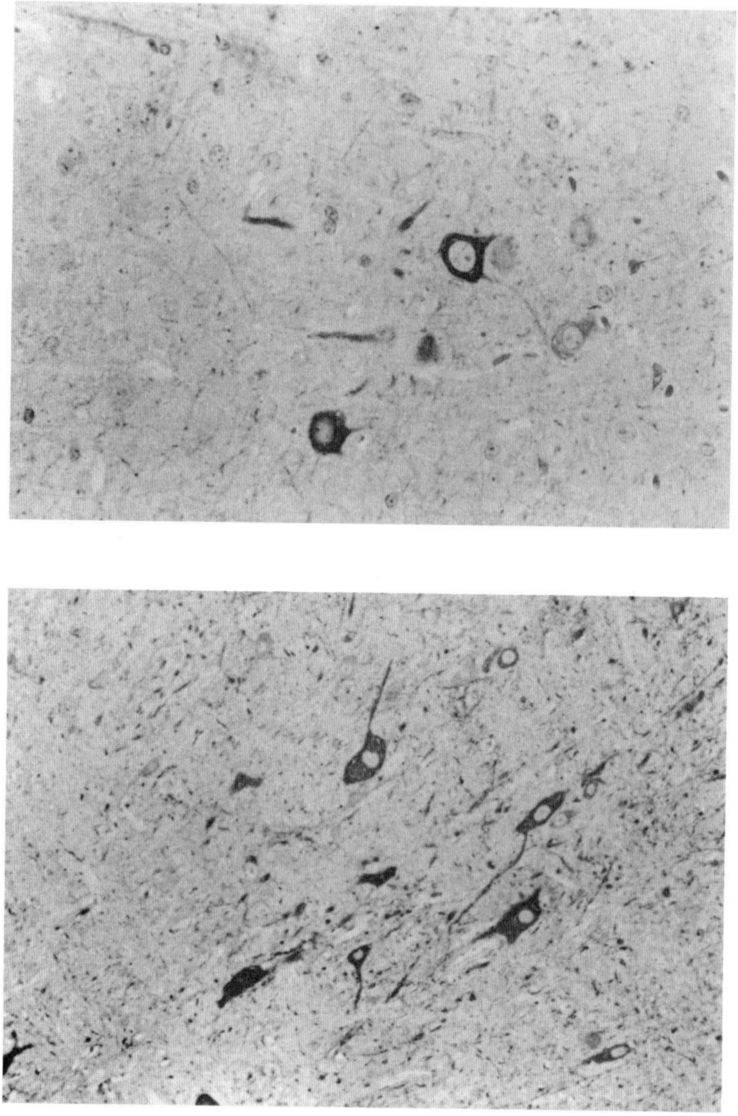

FIGURE 5. Immunohistochemical staining of human tau in the brain of a mouse transgenic for the longest human brain tau isoform. The anti-tau antiserum 133 was used. *Upper panel*: Section through the cerebral cortex. *Lower panel*: Section through the brainstem. Note the strong somatodendritic staining of some nerve cells; it resembles the pre-tangle lesions of Alzheimer's disease.

and protein phosphatases have been identified by virtue of their *in vitro* phosphorylation of tau at some of the sites that are hyperphosphorylated in Alzheimer's disease. It appears likely that the brain contains additional protein kinases that can phosphorylate tau in this manner. It is unknown which enzymes phosphorylate tau *in vivo* in either the normal or the Alzheimer's disease brain. It is therefore essential to relate these and future *in vitro* experiments to the mechanisms that lead to tau phosphorylation *in vivo*.

Ultimately, an understanding of the mechanisms that lead to the formation of the neurofibrillary lesions of Alzheimer's disease will require an *in vivo* model of PHF formation. We are attempting to reach this goal by producing transgenic mice which express human tau protein in nerve cells under control of the human Thy1 promoter.[26] Expression of human tau is high in a small percentage of nerve cells, where it is present in nerve cell bodies, axons and dendrites (FIG. 5). This contrasts with the axonal localization of tau in normal brain, but is similar to its distribution in the nerve cells that are affected by the neurofibrillary pathology in Alzheimer's disease. Moreover, in these transgenic mice, some nerve cells also stain with phosphorylation-dependent anti-tau antibodies, indicating hyperphosphorylation of tau. The hope is that in due course these and other transgenic lines will develop PHFs. This would constitute an animal model for the neurofibrillary lesions, an essential prerequisite for the testing of compounds aimed at halting or preventing the intracellular pathology of Alzheimer's disease.

REFERENCES

1. GOEDERT, M, M. G. SPILLANTINI, R. JAKES, D. RUTHERFORD & R. A. CROWTHER. 1989. Multiple isoforms of human microtubule-associated protein tau: Sequences and localization in neurofibrillary tangles of Alzheimer's disease. Neuron **3:**519–526.
2. GOEDERT, M., M. G. SPILLANTINI & R. A. CROWTHER. 1992. Cloning of a big tau microtubule-associated protein characteristic of the peripheral nervous system. Proc. Natl. Acad. Sci. USA **89:**1983–1987.
3. COUCHIE, D., C. MAVILIA, I. S. GEORGIEFF, R. K. H. LIEM, M. L. SHELANSKI & J. NUNEZ. 1992. Primary structure of high molecular weight tau protein present in peripheral nervous system. Proc. Natl. Acad. Sci. USA **89:** 4378–4381.
4. BUTLER, M. & M. L. SHELANSKI. 1986. Microheterogeneity of tau microtubule-associated proteins is due to differences in phosphorylation. J. Neurochem. **47:** 1517–1522.
5. BIERNAT, J., E. M. MANDELKOW, C. SCHRÖTER, B. LICHTENBERG-KRAAG, B. STEINER, B. BERLING, H. MEYER, M. MERCKEN, M. VANDERMEEREN, M. GOEDERT & E. MANDELKOW. 1992. The switch of tau protein to an Alzheimer-like state includes the phosphorylation of two serine-proline motifs upstream of the microtubule-binding region. EMBO J. **11:**1593–1597.

6. GOEDERT, M., R. JAKES, R. A. CROWTHER, J. SIX, U. LÜBKE, M. VANDERMEEREN, P. CRAS, J. Q. TROJANOWSKI & V. M.-Y. LEE. 1993. The abnormal phosphorylation of tau protein at serine 202 in Alzheimer disease recapitulates phosphorylation during development. Proc. Natl. Acad. Sci. USA **90:** 5066–5070.
7. GOEDERT, M., R. JAKES, R. A. CROWTHER, P. COHEN, E. VANMECHELEN, M. VANDERMEEREN & P. CRAS. 1994. Epitope mapping of monoclonal antibodies to the paired helical filaments of Alzheimer's disease: Identification of phosphorylation sites in tau protein. Biochem. J. **301:**871–877.
8. GOEDERT, M., R. JAKES & E. VANMECHELEN. Monoclonal antibody AT8 recognizes tau protein phosphorylated at serine 202 and threonine 205. Neurosci. Lett. **189:**167–170.
9. OTVOS, L., L. FEINER, E. LANG, G. I. SZENDREI, M. GOEDERT & V. M.-Y. LEE. 1994. Monoclonal antibody PHF1 recognizes tau protein phosphorylated at serine residues 396 and 404. J. Neurosci. Res. **39:**669–673.
10. SEUBERT, P., M. MAWAL-DEWAN, R. BARBOUR, R. JAKES, M. GOEDERT, G. V. W. JOHNSON, J. M. LITERSKY, D. SCHENK, I. LIEBERBURG, J. Q. TROJANOWSKI & V. M.-Y. LEE. 1995. Detection of phosphorylated Ser^{262} in fetal tau, adult tau and paired helical filament-tau. J. Biol. Chem. **270:**18917–18922.
11. WATANABE, A., M. HASEGAWA, M. SUZUKI, K. TAKIO, M. MORISHIMA-KAWASHIMA, K. TITANI, T. ARAI, K. S. KOSIK & Y. IHARA. 1993. In vivo phosphorylation sites in fetal and adult rat tau. J. Biol. Chem. **268:**25712–25717.
12. GOEDERT, M., M. G. SPILLANTINI, N. J. CAIRNS & R. A. CROWTHER. 1992. Tau proteins of Alzheimer paired helical filaments: Abnormal phosphorylation of all six brain isoforms. Neuron **8:**159–168.
13. LITERSKY, J. M., G. V. W. JOHNSON, R. JAKES, M. GOEDERT, M. LEE & P. SEUBERT. 1995. Tau protein is phosphorylated by cAMP-dependent protein kinase and calcium/calmodulin-dependent protein kinase II within its microtubule-binding domains at Ser262 and Ser356. Submitted for publication.
14. GOEDERT, M., E. S. COHEN, R. JAKES & P. COHEN. 1992. p42 MAP kinase phosphorylation sites in microtubule-associated protein tau are dephosphorylated by protein phosphatase $2A_1$: Implications for Alzheimer's disease. FEBS Lett **312:**95–99.
15. GOEDERT, M., C. M. WISCHIK, R. A. CROWTHER, J. E. WALKER & A. KLUG. 1988. Cloning and sequencing of the cDNA encoding a core protein of the paired helical filament of Alzheimer disease. Proc. Natl. Acad. Sci. USA **85:** 4051–4055.
16. KONDO, J., T. HONDA, H. MORI, Y. HAMADA, R. MIURA, H. OGAWARA & Y. IHARA. 1988. The carboxyl third of tau is tightly bound to paired helical filaments. Neuron **1:**827–834.
17. GREENBERG, S. G. & P. DAVIES. 1990. A preparation of Alzheimer paired helical filaments that displays distinct tau proteins by polyacrylamide gel electrophoresis. Proc. Natl. Acad. Sci. USA **87:**5827–5831.
18. LEE, V.M.-Y. B. J. BALIN, L. OTVOS & J. Q. TROJANOWSKI. 1991. A68: A major subunit of paired helical filaments and derivatized forms of normal tau. Science **251:**675–678.

19. Crowther, R. A., O. F. Olesen, R. Jakes & M. Goedert. 1992. The microtubule binding repeats of tau protein assemble into filaments like those found in Alzheimer's disease. FEBS Lett **309:**199–202.
20. Wille, H., G. Drewes, J. Biernat, E. M. Mandelkow & E. Mandelkow. 1992. Alzheimer-like paired helical filaments and antiparallel dimers formed from microtubule-associated protein tau in vitro. J. Cell Biol. **118:**573–584.
21. Morishima-Kawashima, M., M. Hasegawa, K. Takio, M. Suzuki, K. Titani & Y. Ihara. 1993. Ubiquitin is conjugated with amino-terminally processed tau in paired helical filaments. Neuron **10:**1151–1160.
22. Bramblett, G. T., M. Goedert, R. Jakes, S. E. Merrick, J. Q. Trojanowski & V. M.-Y. Lee. 1993. Abnormal tau phosphorylation at Ser^{396} in Alzheimer's disease recapitulates development and contributes to reduced microtubule binding. Neuron **10:**1089–1099.
23. Yoshida, H. & Y. Ihara. 1993. Tau in paired helical filament is functionally distinct from fetal tau: Assembly incompetence of paired helical filament tau. J. Neurochem. **61:**1183–1186.
24. Hasegawa, M., M. Morishima-Kawashima, K. Takio, M. Suzuki, K. Titani & Y. Ihara. 1992. Protein sequence and mass spectrometric analyses of tau in the Alzheimer's disease brain. J. Biol. Chem. **267:**17047–17054.
25. Morishima-Kawashima, M., M. Hasegawa, K. Takio, M. Suzuki, H. Yoshida, K. Titani & Y. Ihara. 1995. PHF-tau: proline-directed and non-proline-directed phosphorylation. J. Biol. Chem. **270:**823–829.
26. Götz, J., A. Probst, M. G. Spillantini, T. Schäfer, R. Jakes, K. Bürki & M. Goedert. 1995. Somatodendritic localization and hyperphosphorylation of tau protein in transgenic mice expressing the longest human brain tau isoform. EMBO J. **14:**1304–1313.

Molecular Mechanism of Alzheimer's Neurofibrillary Degeneration and Therapeutic Intervention[a]

K. IQBAL[b,d] AND I. GRUNDKE-IQBAL[c]

[b] *Chemical Neuropathology and* [c]*Neuroimmunology Laboratories, New York State Institute for Basic Research in Developmental Disabilities, Staten Island, New York 10314 USA*

ABSTRACT: Microtubule-associated protein tau is abnormally hyperphosphorylated in the brain of patients with Alzheimer's disease (AD) and the abnormal tau is the major protein subunit of paired helical filaments (PHF). The abnormal phosphorylation of tau probably precedes its polymerization into PHF. The abnormal tau does not bind to tubulin, but competes with tubulin in binding to normal tau and thereby inhibits the assembly of microtubules in the affected neurons. The abnormal tau can be dephosphorylated enzymatically and by this way its microtubule assembly promoting activity can be restored. The activities of protein phosphatases might be decreased in the affected neurons in AD brain, allowing the abnormal hyperphosphorylation of tau. Neurofibrillary degeneration can probably be inhibited by increasing the activities of protein phosphatases in the brain of patients with Alzheimer's disease.

INTRODUCTION

Alzheimer's disease (AD) probably has a polyetiology, which includes genetic, environmental, and metabolic factors. Independent of the etiologic agent, histopathologically, AD is characterized by the presence of two brain lesions, paired helical filaments (PHF) in the neurons and β-amyloid in the extracellular space. At present, the exact relationship between PHF and β-amyloid in the pathogenesis of AD is not understood. However, there is growing evidence from a number of laboratories that dementia in AD patients is associated with neurofibrillary degeneration; β-amyloid alone in the absence of Alzheimer's neurofibrillary degeneration does not produce the disease clinically. Thus it is critical to understand the mechanism of Alzheimer's

[a] This work was supported in part by funds from the New York State Office of Mental Retardation and Developmental Disabilities, National Institutes of Health Grants AG05892, AG08076, NS18105, and Zenith Award from the Alzheimer's Association, USA.

[d] *Send correspondence to:* Dr. Khalid Iqbal, NYS Institute for Basic Research in Developmental Disabilities, 1050 Forest Hill Road, Staten Island, NY 10314; Tel: (718)-494-5259; Fax: (718)-494-1080.

neurofibrillary degeneration. One of the most rational therapeutic approaches to AD is to inhibit the neurofibrillary degeneration. In this chapter the molecular pathology and the mechanism of Alzheimer's neurofibrillary degeneration and a strategy to inhibit this type of degeneration are described.

METHODS

Methods have been described elsewhere and only references are given here. These references are monoclonal antibody Tau-1 and polyclonal antibodies to tau[1–3] and amino terminal tau peptide, Peptide-1,[4] isolation of normal and of abnormally phosphorylated microtubule-associated protein tau[2,5], dephosphorylation of abnormally phosphorylated tau/PHF-tau,[3,4,6,7] *in vitro* microtubule assembly and binding of tau to abnormal tau and to tubulin,[8] isolation of protein phosphatases and the enzyme activity assays,[1,9–11] and radioimmunoassay for normal and abnormally phosphorylated tau in brain homogenates.[12]

RESULTS AND DISCUSSION

Molecular Pathology of Alzheimer's Neurofibrillary Changes

The PHF accumulate in neuronal cell bodies as neurofibrillary tangles, in neurites as neuropil threads, and in the dystrophic neurites surrounding wisps or a core of β-amyloid in neuritic (senile) plaques. Many of the neurons with neurofibrillary changes die, leaving behind the ghost tangles in the extracellular space. However, the neurofibrillary degeneration appears to be a very slow process because even after 10-15 years of the disease most of the affected neurons, though markedly devoid of the arborization, can be seen. Furthermore, in some of the hypothalamic nuclei such as nucleus tuberalis lateralis the presence of the cytoskeletal protein pathology but no loss of neurons is seen.[13] Thus the clinical expression of AD might not only be due to the loss of but also the hypofunctional state of the neurons.

In addition to PHF, there is a significant pool (up to 40% of total abnormal tau) of soluble abnormally phosphorylated tau in AD (AD P-tau) brain.[5] The soluble abnormal tau binds to normal tau and becomes sedimentable at 200,000 $\times$ *g*. Like PHF-tau, the soluble abnormal tau is phosphorylated (5-9 moles, phosphates/mole of tau) at all the sites that were seen in PHF-tau using phosphorylation-dependent antibodies.[1,5] Furthermore, immunocytochemical studies have shown the presence of the abnormally phosphorylated tau as amorphous non-PHF deposits in pretangle neurons.[5,14] Thus, these studies indicate that in AD brain the abnormal phosphorylation of tau most likely precedes and is probably involved in its polymerization into PHF.

Mechanism of Alzheimer's Neurofibrillary Degeneration

Ultrastructural studies of brain biopsies have revealed that in certain selected neurons in AD brain the microtubule system is disrupted and replaced by neurofibrillary tangles of PHF.[16] Tau stimulates microtubule assembly by polymerizing with tubulin and maintains the microtubule structure. The levels of tau in homogenates of frontal cortex from AD patients are severalfold higher than in age-matched controls, and this increase is in the form of abnormally phosphorylated protein.[12] Both immunocytochemical[14,15] and biochemical[5,6] studies have shown in AD brain the presence of a significant pool of the abnormally phosphorylated tau that is not polymerized into PHF and not associated with ubiquitin. These studies suggest that the abnormal phosphorylation of tau probably precedes both its polymerization into PHF and the incorporation of ubiquitin.

Microtubule assembly is defective in AD brain. Microtubules can be assembled *in vitro* from the cytosol of normal fresh autopsy brain obtained within 5 h postmortem, but not from identically treated brains of AD cases. The microtubule assembly from the AD brains, however, is induced by the addition of DEAE dextran, a polycation that mimics the effect of tau for microtubule assembly.[6] Because tau in AD brain cytosol is abnormally phosphorylated[5,6] and phosphorylation of tau depresses tau's ability to promote microtubule assembly,[17] it appears that this alteration of tau in the AD brain might have been the cause of the microtubule assembly defect. Both PHF-tau[7] and the non-PHF soluble abnormally phosphorylated tau[8] when dephosphorylated with alkaline phosphatase stimulate *in vitro* microtubule assembly. While normal tau promotes the GTP binding to the exchangeable site in the β-subunit of tubulin, the PHF-tau does it only after prior dephosphorylation of the abnormal sites.[18] Furthermore, the soluble abnormally phosphorylated tau isolated from AD brain binds to normal tau and not to tubulin, and inhibits the microtubule assembly.[8] These findings suggest that the neurofibrillary degeneration is probably caused by the binding of the abnormally phosphorylated tau to normal tau and thereby disrupting the microtubules in the affected neurons (see FIG. 1).

For a neuron to function it must be able to transport materials between its cell body and synapses and integrity of the microtubule system is essential for this axonal transport. A breakdown of the microtubule system probably leads to a compromised axonal transport and eventually a retrograde degeneration of the affected neurons. The retrograde degeneration in AD brain has been confirmed both by Golgi stain studies, showing massive loss of the neuronal arborization,[19] and by the presence of thickening of the distal ends of the neurites, an early stage of neuropil threads, in pretangle neurons stained for the abnormally phosphorylated tau.[15]

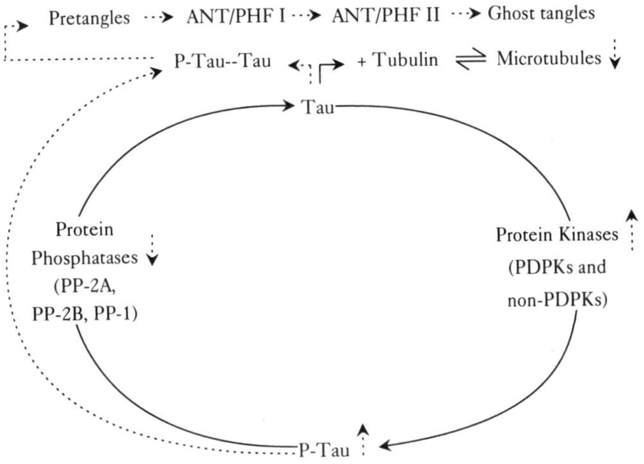

FIGURE 1. A hypothetical scheme showing the mechanism of neurofibrillary degeneration in Alzheimer disease. Tau is phosphorylated by several protein kinases. Because of a decrease (↓) in the activities of protein phosphatases that normally dephosphorylate tau, an imbalance of phosphorylation/dephosphorylation occurs, leading to the abnormal hyperphosphorylation of tau (↑). The latter 1) does not bind to tubulin to promote assembly of microtubules, 2) binds to normal tau and inhibits the microtubule assembly, and 3) becomes stabilized and polymerizes into PHF. The affected neurons degenerate retrogradely both as a result of the breakdown of the microtubule system, and because of the accumulation of PHF as AD neurofibrillary tangles (ANT) filling the entire cell cytoplasm, leaving behind ghost tangles in the extracellular space.

Alzheimer's disease changes are indicated by hatched lines and arrows. The increase in protein kinase (proline directed protein kinases, PDPKs, and non-PDPKs) activity(s) is as yet not supported by any data.

Role of Protein Kinases and Protein Phosphatases in the Abnormal Phosphorylation of Tau

The state of phosphorylation of substrate proteins depends on the relative activities of protein kinases and phosphoprotein phosphatases. Eleven of twenty-one abnormal phosphorylation sites of the AD hyperphosphorylated tau identified to date are canonical sites for the proline-directed protein kinases (PDPKs), suggesting that more than one protein kinase is likely to be involved in the abnormal phosphorylation. At present, the nature of these protein kinases is not clearly understood.

The activities of the phosphoseryl/phosphothreonyl protein phosphatases (PP)-1 and -2A towards phosphorylase kinase, and phosphotyrosyl phosphatase towards poly (Glu, Tyr) 4:1 are decreased in brains of AD cases.[9] Furthermore, by immunocytochemical studies it has been observed that PP-1, PP-2A, PP-2B and cytosolic phosphotyrosyl protein phosphatase PTP-1B are all present in pyramidal neurons in human brain.[20] To date, all the tau abnormal phosphorylation sites identified are Ser or Thr; no phosphorylation of Tyr has been observed. However, a decrease in phosphotyrosine phosphatase in AD brain might contribute to hyperphosphorylation of tau by keeping MAP kinase activated for extended periods; dephosphorylation of MAP kinase at either Ser/Thr or Tyr inactivates it.[21] Furthermore, treatment of cultured neuroblastoma cells with inhibitors of PP-2A and PP-2B leads to increased phosphorylation of tau.[22]

Studies on site-specific dephosphorylation by protein phosphatases have revealed that the soluble abnormally phosphorylated tau isolated from AD brain is rapidly dephosphorylated at the abnormal sites Ser 46, Ser 199/Ser 202, Ser 235, and Ser 396/Ser 404 by protein phosphatase PP-2B, at all the above sites except Ser 235 by PP-2A, and at only Ser 199/Ser 202 and Ser 396/Ser 404 by PP-1.[1,10,11] The activities of all the three phosphatases, i.e. PP-2B, PP-2A and PP-1 towards the abnormally phosphorylated tau are markedly increased by the presence of Mn^{2+}. Dephosphorylation of the abnormal tau by PP-2C at none of the above sites has been detected.[11]

Therapeutic Approach to Neurofibrillary Degeneration

Since the abnormal phosphorylation of tau appears to be the cause of the neurofibrillary degeneration in AD two approaches to prevent and/or arrest neurofibrillary degeneration are i) to inhibit the tau kinases and ii) to activate the tau phosphatases involved. More than one kinase are involved in the abnormal phosphorylation of tau and inhibition of these kinases might jeopardize several other cellular functions directly or indirectly dependent on these kinase activities. On the other hand a 10–20% activation of tau phosphatases might be all that is required to correct the tau phosphorylation/dephosphorylation imbalance. Increase of phosphatase activities are less likely to disturb other cellular functions because enzymes are known to be generally present in excess of their use in tissues. An activator like Mn^{2+} which increases the activity of all the three phosphatases, PP-2B, PP-2A, and PP-1 towards the abnormally phosphorylated tau might prevent and arrest the neurofibrillary degeneration (see refs. 1 and 10).

ACKNOWLEDGMENTS

Secretarial support was provided by Joanne Lopez and Padmini Reginald.

REFERENCES

1. Gong, C.-X., T. J. Singh, T. J. Grundke-Iqbal & K. Iqbal. 1994. Alzheimer disease abnormally phosphorylated τ is dephosphorylated by protein phosphatase-2B (calcineurin). J. Neurochem. **62:**803–806.

2. GRUNDKE-IQBAL, I., K. IQBAL, M. QUINLAN, Y.-C. TUNG, M. S. ZAIDI & H. M. WISNIEWSKI. 1986. Microtubule-associated protein tau a component of Alzheimer paired helical filaments. J. Biol. Chem. **261**:6084–6089.
3. GRUNDKE-IQBAL, I., K. IQBAL, Y.-C. TUNG, M. QUINLAN, H. M. WISNIEWSKI & L. I. BINDER. 1986. Abnormal phosphorylation of the microtubule-associated protein tau in Alzheimer cytoskeletal pathology. Proc. Natl. Acad. Sci. USA **83**:4913–1917.
4. IQBAL, K., I. GRUNDKE-IQBAL, A. J. SMITH, L. GEORGE, Y.-C. TUNG & T. ZAIDI. 1989. Identification and localization of a tau peptide to paired helical filaments of Alzheimer disease. Proc. Natl. Acad. Sci. USA **86**:5646–5650.
5. KÖPKE, E., Y.-C. TUNG, S. SHAIKH, A. DEL C. ALONSO, K. IQBAL & I. GRUNDKE-IQBAL. 1993. Microbutule associated protein tau: Abnormal phosphorylation of a non-paired helical filament pool in Alzheimer disease. J. Biol. Chem. **268**:24374–24384.
6. IQBAL, K., I. GRUNDKE-IQBAL, T. ZAIDI, P. A. MERZ, G. Y. WEN, S. S. SHAIKH, H. M. WISNIEWSKI, I. ALAFUZOFF & B. WINBLAD. 1986. Defective brain microtubule assembly in Alzheimer's disease. Lancet **2**:421–426.
7. IQBAL, K., T. ZAIDI, C. BANCHER & I. GRUNDKE-IQBAL. 1994. Alzheimer paired helical filaments: Restoration of the biological activity by dephosphorylation. FEBS Lett. **349**:104–108.
8. ALONSO, A., C. DEL, T. ZAIDI, I. GRUNDKE-IQBAL & K. IQBAL. 1994. Role of abnormally phosphorylated tau in the breakdown of microtubules in Alzheimer disease. Proc. Natl. Acad. Sci. USA **91**:5562–5566.
9. GONG, C.-X., T. J. SINGH, I. GRUNDKE-IQBAL & K. IQBAL. 1993. Phosphoprotein phosphatase activities in Alzheimer disease. J. Neurochem. **61**:921–927.
10. GONG, C.-X., I. GRUNDKE-IQBAL & K. IQBAL. 1994. Dephosphorylation of Alzheimer disease abnormally phosphorylated tau by protein phosphatase-2A. Neuroscience **61**:765–772.
11. GONG, C.-X., I. GRUNDKE-IQBAL, Z. DAMUNI & K. IQBAL. 1994. Dephosphorylation of microtubule-associated protein tau by protein phosphatase-1 and -2C and its implication in Alzheimer disease. FEBS Lett. **341**:94–98.
12. KHATOON, S., I. GRUNDKE-IQBAL & K. IQBAL. 1992. Brain levels of microtubule-associated protein tau are elevated in Alzheimer's disease: a Radioimmuno-slot blot assay for nanograms of the protein. J. Neurochem. **59**:750–753.
13. SWAAB, D. F., I. GRUNDKE-IQBAL, K. IQBAL, H. P. H. KREMER, R. RAVID & J. A. P. VAN DE NES. 1992. Tau and ubiquitin in the human hypothalamus in aging and Alzheimer's disease. Brain Res. **590**:239–249.
14. BANCHER, C., C. BRUNNER, H. LASSMANN, H. BUDKA, K. JELLINGER, G. WICHE, F. SEITELBERGER, I. GRUNDKE-IQBAL, K. IQBAL & H. M. WISNIEWSKI. 1989. Accumulation of abnormally phosphorylated tau precedes the formation of neurofibrillary tangles in Alzheimer's disease. Brain Res. **477**:90–99.
15. BRAAK, E., H. BRAAK & E. M. MANDELKOW. 1994. A sequence of cytoskeleton changes related to the formation of neurofibrillary tangles and neuropil threads. Acta Neuropathol. (Berl.) **87**:554–567.
16. DUSTIN, P. & J. FLAMENT-DURAND. 1982. Disturbances of axoplasmic transport in Alzheimer's disease. *In* Axoplasmic Transport in Physiology and Pathol-

ogy. D. G. Weiss, & A. Gorio, Eds. :131–136. Berlin, Heidelberg, New York: Springer-Verlag.
17. LINDWALL, G. & R. D. COLE. 1984. Phosphorylation affects the ability of tau protein to promote microtubule assembly. J. Biol. Chem. **259**:5301–5305.
18. KHATOON, S., I. GRUNDKE-IQBAL & K. IQBAL. 1995. Guanosine triphosphate binding to β-subunit of tubulin in Alzheimer's disease brain: Role of microtubule-associated protein τ J. Neurochem. **64**:777–787.
19. SCHIEBEL, M. E., R. D. LINDSAY, U. TOMIYASU & A. B. SCHEIBEL. 1976. Progressive dendritic changes in the aging human limbic system. Exp. Neurol. **53**:420–430.
20. PEI, J.-J., E. SERSEN, K. IQBAL & I. GRUNDKE-IQBAL. 1994. Expression of protein phosphatases PP-1, PP-2A, PP-2B and PTP-1B and protein kinases MAP kinase and $P34^{cdc2}$ in the hippocampus of patients with Alzheimer disease and normal aged individuals. Brain Res. **655**:70–76.
21. PELECH, S. L. & T. S. SANGHERA. 1992. MAP kinases charting the regulatory pathways. Science **257**:1355–1356.
22. TANAKA, T., K. IQBAL, E. TRENKNER & I. GRUNDKE-IQBAL. 1995. Phosphatase inhibitors increase tau phosphorylation and induce cytotoxicity in SH-SY5Y neuroblastoma cells. Soc. Neurosci. Abst. **21**:255.

Apolipoprotein E

Structure, Function, and Possible Roles in Alzheimer's Disease[a]

R.W. MAHLEY,[b–e] B.P. NATHAN,[b] AND R.E. PITAS[b,c]

[b] *Gladstone Institute of Cardiovascular Disease, Cardiovascular Research Institute,* [c]*Department of Pathology,* [d] *Department of Medicine, University of California, San Francisco, California 94141-9100 USA*

ABSTRACT: Apolipoprotein (apo) E is associated with the two characteristic neuropathologic lesions of Alzheimer's disease—extracellular neuritic plaques representing deposits of amyloid beta (Aβ) peptide and intracellular neurofibrillary tangles representing filaments of a microtubule-associated protein called tau. Incubation of the apoE4 isoform with the Aβ peptide *in vitro* results in the formation of a dense, stable network of very long monofibrils, while incubation of apoE3 with the Aβ peptide results in the formation of a less dense, less stable network. The more complex nature of the plaques formed with the Aβ peptide in the presence of apoE4 *in vivo* may impair the normal clearance process and enhance plaque formation. Alternatively or additionally, apoE may alter the cytoskeletal structure and function and, under certain conditions, may promote the formation of the neurofibrillary tangles. Our studies have demonstrated that apoE3 and apoE4 exert differential effects on neuronal growth (*i.e.,* neurite extension and branching) *in vitro.* When combined with a source of lipid, apoE3 stimulated neurite extension in peripheral nervous system neurons (dorsal root ganglia), whereas apoE4 inhibited it. Similar results were obtained with central nervous system neurons (murine neuroblastoma Neuro-2a cells). Addition of free apoE3 or apoE4 without β-VLDL had no effect on neurite outgrowth. There was also differential accumulation of apoE3 and apoE4 by the neuroblastoma cells: apoE3 accumulated within cell bodies and neurites to a greater extent than apoE4. Thus, apoE3 may facilitate cytoskeletal activity, whereas apoE4 may inhibit it, which would be detrimental during synaptic remodeling.

INTRODUCTION

Apolipoprotein (apo) E, a 34,000 molecular weight protein that is the product of a single gene on chromosome 19 (see ref. 1 for review), exists in

[a] This research was funded in part by NIH Program Project Grant HL41633.

[e] *Send correspondence to:* Dr. Robert W. Mahley, Gladstone Institute of Cardiovascular Disease, P.O. Box 419100, San Francisco, CA 94141-9100, Tel: (415) 826-7500; Fax: (415) 285-5632.

three major forms of which apoE3 is the most common form. Apolipoprotein E2, which is associated with type III hyperlipoproteinemia,[1] differs from apoE3 at residue 158, where cysteine substitutes for the normally occurring arginine. Apolipoprotein E4, which has been linked to Alzheimer's disease, differs from apoE3 at residue 112, where arginine substitutes for cysteine.

Through its function in redistributing lipids among the cells of different organs, apoE plays a critical role in lipid metabolism (see ref. 1 for review). While apoE exerts this global transport mechanism in chylomicron and very low density lipoprotein (VLDL) metabolism, it also functions in the local transport of lipids by redistributing them among cells within a tissue. Cells with excess cholesterol and other lipids may release these substances to apoE·lipid complexes or to high density lipoproteins containing apoE, which can transport the lipids to cells requiring them for proliferation or repair. The apoE on these lipoprotein particles mediates their interaction and uptake via the low density lipoprotein (LDL) receptor or the LDL receptor–related protein (LRP).

A series of observations made over several years suggests a neurobiological role for apoE. First, apoE mRNA is abundant in the brain, where it is synthesized and secreted primarily by astrocytes.[2–4] Second, apoE-containing lipoproteins are found in the cerebrospinal fluid and appear to play a major role in lipid transport in the central nervous system.[5] Third, apoE plus a source of cholesterol promotes marked neurite extension in dorsal root ganglion cells in culture.[6] Fourth, apoE levels dramatically increase after peripheral nerve injury.[7,8] Apolipoprotein E appears to participate both in the scavenging of lipids generated after axon degeneration and in the redistribution of these lipids to sprouting neurites for axon regeneration and later to Schwann cells for remyelination of the new axons.[9,10]

Most recently, apoE has been implicated in the pathogenesis of Alzheimer's disease.[11,12] Apolipoprotein E is associated with the two characteristic neuropathologic lesions of Alzheimer's disease—extracellular neuritic plaques representing deposits of amyloid beta (Aβ) peptide and intracellular neurofibrillary tangles representing filaments of a microtubule-associated protein called tau (for review, see refs. 13-18). The occurrence of the apoE4 isoform is strongly linked with late-onset Alzheimer's disease.[11,12] It is unknown whether apoE4 alters or accelerates the formation of extracellular amyloid plaques *in vivo.* The Aβ peptide and apoE4 interact readily *in vitro* to form a very stable complex.[19,20] Apolipoprotein E3, on the other hand, displays a less avid interaction with the Aβ peptide.

Morphological studies using negative staining electron microscopy have revealed interesting differences in the ultrastructures of apoE·Aβ peptide complexes formed with apoE3 versus apoE4.[21] When apoE4 and the Aβ peptide were incubated together, a dense complex composed of very long monofibrils was observed. In contrast, much of the Aβ peptide appeared unreactive in the presence of apoE3, and the monofibrillar complex was less dense. Immunogold labeling demonstrated that both apoE3 and apoE4 associated with the monofibrils each formed with the Aβ peptide. The apoE was uniformly distributed along the fibrils or was insinuated within the structure of the fibrils.[21]

While the apoE4-initiated network of monofibrils was massive, the apoE3-initiated network was simpler. The more complex nature of the plaque formed with the Aβ peptide in the presence of apoE4 *in vivo* could impair the normal clearance process and in fact enhance plaque formation. In contrast, apoE3 may allow a more normal clearance of the Aβ peptide and thus impair plaque formation, or at least the extent of deposition.

The neurofibrillary tangles, which are paired helical filaments of hyperphosphorylated tau (see refs. 15, 17, and 18 for review), accumulate in the cytoplasm of neurons. Tau is a microtubule-associated protein which stabilizes the neuronal microtubules. Tau is a phosphoprotein; however, hyperphosphorylation impairs tau's ability to interact with the microtubules.

It has been shown that tau interacts with apoE3 *in vitro,* but not with apoE4.[22] This observation suggests that the interaction of apoE3 with tau prevents tau's hyperphosphorylation, thus allowing it to function normally in stabilization of microtubular structure and function.[17,22] In the presence of apoE4, tau could become hyperphosphorylated and thus inactive, which could promote the formation of neurofibrillary tangles. This postulate remains to be proven.

RESULTS AND DISCUSSION

In neurons the cytoskeleton functions in neurite extension and retraction. Therefore, our studies have focused on the different effects of apoE3 and apoE4 on neurite extension and branching. Our postulate is that apoE modulates the intracellular cytoskeletal apparatus and alters neurite extension and branching. Understanding how the various isoforms of apoE alter the cytoskeleton may shed light on the process of neurofibrillary tangle formation and suggest how apoE may modulate the remodeling of synaptic connections later in life.

Incubation of dorsal root ganglion neurons in culture with β-migrating very low density lipoproteins (β-VLDL) or cholesterol alters the neurite growth of these cells compared to that of cells grown in media alone.[6] In the presence of a source of lipid (β-VLDL or free cholesterol), neurite outgrowth is greatly enhanced, specifically due to extensive branching (with little or no neurite extension). When rabbit apoE (equivalent to human apoE3 with respect to the occurrence of cysteine at residue 112) was added to this system, however, the growth pattern changed markedly. Incubation of dorsal root ganglion neurons with rabbit apoE3 and either β-VLDL or cholesterol resulted in very significant neurite extension.[6] A source of cholesterol appears to enhance membrane biosynthesis, whereas the addition of rabbit apoE with a source of lipid results in long neuritic extensions and a trimming back of the branches. The elongation of neurites in cultured neurons may be equivalent to a target-directed extension of neurites to form synaptic connections or to innervate end organs.

A comparison of the effects of human apoE3 versus human apoE4 showed pronounced differential isoform-specific effects on neurite outgrowth.[23] Apolipoprotein E3 plus β-VLDL resulted in an increase in neurite extension, while apoE4 plus β-VLDL resulted in a marked decrease in both neurite branching and extension. Dorsal root ganglion neurons incubated with apoE4 plus β-VLDL displayed very short, stunted neurites.[23] This was not a toxic effect of apoE4 since replacement of the apoE4-containing media with fresh apoE4-lacking media restored the ability of the neurons to produce neuritic extensions. Furthermore, the apoE3- and apoE4-specific effects were blocked by addition of an antibody against the receptor-binding domain of apoE or reductive methylation of critical lysine residues, suggesting that the effect of apoE was receptor-mediated.

Murine neuroblastoma cells (Neuro-2a) from the central nervous system were used to compare the effects of apoE on the peripheral nervous system neurons described above with those on cortical neurons. Similar results were obtained. When combined with a source of lipid, apoE3 stimulated neurite extension, whereas apoE4 inhibited it.[24] Addition of free apoE3 or apoE4 without β-VLDL had no effect on neurite outgrowth. These results further suggest that the effect of apoE on neurons requires the lipoprotein receptor-mediated uptake of apoE or a combination of apoE plus a source of lipid. Free of lipid, apoE does not bind to either the LDL receptor or the LRP.

Apolipoprotein E3 or E4 plus β-VLDL was incubated with Neuro-2a cells for 48 h, and the intracellular apoE was localized by immunofluorescence using confocal microscopy.[24] By optical sectioning through the neuronal cell body and neurites, it was possible to visualize the apoE on the cell surface and throughout the cell. Apolipoprotein E3 accumulated in high concentration in the cell bodies around the nucleus and diffusely throughout the neurites within the intracellular compartment. Although the apoE3 clearly is present within the intracellular compartment of the cell, it is not yet possible to ascertain whether it occurs within the cytoplasm, in association with the cytoskeletal elements, or within vesicles. On the other hand, apoE4 accumulated in the cell bodies and neurites to a much lesser degree; it was localized near the cell surface of the cell bodies and to a lesser extent within the short neurites. Although we have demonstrated that apoE3 and apoE4 in the presence of β-VLDL are taken up by the cells to a similar extent, apoE3 appears to be retained, whereas apoE4 is not.

SUMMARY

In dorsal root ganglion or neuroblastoma cells, apoE3 plus a source of lipid supports and facilitates neurite extension. It is envisioned that apoE3 stabilizes the cytoskeleton and supports neurite elongation. Apolipoprotein E3 accumulates widely in the cell bodies and neurites and may directly or indirectly modulate cytoskeletal activity, thus facilitating neurite extension. Apolipoprotein E4, on the other hand, does not appear to support neurite extension, does not accumulate within neurons, and may even destabilize the cytoskeleton, possibly the microtubules. To the best of our knowledge, individuals with apoE4 have normal neuronal development early in life. However, apoE4 may exert its detrimental effects later in life, by not allowing or supporting remodeling of synaptic connections. This effect may be important in the pathogenesis of Alzheimer's disease. Alternatively, apoE4 may contribute to Alzheimer's disease by altering amyloid plaque formation. Apolipoprotein E4 may participate in the formation of dense, complicated plaques of $A\beta$ peptide. Such plaques may be toxic and thus impact the development of Alzheimer's disease via this mechanism. At the present time, the pathway whereby apoE affects the development of Alzheimer's disease remains speculative.

REFERENCES

1. MAHLEY, R. W. 1988. Apolipoprotein E: Cholesterol transport protein with expanding role in cell biology. Science **240:**622–630.
2. ELSHOURBAGY, N. A., W. S. LIAO, R. W. MAHLEY & J. M. TAYLOR. 1985. Apolipoprotein E mRNA is abundant in the brain and adrenals, as well as in the liver, and is present in other peripheral tissues of rats and marmosets. Proc. Natl. Acad. Sci. USA **82:**203–207.
3. BOYLES, J. K., R. E. PITAS, E. WILSON, R. W. MAHLEY & J. M. TAYLOR. 1985. Apolipoprotein E associated with astrocytic glia of the central nervous system and with nonmyelinating glia of the peripheral nervous system. J. Clin. Invest. **76:**1501–1513.
4. PITAS, R. E., J. K. BOYLES, S. H. LEE, D. FOSS & R. W. MAHLEY. 1987. Astrocytes synthesize apolipoprotein E and metabolize apolipoprotein E-containing lipoproteins. Biochim. Biophys. Acta. **917:**148–161.
5. PITAS, R. E., J. K. BOYLES, S. H. LEE, D. Y. HUI & K. H. WEISGRABER. 1987. Lipoproteins and their receptors in the central nervous system: Characterization of the lipoproteins in cerebrospinal fluid and identification of apolipoprotein B,E(LDL) receptors in the brain. J. Biol. Chem. **262:**14352–14360.
6. HANDELMANN, G. E., J. K. BOYLES, K. H. WEISGRABER, R. W. MAHLEY & R. E. PITAS. 1992. Effects of apolipoprotein E, β-very low density lipoproteins, and cholesterol on the extension of neurites by rabbit dorsal root ganglion neurons in vitro. J. Lipid Res. **33:**1677–1688.

7. MÜLLER, H. W., P. J. GEBICKE-HÄRTER, D. H. HANGEN & E. M. SHOOTER. 1985. A specific 37,000-dalton protein that accumulates in regenerating but not in nonregenerating mammalian nerves. Science **228:**499–501.
8. IGNATIUS, M. J., P. J. GEBICKE-HÄRTER, J. H. P. SKENE, J. W. SCHILLING, K. H. WEISGRABER, R. W. MAHLEY & E. M. SHOOTER. 1986. Expression of apolipoprotein E during nerve degeneration and regeneration. Proc. Natl. Acad. Sci. USA **83:**1125–1129.
9. BOYLES, J. K., C. D. ZOELLNER, L J. ANDERSON, L. M. KOSIK, R. E. PITAS, K. H. WEISGRABER, D. Y. HUI, R. W. MAHLEY, P. J. GEBICKE-HAERTER, M. J. IGNATIUS & E. M. SHOOTER. 1989. A role for apolipoprotein E, apolipoprotein A-I, and low density lipoprotein receptors in cholesterol transport during regeneration and remyelination of the rat sciatic nerve. J. Clin. Invest. **83:** 1015–1031.
10. IGNATIUS, M. J., E. M. SHOOTER, R. E. PITAS & R. W. MAHLEY. 1987. Lipoprotein uptake by neuronal growth cones in vitro. Science **236:**959–962.
11. SAUNDERS, A. M., W. J. STRITTMATTER, D. SCHMECHEL, P. H. ST. GEORGE-HYSLOP, M. A. PERICAK-VANCE, S. H. JOO, B. L. ROSI, J. F. GUSELLA, D. R. CRAPPER-MACLACHLAN, M. J. ALBERTS, C. HULETTE, B. CRAIN, D. GOLDGABER & A. D. ROSES. 1993. Association of apolipoprotein E allele $\epsilon 4$ with late-onset familial and sporadic Alzheimer's disease. Neurology **43:**1467–1472.
12. CORDER, E. H., A. M. SAUNDERS, W. J. STRITTMATTER, D. E. SCHMECHEL, P. C. GASKELL, G. W. SMALL, A. D. ROSES, J. L. HAINES & M. A. PERICAK-VANCE. 1993. Gene dose of apolipoprotein E type 4 allele and the risk of Alzheimer's disease in late onset families. Science **261:**921–923.
13. MCKHANN, G., D. DRACHMAN, M. FOLSTEIN, R. KATZMAN, D. PRICE & E. M. STADLAN. 1984. Clinical diagnosis of Alzheimer's disease: Report of the NINCDS-ADRDA Work Group under the auspices of Department of Health and Human Services Task Force on Alzheimer's Disease. Neurology **34:** 939–944.
14. SELKOE, D. J. 1991. The molecular pathology of Alzheimer's disease. Neuron **6:**487–498.
15. CROWTHER, R. A. 1993. Tau protein and paired helical filaments of Alzheimer's disease. Curr. Opin. Struct. Biol. **3:**202–206.
16. ROSES, A. D. 1994. The Alzheimer diseases. Curr. Neurol. **14:**111–141.
17. WEISGRABER, K. H., A. D. ROSES & W. J. STRITTMATTER. 1994. The role of apolipoprotein E in the nervous system. Curr. Opin. Lipidol. **5:**110–116.
18. WEISGRABER, K. H., R. E. PITAS & R. W. MAHLEY. 1994. Lipoproteins, neurobiology, and Alzheimer's disease: Structure and function of apolipoprotein E. Curr. Opin. Struct. Biol. **4:**507–515.
19. STRITTMATTER, W. J., A. M. SAUNDERS, D. SCHMECHEL, M. PERICAK-VANCE, J. ENGHILD, G. S. SALVESEN & A. D. ROSES. 1993. Apolipoprotein E: High avidity binding to β-amyloid and increased frequency of type 4 allele in late-onset familial Alzheimer disease. Proc. Natl. Acad. Sci. USA **90:**1977–1981.
20. STRITTMATTER, W. J., K. H. WEISGRABER, D. Y. HUANG, L.-M. DONG, G. S. SALVESEN, M. PERICAK-VANCE, D. SCHMECHEL, A. M. SAUNDERS, D. GOLDGABER & A. D. ROSES. 1993. Binding of human apolipoprotein E to synthetic

amyloid β peptide: Isoform-specific effects and implications for late-onset Alzheimer disease. Proc. Natl. Acad. Sci. USA **90:**8098–8102.

21. SANAN, D. A., K. H. WEISGRABER, S. J. RUSSELL, R. W. MAHLEY, D. HUANG, A. SAUNDERS, D. SCHMECHEL, T. WISNIEWSKI, B. FRANGIONE, A. D. ROSES & W. J. STRITTMATTER. 1994. Apolipoprotein E associates with β amyloid peptide of Alzheimer's disease to form novel monofibrils. Isoform apoE4 associates more efficiently than apoE3. J. Clin. Invest. **94:**860–869.
22. STRITTMATTER, W. J., K. H. WEISGRABER, M. GOEDERT, A. M. SAUNDERS, D. HUANG, E. H. CORDER, L.-M. DONG, R. JAKES, M. J. ALBERTS, J. R. GILBERT, S.-H. HAN, C. HULETTE, G. EINSTEIN, D. E. SCHMECHEL, M. A. PERICAK-VANCE & A. D. ROSES. 1994. Hypothesis: Microtubule instability and paired helical filament formation in the Alzheimer disease brain are related to apolipoprotein E genotype. Exp. Neurol. **125:**163–171.
23. NATHAN, B. P., S. BELLOSTA, D. A. SANAN, K. H. WEISGRABER, R. W. MAHLEY & R. E. PITAS. 1994. Differential effects of apolipoproteins E3 and E4 on neuronal growth in vitro. Science **264:**850–852.
24. NATHAN, B. P., S. BELLOSTA, R. W. MAHLEY & R. E. PITAS. 1994. Apolipoprotein E3- and E4-induced differences in neurite outgrowth are associated with differences in the subcellular localization of apolipoprotein E. Soc. Neurosci. **20:** 1033. (Abstr.)

Morphological, Biochemical, and Genetic Support for an Apolipoprotein E Effect on Microtubular Metabolism[a]

A. D. ROSES,[b,c,g] AND (in alphabetical order) G. EINSTEIN,[b,c]
J. GILBERT,[b,c] M. GOEDERT,[d] S-H. HAN,[b,c] D. HUANG,[b,c]
C. HULETTE,[b,c] E. MASLIAH,[e] M. A. PERICAK-VANCE,[b,c]
A. M. SAUNDERS,[b] D. E. SCHMECHEL,[b,c]
W. J. STRITTMATTTER,[b,c]
K. H. WEISGRABER,[f] AND P.-T. XI[b,c]

[b] *Department of Medicine (Neurology) and* [c] *Department of Neurobiology, Joseph and Kathleen Bryan Alzheimer's Disease Research Center (Box 2900) Duke University Medical Center, Durham, North Carolina 27710 USA*

[d] *Medical Research Council Laboratory of Molecular Biology, Cambridge CB2 2QH, England*

[e] *Department of Neurosciences, University of California at San Diego, La Jolla, California 92093-0624 USA*

[f] *Gladstone Institute of Cardiovascular Disease, Cardiovascular Research Institute, Department of Pathology, University of California, San Francisco, California 94140-9100 USA*

ABSTRACT: There are two distinct viewpoints on the association of the inheritance of apolipoprotein E (APOE) alleles and the age of onset distribution of Alzheimer's disease (AD): genetic and phenotypic expression. There have been multiple corroborations of the APOE-ϵ4 association with Alzheimer's disease in populations around the world in clinic based studies as well as emerging epidemiological studies. The genetic data do not imply mechanism of pathogenesis. The phenotypic expression of AD has been based in theories based on amyloid plaques or neurofibrillary tangles. ApoE protein interacts with both β-amyloid

[a] Supported by a National Institutes of Health Leadership and Excellence in Alzheimer's Disease Award 5R35 AG-07922 (A.D.R.), National Institutes of Health Alzheimer's Disease Research Center 5P50 AG-05128 (A.D.R.), numerous contributions to the Joseph and Kathleen Bryan Alzheimer's Disease Research Center, the Bader Foundation (A.D.R.), NIH Program Project Grant HL-41633 and the L. K. Whittier Foundation, South Pasadena, CA (K.H.W.), grants AGO5131, AG10689, funding from the UCSD Neurosciences Educational and Research Foundation, the Sam and Rose Stein Institute for Research on Aging (E.M.) and the Medical Research Council, London (M.G.).

[g] *Send correspondence to:* Dr. Allen D. Roses, Departments of Medicine (Neurology) and Neurobiology, Joseph and Kathleen Bryan Alzheimer's Disease Research Center, Box 2900, Duke University Medical Center, Durham, N.C. 27710; Tel: (919) 684-6274; Fax: (919) 684-6514.

and tau in an isoform-specific manner. The interaction with tau had been thought to be an *in vitro* artifact, since apoE had not been previously localized to the neuronal cytoplasm. Immuno-EM studies have localized apoE in neuronal cytoplasm. ApoE3 interacts with both tau and MAP2c at the microtubule binding repeat domain under conditions in which apoE4 is less tightly bound. These data further support a hypothesis that apoE3 (and apoE2) protect the microtubule binding domain of tau from binding to itself to form paired helical filaments and neurofibrillary tangles, while protecting the site for microtubule stabilizing interactions with β-tubulin. These data are supported by recent data from APOE knock-out mice demonstrating dendritic alterations leading to synaptic simplification similar to that observed in AD.

INTRODUCTION

Disease-oriented research usually begins with clinical descriptions and nosology, then progresses to pathological correlations. Historically, the next step has been biochemical subtyping. Recently, with the development of molecular mapping strategies, relevant genetic factors can be identified independent of the known descriptive pathology, by using genetic linkage strategies that locate the regions on chromosomes containing relevant genes. Research in AD had focused on clinical definitions and pathological differences. With the biochemical analyses of neuritic plaques and neurofibrillary tangles (NFTs), hypotheses have been centered on amyloid and tau.[1,2] Genetic epidemiological methods were used to link two early-onset Alzheimer's disease (AD) genes that appear to segregate as autosomal dominant traits.[3,4] Several specific mutations of the β-amyloid precursor protein gene (APP) on chromosome 21 were identified in one early-onset form of AD in approximately 15 known families worldwide.[4] However the gene for most early-onset disease families segregates as the presenilin 1 gene on chromosome 14.[3,5] Using a new strategy for application of nonparametric and standard linkage methods to determine linkage of late-onset AD on chromosome 19, apolipoprotein E (APOE, gene; apoE, protein) was identified as a major susceptibility gene contributing to the age of onset distribution.[6–9] Variations in the APOE genotype can account for more than two decades difference in the mean age of onset, with APOE4/4 having the earliest onset and the greatest risk.[10] APOE2/3 has the least risk and the oldest age of onset distribution.[10] TABLE 1 is a classification of the currently known genetic etiologies of the Alzheimer's diseases.[11]

APOE ALLELES AND ALZHEIMER'S DISEASE SUSCEPTIBILITY

Genetic linkage data identified a locus for late-onset AD on chromosome 19 in 1991.[8] In 1993, Strittmatter *et al.* identified a genetic association of APOE-ϵ4 (APOE4) with late-onset familial AD.[7] APOE had been mapped to this region of chromosome 19 for almost a decade, but was never known to be genetically or functionally related to AD. Saunders *et al.* found that

TABLE 1. Genetic Classification of the Dementias

I. THE ALZHEIMER'S DISEASES			
	Type	Chromosome	Gene
A.	Early-onset familial, autosomal dominant mutations, ***AD1***	21	APP[4]
B.	Late-onset familial and sporadic associated, susceptibility gene, ***AD2***	19	APOE[6,7] genotype
C.	Early-onset familial, autosomal dominant, ***AD3***	14	*presenilin I* (S182)[31]
D.	Early-onset familial, autosomal dominant (Volga-German founder and other families), ***AD4***	1	*presenilin II* (E5.1, STM2)[32–34]
E.	Other late-onset susceptibility genes	Several reported, but unconfirmed	

the APOE4 association was also present using DNA from 176 sporadic, pathologically verified AD cases.[6] Namba *et al.* reported in 1991 that apoE was immunoreactive with amyloid plaques, vascular amyloid, and neurofibrillary tangles.[12] Wisniewski and Frangione subsequently reported that apoE was bound to the amyloids of many diseases.[13] Strittmatter *et al.* confirmed the Namba *et al.* data in AD, but also noted apoE immunoreactivity in some neurons without the presence of NFTs.[7] Diedrich *et al.* had earlier noted that mRNA for APOE was increased in astrocytes in AD but found no evidence that neurons expressed the mRNA for apoE.[14] Thus, the unexpected presence of apoE-like immunoreactivity in neurons raised the possibility that apoE could be involved in intraneuronal metabolism. Striking differences in the apoE isoform-specific *in vitro* interactions with tau and other microtubule-associated proteins catalyzed a series of multidisciplinary investigations.[15,16]

The probability that the linkage of APOE to AD was due to a nearby gene, or an example of linkage disequilibrium, was raised immediately. However, within 18 months of the initial publication there were more than 50 published verifications of the association of APOE4 and AD in the literature. No evidence for linkage disequilibrium with any nearby gene or marker has been reported for any data set of unrelated families. Only one published report failed to find an association of APOE4 and sporadic AD.[17] This report ascertained only a small number of patients who were 75 years or older. This biased ascertainment missed the main fraction of age of onset effects of APOE4 inheritance and was subsequently corrected.[11a]

APOE ALLELES AND AGE OF ONSET DISTRIBUTION

Corder *et al.* analyzed the distribution of age of onset in all members over 50 years old in 42 AD families and found an APOE4 dose-related increase in risk and decrease in the mean age of onset.[9] Adding sporadic AD patients and age-matched controls to the analysis, Corder *et al.* also found that the inheritance of an APOE2 allele decreased the risk and increased the age of onset.[10] The opposite phenotypic effect of inheriting one allele that was asso-

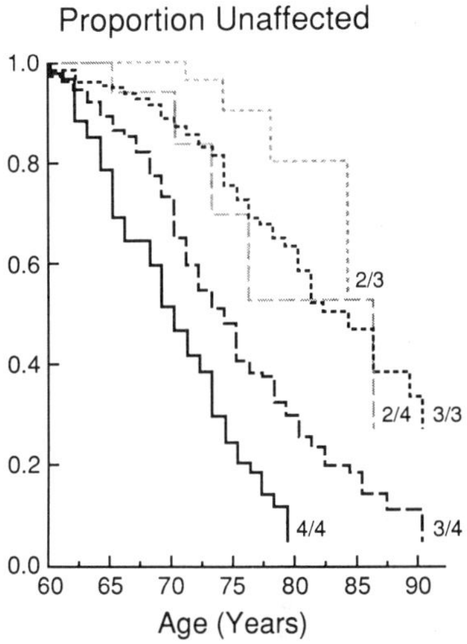

FIGURE 1. APOE genotype specific risk of remaining unaffected. Age of onset for subjects with each indicated APOE genotype. Onset curves were estimated by Kaplan-Meier product limit distributions. These data do not represent a population-based epidemiological survey, are limited to late-onset Alzheimer's disease and should not be generalized for all populations at risk. The figure is meant to illustrate the genotype-specific age of onset distributions. Proper epidemiologically based distribution will be needed before risks can be estimated in the general population. (Reprinted from Roses,[19] with permission.)

ciated with disease and another than spared disease at the same APOE locus made the possibility of linkage disequilibrium with two separate segregating genes vanishingly small. FIGURE 1 illustrates the relationship of APOE genotype to age of onset.[10,11]

APOE IS LOCATED IN THE CYTOPLASM OF NEURONS

It should be emphasized that the genetic data do not predict the specific mechanisms involved in the pathogenesis of the disease for either the APOE or APP genes.[18] Testing the phenotypic effects of inheriting different APOE alleles proceeded along three principal pathways: 1) the popular amyloid hypothesis, 2) interactions with tau and NFTs, and 3) additional genetic approaches. The relationship of the APOE genetic data to the amyloid hypothesis has recently been summarized from two opposing viewpoints.[2,19] The interaction of tau with apoE was criticized because apoE was not known to be in neuronal cytoplasm where tau is located.[2] It was stated that "a novel mechanism for the escape of the endocytosed, membrane-surrounded ApoE molecular from lysosomal degradation and its transport into the cytoplasm would need to be found" to explain apoE interacting with tau in neuronal cytoplasm.[2] FIGURE 2 illustrates the electron microscopic localization of apoE immunoreactivity inside a neuron, clearly demonstrating the cytoplasmic and

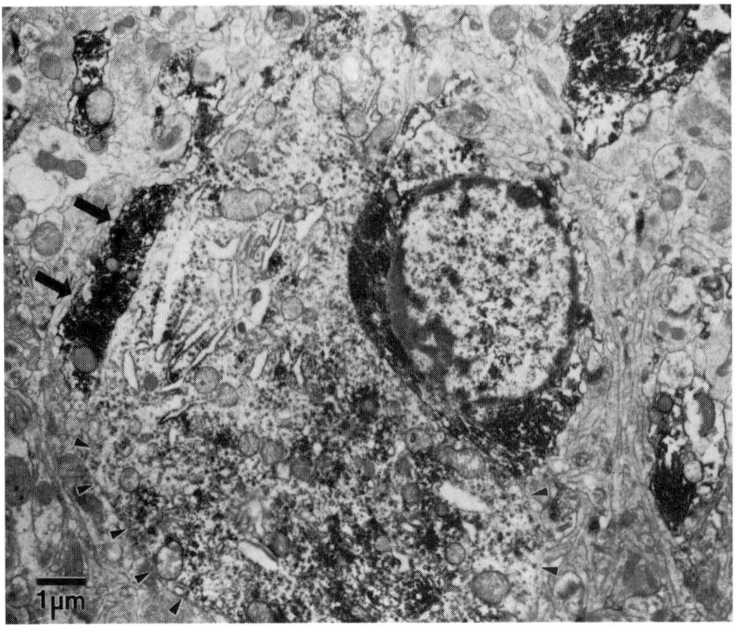

FIGURE 2. ApoE is in the cytoplasm of human neurons. Electron microscopic analysis of immunoreacted vibratome section from surgical specimen of lateral temporal lobe from a young (<30 years old) epilepsy patient. Fixation 2% paraformaldehyde/0.2% glutaraldehyde. The densely immunoreactive cell in right center with relatively condensed nuclear chromatin is assumed to represent a satellite glial cell due to its close apposition to a large, less strongly immunoreactive region (arrowheads indicate border) filling much of the photograph. This region is presumed to represent proximal dendrite or perinuclear soma of a large cortical neuron given the size and shape of this process and the intensity of immunoreaction. On the left side of this lightly immunoreactive cell is a strongly immunoreactive process *(arrows)*, which is also opposed to the presumed neuron. The overall appearance suggests the combined profile of satellite glial cell and neuronal cell body. (Reprinted from Han *et al.*[20] with permission.)

micro-organelle localization of apoE.[20,21] Subfractionation experiments using human brain tissue and neuronal tissue culture systems are currently underway to verify the immunological data biochemically.

That "a novel mechanism of escape" is necessary assumes that endosomal transport is the only mechanism available for neurons. There are three known receptors that are involved in apoE-containing lipoprotein transport into cells: the low density lipoprotein receptor (LDL), the low density lipoprotein-related receptor (LRP), and the newly described very low density lipoprotein receptor (VLDL).[22–24] Neurons, like macrophages, are very rich in surface LRP receptors. The LRP receptor is known to have a mechanism to allow transposition of peptides into the cytoplasm without going through endosomal transport. Rebeck *et al.* demonstrated that LRP receptor immunoreactivity was increased in plaques in Alzheimer disease.[23] Ikeda *et al.* (unpublished data) have also demonstrated that a newly described VLDL receptor is also localized to plaques in human brain.[24] Immunoelectron microscopy experiments are in progress to examine the intraneuronal locations of LRP and the VLDL receptor. Studies of the functional mechanisms associated with the VLDL receptor are also in progress. Suffice it to say that there are recognized mechanisms for possible transposition of apoE into neurons. In fact, one might view one of the main functions of apoE to transit membranes when other molecules do not. However, the fact that the specific mechanism may be currently undefined does not rule out a hypothesis that requires apoE in the cytoplasmic compartment of neurons. No matter what the mechanism, apoE is located in the neuronal cytoplasm. That fact makes moot the argument that it cannot be there. It is remotely possible that APOE mRNA may be expressed at some low level in neurons. Either way, apoE is in the right place to interact with microtubule-associated proteins. It was also stated that "There has been no previous evidence that tau requires ApoE in order to perform its normal function as a microtubule-associated protein."[2] We would correct that criticism to propose that apoE is not needed for tau to perform its normal function, but rather apoE3 or apoE2 may enhance that function over time by protecting tau from forming PHF.

ADDITIONAL EXPERIMENTAL SUPPORT FOR APOE/TAU INTERACTIONS

In vitro experiments have demonstrated that apoE3 binds to non-phosphorylated tau at much lower concentrations that does apoE4.[15,25] The endpoints of these experiments, formation of an SDS insoluble complex, are probably not physiologic. Kinetics, affinity and binding position studies of apoE, tau, MAP2c and β-tubulin interactions are currently in progress using label-free analysis of proteins in free solution binding to immobilized ligand at a hydrophilic sensor surface.

Isoform-specific interactions of apoE3 occur with all four microtubule-binding repeats of tau. Huang *et al.* have synthesized peptides for each repeat of the microtubule-binding domain and found that apoE3 binds each repeat at concentrations two orders of magnitude less than apoE4.[26] Replacement of the unique cysteine residue of microtubule-binding repeat 2 with an alanine residue diminishes, but does not abolish, apoE3 binding. Therefore it appears that the apoE specific interactions are not dependent on the formation of apoE3/tau repeat 2 disulfide bonds.

APOE INTERACTIONS WITH MAP2C

Huang *et al.* have also established a similar interaction of the microtubule-binding repeat domains of MAP2c, supporting the notion that apoE3 can protect the same binding domains of MAP2c that are responsible in tau for the formation of paired helical filaments (PHF).[15] Apparently MAP2c simply does not form PHFs or NFTs, just as tau from most other species does not seem to form NFTs. NFTs are composed mainly of highly phosphorylated PHFs.[1,27] We have proposed that a similar functional mechanism exists for tau and MAP2c, but that apoE3 [or apoE2] provides the added benefit of protecting tau from forming PHFs.[16] The hyper-phosphorylation and aggregation of PHF-tau into NFTs may reflect a bias introduced by not protecting the tau binding domains over many years and allowing more opportunity for tau/tau interactions. Experiments are underway examining the interactions of tau and MAP2c with β-tubulin in the presence of apoE3 or apoE4. This model would predict that the apoE3/tau interactions would not functionally interfere with, and may enhance, interactions with β-tubulin.

TAU PHOSPHORYLATION

There is a single phosphorylation site within the tau microtubule-binding domain. Ser^{262} is located within repeat 1 and can be phosphorylated *in vitro* and *in vivo*.[28] Strittmatter *et al.* have demonstrated that *in vitro* phosphorylation of peptide repeat 1 abolishes its ability to bind apoE3.[26] Experiments are in progress to test whether apoE3 binding can block ser^{262} phosphorylation. Thus apoE3 may be involved in the phosphorylation metabolism involving the only tau phosphorylation site within the microtubule-binding domain. The relationship of these data to the formation of PHF-tau and to interactions with β-tubulin are currently being studied.

APOE KNOCK-OUT MICE

The results of genetic experiments using knock-out and transgenic technology cannot always be predicted accurately. Negative experiments provide little useful data. The last seven years have witnessed a large number of investigators using the APP gene or its derivatives to try to create a model of neuritic amyloid plaques. Knocking out the gene for tau does little to the animals, particularly regarding its purported function for axonal growth elongation and branching.[29] Apparently other microtubule-associated proteins, such as MAP2, can replace the function of tau.

APOE knock-out mice were constructed as an excellent model of atherosclerosis, with little attention to the brain. Masliah *et al.* studied aged APOE knock-out mice and found that the synaptic density in homozygous knock-out mice was significantly reduced before one year of age.[30] Perhaps the most striking data that support the role of apoE in microtubular stability in proximal dendrites comes from electron microscopic analysis of the APOE

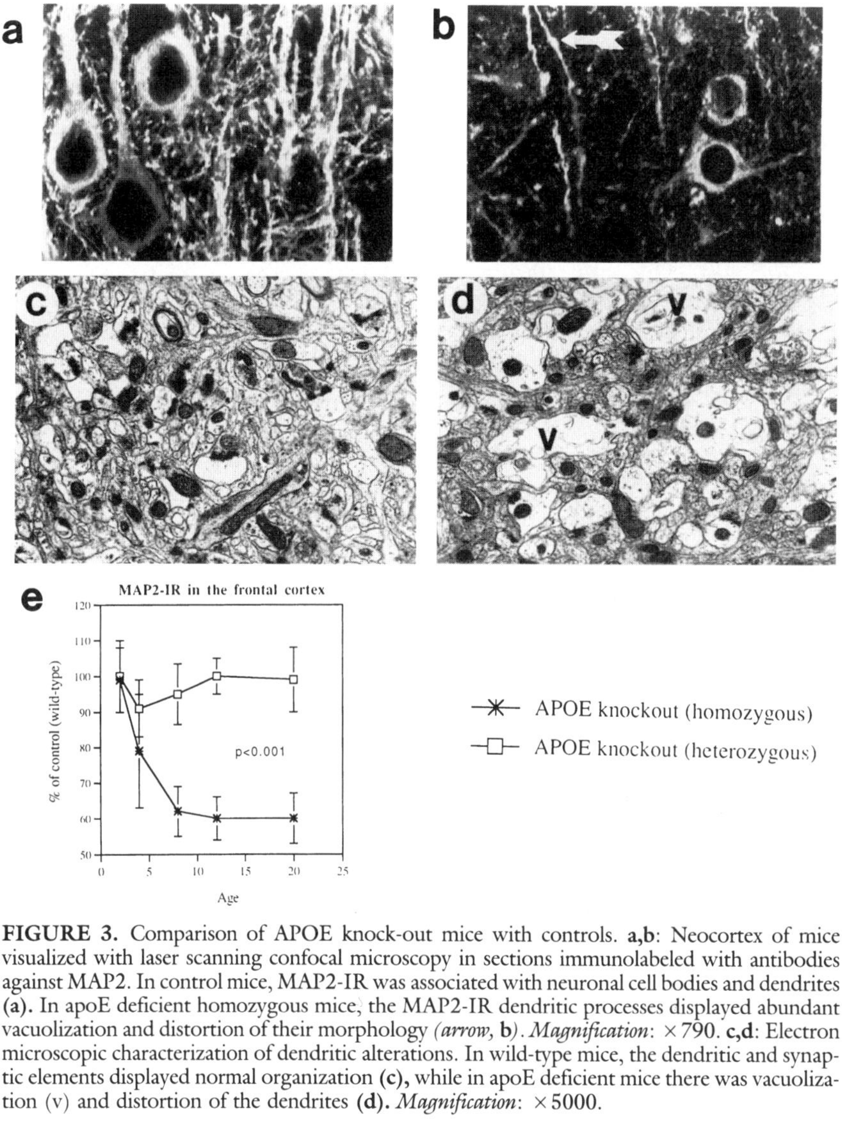

FIGURE 3. Comparison of APOE knock-out mice with controls. **a,b**: Neocortex of mice visualized with laser scanning confocal microscopy in sections immunolabeled with antibodies against MAP2. In control mice, MAP2-IR was associated with neuronal cell bodies and dendrites **(a).** In apoE deficient homozygous mice, the MAP2-IR dendritic processes displayed abundant vacuolization and distortion of their morphology *(arrow,* **b**). *Magnification*: ×790. **c,d**: Electron microscopic characterization of dendritic alterations. In wild-type mice, the dendritic and synaptic elements displayed normal organization **(c),** while in apoE deficient mice there was vacuolization (v) and distortion of the dendrites **(d).** *Magnification*: ×5000.

knock-out mice (FIG. 3,a,b,e). Masliah *et al.* have also demonstrated vacuolization of dendrites with fragmented microtubule-like structures in homozygous mice by eight months of age, at the time when dendritic simplification analyses become sensitive enough to detect significant differences.[30] Additional immuno-EM studies are also in progress, but provide striking localization for dendritic pathology and microtubule-like fragmentation (FIG. 3,c,d).

Perhaps having no apoE is much worse than having apoE4—which is worse than apoE3 or apoE2.

That damage to dendrites should result from knocking out a gene that is not known to be expressed in neurons suggests that there is an important function for the small amounts of apoE that get into neuronal cells. This is consistent with the *in vitro* data suggesting that apoE3 binds both tau and MAP2c as an important co-factor related to dendritic integrity. The critical functions of apoE3 and apoE4 may be similar *in vivo* in humans. Differential metabolic problems with apoE/microtubule-associated protein interactions may become manifest over time or as a function of maintenance and repair of microtubule systems.

ENVIRONMENTAL STRESS IN HUMANS

Several groups using a variety of similar quantitative techniques had previously found synaptic density to be reduced in AD. FIGURE 4 illustrates the appearance of representative granule cells in the hippocampus of a 70-year-old nondemented control and a 70-year-old patient. The synaptic arborizations are less dense and shrunken, yet the cell body is still there and the cell had not died.[31] The fact that APOE mice may suffer the same accelerated fate provides a remarkable, positive coincidence.

Environmental stress may also serve to differentiate genetic differences in repair mechanisms. In a series of volunteers who were undergoing open-heart surgery we tested cognitive function pre-operatively and postoperatively.[32] Functional impairments in several cognitive tests were observed post-opera-

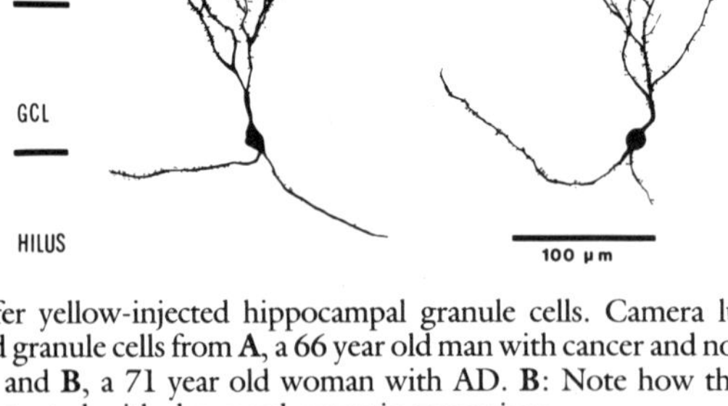

FIGURE 4. Lucifer yellow-injected hippocampal granule cells. Camera lucida drawings of Lucifer Yellow-filled granule cells from **A**, a 66 year old man with cancer and no neuropsychological evidence of AD and **B**, a 71 year old woman with AD. **B**: Note how the dendritic tree of the granule cells is stunted with decreased synaptic corrections.

tively. When the patients were subsequently analyzed as a function of whether they carried an APOE4 allele, those patients with an APOE4 allele were more psychologically impaired. Independent of whether these data are due to the anesthetic stress, the cardiopulmonary bypass stress, or other factors, they suggest that apoE4 may not protect the brain from metabolic stress as well as apoE3 and apoE2. These data and recent reports that apoE4 is the risk factor associated with death from head injury and intracerebral hemorrhage provide an impetus for further stress related studies.[33–35]

CONCLUSION

We are at the beginning of a new era of understanding neuronal metabolism. In a reversal of the usual pattern of basic science discoveries anticipating and fertilizing clinical research, the genetic study of AD has identified a new role for apoE in neurobiology. Extensive prior basic science studies of apoE have catalyzed our ability to perform meaningful experiments. The involvement of apoE in neuronal metabolism will no doubt have implications beyond the isoform-specific mechanisms leading to AD. A role for apoE in lipid peroxidation is also suggested by the localization of apoE in peroxisomes and other intraneuronal micro-organelles. There appears to be a confluence of possible mechanisms for Alzheimer's disease, amytrophic lateral sclerosis (superoxide dismutase), Parkinson disease (mitochondrial metabolism), and perhaps other degenerative neurological diseases through oxygen and lipid peroxidase metabolic pathways.

REFERENCES

1. GOEDERT, M., C. M. WISCHIK, R. A. CROWTHER, *et al.* 1988. Cloning and sequencing of the cDNA encoding a core protein of the paired helical filament of Alzheimer disease: Identification as the microtubule-associated protein tau. Proc. Natl. Acad. Sci. USA **85**(11):4051–4055.
2. SELKOE, D. J. 1994. Alzheimer's disease: A central role for amyloid. J. Neuropathol. Exp. Neurol. **53:**438–447.
3. SCHELLENBERG, G. D., T. D. BIRD, E. M. WIJSMAN, *et al.* 1992. Genetic linkage evidence for a familial Alzheimer's disease locus on chromosome 14. Science **258:**668–671.
4. GOATE, A., H. M. CHARTIER, M. MULLAN, *et al.* 1991. Segregation of a missense mutation in the amyloid precursor protein gene with familial Alzheimer's disease [see comments]. Nature **349**(6311):704–706.
5. SHERRINGTON, R., E. I. ROGAEV, Y. LIANG, *et al.* 1995. Cloning of a gene bearing missense mutations in early-onset familial Alzheimer's disease. Nature **375:**754–760.

6. SAUNDERS, A. M., W. J. STRITTMATTER, D. SCHMECHEL, *et al.* 1993. Association of apolipoprotein E allele epsilon 4 with late-onset familial and sporadic Alzheimer's disease. Neurology **43**(8):1467–1472.
7. STRITTMATTER, W. J., A. M. SAUNDERS, D. SCHMECHEL, *et al.* 1993. Apolipoprotein E: High-avidity binding to beta-amyloid and increased frequency of type 4 allele in late-onset familial Alzheimer disease. Proc. Natl. Acad. Sci. USA **90**(5):1977–1981.
8. PERICAK, V. M., J. L. BEBOUT, P. J. GASKELL, *et al.* 1991. Linkage studies in familial Alzheimer disease: Evidence for chromosome 19 linkage. Am. J. Hum. Genet. **48**(6):1034–1050.
9. CORDER, E. H., A. M. SAUNDERS, W. J. STRITTMATTER, *et al.* 1993. Gene dose of apolipoprotein E type 4 allele and the risk of Alzheimer's disease in late onset families [see comments]. Science **261**(5123):921–923.
10. CORDER, E. H., A. M. SAUNDERS, N. J. RISCH, *et al.* 1994. Protective effect of apolipoprotein E type 2 allele for late onset Alzheimer disease. Nat. Genet. **7**:180–184.
11. ROSES, A. D., W. J. STRITTMATTER, V. M. PERICAK, *et al.* 1994. Clinical application of apolipoprotein E genotyping to Alzheimer's disease [letter]. Lancet **343**(8912):1564–1565.
11a. BASUN, H., M. GRUT, B. WINBLAD, *et al.* 1995. Apolipoprotein epsilon 4 allele and disease progression in patients with late-onset Alzheimer's disease. Neurosci. Lett. **183**:32–34.
12. NAMBA, Y., M. TOMONAGA, H. KAWASAKI, *et al.* 1991. Apolipoprotein E immunoreactivity in cerebral amyloid deposits and neurofibrillary tangles in Alzheimer's disease and kuru plaque amyloid in Creutzfeldt-Jakob disease. Brain Res. **541**(1):163–166.
13. WISNIEWSKI, T. & B. FRANGIONE. 1992. Apolipoprotein E: A pathological chaperone protein in patients with cerebral and systemic amyloid. Neurosci. Lett. **135**(2):235–238.
14. DIEDRICH, J., S. WIETGREFE, A. HAASE, *et al.* 1991. Identifying and mapping changes in gene expression involved in the neuropathology of scrapie and Alzheimer's disease. [Review]. Curr. Top. Microbiol. Immunol. **172**(259):259–274.
15. HUANG, D. Y., M. GOEDERT, R. JAKES, *et al.* 1994. Isoform-specific interactions of apolipoprotein E with the microtubule associated protein MAP2c: Implications for Alzheimer's disease Neurosci. Lett. In press.
16. STRITTMATTER, W. J., K. H. WEISGRABER, M. GOEDERT, *et al.* 1994. Hypothesis: Microtubule instability and paired helical filament formation in the Alzheimer disease brain are related to apolipoprotein E genotype. [Review]. Exp. Neurol. **125**(2):163–171.
17. LANNFELT, L., L. LILIUS, M. NASTASE, *et al.* 1994. Lack of association between apolipoprotein E allele epsilon 4 and sporadic Alzheimer's disease. Neurosci. Lett. **169**(1–2):175–178.
18. ROSES, A. D. 1994. The Alzheimer diseases. *In* Current Neurology. S. H. Appel, Ed. 111–141. Chicago: Mosby-Year Book.
19. ROSES, A. D. 1994. Apolipoprotein E affects the rate of Alzheimer disease expression: β-amyloid burden is a secondary consequence dependent on APOE genotype and duration of disease. J. Neuropathol. Exp. Neurol. **53**:429–437.

20. HAN, S.-H., G. EINSTEIN, K. H. WEISGRABER, *et al.* 1994. Apolipoprotein E is localized to the cytoplasm of human cortical neurons: A light and electron microscopic study. J. Neuropathol. Exp. Neurol. **53:**535–544.
21. HAN, S.-H., C. HULETTE, A. M. SAUNDERS, *et al.* 1994. Apolipoprotein E is present in hippocampal neurons without neurofibrillary tangles in Alzheimer's disease and age-matched controls. Exp. Neurol. **128:**13–26.
22. POIRIER, J., A. BACCICHET, D. DEA, *et al.* 1993. Cholesterol synthesis and lipoprotein reuptake during synaptic remodelling in hippocampus in adult rats. Neuroscience **55**(1)**:**81–90.
23. REBECK, G. W., J. S. REITER, D. K. STRICKLAND, *et al.* 1993. Apolipoprotein E in sporadic Alzheimer's disease: Allelic variation and receptor interactions. Neuron **11**(4)**:**575–580.
24. SAKAI, J., A. HOSHINO, S. TAKAHASHI, *et al.* Structure, chromosome location, and expression of the human very low density lipoprotein receptor gene. J. Biol. Chem. **269**(3)**:**2173–2182.
25. STRITTMATTER, W. J., A. M. SAUNDERS, M. GOEDERT, *et al.* 1994. Isoform-specific interactions of apolipoprotein E with tau and phosphorylated tau: Implications for Alzheimer disease. Proc. Natl. Acad. Sci. USA **91:**11183–11186.
26. HUANG, D., K. H. WEISGRABER, M. GOEDERT, *et al.* 1995. ApoE3 binding to tau tandem repeat I is abolished by tau $serine_{262}$ phosphorylation. Neurosci. Lett. **192:**209–212.
27. LEE, V. M., B. J. BALIN, L. J. OTVOS, *et al.* 1991. A68: A major subunit of paired helical filaments and derivatized forms of normal Tau. Science **251**(4994)**:**675–678.
28. MANDELKOW, E. M., J. BIERNAT, G. DREWES, *et al.* 1993. Microtubule-associated protein tau, paired helical filaments, and phosphorylation [Review]. Ann. N.Y. Acad. Sci. **695**(209)**:**209–216.
29. HARADA, A., K. OGUCHI, S. OKABE, *et al.* 1995. Altered microtubule organization in small-calibre axons of mice lacking tau protein. Nature **369**(6480)**:** 488–491.
30. MASLIAH, E., M. MALLORY, N. GE, *et al.* 1995. Neurodegeneration in the central nervous system of apoE-deficient mice. Exp. Neurol. In press.
31. EINSTEIN, G., R. BURANOSKY & B. J. CRAIN. 1994. Dendritic pathology of granule cells in Alzheimer's disease is unrelated to neuritic plaques. J. Neurosci. **14**(8)**:**5077–5088.
32. NICOLL, J. A. R., G. W. ROBERTS & D. I. GRAHAM. 1995. Apolipoprotein E e4 allele is associated with deposition of amyloid β-protein following head injury. Nature Med. **1:**135–137.
33. ROSES, A. D. & A. M. SAUNDERS. 1995. Head injury, amyloid β and Alzheimer's disease. Nature Med. **1:**603–604.
34. NEWMAN, M. F., N. D. CROUGHWELL, J. A. BLUMENTHAL, *et al.* 1995. Predictors of cognitive decline after cardiac operation. Ann. Thorac. Surg. **59:** 1326–1330.
35. ALBERTS, M. J., C. GRAFFAGNINO, C. MCCLENNEY, *et al.* 1995. ApoE genotype and survival from intracerebral hemorrhage. Lancet **346:**575.

Clinical and Neuropathological Correlates of Apolipoprotein E Genotype in Alzheimer's Disease

Window on Molecular Epidemiology[a]

B.T. HYMAN,[b] T. GOMEZ-ISLA, H. WEST, M. BRIGGS, H. CHUNG, J.H. GROWDON, AND G.W. REBECK

Neurology Service, Massachusetts General Hospital and Harvard Medical School Boston, Massachusetts 02114 USA

ABSTRACT: Inheritance of the apolipoprotein E (apo E) $\epsilon 4$ allele has recently been found to be associated with Alzheimer's disease. We have studied the clinical and neuropathological correlates of apolipoprotein E genotype in a large group of Alzheimer's patients. The primary influence on clinical presentation is a shift towards earlier age of onset in individuals who have the apo E $\epsilon 4$ gene: no change in clinical course was observed. In neuropathological studies, we find that the major influence of apo E $\epsilon 4$ is on increased Aβ deposition. These results led to a model of the biological interaction between the apo E protein and Alzheimer's disease.

INTRODUCTION

The recent observation that inheritance of the apolipoprotein E $\epsilon 4$ allele is a risk factor for Alzheimer's disease (AD) provides an opportunity to study subgroups of Alzheimer's patients based on their genetic predisposition to the disease. The general idea of examining phenotypes of individuals with various genetic predisposition to disease is referred to as molecular epidemiology and can be used to gain insight into the underlying biological processes that predispose toward disease. With this in mind, we have examined the clinical characteristics and neuropathological data from individuals with AD who had various apolipoprotein E genotypes.

Apolipoprotein E is inherited in the general population in 3 major alleles: apo E $\epsilon 2$, apo E $\epsilon 3$, and apo E $\epsilon 4$ (alleles) (apo E2, apo E3, apo E4 protein).

[a] Supported by the Massachusetts ADRC and by NIH grants AGO8487, AG12406, and AGO5134. HLW is a Howard Hughes Medical Institute Medical Student Fellow.

[b] *Send correspondence to:* Dr. Bradley T. Hyman, Neurology Service/WRN 408, Massachusetts General Hospital, Boston, MA 02114; Tel: (617) 726-8643; Fax: (617) 724-1480.

Apo E ϵ3 is the most common, accounting for approximately .78 of all alleles. Apo E ϵ4 is present in .14 of all alleles, and apo E ϵ2 in about .08. Because everyone inherits 2 alleles, there are 6 possible combinations: apo E ϵ4/4, ϵ3/4, ϵ2/4, ϵ3/3, ϵ2/3, and ϵ2/2. The most common combination in the general population is apo E ϵ3/3. Homozygosity for apo E ϵ2 (<1%) or apo E ϵ4 (2%) is unusual in the general population. Strittmatter and colleagues[1] demonstrated that, among late onset familial AD patients, the incidence of the apo E ϵ4 allele was markedly elevated. This observation was rapidly confirmed and generalized to include essentially all so-called sporadic Alzheimer's patients.[2-4] In addition, recent data suggest that inheritance of the apo E ϵ2 allele is under represented in AD populations, even after taking into account the over representation of the apo E ϵ4 allele.[5,6] The latter observation is of importance because it strongly suggests that the biological influence of the apo E ϵ4 effect is due to inheritance of the apo E4 protein, rather than due to a genetic disequilibrium implicating another (perhaps unknown) protein coded for by a gene adjacent to the apolipoprotein E locus on chromosome 19. These genetic observations have now been replicated by several dozen groups around the world (see Roses[7] for review). Taken together, these studies in hundreds of patients suggest that the apo E ϵ4 allele frequency is increased in AD about 3-fold, from 0.14 in the general population to about 0.38. The effect on the apo E ϵ4 homozygotes is even more marked. In the general population apo E ϵ4/4 accounts for about 1–2% of individuals, whereas in a typical Alzheimer's clinic this genotype is present in 15–18% of all patients.

These observations, derived from genetic analysis, strongly implicate a biological effect of apolipoprotein E in the Alzheimer's disease process. Initial data suggested that the apo E protein can bind to Aβ[1] and, in fact, apo E had previously been observed to be associated with senile plaques[8,9] by immunohistochemical techniques. Schmeckel *et al.*[10] and Rebeck *et al.*[3] examined large groups of AD brains and concluded that there was increased amyloid deposition in those individuals who had inherited apo E ϵ4. From a clinical perspective, the apo E ϵ4 allele appears to be associated with an earlier age of disease onset.[11–13] We have now evaluated the clinic population of the Massachusetts Alzheimer Disease Research Center to further refine both the clinical and neuropathological phenotypes associated with inheritance of apo E ϵ4.

CLINICAL AND NEUROPATHOLOGICAL CORRELATES OF APO E ϵ4

Over 200 individuals have been examined to date. In both our clinical population and in our autopsy-proven postmortem study series, the apo E

$\epsilon 4$ allele is markedly over represented in Alzheimer disease. In these groups the apo E $\epsilon 4$ allele frequency is 0.38, agreeing closely with our preliminary observations[3] as well as with multiple studies from around the world.[7]

In this large group of patients we also find that the age of onset of dementia decreases in accord with increasing numbers of apo E $\epsilon 4$ alleles. Individuals who are apo E $\epsilon 4/4$ have an age of onset of approximately 68 years whereas apo E $\epsilon 3/3$ individuals have an average age of onset of about 74, and heterozygote apo E $\epsilon 3/4$ individuals are intermediate. However, it is important to note that these are averages with rather broad distributions so that age of onset by itself is not sufficient to strongly predict genotype, and vice versa. We have also examined the overall rate of progression of individuals with the various genotypes. It is well known that individuals with AD progress at markedly differing rates, with some individuals very rapidly failing while others have a much more benign course. In fact, in our clinic population, approximately 20% of individuals show essentially no change in neuropsychometric and behavioral measures over a 12 month period, whereas about 20% have a much faster rate than average. Our initial hypothesis was that those individuals who had inherited the apo E $\epsilon 4$ allele would tend to have a faster progression of illness. However, comparing the rates of change on the Blessed dementia scale, a brief cognitive battery assessing information, memory and concentration, each subgroup progressed at the rate of 3-4 points per year with broad standard deviations regardless of genotype. Similarly, comparison of rate of progression as assessed by the activities of daily living scale (an assessment of functional decline) also revealed no differences among groups defined by genotype. A finer grain analysis, based on particular cognitive profiles, is currently underway.

We have also assessed whether the neuropathological profile of individuals with various genotypes differ. Our initial study[3] suggested increased amyloid burden (percent cortical surface covered by Aβ) in individuals with apo E $\epsilon 4/4$ (about 9%) greater than apo E $\epsilon 3/4$ (about 7%) greater than apo E $\epsilon 3/3$ (about 5%). We have now studied over 30 individuals, and this trend is maintained. Examination of the size distribution of senile plaques shows no difference across genotypes, suggesting that the effect is due primarily to increased numbers of successful nucleation sites for new plaques.[14] This is consistent with the pathological chaperone hypothesis of Wisniewski and Frangione[9] and the observation of increased amyloid fibril formation in the presence of apo E4.[15–17]

In collaboration with Dr. Vonsattel at Massachusetts General Hospital, we have recently examined the genotype of individuals with AD who have marked congophilic angiopathy. Congophilic angiopathy was rated on a scale of 0–3, with 0 being minimal vascular changes and 3 being severe. Approximately 80 cases were ranked, with nearly equal numbers in each group. Apo E genotype was then analyzed. We asked whether genotype would correlate with congophilic angiopathy score. Surprisingly, over 75% of those cases that were independently ranked as "severe" congophilic angiopathy had the apo E $\epsilon 4$ allele, whereas under 25% of those cases ranked "mild" had inherited the apo E $\epsilon 4$ allele. Thus, amyloid angiopathy also shows a striking difference among apo E genotypes.[18]

Strittmatter *et al.*[19] have shown that apo E protein interacts with tau protein in an isoform specific manner. This has led to the hypothesis that the relevant biological interaction of apo E protein is in stabilizing the cytoskeleton rather than in interacting with Aβ or APP in the extracellular domain. This is a controversial perspective, because apo E is known to be secreted from astrocytes and not synthesized by neurons. Thus, interactions of apo E with cytoskeletal elements within the cell would appear to be improbable. Nonetheless, some data from Roses's group does suggest that apo E is present intracellularly based on immunoelectron microscopy of neurons.[20] This hypothesis would imply that apo E $\epsilon 4$ allele individuals would have a predisposition towards increased tangle formation. We directly tested this idea by assessing the number of tangles in the superior temporal sulcus high order association cortex in individuals with Alzheimer disease of various genotypes. Our earlier studies had suggested that the number of tangles in this area increased monotonically with increasing severity and duration of illness, so that this confounding factor was taken into account in the analysis. Our current data suggest that there is no difference among individuals with different genotypes in terms of the number of tangles.[13]

Thus, our neuropathological studies suggest an association of apo E $\epsilon 4$ with increased amyloid burden, but no influence on neurofibrillary tangle formation.[13]

This fits well with the clinical studies noted above, in that neurofibrillary tangle formation does appear to be related to clinical progression of the illness.[21] These data imply that the biological effect of apo E is most likely an influence on amyloid deposition.

BIOLOGICAL ROLE OF APOLIPOPROTEIN E IN THE BRAIN

Apo E is a 299 amino acid protein with two major domains: a ligand binding domain that contains several amphipathic alpha helices and a receptor binding domain that interacts with 4 specific receptors. Apo E is known to be synthesized and secreted by astrocytes and upregulated following neuronal injury. It has been suggested that this is in part to clear the myelin breakdown products and other hydrophobic debris of neuronal damage (see Mahley[22] for review). It is also possible that apo E has a direct effect promoting neuronal regeneration and neurite outgrowth.[23] As noted above, apo E immunoreactivity is associated with senile plaques.[3,9,24] Some groups have also found apo E associated with neurofibrillary tangles.[19] Isoform specific interactions of apo E with Aβ have been reported,[1] but different groups have obtained diametrically opposing isoform specific interactions.[25]

We have also begun to study the potential biological roles of apo E in the central nervous system by examining the distribution of apo E receptors. There are four known apo E receptors. The first cloned was the LDL receptor, initially studied by Brown and Goldstein in the context of studying lipid metabolism. Three other receptors have been cloned based primarily on homology to this receptor. The first, the LDL receptor related protein (also cloned as the $\alpha 2$ macro globulin receptor) is referred to as LRP.[26] This multifunctional receptor is much larger than the LDL receptor and has multiple binding sites for at least 7 different ligands. The last two receptors, GP330 and VLDL-receptor (VLDL-r) are more recently cloned and appear biochemically to share many of the properties of the multifunctional LRP receptor rather than the LDL receptor (see Herz[27] for review).

We have found that all four receptors are expressed in brain. The LDL receptor is primarily on astrocytes although strong staining of the neuropil suggests that it is possibly present on neuronal processes as well. LRP is present on a subset of neurons, and on reactive but not resting astrocytes. GP330 appears not to be present in the cortex but has been reported to be present in choriod plexus. VLDL receptor is present primarily on microglial cells. These initial observations, carried out in collaboration with Dr. Dudley Strickland at the American Red Cross Holland Laboratory, suggest some of the complexity of the apo E receptor profiles in the brain. Different receptors appear on different cell types and appear to be under different regulatory influences.

Of particular interest, LRP is also present on senile plaques.[3] This may be a result of interactions with apo E that is bound on plaques or via other ligands. In fact, our recent studies[24] suggest that all 7 known LRP ligands are also senile plaque constituents. This diverse group of proteins includes $\alpha 2$ macroglobulin, urokinase plasminogen activator, tissue plasminogen activator, PAI-1 protease inhibitor, lactoferrin, lipoprotein lipase, and apo E. Thus, LRP by virtue of its presence on senile plaques and its association with a variety of plaque constituent proteins, and VLDL receptor, by virtue of its presence on microglia, appear to be most closely associated with Alzheimer pathology. We hypothesize that these receptors might interact with small aggregates of Aβ perhaps in association with apo E or other ligands. This would imply a complex set of interactions that could influence the relative likelihood of Aβ deposition and stabilization. How various apo E isoforms interact with these more recently discovered receptors remains largely unknown especially in the context of putative apo E/Aβ complexes. Nonetheless, because these receptors have been studied extensively with regard to their role in lipid metabolism, they provide an appealing target for therapeutic intervention.

ACKNOWLEDGMENTS

We thank Dr. Joseph Locasio for assistance with statistical analyses and Dr. Rudolph Tanzi for helpful discussion.

REFERENCES

1. STRITTMATTER, W. J., K. H. WEISGRABER, D. HUANG, L. M. DONG, G. S. SALVESEN, M. PERICAK-VANCE, D. SCHMECHEL, A. M. SAUNDERS, D. GOLDGABER & A. D. ROSES. 1993. Binding of human apolipoprotein E to βA4 peptide: Isoform specific effects and implications for late onset Alzheimer disease. Proc. Natl. Acad. Sci. USA **90:**8098–8102.
2. SAUNDERS, A. M., W. J. STRITTMATTER, D. SCHMECHEL, P. H. ST GEORGE-HYSLOP, M. A. PERICAK-VANCE, S. H. JOO, B. L. ROSI, J. F. GUSELLA, D. R. CRAPPER-MACLACHLAN, M. J. ALBERTS, C. HULETTE, B. CRAIN, D. GOLDGABER & A. D. ROSES. 1993. Association of apolipoprotein E allele $\epsilon 4$ with late-onset familial and sporadic Alzheimer's disease. Neurology **43:**1467–1472.
3. REBECK, G. W., J. S. REITER, D. K. STRICKLAND & B. T. HYMAN. 1993. Apolipoprotein E sporadic Alzheimer's disease: Allelic variation and receptor interactions. Neuron **11:**575–580.
4. POIRIER, J., J. DAVIGNON, D. BOUTHILLIER, S. KOGAN, P. BERTRAND & A. GAUTHIER. 1993. Apolipoprotein E polymorphism and Alzheimer's disease. Lancet **342:**697–699.
5. WEST, H. E., G. W. REBECK & B. T. HYMAN. 1994. Frequency of the apolipoprotein E $\epsilon 2$ allele is diminished in sporadic Alzheimer disease. Neurosci. Lett. **175:**46–48.
6. CORDER, E. H., A. M. SAUNDERS, N. J. RISCH, W. J. STRITTMATTER, D. E. SCHMECHEL, D. C. GASKELL, J. B. RIMMLER, P. A. LOCKE, M. CONNEALLY, K. E. SCHMADER, G. W. SMALL, A. D. ROSES, J. L. HAINES & M. A. PERICAK-VANCE. 1994. Apolipoprotein E type 2 allele decreases the risk of late onset Alzheimer disease. Nature Genetics **7:**180–184.
7. ROSES, A. D. 1994. Apolipoprotein E affects the rate of Alzheimer disease expression: β-amyloid burden is a secondary consequence dependent on APOE genotype and duration of disease. J. Neuropathol. Exp. Neurol. **53:**429–437.
8. NAMBA, Y., M. TOMONAGA, H. KAWASAKI, E. OTOWO & K. IKEDA. 1991. Apolipoprotein E immunoreactivity in cerebral amyloid deposits and neurofibrillary tangles in Alzheimers disease and kuru plaque amyloid in Greutzfeldt-Jahob disease. Brain Res. **541:**163–166.
9. WISNIEWSKI, T. & B. FRANGIONE. 1992. Apolipoprotein E: A pathological chaperone protein in patients with cerebral and systemic amyloid. Neurosci. Lett. **135:**235–238.
10. SCHMECHEL, D. E., A. M. SAUNDERS, W. J. STRITTMATTER, B. CRAIN, C. HULETTE, S. H. JOO, M. A. PERICACK-VANCE, D. GOLDGABER & A. D. ROSES. 1993. Increased amyloid β-peptide deposition as a consequence of apolipopro-

tein E genotype in late-onset Alzheimer's disease. Proc. Natl. Acad. Sci. USA **90:**9649–9653.

11. CORDER, E. H., A. M. SAUNDERS, W. J. STRITTMATTER, D. E. SCHMECHEL, P. C. GASKELL, G. W. SMALL, A. D. ROSES, J. L. HAINES & M. A. PERICAK-VANCE. 1993. Gene dose of apolipoprotein E type 4 allele and the risk of Alzheimer's disease in late onset families. Science **261:**921–923.
12. TSAI, M.-S., E. O. TANGALOS, R. C. PETERSEN, G. E. SMITH, D. J. SCHAID, E. KOKMEN, R. J. IVNEK & N. THIBODEAU. 1994. Apolipoprotein E: Risk factor for Alzheimer's disease. Am. J. Hum. Genet. **54:**643–649.
13. GOMEZ-ISLA, T., H. L. WEST, G. W. REBECK, S. D. HARR, J. H. GROWDON, J. T. LOCASIO, T. T. PERLS, L. A. LIPSITZ & B. T. HYMAN. 1995. Clinical and pathological correlates of apolipoprotein E $\epsilon 4$ in Alzheimer's disease. Ann. Neurol. In press.
14. HYMAN, B. T., H. L. WEST, G. W. REBECK, S. V. BULDYREV, R. N. MANTEGNA, M. UKLEJA, S. HAVLIN & H. E. STANLEY. 1995. Quantitative analysis of senile plaques in Alzheimer's disease: Observation of log-normal size distribution and molecular epidemiology of differences associated with apolipoprotein E genotype and trisomy 21 (Down syndrome). Proc. Natl. Acad. Sci. USA **92:** 3586–3590.
15. MA, Y., A. YEE, A. Y. H. BREWER, JR., S. DAS & H. POTTER. 1994. Amyloid-associated proteins α_1-antichymotrypsin and apolipoprotein E promote assembly of Alzheimer β-protein into filaments. Nature **372:**92–94.
16. SANAN, D. A., K. H. WEISGRABER, S. J. RUSSELL, R. W. MAHLEY, D. HUANG, A. SAUNDERS, D. SCHMECHEL, T. WISNIEWSKI, B. FRANGIONE, A. D. ROSES & W. J. STRITTMATTER. 1994. Apolipoprotein E associates with β amyloid peptide of Alzheimer's disease to form novel monofibrils. J. Clin. Invest. **94:** 860–869.
17. WISNIEWSKI, T., E. M. CASTANO, A. GOLABEK, T. VOGEL & B. FRANGIONE. 1994. Acceleration of Alzheimer's fibril formation by apolipoprotein E in vivo. Am. J. Pathol. **145:**1030–1035.
18. GREENBERG, S. M., G. W. REBECK, J. P. G. VONSATTEL, T. GOMEZ-ISLA & B. T. HYMAN. 1995. Apolipoprotein E $\epsilon 4$ and cerebral hemorrhage associated with amyloid angiopathy. Ann. Neurol. **38:**254–259.
19. STRITTMATTER, W. J., K. H. WEISGRABBER, M. GOEDERT, A. M. SAUNDERS, D. HUANG, E. H. CORDER, L.-M. DONG, R. JAKES, M. J. ALBERTS, J. R. GILBERT, S.-H. HAN, C. HULETTE, G. EINSTEIN, D. E. SCHMECHEL, M. A. PERICAK-VANCE & A. D. ROSES. 1994. Hypothesis: microtubule instability and paired helical filament formation in the Alzheimer disease brain are related to apolipoprotein E genotype. Exp. Neurol. **125:**163–171.
20. HAN, S.-H., G. EINSTEIN, K. H. WEISGRABER, W. J. STRITTMATTER, A. M. SAUNDERS, M. PERICAK-VANCE, A. D. ROSES & D. E. SCHMECHEL. 1994. Apolipoprotein E is localized to the cytoplasm of human cortical neurons: A light and electron microscopic study. J. Neuropathol. Exp. Neurol. **53:** 535–544.
21. ARRIAGADA, P. V., J. H. GROWDON, E. T. HEDLEY-WHYTE & B. T. HYMAN. 1992. Neurofibrillary tangles but not senile plaques parallel duration and severity of Alzheimer disease. Neurology **42:**631–639.

22. MAHLEY, R. W. 1988. Apolipoprotein E: Cholesterol transport protein with expanding role in cell biology. Science **240:**622–630.
23. NATHAN, B. P., S. BELLOSTA, D. A. SANAN, K. H. WEISGRABER, R. W. MAHLEY & R. E. PITAS. 1994. Differential effects of apolipoproteins E3 and E4 on neuronal growth in vitro. Science **264:**850–852.
24. REBECK, G. W., S. D. HARR, D. K. STRICKLAND & B. T. HYMAN. 1995. Multiple, diverse senile plaque-associated proteins are ligands of an apolipoprotein E receptor, the α_2-macroglobulin receptor/low-density-lipoprotein receptor-related protein. Ann Neurol. **36:**211–227.
25. LADU, M. J., M. T. FALDUTO, A. M. MANELLI, C. A. REARDON, G. S. GETZ & D. E. FRAIL. 1994. Isoform-specific binding of apolipoprotein E to β-amyloid. J. Biol. Chem. **269:**23403–23406.
26. STRICKLAND, D. K., J. D. ASHCOM, S. WILLIAMS, W. H. BURGESS, M. MIGLIORINI & W. S. ARGRAVES. 1990. Sequence identity between the α2-macroglobulin receptor and low density lipoprotein receptor-related protein suggests that this molecule is a multifunctional receptor. J. Biol. Chem. **265:**17401–17404.
27. HERZ, J. 1993. The LDL-receptor-related protein–portrait of a multifunctional receptor. Curr. Opinion Lipidol. **4:**107–113.

Apolipoprotein E, Plaques, Tangles and Cholinergic Dysfunction in Alzheimer's Disease[a]

U. BEFFERT AND J. POIRIER

Department of Neurology and Neurosurgery, McGill University, The McGill Centre for Studies in Aging, Douglas Hospital Research Centre, Verdun, PQ, Canada

ABSTRACT: Apolipoprotein E is a plasma cholesterol and phospholipid transporter which plays a central role in lipoprotein metabolism in the brain. Apolipoprotein E is a polymorphic protein with three common alleles in the general population, designated $\epsilon 2$, $\epsilon 3$ and $\epsilon 4$ coding for proteins ApoE2, ApoE3 and ApoE4, respectively. Recent findings have demonstrated a significant relationship between the $\epsilon 4$ allele and late onset familial and sporadic Alzheimer's disease. We examined several classical neuropathological hallmarks of Alzheimer's disease to determine whether they might be related to apolipoprotein E genotype: the presence of intracellular neurofibrillary tangles, extracellular senile plaques, and the attenuation of choline acetyltransferase activity. Significant correlations were found between $\epsilon 4$ allele copy number and senile plaque density in the frontal, parietal and fusiform cortical areas. Similarly, significant correlations were also found with increased neurofibrillary tangle number in the frontal and fusiform cortex. Interestingly, there was an inverse correlation between the $\epsilon 4$ allele with temporal cortical choline acetyltransferase activity. To further define the specific function of ApoE4, cultured rat hippocampal neurons were used to investigate interactions involving β-amyloid protein. In this model, ApoE4 (but not ApoE2) was able to reverse the neuroprotective effects of β-amyloid. ApoE3 was demonstrated to increase the internalization of β-amyloid peptide into these neurons. Taken together, these results support the involvement of ApoE4 in the pathogenesis of Alzheimer's disease and also provide some explanations for the possible function of this protein.

INTRODUCTION

Apolipoprotein E (ApoE) is a lipophilic plasma protein involved in cholesterol and phospholipid transport. ApoE synthesis occurs primarily in the liver,

[a] This work was sponsored by the Medical Research Council of Canada, the Fonds d'Aide aux Chercheurs à la Recherche, the Fonds de la Recherche en Santé du Québec, and the National Institute on Aging (J.P.).

[b] *Send correspondence to:* Dr. Judes Poirier, McGill Centre for Studies in Aging, 6875 LaSalle Blvd., Verdun, Quebec, Canada, H4H 1R3; Tel: (514) 762-3048 ext. 23936; Fax: (514) 762-3034.

as well as in macrophages and adrenals, but also in significant amounts in the brain.[1] In the central nervous system, the function of ApoE appears to involve the mobilization and redistribution of cholesterol and phospholipid during membrane remodeling associated with synaptic plasticity.[2–4] The absence of other major plasma apolipoproteins such as ApoA1 and ApoB further emphasizes the potential role of ApoE in the brain.[5]

Human ApoE is secreted as a 299 amino acid, 34 kD protein encoded by a four exon gene (3.6 kb in length) located on the long arm of chromosome 19. Three major ApoE isoforms (E2, E3, E4) exist due to polymorphisms at residues 112 and 158, resulting in six ApoE phenotypes in the population: E2/2, E3/3 and E4/4 homozygotes, and E2/3, E2/4 and E3/4 heterozygotes. Functionally, ApoE3 and E4 appear to have a higher affinity for the low density lipoprotein (LDL) receptor than does ApoE2. Furthermore, ApoE4 associated lipoproteins are cleared more efficiently from plasma than ApoE3 or ApoE2.[6]

Astrocytes are responsible for the synthesis and secretion of ApoE in the deafferented zone of the hippocampus following lesions of the entorhinal cortex.[3,4] Following the post-lesion peak of ApoE synthesis, a cell-specific induction of the ApoE/ApoB (LDL) receptor occurs in the granule cell layer of the dentate gyrus which closely correlates with the cholinergic synaptogenesis due to the loss of entorhinal cortex projections.[2] During this same period, hippocampal cholesterol synthesis decreases by more than 60% during the early phase of cholinergic reinnervation.[2] These compensatory changes following entorhinal cortex lesions indicate a role for ApoE as part of a cholinergic response to the loss of entorhinal cortex neurons, a process which is also observed in Alzheimer's disease (AD).[7]

Recently, the gene dose of ApoE4 was shown to be a risk factor for late onset familial[8] and sporadic forms of AD,[9] *i.e.*, as E4 allele number increases, the age of onset of AD decreases. Furthermore, ApoE immunoreactivity is associated with β-amyloid (Aβ) in extracellular senile plaques, intracellular neurofibrillary tangles, and at sites of cerebral vessel congophilic angiopathy in Alzheimer's disease brains.[10] The presence of ApoE in plaques and tangles also suggests a potential role for ApoE in the development of these lesions. The association of the ApoE4 isoform with AD further implies that the function of this protein may be important to the pathogenesis of the disorder.

To investigate the potential role of ApoE in the pathology of AD, we examined the degree of amyloid plaque formation, neurofibrillary tangles and choline acetyltransferase (ChAT) activity of ApoE-genotyped sporadic late-onset AD individuals. We also examined interactions between the major ApoE isoforms and the Aβ protein in a primary neuronal cell culture model.

METHODS

Case Selection and ApoE Genotyping

Frozen tissue from 84 autopsied cases of sporadic AD (35 females, 77.3 ± 8.7 years: 49 males, 76.1 ± 9.5) and from 14 controls (8 f, 71.6 ± 3.6 years: 6 m, 66.4 ± 3.1) were obtained from the Douglas Hospital Brain Bank in Verdun, Canada. The average postmortem delay was 17.2 ± 1.3 and 20.0 ± 4.6 hours for AD and control subjects, respectively. High molecular weight DNA for genotype analysis was isolated from frozen cerebellum or temporal cortex. ApoE genotype was determined by allele-specific extension of purified brain DNA.

Neuropathological Analyses

Neurofibrillary tangles (NFT) and senile plaques (SP) indices were determined as previously described.[11] The criteria used are consistent with those used in the classification of Khachaturian.[12]

Choline Acetyltransferase Activity

Hippocampal and temporal cortex tissue was homogenized and incubated for 15 min in buffer containing [^{14}C]-acetyl-CoA (NEN, MA) as previously described using the method of Tucek for determination of ChAT activity.[13]

Hippocampal Cultures and Lactate Dehydrogenase Activity

Neuronal cells form hippocampi of embroyonic day 18/19 Sprague-Dawley rats were dissociated and cultured in Dulbecco's Modified Eagle Medium supplemented with a growth supplement (B27; Gibco). Cells were plated at a density of approximately 200,000 cells/35 mm^2 well and cultured for six days prior to treatment and/or collection. Neurons were treated with combinations of ApoE2 and ApoE4 liposomes and Aβ protein for a period of 24 hours. Media from the treated neurons was used to assess cell viability via lactate dehydrogenase (LDH) activity which was measured via the Sigma kit sample start procedure.

Western Blot Analysis

Ten μg of hippocampal rat neuronal homogenate pretreated with neuraminidase (Pharmacia, Canada) was loaded on a SDS polyacrylamide gel. The

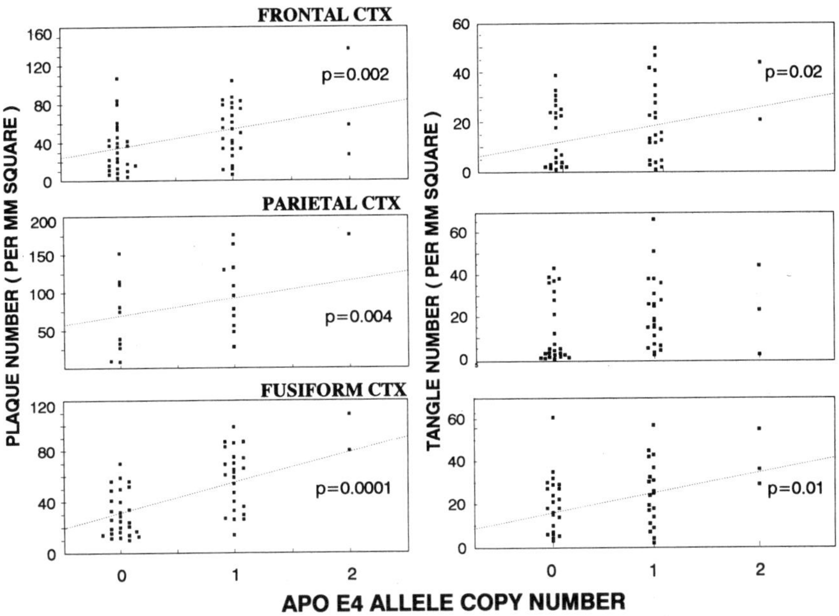

FIGURE 1. Senile plaque and neurofibrillary tangle number per mm square vs ApoE4 allele copy number as measured in three different cortical areas.

proteins were then transferred onto a nitrocellulose membrane. Detection was performed using a polyclonal antibody against fragment 1-42 of the beta amyloid peptide diluted 1:1000 (generous gift of Dr. D. Selkoe). Adsorption of the antibody with the purified peptide completely blocked the Aβ band at 4-6 kD. Visualization was performed using the enhanced chemiluminescence detection kit (Amersham, Canada). Quantification of autoradiographic signals was done using the MCID image analysis system (Ste. Catherine, Canada) equipped with 1D-gel analysis software.

Statistical analysis was performed using SYSTAT (Evanston, IL) statistical software Multivariate general linear models were used for the correlational analyses whereas statistical differences for ChAT values in the different genotypes were evaluated via Student's unpaired *t*-test.

RESULTS

In human, three microscopic fields of the cerebral cortex were chosen to determine senile plaque and neurofibrillary tangle number in control and AD patients. These values were then correlated with ϵ4 allele copy number (FIG. 1). The number of senile plaques and neurofibrillary tangles increased with

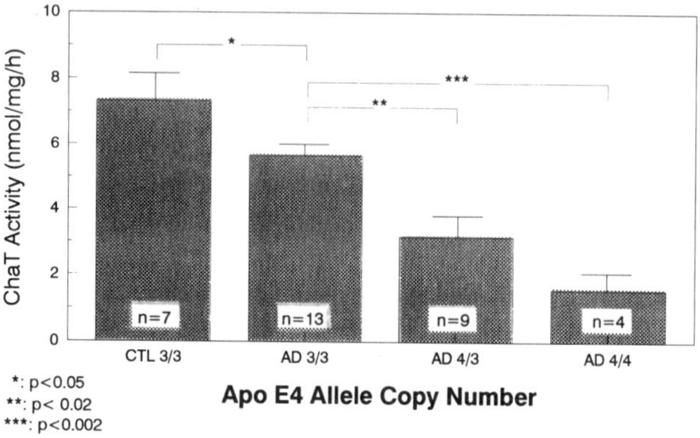

FIGURE 2. Choline acetyltransferase activity as measured in the temporal cortex plotted against ApoE4 allele copy number of control and AD individuals.

the number of ϵ4 allele. FIGURE 1 (*left*) demonstrates that as ϵ4 allele copy number increases, the density of senile plaques in the frontal (fc), parietal (pc) and fusiform cortices (fu) increases ($p = 0.002$ for fc, $p = 0.004$ for pc, and $p = 0.0001$ for fu). Similar results were obtained for the density of neurofibrillary tangles in these same brain regions, although the correlation was not as significant (FIG. 1-*right*: $p = 0.02$ for fc and $p = 0.01$ for fu).

Another well-established hallmark of AD, choline acetyltransferase activity, was measured in the temporal cortex and examined in relation to ApoE genotype. FIGURE 2 demonstrates a significant decrease in ChAT activity with respect to increasing ϵ4 allele copy number of age-matched control and AD individuals. Pronounced reductions in ChAT activity was observed in ApoE4 carriers as compared to controls. In the temporal cortex, ChAT values of 7.48 ± 0.74 (control ApoE3/3), 5.65 ± 0.33 (AD ApoE3/3), 2.91 ± 0.66 (AD ApoE4/3) and 1.56 ± 0.47 (AD ApoE4/4) nmol Ach/mg protein/h were obtained. The control ApoE3/3 group was shown to be significantly different from all the AD groups whereas AD ApoE3/3 ChAT levels were also significantly higher than both the AD ApoE3/4 ($p < 0.002$) and the ApoE4/4 group ($p < 0.02$).

In rat primary hippocampal neurons, lactate dehydrogenase activity was used to evaluate the effect of Aβ on cell viability in the presence of different ApoE isoforms (FIG. 3). Aβ (1 μg/ml) significantly decreased the amount of LDH activity compared to controls ($p = 0.004$). In the presence of ApoE2 micelles (1 μg/ml), Aβ was also able to inhibit LDH activity ($p = 0.01$). ApoE4 micelles (1 μg/ml) caused the opposite effect though in the presence of Aβ, increasing LDH activity significantly versus control values ($p = 0.02$).

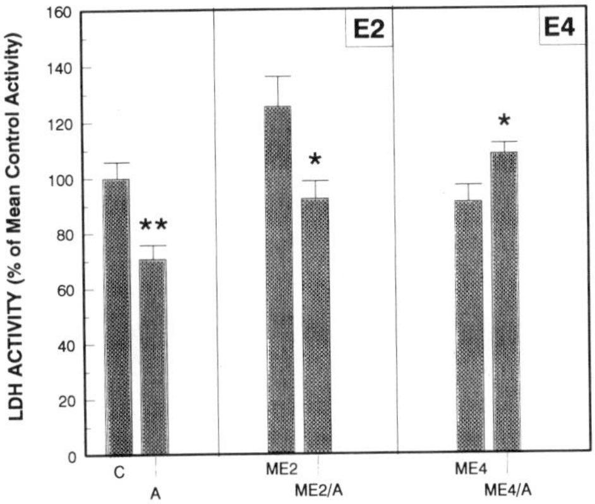

FIGURE 3. Lactate dehydrogenase activity toxicity assay using primary hippocampal neurons from rat. Key: C = control, A = 1 μg/ml β-amyloid 1-42, ME2 and ME4 = 1 μg/ml Apolipoprotein E2 and E4 micelles (liposomes).

Since ApoE was previously shown to bind synthetic Aβ, we examined via Western blot (for Aβ protein) the potential role of ApoE to facilitate the internalization of Aβ into neuronal cells (FIG. 4). Using primary hippocampal neurons from rat, ApoE3 demonstrated a significant increase of Aβ internalization (lanes 1 through 4) as compared to either Aβ or ApoE3 alone (as represented by cell homogenate).

DISCUSSION

The apolipoprotein E ε4 allele has been shown to be genetically associated with the late onset familial and sporadic forms of Alzheimer's disease.[9] Risk for AD increases significantly with increasing ApoE4 gene dose. Histologically identified senile plaque and neurofibrillary tangle density were both found to increase in several regions of the cerebral cortex of sporadic late-onset AD individuals as a function of ApoE4 gene dose. A similar finding which examined the hippocampus and middle frontal gyrus of the cerebral cortex of patients with AD reported a significant association of Aβ deposition and the dosage of ApoE4.[14] Our results imply that the E4 isoform of apolipoprotein E may function to accelerate the formation of senile plaques (perhaps via increased sequestration of Aβ) and neurofibrillary tangles, two classical hallmarks of AD.

Loss of cholinergic neurons and choline acetyltransferase activity are also well established hallmarks of AD. We observed a significant reduction of ChAT activity in the temporal cortex of control and AD individuals to be associated with ε4 copy number (*i.e.*, as ε4 allele copy number increased, ChAT activity decreased). The reduction of ChAT activity could be secondary

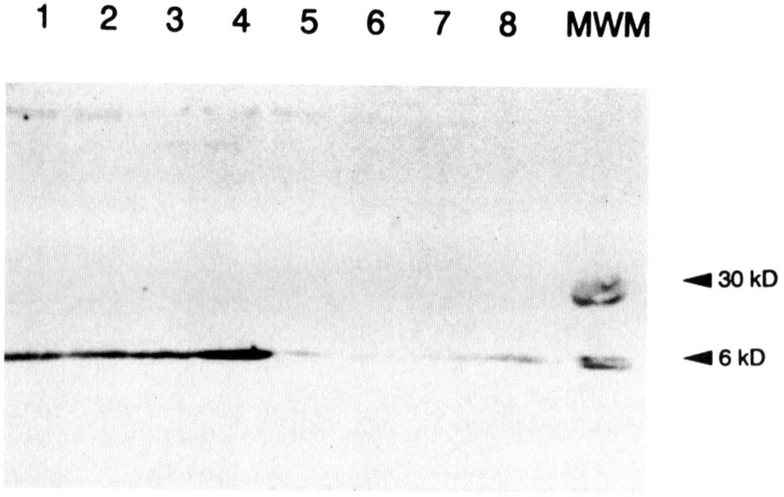

1: Apo E - Liposomes (10 ug/mL) + Amyloid (1 ug/mL)
2: Apo E - Liposomes (5 ug/mL) + Amyloid (1 ug/mL)
3: Apo E - Liposomes (3 ug/mL) + Amyloid (1 ug/mL)
4: Apo E - Liposomes (1 ug/mL) + Amyloid (1 ug/mL)
5: Liposomes
6: Apo E - Liposomes (1 ug/mL)
7: Amyloid (1-42, 1 ug/mL)
8: Apo E (1 ug/mL)
MWM: Molecular Weight Markers

FIGURE 4. β-amyloid Western blot of 10 μg cell homogenate (representing internalized portion) of rat primary hippocampal neurons treated with ApoE3 liposomes (micelles) and β-amyloid in the indicated concentrations.

to the loss of cholinergic neurons in the basal forebrain and may be interpreted as evidence for poor compensation of cholinergic fibers in ApoE4 individuals.

Further evidence of an altered function for the ApoE4 protein was demonstrated in an in vitro cell culture model. When hippocampal neurons in culture were exposed to physiological concentrations of Aβ protein, LDH activity decreased (*i.e.*, cell viability increased), supporting the reported neurotrophic role for the peptide at this concentration (1 μg/ml).[15] This effect was abolished in the presence of ApoE4 but not ApoE2 reconstituted liposomes, suggesting that a combination of ApoE4 and Aβ leads to increased neuronal toxicity while ApoE2 in the presence of Aβ actually increases cell viability over control values. Thus, a critical role for the ApoE4 protein in AD may be a direct interaction with Aβ which leads to decreased amyloid trophicity, while ApoE2 may afford a similar interaction but with opposite, neurotrophic results.

A direct interaction of ApoE with Aβ is further substantiated by the increased amount of amyloid staining shown by Western blot of neuronal cultures in the presence of these two proteins. In the absence of ApoE3 liposomes, little or no Aβ is internalized, while the presence of ApoE3 significantly increases the internalization of Aβ. Apolipoprotein E is thus capable of not only binding and interacting with Aβ, but also of internalizing Aβ. Increased internalization of Aβ by ApoE may lead to increased storage of amyloid and eventually to increased cell death and senile plaque formation.

In summary, we report possible functions of apolipoprotein E in regard to the etiology of Alzheimer's disease. We extend clear evidence for a genetic link between ApoE4 to several established hallmarks of AD. Also, increased toxicity due to the interaction of the ApoE4 isoform with Aβ may offer an explanation for this allele being associated with AD.

REFERENCES

1. Elshourbagy, N. A., W. S. Liao, R. W. Mahley & J. M. Taylor. 1985. Apolipoprotein E mRNA is abundant in the brain and adrenals, as well as in the liver, and is present in other peripheral tissues of rats and marmosets. Proc. Natl. Acad. Sci. USA **82:**203–207.
2. Poirier, J., A. Baccichet, D. Dea & S. Gauthier. 1993. Cholesterol synthesis and lipoprotein reuptake during synaptic remodelling in hippocampus in adult rats. Neuroscience **55:**81–90.
3. Poirier, J., M. Hess, P. C. May & C. E. Finch. 1991. Astrocytic apolipoprotein E mRNA and GFAP mRNA in hippocampus after entorhinal cortex lesioning. Brain Res. Mol. Brain. Res. **11:**97–106.
4. Poirier, J., M. Hess, P. C. May & C. E. Finch. 1991. Cloning of hippocampal poly(A) RNA sequences that increase after entorhinal cortex lesion in adult rat. Brain Res. Mol. Brain Res. **9:**191–195.
5. Roheim, P. S., M. Carey, T. Forte & G. L. Vega. 1979. Apolipoproteins in human cerebrospinal fluid, Proc. Natl. Acad. Sci. USA **76:**4646–4649.
6. Weisgraber, K. H., T. L. Innerarity & R. W. Mahley. 1982. Abnormal lipoprotein receptor-binding activity of the human E apoprotein due to cysteine-arginine interchange at a single site. J. Biol. Chem. **257:**2518–2521.
7. Hyman, B. T., G. W. Van Horsen, A. R. Damasio & C. L. Barnes. 1984. Alzheimer's disease: Cell-specific pathology isolates the hippocampal formation. Science **225:**1168–1170.
8. Corder, E. H., A. M. Saunders, W. J. Strittmatter, D. E. Schmechel, P. C. Gaskell, G. W. Small, A. D. Roses, J. L. Haines & M. A. Pericak-Vance. 1993. Gene dose of apolipoprotein E type 4 allele and the risk of Alzheimer's disease in late onset families. Science **261:**921–923.
9. Poirier, J., J. Davignon, D. Bouthillier, S. Kogan, P. Bertrand & S. Gauthier. 1993. Apolipoprotein E polymorphism and Alzheimer's disease. Lancet **342:**697–699.
10. Namba, Y., M. Tomonaga, H. Kawasaki, E. Otomo & K. Ikeda. 1991.

Apolipoprotein E immunoreactivity in cerebral amyloid deposits and neurofibrillary tangles in Alzheimer's disease and kuru plaque amyloid in Creutzfeldt-Jakob disease. Brain Res. **541**:163–166.

11. ETIENNE, P. T., ROBITAILLE, P. WOOD, S. GAUTHIER, N. P. NAIR & R. QUIRION. 1986. Nucleus basalis neuronal loss, neuritic plaques and choline acetyltransferase activity in advanced Alzheimer's disease. Neuroscience **19:** 1279–1291.
12. KHACHATURIAN, Z. S. 1985. Diagnosis of Alzheimer's disease. Arch. Neurol. **42**:1097–1105.
13. TUCEK, S. 1978. Choline acetyltransferase. *In* Acetylcholine Synthesis in Neurons: 29–34. London: Chapman and Hall Ltd.
14. SCHMECHEL, D. E., A. M. SAUNDERS, W. J. STRITTMATTER, B. J. CRAIN, C. M. HULETTE, S. H. JOO, M. A. PERICAK-VANCE, D. GOLDGABER & A. D. ROSES. 1993. Increased amyloid beta-peptide deposition in cerebral cortex as a consequence of apolipoprotein E genotype in late-onset Alzheimer disease, Proc. Natl. Acad. Sci. USA **90**:9649–9653.
15. KOO, E. H., L. PARK & D. J. SELKOE. 1993. Amyloid beta-protein as a substrate interacts with extracellular matrix to promote neurite outgrowth. Proc. Natl. Acad. Sci. USA **90**:4748–4752.

Regulation of APP Processing

Potential for the Therapeutical Reduction of Brain Amyloid Burden[a]

R. M. NITSCH,[b–d] R. J. WURTMAN,[b] AND J. H. GROWDON[c]

[b] *Department of Brain and Cognitive Sciences, Massachusetts Institute of Technology, Cambridge, Massachusetts 02139 USA*

[c] *Department of Neurology, Massachusetts General Hospital, Harvard Medical School, Boston, Massachusetts 02114 USA*

ABSTRACT: The role of brain amyloid in the pathogenesis of Alzheimer's disease (AD) is discussed controversially, but combined genetic and biochemical evidence points to a central role of the gene encoding the amyloid precursor APP in at least some forms of AD. This article proposes that preventing brain amyloid formation is a rational concept for drug treatment of AD. We suggest that pharmacologically active ligands for specific cell surface receptor subtypes—normally stimulated by neurotransmitters, growth factors, and cytokines—constitute a class of chemicals that might be useful to accelerate processing of APP into non-amyloidogenic, and biologically active, derivatives. This class of agents includes muscarinic m1 and m3 agonists, serotoninergic 5-HT$_{2a}$ and 5-HT$_{2c}$ agonists, glutamatergic mGluR1 agonists, as well as agonists for bradykinin and vasopressin receptors.

INTRODUCTION

Rational therapy of disease is targeted at correcting, or preventing, pathology based upon the understanding of the biology of molecules involved in the pathological characteristics of the disease. Alzheimer's disease (AD) is a syndrome with multifactorial, and possibly multigenic, etiologies that cause a common clinicopathological phenotype characterized by progressive dementia, amyloid deposition, tangle formation, and neuronal degeneration. The significance of amyloid in the pathogenesis of AD is discussed controversially but it is widely accepted that formation of brain amyloid is an early

[a] Funded by grants from NIA, NIMH, BMFT, the Center for Brain Sciences and Metabolism, and the Hoffman Fellowship in Alzheimer's Disease at MGH.

[d] *Send correspondence to:* Dr. Roger M. Nitsch, Center for Molecular Neurobiology, University of Hamburg, UKE 22, Martinistr. 52, 20246 Hamburg, Germany; Tel: 49-40-4717-4746; Fax: 49-40-4717-4774. E-mail nitsch@zuinh.zuinh.uni-hamburg.de.

and prominent feature of a vast majority of, if not all, clinically diagnosed AD cases.[1–3]

APP GENE MUTATIONS LINKED TO THE CLINICOPATHOLOGICAL PHENOTYPE OF AD CAUSE MISPROCESSING OF APP

For selected etiologic subgroups of AD, combined genetic and biochemical evidence points to a direct and causal role the gene encoding the biological amyloid precursor APP, in the cascade of pathological events. An increasing number of mutations within the APP gene is linked to the familial phenotype of AD. Notably, these mutations are clustered within, or close to, the region encoding APP's β-amyloid (Aβ) domain. In tissue culture systems, some of these mutations produce significant changes in proteolytic processing patterns of APP. As an example, the double mutation at the N-terminus of the Aβ domain linked to the Swedish variant of familial AD causes 5– to 8–fold increases in the rate of Aβ production.[4–6] Because the kinetics of aggregation of soluble Aβ into fibrillar amyloid are a function of Aβ concentration,[7] it is possible that increased Aβ production leads to increased amyloid formation (provided that clearance of Aβ remains unchanged). The missense mutations at codon 717 of APP close to the C-terminus of the Aβ domain cause the formation of longer, C-terminally extended forms of Aβ.[8] Such extended forms of Aβ aggregate more readily as compared to the shorter forms derived from the wild-type APP sequence, and they can further accelerate fibrillization of shorter forms of Aβ.[9] It is thus possible that the 717 mutations are associated with increased amyloid formation. Additional Aβ species with intermediate lengths are generated as a result of a point mutation at codon 692 within the Aβ domain.[10] This mutation causes severe congophilic amyloid angiopathy, cerebral hemorrhage, and premature death in humans.[11] Together, these initial observations indicate that disease-causing mutations at strategic positions within, or close to, the Aβ domain of APP can be associated with increases in the quantity, and changes in the quality, of Aβ molecules that are pathologically relevant for amyloid formation.

Besides mutations of the APP gene, markers on chromosomes 19, 14, and 1 have been linked to the familial AD phenotype. The genes accounting for linkage are identified[27,28] and it remains a central question whether their products bear a relationship to the mechanisms involved in the expression or processing of APP, or in the removal of amyloidogenic APP derivatives. Preliminary tissue culture experiments suggest some of them do: Fibroblasts obtained from patients with familial AD linked to chromosome 14 produced 3 to 7 times as much Aβ per APP synthesis rate as fibroblasts obtained from unaffected family members.[12]

Combined, the genetic and the biochemical data available to date strongly suggest that abnormal structure of the APP gene can be sufficient to produce both the clinicopathological phenotype of AD, and alterations in proteolytic APP processing resulting in quantitatively or qualitatively altered Aβ formation.

PROCESSING OF APP INTO AMYLOIDOGENIC DERIVATIVES MAY IMPAIR ITS NORMAL FUNCTION

An additional, if hypothetical, dimension of APP misprocessing in brain could be associated with impaired normal biological function of APP. Although not fully understood, initial evidence suggests developmental regulation and correlation with synaptogenesis,[13] potent roles in neurite outgrowth,[14,15] and functions in cell adhesion.[16] Additionally, APP^S can prevent neuronal damage induced by excess glutamate, or by glucose deprivation.[17] Targeted disruption of exon 2 of the APP gene in mice results in lower levels of a shortened molecule that lacks part of the ectodomain, and is associated with a phenotype characterized by impaired learning and memory behavior.[18]

Tissue culture studies on the regulation of secretory APP processing have revealed a reciprocal relationship between the formation of the α-secretase product APP^S and that of Aβ, indicating that production of more Aβ is paralleled by reduced amounts of secreted APP^S.[19] Provided that mismetabolism of APP in brain is associated with physiologically relevant decreases in the production of biologically active APP derivatives, impaired APP function might contribute to cellular malfunction in AD.

AMYLOID AS A TARGET FOR RATIONAL DRUG THERAPY OF ALZHEIMER'S DISEASE

Regardless of whether amyloid formation or APP mismetabolism occurs in the beginning, during, or at the end of the pathological cascade that leads to neurodegeneration—or during all stages—it provides a well-characterized target for drug therapy. If the "amyloid cascade hypothesis" turns out to be correct, therapies aimed at preventing amyloid formation will be obviously useful, because the disease-causing agent is being removed from brain cells which may benefit from reduced amyloid formation. Should it turn out that brain amyloid is an epiphenomenon that bears little importance for the etiology of AD, it will still remain to be tested whether preventing the formation of the pathognomic structures can affect the clinical phenotype of the syndrome. Thus, it appears useful to develop and to test treatments that are designed to lower brain amyloid burden even before a complete understanding of the multifactorial etiology of AD has emerged.

POSSIBLE ROUTES TO LOWER BRAIN AMYLOID BURDEN

It is proposed that brain amyloid burden can be lowered by the following options:

1. Inhibition of Aβ production. Aβ is generated from APP by two proteolytic cleavage events within the ecto- and transmembrane domains. Inhibition of the enzymes involved in these cleavages likely reduces brain concentrations of Aβ and possibly slow the formation of amyloid. Primary targets for such protease inhibitors are the β- and γ-secretase activities. Inhibition of their putative activators provides an additional possibility to reduce their activities.
2. Simulation of non-amyloidogenic APP secretase activities. Non-amyloidogenic APP processing including the formation of APP^S by α-secretase processing most likely precludes the formation of amyloid from APP.
3. Acceleration of the clearance or the degradation of soluble Aβ peptides.
4. Prevention of the aggregation of soluble Aβ peptides into amyloid fibrils.
5. Inhibition of such potential amyloidotrophic factors as apolipoprotein E, α1-antichymotrypsin, zinc, and free oxygen radicals.
6. Removal of amyloid plaques.

Of the above, only the first two possibilities aim at reducing the formation of amyloidogenic APP derivatives.

CELL-SURFACE RECEPTORS CAN INCREASE APP^S SECRETION

Although the identity of α-secretase is unknown, much is known about the factors that regulate its activity. Early studies have shown that phorbol esters accelerate α-secretase processing,[20] and subsequent experiments revealed that cell-surface receptors coupled to G protein- and phospholipase-signaling readily stimulate α-secretase activity, and cause a rapid increase in the formation of secreted APP^S.[21] Receptor molecules that regulate APP processing in this manner include muscarinic m1 and m3,[22] serotoninergic 5-HT_{2a} and 5-HT_{2c}, glutamatergic mGluR1, and peptidergic receptors for

TABLE 1. First Messengers and Receptors Known to Increase APP^S Secretion

First Messenger	Receptor Subtype	APP^S Secretion
Acetylcholine	m1, m3	increase
Serotonin	$5\text{-}HT_{2a}$, $5\text{-}HT_{2c}$	increase
Glutamate	mGluR1	increase
Bradykinin		increase
Vasopressin		increase
NGF		increase
EGF		increase
Interleukin 1		increase
Thrombin		increase

bradykinin and vasopressin (TABLE 1). In addition to cell-surface receptors coupled to G protein signaling, several tyrosine kinase-coupled receptors can rapidly accelerate the formation and secretion of APP^S. For instance, in PC-12 cells, NGF and EGF stimulate APP^S secretion within minutes. Moreover, such cytokines as IL-1 also increase secretory APP processing.[23]

This preliminary inventory suggests that APP processing in brain may be regulated by many different neurotransmitter systems. In intact brain tissue slices, release of endogenous neurotransmitters induced by electrical depolarization is paralleled by a tetrodotoxon-sensitive and frequency-dependent increase in APP^S secretion.[24] Together, these observations support the idea that the regulation of APP processing in brain cells is coupled to their electrical activity, conceivably via first messengers, cell surface receptors and transmembrane signaling. Similar stimulatory effects on APP^S secretion may occur in many different cell types expressing distinct surface receptor molecules. Although the effects of neuronal stimulation were rather specific with respect to the release of other transmembrane and cytosolic proteins, our current data indicate that the processing of other members of the larger family of APP-like proteins (APLP) can also be regulated by the activity of cell-surface receptors. In particular, cleavage and secretion of APLP2 is readily activated by serotoninergic receptor subtypes and by electrical depolarization. It is likely that future studies will reveal even more cell surface receptor types with functions in the regulation of proteolytic processing of members of the APP gene family.

ACTIVATION OF CELL-SURFACE RECEPTORS CAN REDUCE Aβ FORMATION

It is widely accepted that α-secretase processing precludes the formation of amyloidogenic derivatives from APP. Although the ~3 kD peptide generated from the cell-associated ~10 kD C-terminal α-secretase product (p3; Aβ17-40; ref.26) can possibly form amyloid *in vitro,* the vast majority of Aβ molecules found in brain amyloid begin at its N-terminus around residue Asp1, suggesting that p3 does not contribute to amyloid formation in brain, perhaps because it is more efficiently removed than Aβ. Thus, activation of α-secretase appears to be a useful possibility to reduce amyloid formation in brain. To address directly whether activation of α-secretase reduces Aβ formation, we co-transfected cells with both muscarinic m1 receptors known to stimulate α-secretase, and with APP to overproduce Aβ. Stimulation of the m1 receptors with carbachol decreased Aβ secretion to approximately 40% of the unstimulated baseline levels.[19] These data indicate that activation of α-secretase processing can be associated with decreases in Aβ formation, and they suggest the possible consequence of reduced rates of amyloid formation. Activation of m1 receptors also reduced the abnormally high production of Aβ caused by the Swedish double mutation, suggesting that processing of both normal APP and APP with this mutation can be regulated by muscarinic m1 receptor activation.[19]

Future clinical studies will show whether treatments designed to reduce Aβ formation can reduce brain amyloid burden, slow neurodegeneration, and halt progression of the disease.

REFERENCES

1. HYMAN, B. T. & R. D. TERRY. 1994. Apolipoprotein E, *Aβ,* and Alzheimer's disease. An editorial comment. J. Neuropathol. Exp. Neurol. **53:**427–428.
2. ROSES, A. D. 1994. Apolipoprotein E affects the rate of Alzheimer's disease expression: β-amyloid burden is a secondary consequence dependent on APOE genotype and duration of disease. J. Neuropathol. Exp. Neurol. **53:**429–437.
3. SELKOE, D. J. 1994. Alzheimer's disease: A central role for amyloid. J. Neuropathol. Exp. Neurol. **53:**438–447.
4. CITRON, M., T. OLTERSDORF, C. HAASS, *et al.* 1992. Mutation of the β-amyloid precursor protein in familial Alzheimer's disease increases β-protein production. Nature **360:**672–674.
5. CAI, X.-D., T. E. GOLDE & S. G. YOUNKIN. 1993. Release of excess amyloid β-protein from a mutant amyloid β-protein precursor. Science **259:**514–516.
6. CITRON, M., C. VIGO-PELFREY, D. B. TEPLOW, *et al.* 1994. Excessive production of amyloid β-protein by peripheral cells of symptomatic and presymptomatic patients carrying the Swedish familial Alzheimer's disease mutation. Proc. Natl. Acad. Sci. USA **91:**11993–11997.
7. JARRETT, J. T. & P. T. LANSBURY, JR. 1993. Seeding "one-dimensional crystallization" of amyloid: A pathogenic mechanism in Alzheimer's disease and scrapie. Cell **73:**1055–1058.
8. SUZUKI, N., T. T. CHEUNG, X.-D. CAI, *et al.* 1994. An increased percentage

of long amyloid β protein secreted by familial amyloid β protein precursor (βAPP717) mutants. Science **264:**1336–1340.
9. JARRETT, J. T., E. P. BERGER & P. T. LANDSBURY, JR. 1993. The carboxy-terminus of the beta amyloid protein is critical for the seeding of amyloid formation: Implications for the pathogenesis of Alzheimer's disease. Biochemistry **32:** 4693–4697.
10. HAASS, C., A. Y. HUNG, D. J. SELKOE & D. B. TEPLOW. 1994. Mutations associated with a locus for familial Alzheimer's disease result in alternative processing of amyloid β-protein precursor. J. Biol. Chem. **269:**1–8.
11. LEVY, E., M. D. CARMAN, I. J. FERNANDEZ-MARID, *et al.* 1990. Mutation of Alzheimer's disease amyloid gene in hereditary cerebral hemorrhage, Dutch-type. Science **248:**1124–1126.
12. SUZUKI, N., T. T. CHEUNG, X.-D. CAI, *et al.* 1994. Factors affecting production of secreted amyloid β-protein (Aβ) ending at Aβ42,43. Soc. Neuroscience Abstr. **20:**437.
13. MOYA, K. L., L. I. BENOWITZ, G. E. SCHNEIDER & B. ALLINQUANT. 1994. The amyloid precursor protein is developmentally regulated and correlated with synaptogenesis. Dev. Biol. **161:**597–603.
14. MILWARD, E. A., R. PAPADOPOULOS, S. J. FULLER, *et al.* 1992. The amyloid protein precursor of Alzheimer's disease is a mediator of the effects of nerve growth factor on neurite outgrowth. Neuron **9:**129–137.
15. KOO, E. H., L. PARK & D. J. SELKOE. 1993. Amyloid β-protein as a substrate interacts with extracellular matrix to promote neurite outgrowth. Proc. Natl. Acad. Sci. USA **90:**4748–4752.
16. SCHUBERT, D., L.-W. JIN, T. SAITOH & G. COLE. 1989. The regulation of amyloid β protein precursor secretion and its modulatory role in cell adhesion. Neuron **3:**689–694.
17. MATTSON, M., B. CHENG, A. CULWELL, F. ESCH, I. LIEBERBURG & R. RYDEL. 1993. Evidence for excitoprotective and intraneuronal calcium-regulating roles for secreted forms of the β-amyloid precursor protein. Neuron **10:**243–254.
18. MÜLLER, U., N. CRISTINA, Z.-W. LI, *et al.* 1994. Behavioral and anatomical deficits in mice homologous for a modified β-amyloid precursor protein gene. Cell **79:**755–765.
19. HUNG, A. Y., C. HAASS, R. M. NITSCH, *et al.* 1993. Activation of protein kinase C inhibits cellular production of the amyloid β-protein. J. Biol. Chem. **268:**22959–22962.
20. BUXBAUM, J. D., S. E. GANDY, P. CICCHETTI, *et al.* 1990. Processing of Alzheimer β/A4 amyloid precursor protein: modulation by agents that regulate protein phosphorylation. Proc. Natl. Acad. Sci. USA **87:**6003–6006.
21. NITSCH, R. M. & J. H. GROWDON. 1994. Role of neurotransmission in the regulation of amyloid β-protein precursor processing. Biochem. Pharmacol. **47:** 1275–1284.
22. NITSCH, R. M., B. E. SLACK, R. J. WURTMAN & J. H. GROWDON. 1992. Release of Alzheimer amyloid precursor derivatives stimulated by activation of muscarinic acetylcholine receptors. Science **258:**304–307.
23. BUXBAUM, J. D., M. OISHI, H. I. CHEN, *et al.* 1992. Cholinergic agonists and

interleukin 1 regulate processing and secretion of the Alzheimer β/A4 amyloid protein precursor. Proc. Natl. Acad. Sci. USA **89:**10075–10078.

24. NITSCH, R. M., S. A. FARBER, J. H. GROWDON & R. J. WURTMAN. 1993. Release of amyloid β-protein precursor derivatives from hippocampal slices by electrical depolarization. Proc. Natl. Acad. Sci. USA **90:**5191–5193.
25. FARBER, S. A., J. G. SCHULZ, R. J. WURTMAN & R. M. NITSCH. 1995. Regulation of amyloid β-protein precursor secretion in rat brain. J. Neurosci. In press.
26. HAASS, C., A. Y. HUNG, M. SCHLOSSMACHER, D. B. TEPLOW & D. J. SELKOE. 1993. β-amyloid peptide and a 3-kDa fragment are derived by distinct cellular mechanisms. J. Biol. Chem. **268:**3021–3024.
27. SHERRINGTON, R., E.I. ROGAEV, Y. LIANG, *et al.* 1995. Cloning of a gene bearing missense mutations in early-onset familial Alzheimer's disease. Nature **375:**754–760.
28. LEVY-LAHAD, E., W. WASCO, P. POORKAJ, *et al.* 1995. Candidate gene for the chromosome 1 familial Alzheimer's disease locus. Science **269:**973–977.

A Novel Brain Cysteine Protease Forms an SDS Stable Complex with the β-Amyloid Precursor Protein[a]

TIEN CHANG[b] AND CARMELA R. ABRAHAM[b–d]

[b] *Department of Medicine and* [c]*Department of Biochemistry Boston University School of Medicine, Boston Massachusetts 02118 USA*

ABSTRACT: Alzheimer's disease (AD) brain accumulates β-protein (Aβ) a peptide proteolytically derived from the β-amyloid precursor protein (APP). The abnormal production and aggregation of Aβ have been implicated in the pathogenesis of the disease. The mechanism of production of Aβ *in vivo* is not yet clear; but endoproteases capable of degrading APP are likely to be involved in the process. We have isolated a protease from AD brain by following its activity in digesting a synthetic peptide of 10 amino acids derived from the APP sequence flanking the N-terminus of Aβ. The protease was purified by a fractionation scheme including ammonium sulfate precipitation and column chromatography using hydrophobic interaction, anion exchange, affinity, hydroxyapatite and size exclusion gels. The purity of the final product was assessed on a silver stained SDS gel by the presence of a single band. Microsequencing was performed following trypsin digestion of the sample. Internal peptide sequences were found to have sequence homology to cysteine proteases in the database. The enzyme requires DTT for activity and can be inhibited by specific inhibitors of cysteine but not serine proteases. The purified enzyme has a pI of 5.0 and a native tetrameric structure with subunits of 48 kD each. The enzyme is capable of digesting APP and generating a short peptide recognizable by antibodies specific to the C-terminus of APP. Interestingly, the purified protease also forms heat- and SDS-stable complexes with APP.

INTRODUCTION

The amyloid precursor protein (APP) is involved in the pathogenesis of AD because mutations in its gene are linked to familial AD and because its fragment, Aβ, accumulates in plaques and blood vessel walls in AD brains.[1] Aβ is neurotoxic in cell culture,[2] especially in its aggregated form.[3] Under-

[a] This research was supported by the Zenith Award from the Alzheimer Association and NIH grant AG09905.

[d] *Send correspondence to:* Dr. Carmela R. Abraham, K-5, Boston University School of Medicine, 80 E. Concord Street, Boston, MA 02118; Tel: (617) 638-4308; Fax: (617) 638-5226; Email: cabraham@acs.bu.edu.

standing the pathways of APP degradation and Aβ production is an important step in elucidating early events in the development of AD. Several proteolytic pathways have been described. In one pathway APP is cleaved at position Lys16 of Aβ, thus generating secreted APP[4,5] and a 10 kD membrane-bound fragment which is then internalized and degraded in lysosomes. The enzyme responsible for such a cleavage has been termed α-secretase. Activity of an α-secretase precludes Aβ formation. In a second pathway, APP is cleaved at the N-terminus of Aβ by a protease(s) termed β-secretase.[6] This cleavage is amyloidogenic because Aβ can be produced from the resulting C-terminal fragment by the activity of a γ-secretase, a protease(s) that cleaves around the C-terminus of Aβ. In a third pathway, full length APP and other C-terminus derivatives are internalized and degraded in lysosomes.[7–9] Lastly, several groups reported that small amounts of full length APP are found in the media of cultured cells.[10–12] In such a scenario, Aβ could be produced extracellularly by secreted proteases. From recent data it is plausible that more than one β-secretase or γ-secretase exist since the N and C-termini of Aβ appear heterogeneous.[13] Nevertheless, the most common form of the soluble Aβ sequence starts at aspartic acid. For the identification of a putative β-secretase, we have used a synthetic peptide (see below) and a recombinant full length human APP. Using these two substrates, we have previously identified three novel proteases. Here we describe the identification, purification and characterization of a novel cysteine protease from AD brain.

METHODS

Protease activities were followed by their ability to hydrolyze the synthetic peptide P1 (HSEVKMDAEF) which flanks the N-terminus of Aβ. A convenient assay can be performed by incubating ^{125}[I] labeled P1 with protease fractions at 37°C, after which the reaction mixture is spotted on a thin layer chromatography (TLC) plate to separate the P1 and the reaction products based on the different mobility of the peptides on the TLC plate.[14] We are able to distinguish peptide fragments generated by proteolytic activities cutting either after Lys or Met residues. We followed the strongest proteolytic activity cleaving after the Lys at -1 position of Aβ. Alzheimer's disease brain was homogenized in Tris buffer at neutral pH, the homogenate was then centrifuged at 10,000 × g. The supernatant was fractionated by ammonium sulfate, a series of liquid column chromatography techniques including Phenyl-Sepharose, DEAE anion exchange, Blue-dye affinity, hydroxyapatite and size exclusion gels. At each step of the purification, the fractions were analyzed for their ability to generate the HSEVK peptide fragment. Active fractions were analyzed by SDS-PAGE to monitor the degree of purification before being pooled together. Purified protein was then characterized for its biochemical and biophysical properties; *i.e.,* size exclusion column to determine the native molecular weight, SDS-PAGE to determine molecular weight at denatured state. The isoelectric point was determined by running the enzyme on a pH gradient gel of 3 to 10. We then tested the effects of protease inhibitors specific for different classes of proteases. Lastly, the purified enzyme was dialyzed against a volatile buffer, lyophilized and submitted to amino acid sequence analysis.

RESULTS

We have purified to homogeneity a protease capable of hydrolyzing P1 after the Lys residue (Lys cut). This protease has apparent native molecular weight of 190,000 as determined by size exclusion column chromatography. Following SDS-PAGE, a single band is observed at apparent molecular mass of 48 kD. A single band is also observed on the isoelectric focusing gel at pH 5.0. We concluded that the native protein consists of four identical subunits as a tetramer. The tetrameric structure is stable up to 60°C and likely to be the active form of the enzyme. We also observed a mobility shift of the purified protease on SDS-PAGE with and without β-mercaptoethanol, indicating the presence of cysteine-cysteine disulfide bridge(s). DTT is required for the proteolytic activity. The protease is strongly inhibited by E-64, an irreversible specific cysteine protease inhibitor; yet it is not inhibited by the serine protease inhibitor DFP at a concentration cathepsin G activity is completely abolished. So far we have obtained 5 internal amino acid sequences, a 13-residue sequence was found to have a 9-residue match to a cysteine protease in the database. Therefore, it is likely we have purified a cysteine protease. The ability of the purified protease to digest APP was assayed by mixing with the human recombinant APP751 at 37°C. The reaction mixture was then separated by SDS-PAGE, and the proteins were blotted onto PDVF immobilon filter for Western blot analysis. On the Western blot we observed low molecular weight bands (> C-100) which can be recognized by polyclonal antibodies against the C-terminus of APP. Interestingly, while the APP band completely disappeared, a smear pattern appeared above the position of the APP band which continued to the top of the gel. It appears that the newly purified protease is capable of forming stable complexes with the APP or its digested product fragment(s). The complexes are stable under the denaturing conditions of SDS-PAGE.

DISCUSSION

In the past several years the search for proteases that degrade APP has become a very active area of investigation in many laboratories (for review see ref. 13). We have focused our attention on proteases that play the role of a β-secretase. In our opinion, a protease has to fulfill three criteria in order to be considered a β-secretase. First, the site of the cleavage has to be the KM or MD bond of HSEVKMDAEF (P1). Secondly, it has to generate amyloidogenic fragments from APP. Third, modulation of the protease activity has to correlate with a change in the pattern of APP degradation and/or Aβ production, *i.e.,* using antisense experiments to knock out the expression of the protease gene in cell culture, or by inhibiting the protease activity by specific inhibitors.

The first protease identified by us was a calcium-activated serine protease (CASP).[15,16] It cleaves between Lys and Met in the synthetic peptide and it can generate amyloidogenic fragments which comigrate with C-100 (the last 100 amino acids of APP starting at Met-1 of Aβ). Judging from its activity on P1, its inhibitory profile and from its cross reactivity with cathepsin G antibodies, CASP appears to be a brain cathepsin G–like protease.

A second protease purified is a zinc metalloprotease homologous to the rat metalloendopeptidase E.C.3.4.24.15. (MP24.15). The metalloprotease cleaves between Met and Asp and is also able to generate amyloidogenic fragments.[17] Human neuroblastoma, but not glioblastoma, cells secrete into the medium an 80 kD protease which can be recognized by monoclonal antibodies against MP24.15 on Western blots. In homogenates of human brain and various monkey organs including brain, the antibodies immunostain two bands of 80 and 55 kD each. Recently, it has been shown that the 55 kD protein is also a protease that can be inhibited by a specific inhibitor of MP24.15.[18] Surprisingly, the MP24.15 monoclonal antibodies strongly labeled neurofibrillary tangles, dystrophic neurites in senile plaques and neuropil threads in AD brain. The staining on Western blots and on sections was completely abolished following absorption of the antibodies with the peptide used as immunogen.[19]

Here we describe the purification and characterization of a novel brain cysteine protease (NCP). A cysteine protease is most likely to be active in an intracellular compartment where it can maintain a reduced state, *i.e.,* cytosol or lysosomes. NCP is active at neutral pH, thus it may be active in both cytosol or lysosomes. Some lysosomal cysteine proteases are active over a wide range of pHs, for example cathepsin H is active between pH 3-7.[20] Subcellular fractionation should aid in the localization of NCP activity. What we find particularly interesting is the tight binding of NCP to APP *in vitro.* In further studies we will investigate the APP domain to which NCP binds, under what conditions they bind and whether this binding occurs *in vivo.*

It is likely that more than one β-secretase exists. Since APP processing has been studied in such a variety of primary cultured cells, cell lines from different organs and species, cells transfected with APP or untransfected cells, we witness very heterogeneous N-termini of Aβ. It is also possible that in AD cases harboring mutations in the APP gene, in or close to Aβ, other proteases may act as α, β or γ secretases. Finding the proteases involved in the amyloidogenic processing of Aβ may help in the design of specific inhibitors of this pathway.

REFERENCES

1. SELKOE, D. J. 1994. Alzheimer's disease: A central role for amyloid. J. Neuropathol. Exp. Neurol. **53**:438–447.
2. YANKNER, B. A., L. K. DUFFY & D. A. KIRSCHNER. 1990. Neurotrophic and neurotoxic effects of amyloid β protein: Reversal by tachykinin neuropeptides. Science **250**:279–282.
3. PIKE, C. J., D. BURDICK, A. J. WALENCEWICZ, C. G. GLABE & C. W. COTMAN. 1993. Neurodegeneration induced by β-amyloid peptides *in vitro*: The role of peptide assembly state. J. Neurosci. **13**:1676–1687.
4. ESCH, F. S., P. S. DEIM, E. BEATTIE, R. W. BLACHER, A. R. CULWELL, T. OLTERSDORF, D. MCCLURE & P. J. WARD. 1990. Cleavage of amyloid β-peptide during constitutive processing of its precursor. Science **248**: 1122–1124.
5. SISODIA, S. S., E. H. KOO, K. BEYREUTHER, K., A. UNTERBECK & D. L. PRICE. 1990. Evidence that β-amyloid protein in Alzheimer's disease is not derived by normal processing. Science **248**:492–495.
6. SEUBERT, P., T. OLTERSDORF, M. G. LEE, R. BARBOUR, C. BLOMQUIST, D. L. DAVID, K. BRYANT, L. C. FRITZ, D. GALASKO, L. J. THAL, I. LIEBERBURG & D. B. SCHENK. 1993. Secretion of β-amyloid precursor protein cleaved at the amino terminus of the β-amyloid peptide. Nature **361**:260–263.
7. COLE, G. M., T. V. HUYNH & T. SAITOH. 1989. Evidence for lysosomal processing of β-amyloid precursor in cultured cells. Neurochem. Res. **14**:933–939.
8. GOLDE, T. E., S. ESTUS, L. H. YOUNKIN, D. J. SELKOE & S. G. YOUNKIN. 1992. Processing of the amyloid precursor protein to potentially amyloidogenic derivatives. Science **255**:728–730.
9. KOO, E. H. & S. SQUAZZO. 1994. Evidence that production and release of amyloid β-protein involves the endocytic pathway. J. Biol. Chem. **269**: 17386–17389.
10. RIPELLINO, J. A., D. VASSILACOPOULOU & N. K. ROBAKIS. 1994. Solubilization of full-length amyloid precursor proteins from PC12 cell membranes. J. Neurosci. Res. **39**:211–218.
11. BASHIN, R., L. GREGORI, I. MOROZOV & D. GOLDGABER. 1994. A processing pathway for amyloid β-protein precursor generates a secreted form with an intact carboxyl terminus. Amyloid **1**:221–231.
12. CONN, K., G. PAPASTOITSIS, B. MECKELEIN & C. R. ABRAHAM. 1994. Identification of full length β-amyloid precursor protein in human neuronal and non-neuronal cell culture supernatant: A possible extracellular source for the generation of Aβ. Amyloid **1**:232–239.
13. EVIN, G., K. BEYREUTHER & C. MASTERS. 1994. Review: Alzheimer's disease amyloid precursor protein (AβPP): Proteolytic processing, secretases and βA4 amyloid production. Amyloid **1**:263–280.

14. ABRAHAM, C. R., A. BEN MEIR & P. TEMPST. 1990. Facile and sensitive assay for monitoring proteolytic activities with defined specificities: Studies on amyloid β-protein processing in Alzheimer's disease. Peptide Res. **3**:211–215.
15. ABRAHAM, C. R., J. DRISCOLL, H. POTTER, W. E. VAN NOSTRAND & P. TEMPST. 1991. A calcium-activated protease from Alzheimer's disease brain cleaves at the N-terminus of the amyloid β-protein. Biochem. Biophys. Res. Commun. **174**:790–796.
16. RAZZABONI, B. L., G. PAPASTOITSIS, E. H. KOO & C. R. ABRAHAM. 1992. A calcium stimulated serine protease from monkey brain degrades the β-amyloid precursor protein. Brain. Res. **589**:207–216.
17. PAPASTOITSIS, G., R. SIMAN, R. SCOTT & C. R. ABRAHAM. 1994. Indentification of a metalloprotease from Alzheimer's disease brain able to degrade the β-amyloid precursor protein and generate amyloidogenic fragments. Biochemistry **33**:192–199.
18. SONNENBERG-REINES, J., A. B. BROWN, D. M. TUMMOLO, M. A. SPRUYT & J. S. JACOBSEN. 1994. Comparison of a brain purified metalloprotease with human cathepsins G and D: Evaluation of sequence specificity using peptide substrates and regional selectivity using purified brain APP. Neurosci. Abstr. **20**:1484.
19. ABRAHAM, C. R., K.-J. CONN, M. PIETROPAOLO, A. AMARATUNGA, R. E. FINE & B. MECKELEIN. 1994. Localization of a human metalloprotease capable of generating amyloidogenic fragments from the amyloid precursor protein. Neurosci. Abstr. 612.3, **20**:1485.
20. BARRET, A. J. & H. KIRSCHKE. 1981. Cathepsin B, cathepsin H and cathepsin L. Methods Enzymol. **80**:535.

M1 Agonists for the Treatment of Alzheimer's Disease

Novel Properties and Clinical Update[a]

A. FISHER,[b,e] E. HELDMAN,[b] D. GURWITZ,[b] R. HARING,[b] Y. KARTON,[b] H. MESHULAM,[b] Z. PITTEL,[b] D. MARCIANO,[b] R. BRANDEIS,[b] E. SADOT,[c] Y. BARG,[c] R. PINKAS-KRAMARSKI,[c] Z. VOGEL,[c] I. GINZBURG,[c] T. A. TREVES,[d] R. VERCHOVSKY,[d] S. KLIMOWSKY,[d] AND A.D. KORCZYN[d]

[b] *Israel Institute for Biological Research, Ness-Ziona 74100 Israel*

[c] *Department of Neurobiology, The Weizmann Institute of Science, Rehovot 76100 Israel*

[d] *Tel Aviv Medical Center, Department of Neurology, Tel Aviv, Israel*

ABSTRACT: The *AF series* compounds, AF102B and congeners of AF150(S), are functionally selective agonists for m1 muscarinic receptors (m1AChRs). This is shown in stable transfected CHO and PC12 cells (PC12M1) with m1-m5AChRs and m1AChRs, respectively. AF102B and AF150(S) are partial agonists, but AF150, AF151, and AF151(S) are full agonists in stimulating phosphoinositides hydrolysis or arachidonic acid release in these cells. Yet, all these compounds behave as antagonists when compared with carbachol in elevating cAMP levels. In PC12M1 cells, unlike carbachol, the AF *series* compounds induce only minimal to moderate neurite outgrowth. Yet, these agonists synergize strongly with NGF, which by itself mediates only a mild response. Stimulation of m1AChRs by AF102B, AF150(S) and AF151(S) in PC12M1 cells enhances secretion of β/A4 amyloid precursor protein derivatives (APPs). The enhanced APPs secretion induced by AF102B is potentiated by NGF. AF102B also stimulates APPs secretion from rat cortical slices. Stimulation of m1AChR in PC12M1 cells with carbachol or AF102B decreases tau phosphorylation as indicated by specific tau-1 mAb and alkaline phosphatase treatment. Due to the above mentioned properties m1 agonists may be of unique value in delaying the progression of Alzheimer's disease (AD). The *AF series* compounds show a wide safety margin and improve memory and learning deficits in animal models for AD. There is a dearth of clinical reports on m1 agonists. These include studies on AF102B and xanomeline, another m1 selective agonist. We tested AF102B in escalating doses of 20, 40, 60 mg, tid, po, (each dose for 2 weeks) for a total of 10 weeks. This was a single-blind placebo-controlled, parallel-group study in patients with prob-

[a] Supported in part by Snow Brand Milk Products, Japan.

[e] *Send correspondence to:* Dr. Abraham Fisher, Israel Institute for Biological Research, P.O. Box 19, Ness-Ziona 74100 Israel; Tel: (972)-8-381-603; Fax: (972)-8-401-094.

able AD. AF102B was significantly effective at 40 and 60 mg, tid in the ADAS, ADAS-*cognitive* and ADAS-*word recognition scales.*

INTRODUCTION

Much interest is focused recently on the design of selective M1 muscarinic agonists (reviewed by Fisher and Barak[1]). This can be attributed to:

1) *The heterogeneity in muscarinic acetylcholine (ACh) receptors (mAChRs)*. Five structurally different human mAChR subtypes (m1-m5) genes have been cloned and expressed in suitable cell systems (reviewed by Hulme *et al.*[2]). The five mAChR subtypes belong to the family of cell surface receptors coupled to guanosine-triphosphate-binding proteins (G-proteins).[2] When expressed in mammalian cells, these mAChRs subtypes mediate a variety of signal transduction pathways.[2] Additional activities associated with mAChR subtypes were uncovered recently (*vide infra*).

2) *The presynaptic cholinergic deficit in Alzheimer's disease (AD)* (reviewed by Court and Perry[3]). The lack of ACh has been linked with cognitive impairments in AD patients. However, postsynaptic M1AChRs, facilitating cellular excitation and cognitive processes relevant to AD, are relatively unchanged in AD.[3] Activation of M1AChRs could be considered a direct and rational strategy to treat AD patients (reviewed by Fisher and Barak[1]). Since abnormalities in AD may also occur along various signal transduction pathways,[4] we suggested that such selective M1 (or m1) agonists may also activate hypofunctional signalling pathways in AD.[5] Mismetabolism of amyloid precursor proteins (APPs) may induce AD.[6] Interestingly, activation of m1 and m3 mAChRs enhances secretion of APPs, *in vitro*.[7] Finally, activation of m1AChRs induces neurotrophic-like activities[5,8] and decreases *tau* phosphorylation (E. Sadot, D. Gurwitz, J. Barg, I. Ginzburg and A. Fisher, submitted for publication). To address some of these new findings, we describe below novel properties of some compounds from the *AF series*[9] (FIG. 1). We also survey a study with AF102B in AD patients in Israel and some clinical data on xanomeline (FIG. 1), another recently reported M1 (or m1) agonist.[10]

RESULTS AND DISCUSSION

AF102B was the first reported selective M1 agonist (reviewed by Fisher and Barak[1]). It can act as a full agonist, a partial agonist, or an antagonist depending on the tissue, the mAChR subtype and the functional assays studied. Thus, AF102B can be considered as a selective M1 (or m1) agonist when defined through functional assays. Selectivity and efficacy in binding studies (± GppNHp, a non-hydrolyzable GTP analog) were assayed using rat brain

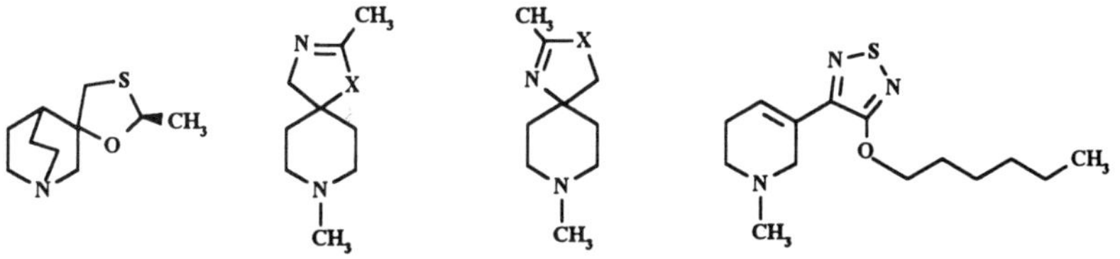

AF102B **AF150 X=O** **AF151 X=O**
AF150(S) X=S **AF151(S) X=S**

The AF series **Xanomeline**

FIGURE 1. The m1, m2, m3, m4 AChRs fit the pharmacological definition of the M1, M2, M3 and M4 AChRs, respectively.[2] We use the term m1-m5 and M1-M3 agonists (or antagonists) for ligands defined using the cloned m1-m5 and M1-M3 pharmacologically characterized mAChRs, respectively.

regions rich in M1AChRs (*e.g.*, cerebral cortex) *vs.* brain regions rich in M2AChRs (*e.g.*, cerebellum). The results showed that AF150, AF151, AF150(S) and AF151(S) were more efficacious M1 agonists than AF102B.[9]

Ligand-mediated Selective Signaling via Distinct G-proteins

The *AF series* compounds (FIG. 1) are full agonists when assayed for elevating $[Ca^{2+}]_i$ in Chinese hamster ovary (CHO) cells stably transfected with cloned m1AChR. In CHO and in rat pheochromocytoma (PC12) cells stably transfected with mlAChR (PC12M1), (*re*: Pinkas-Kramarski *et al.*[8]), AF102B and AF150(S) are partial agonists (30–50% vs. carbachol, CCh), while AF150, AF151, and AF151(S) are full agonists in stimulating phosphoinositides (PI) hydrolysis or arachidonic acid release. Yet, all these compounds behave as antagonists when compared with CCh in elevating cAMP levels.[5,9] These data imply that selective muscarinic ligands may activate distinct sets of G-proteins and that drug selectivity may extend beyond the ligand recognition site.[5]

Neurotrophic-like Effects in PC12M1 Cells

Oxotremorine induces neurite outgrowth, which is synergized with nerve growth factor (NGF) in PC12M1 cells.[8] AF102B (up to 100 μM) induced only minimal neurite outgrowth. Yet, AF102B strongly synergized (EC_{50} = μM) with 2 nM NGF, a concentration which by itself mediated only a mild response. Similar results were obtained with AF150(S) and AF151(S) which were also synergistic ($EC_{50} \sim 1\mu M$) with NGF (2 nM). The synergism of AF102B and NGF was not observed in non-transfected PC12 cells, and was completely blocked in PC12M1 cells by 10 μM atropine.[5,11] The mechanism underlying neurotrophic-like effects by these muscarinic agonists, which is dependent on the presence of NGF, probably indicates that these compounds exert neurotrophic activities in conjunction with some signal(s) mediated *via* NGF receptors. Some of the neurotrophic-like effects induced by such agonists may involve increased release of APPs following activation of m1AChR (*vide infra*). Notably, the secreted forms of APPs, are known to regulate neurite outgrowth and to promote neuronal survival (reviewed by Mattson *et al.*[6]).

PC12M1 Cells and Rat Cortical Slices: m1AChR-stimulated APPs Secretion

Stimulation of m1AChRs by AF102B in PC12M1 cells enhances secretion of APPs to the culture medium, and lowers the level of membrane associated APPs. Notably, while being a partial agonist (30% *vs.* CCh) in stimulating PI hydrolysis, AF102B is as efficacious as the full agonist CCh for stimulating APPs secretion in PC12M1 cells, as judged from the decline in membrane-associated APPs.[12] Similar results were found with AF150(S) and AF151(S). The enhanced APPs secretion induced by AF102B is potentiated by NGF.[13] Increased APPs secretion by CCh (100 μM) or AF102B (1 μM) from rat cortical slices was also found by us recently (Z. Pittel, J. Barg, E. Heldman and A. Fisher, unpublished observations).

m1AChR-stimulated tau Dephosphorylation in PC12M1 Cells

Hyperphosphorylated *tau* proteins are the principal fibrous component of the neurofibrillary tangle pathology in AD.[14] Stimulation of m1AChR in PC12M1 cells with CCh or AF102B decreased *tau* phosphorylation as indicated by specific *tau* mAb (tau-1) and alkaline phosphatase treatment. The maximal effect was observed at 6 hours, and declined later. Unlike for CCh, the decreased *tau* phosphorylation by AF102B was maintained for at least four days (the longest period studied). When PC12M1 were treated for 24 hours with NGF together with the muscarinic agonists, an almost additive effect was observed as compared to the effect of NGF and the agonists, separately (E. Sadot, D. Gurwitz, J. Barg, I. Ginzburg and A. Fisher, submitted for publication).

Studies in Animal Models That Mimic the Cholinergic Hypofunction in AD

AF102B, AF150, AF150(S) and AF151 improved memory and learning deficits in a variety of animal models for AD. These agonists were without adverse central or peripheral side-effects at the effective doses and showed a relatively wide safety margin (reviewed by Fisher and Barak[3]; Brandeis *et al.*).[14]

Clinical Results with m1 Agonists in AD Patients

Few clinical results with functionally selective M1 (or m1) agonists were reported so far. In this context we survey below some studies with AF102B and xanomeline [another putative and highly potent M1 (or m1) agonist *in vitro*[10]].

AF102B: Clinical trials were already performed on AF102B in Japan, Israel and the USA. In Israel, we have evaluated AF102B in a 10 weeks, single-blind placebo-controlled, parallel-group study in patients of both sexes diagnosed with probable AD.[16] This study evaluated AF102B vs. placebo in escalating doses of 20, 40 and 60 mg, tid, po, each dose for 2 weeks, with a two weeks placebo *lead in* and a 2 weeks placebo *washout.* A significant effect in favor of AF102B at doses of 40 mg and 60 mg, tid, vs. placebo was found in the Alzheimer's Disease Assessment Scale (ADAS) and its subscale, the ADAS-*cognitive* scale. Notably, the ADAS-*cognitive* was used in studies on tacrine.[17] A dose-related improvement could be detected in the ADAS-subscales (*e.g. cognitive* and *word recognition*), *e.g.* 60 mg, tid, was more effective than the lower dose of 40 mg. The dose of 60 mg, tid, was also efficacious in moderately and more severely demented patients. Improvement was also noted in the drug group by the caregiver(s) at the two doses of 40 and 60 mg, tid. AF102B was well tolerated by the patients with only one dropout due to adverse effect as compared with more dropouts in the placebo group. The only significant adverse effects of AF102B (60 mg, tid) were of cholinergic nature, including diaphoresis (20%) and excessive salivation (7% of the patients).

Xanomeline: This is a very potent M1 (or m1) agonist, *in vitro* with little preclinical data reported *in vivo.*[10] A recent report describes safety and tolerance in AD patients following treatment with xanomeline up to a dose of 115 mg, tid.[18] From the report it can be deduced that xanomeline is producing a range of effects, most probably cholinergic in nature. Some of the side-effects (*e.g.* diarrhea, nausea, vomiting) were quite pronounced.[18] This clinical profile indicates that the activity of this compound may not be mediated solely through m1AChRs. A study in mild and moderate AD patients is to be published.[19] The pharmacokinetic profile of xanomeline in healthy subjects indicated a rapid formation of at least eight metabolites, some of which were present at concentrations significantly greater than the parent drug.[20] The existence of a metabolite(s) which could contribute to the drug pharmacology and/or side effects could not be ruled out.

CONCLUSIONS

The new findings concerning activation of m1AChRs may have both basic and clinical relevance. The functional selectivity is not inherent to the agonists mentioned above, but rather to a combination of agonist-m1AChR-G-proteins complexes. Selective responses are achieved when the agonist-mAChR complex activates only certain G-proteins, which in turn activate distinct signal transduction pathways. The AF series compounds (FIG. 1) exhibit such an activity. The concept of "*ligand-mediated selective signaling*",[5] *e.g.*, activation of only distinct G-protein subset(s) (*e.g.* Gq, Go but not Gs), might be of clinical significance, since altered signal transduction *via* Gs might be relevant in the pathophysiology of AD.[4] This may be linked with the enhanced secretion of APPs following stimulation of m1AChRs by the *AF series* compounds and other functionally m1 selective agonists. Consequently, M1 (or m1) agonists may be of value in preventing amyloid formation by selectively promoting the "α-secretase" processing pathway in AD (suggested originally by Nitsch *et al.*[7] and later by Chessell *et al.*[21]). Should neurotrophic-like events induced by M1 (or m1) agonists *in vitro* also occur *in vivo,* these might have important clinical significance and may constitute a novel treatment for AD. Notably, NGF does not cross the blood-brain barrier. A more practical approach would involve modulation of the function of endogenous NGF by a synergistic agent such as an M1 (or m1) agonist. Finally, the decreased phosphorylation of *tau* protein *via* m1AChR deserves particular attention. This suggests for the first time a linkage between the muscarinic signal transduction system(s) and the neuronal cytoskeleton, *via* regulation of phosphorylation of *tau* microtubule-associated protein. Moreover, these studies propose a possible correlation between the cholinergic deficiency and *tau* hyperphosphorylation in AD. It can be speculated that activation of m1AChRs might provide a novel treatment strategy for AD by modifying *tau* processing in the brain and perhaps delaying the formation and accumulation of overphosphorylated *tau*. Thus M1 (or m1) agonists, in addition to the expected use as a cholinergic replacement strategy, might be of unique value in delaying the progression of AD. Hence, m1 agonists might have a more important and complex role in the treatment of AD than previously envisaged. The favorable results with AF102B in AD patients encourage controlled clinical studies for longer trials and with a larger number of patients to evaluate the therapeutic potential of this and other M1 (or m1) agonists in AD.

REFERENCES

1. FISHER, A. & D. BARAK. Progress and perspectives in new muscarinic agonists. 1994. Drug News & Perspectives **7:**453–464.
2. HULME, E. C., N. J. M. BIRDSALL & N. J. BUCKLEY. 1990. Muscarinic receptor subtypes. Annu. Rev. Pharmacol. Toxicol. **30:**633–673.
3. COURT, J. A. & E. K. PERRY. 1991. Dementia: The neurochemical basis of putative transmitter orientated therapy. Pharmac. Ther. **52:**423–443.
4. HARRISON, P. J., A. J. L. BARTON, B. MCDONALD & R. C. A. PEARSON. 1991. Alzheimer's disease: Specific increases in a G-protein subunit (Gsα) mRNA in hippocampal and cortical neurons. Mol. Brain Res. **10:**71–81.
5. GURWITZ, D., R. HARING, E. HELDMAN, C. M. FRASER, M. C. MANOR & A. FISHER. 1994. Discrete activation of transduction pathways associated with acetylcholine m1 receptor by several muscarinic ligands. Eur. J. Pharmacol. (Mol. Pharmacol.) **267:**21–31.
6. MATTSON, M. P., S. W. BARGER, B. CHENG, I. LIEBERBURG, V. L. SMITH-SWINTOSKY & E. R. RUSSELL. 1993. β-Amyloid precursor protein metabolites and loss of neuronal Ca^{2+} homeostasis in Alzheimer's disease. TINS **16:** 409–414.
7. NITSCH, R. N., B. E. SLACK, R. J. WURTMAN & J. H. GROWDON. 1992. Release of Alzheimer amyloid precursor derivatives stimulated by activation of muscarinic acetylcholine receptors. Science **258:**304–307.
8. PINKAS-KRAMARSKI, R., R. STEIN, L. LINDENBOIM & M. SOKOLOVSKY. 1992. Growth factor-like effects mediated by muscarinic receptors in PC12M1 cells. J. Neurochem. **59:**2158–2166.
9. FISHER, A., Y. KARTON, E. HELDMAN, D. GURWITZ, R. HARING, H. MESHULAM, R. BRANDEIS, Z. PITTEL, Y. SEGALL, D. MARCIANO, I. MARKOVITCH, Z. SAMOCHA, E. SHIRIN, M. SAPIR, B. GREEN, G. SHOHAM & D. BARAK. 1993. Progress in medicinal chemistry of novel selective muscarinic agonists. Drug Design and Discovery **9:**221–235.
10. BYMASTER, F. P., D. T. WONG, C. H. MITCH, J. S. WARD, D. O. CALLIGARO, D. D. SCHOEPP, H. E. SHANNON, M. J. SHEARDOWN, P. H. OLESEN, P. D. SUZDAK, M. D. B. SWEDBERG & P. SAUERBERG. 1994. Neurochemical Effects of the M1 Muscarinic Agonist Xanomeline (LY246708/NNC11-0232). J. Pharmacol. Exp. Therap. **269:**282–289.
11. GURWITZ, D., R. HARING, R. PINKAS-KRAMARSKI, R. STEIN & A. FISHER. 1993. Neurotrophic-like effects of AF102B, an M1-selective muscarinic agonist, in PC12 cells transfected with M1 muscarinic receptors. Soc. Neurosci. Abs. **19:**1767.
12. HARING, R., D. GURWITZ, J. BARG, R. PINKAS-KRAMARSKI, E. HELDMAN, Z. PITTEL, H. D. DANENBERG, A. WENGIER, H. MESHULAM, D. MARCIANO, Y. KARTON & A. FISHER. 1995. NGF promotes amyloid precursor protein secretion via muscarinic receptor activation. Biochem. Biophys. Res. Commun. **213:** 15–23.
13. HARING, R., D. GURWITZ, J. BARG, R. PINKAS-KRAMARSKI, E. HELDMAN, Z. PITTEL, A. WENGIER, H. MESHULAM, D. MARCIANO, Y. KARTON & A. FISHER. 1994. Amyloid precursor protein secretion via muscarinic receptors:

reduced desensitization using the M1-selective agonist AF102B. Biochem. Biophys. Res. Commun. **203**:652–658.

14. LEE, V. M.-Y., B. J. BALIN, L. OTVOS & J. Q. TROJANOWSKI. 1991. A68: A major subunit of paired helical filaments and derivatized forms of normal tau. Science **251**:675–678.
15. BRANDEIS, R., M. SAPIR, N. HAFIF, S. ABRAHAM, N. OZ, E. STEIN & A. FISHER. 1995. AF150(S): A new functionally selective M1 agonist improved cognitive performance in rats. Pharmacol. Biochem. Behav. **51**:667–674.
16. FISHER, A., E. HELDMAN, D. GURWITZ, R. BRANDEIS, T. A. TREVES, Y. KARTON, R. VERCHOVSKY, S. KLIMOWSKY, H. MESHULAM, Z. PITTEL, D. MARCIANO & A. D. KORCZYN. Selective signaling via novel muscarinic agonists: Implications for Alzheimer's disease treatments and clinical update. Third Springfield International Symposium on Alzheimer Disease. Springfield, Il, USA, May 11–15, 1994.
17. FARLOW, M., S. I. GRACON, A. HERSHEY, K. W. LEWIS, C. H. SADOWSKY & J. DOLAN-URENO, for the TACRINE STUDY GROUP. 1992. A controlled trial of tacrine in Alzheimer's disease. JAMA **268**:2523–2529.
18. CUTLER, N. R., J. J. SRAMEK, R. D. SEIFERT, J. J. CONRAD, T. S. WARDLE, D. J. HURLEY & L. M. THOREN. Safety and tolerance of xanomeline tartrate an M1-specific cholinergic agonist, in patients with Alzheimer's disease. The Third International Symposium on Alzheimer's and Parkinson's Diseases: Recent Developments, Nov. 1–6, 1993, Chicago, IL, USA.
19. BODICK, N. Xanomeline—a potent and selective M1 agonist: Clinical study in mild and moderate Alzheimer's disease. The Sixth International Symposium on Subtypes of Muscarinic Receptors. Fort Lauderdale, FL, USA. Nov. 9–12, 1994.
20. DELONG, A. F., P. L. BONATE, T. GILLESPIE, D. P. HENRY & J. H. SATTERWHITE. Absorption distribution, metabolism and elimination of radiolabeled xanomeline in healthy male subjects. The Third International Symposium on Alzheimer's and Parkinson's Diseases: Recent Developments, Nov. 1–6, 1993, Chicago, IL, USA.
21. CHESSELL, I. P., P. T. FRANCIS, M. N. PANGALOS, R. C. A. PEARSON & D. M. BOWEN. 1993. Localisation of muscarinic (m1) and other neurotransmitter receptors on corticofugal-projecting pyramidal neurones. Brain Res. **632**: 86–94.

High Affinity Choline Transporter Status in Alzheimer's Disease Tissue from Rapid Autopsy[a]

G. BISSETTE, F. J. SEIDLER,[b] C. B. NEMEROFF,[c] AND T. A. SLOTKIN[b,d]

[b] *Departments of Psychiatry and Pharmacology, Duke University Medical Center, Durham, North Carolina 27710 USA*

[c] *Department of Psychiatry and Behavioral Sciences, Emory University School of Medicine, Atlanta, Georgia 30322 USA*

ABSTRACT: The degeneration of nucleus basalis cholinergic neurons in Alzheimers disease (AD) has led to therapies that attempt to increase the synaptic availability of acetylcholine in the remaining cholinergic nerve terminals and to thereby reverse or slow the progressive dementia accompanying the disease process. The inadequacy of current choline-replacement therapies suggests that utilization of choline may be disordered and the rate-limiting step in acetylcholine synthesis, the high affinity choline transporter, may be involved. An adequate test of this hypothesis requires the use of fresh, unfrozen tissue, as the transporter activity declines rapidly after death. Using tissue acquired within two hours of death, the activity of the high affinity choline transporter was shown to be increased in cortical brain regions from AD patients compared to non-AD controls. Further studies using frozen tissues with similar short postmortem acquisition times, revealed the expression of the high affinity uptake transporter to be increased in AD cortex as well. When the ratio of regional uptake activity or expression to the regional level of choline acetyltransferase was calculated, the increase in choline transporter activity and expression was clearly statistically significant. Further statistical significance in the choline transporter activity of the AD group was achieved when the putamen, a region without marked pathology in AD, was used as an internal standard to control for agonal state differences in the individual patients contributing tissue to this study. These increases in choline transporter expression and activity in AD indicate disordered regulation of this rate-limiting component of acetylcholine synthesis above and beyond that required to compensate for the reduced cholinergic synaptic availability in AD.

INTRODUCTION

The degeneration of nucleus basalis cholinergic neurons with the subsequent reduction in choline acetyltransferase in the cerebrocortical projection

[a] This work was supported by U.S. Public Health Service grants MH-40524 and AG-05128.

[d] *Send correspondence to:* Dr. Theodore Slotkin, Duke University Medical Center, Durham, North Carolina 27710; Tel: (919) 684-2315; Fax: (919) 684-8197.

fields of these neurons was the first specific neurochemical alteration to be discovered in Alzheimer's disease (AD) research in 1976.[1] In the ensuing two decades, attempts have been made to increase the synaptic availability of acetylcholine (ACh) in the brain tissue of AD patients, including loading with the ACh precursor, choline, and inhibition of ACh breakdown by administering acetylcholinesterase inhibitors (See ref. 2 for review). Although this effort has led to the approval of new pharmacotherapies for AD, the results have been disappointing, with few institutionalized patients returning to their homes or resuming their occupations.

These results suggest that, over-and-above a simple deficiency in cholinergic neurotransmission consequent to neuronal death, AD may be typified by disordered regulation of ACh synthesis and metabolism. The rapid hydrolysis of ACh by acetylcholinesterase at death has made direct measurements of ACh levels or turnover impractical, and interpreting any changes in the concentration of the precursor, choline, is confounded by the existence of multiple metabolic pools. The rate-limiting step for synthesis of ACh is not catalyzed by choline acetyltransferase, but rather, is the availability of free choline as provided to the cell by a specific, high-affinity choline transporter.[3] Indeed, the activity of this presynaptic, membrane-bound transporter is regulated by the neuronal firing rate, whereas the activity of choline acetyltransferase is not. Transporter activity is thus a likely target for examining the effects of AD on choline utilization; however, transporter activity, and specifically the impulse activity–related component of activity, decline within the first two to four hours after death, necessitating the use of live or extremely rapidly obtained autopsy material for examination. Previous attempts to measure choline uptake in "fresh" brain tissue from AD patients employed four to six hour post-biopsy samples from patients with AD and reported decreased transporter activity relative to controls,[4] but this is clearly outside the acceptable time period for maintenance of impulse-related changes.

In order to test hypotheses about ACh synthesis and utilization via measurements of transporter activity, the Rapid Autopsy Program was established as a component of the Joseph and Kathleen Price Bryan Alzheimer's Disease Research Center at Duke University Medical Center in 1985. Over several years, a collection of AD and control brains were obtained, the majority of which were retrieved within two hours of death. During the autopsy procedure, one brain hemisphere was dissected and used fresh for measurements of transporter activity (uptake of [^{3}H]choline into synaptosomes[5]) and choline acetyltransferase activity,[6] whereas the remaining hemisphere was frozen and used to measure transporter gene expression by assessing [^{3}H]hemicholinium-3 binding to transporter protein:[7–9] as well as choline acetyltransferase. The present paper reviews the findings of these two previously published studies and relates these data to the other published reports of transporter alterations in AD.

METHODS

The patient population was drawn from tissues collected by the Rapid Autopsy Program associated with the Joseph and Kathleen Bryan Alzheimer's Disease Research Center at Duke University Medical Center. The patients in the first study of transporter activity[8] consisted of 22 cases of pathologically confirmed AD, including one with coexisting Parkinson disease and one with coexisting multi-infarct dementia (15 F, 7 M, age range 53–93y, mean ± SE: 80 ± 2 years). The non-AD controls included three cases with dementia (chronic meningoencephalitis, Pick disease, severe nonspecific neuronal loss and gliosis), two cases of motor disease (amyotrophic lateral sclerosis), 1 case with cerebral infarcts without dementia, and three cases that were neurologically normal (metastatic melanoma, lymphoma and metastatic colon carcinoma). The non-AD group's demographic distribution in the first study was 3 F, 6 M, age range 48–79y, mean ± SE: 69 ± 3 years. In the second study of transporter expression,[7] the 24 pathologically confirmed AD patients included two with coexisting Pick's disease (16 F, 8 M, age range 57–93, mean ± SE: 79 ± 2 years). The 16 individual cases in the non-AD group included one patient with amyotrophic lateral sclerosis, one patient with gliosis, and five cases of cancer without nervous system involvement (5 F, 11 M, age range 48–79y, mean ± SE: 67 ± 2 years). Fourteen of the AD cases used in the second study of the expression of transporter in the frozen hemisphere corresponded to cases where the freshly dissected hemisphere was used in the first study of transporter activity, as were six of the non-AD control cases. The postmortem interval was less than two hours for the tissues used in the measurement of transporter activity (range 30–112 minutes, 57 ± 2 mean ± SE for the AD group and 63 ± 7 minutes for the non-AD control group) and, except for two cases that were less than four hours, was also less than two hours for the second study of transporter expression (57 ± 2 minutes for the AD group and 89 ± 13 minutes for the non-AD control group). The AD diagnosis was applied by the criteria of Khachaturian[10] and the respective group mean senile plaque and neurofibrillary tangle counts per square millimeter for the AD patients in the first study[3] were 50 ± 4 plaques and 4.5 ± 1.3 tangles in frontal cortex and 38 ± 4 plaques and 14 ± 4 tangles in the temporal cortex. The relative numbers of plaques in the control tissues were 2 ± 2 in frontal cortex and 1 ± 1 in the temporal cortex with no neurofibrillary tangles detected in either cortical region in the controls. In the second study,[7] the mean cortical plaque count in the AD group was 44 ± 3 and the mean cortical neurofibrillary tangle count was 9 ± 2. This compared to a group mean of 1 ± 2 senile plaques and no neurofibrillary tangles in the temporal cortex of the controls. None of the patients was receiving anticholinergic or antihistaminic medication. Statistical significance was determined by post-hoc analysis (Fisher's Protected Least Significant

Difference) after two-way ANOVA for diagnostic category and brain region (FIGS. 1 and 2).

RESULTS

The group mean activity of choline acetyltransferase in the AD samples was found to be reduced in all four cortical regions examined in the first study[8] relative to the non-AD controls. This decrease in choline acetyltransferase activity was not seen in the caudate, putamen or hippocampus. In contrast, the rate of choline uptake by the synaptosomal preparation was *increased* in the cortical AD samples and the ratio of choline uptake to choline acetyltransferase activity, a measure of uptake per nerve terminal, was increased even more. These changes were not seen in the subcortical regions. Because each individual case has unique parameters during the agonal state as well as differences in age and proximate cause of death, the putamen was chosen as an internal standard that would share these other variables. When regional choline acetyltransferase activity and transporter activity were compared using putamen values as a denominator, the statistical significance of the increased transporter and decreased choline acetyltransferase activity in the AD group cortical samples was further enhanced, to values as much as 400% above control.

The increases in choline uptake in regions affected by AD could represent nonspecific metabolic activation of the tissue, or alternatively, could represent specific upregulation of choline metabolism. Accordingly, we examined whether high-affinity choline transporter expression was affected similarly.[7] The relative number of transporter sites as assessed by [^{3}H]hemicholinium-3 binding was significantly increased in the frontal cortex of the AD group, without changes in transporter affinity: putamen transporter expression was not affected. As found before, choline acetyltransferase was significantly lowered in the AD cortical regions, commensurate with loss of cholinergic terminals. When the ratio of transporter binding relative to choline acetyltransferase activity was calculated (transporter sites per nerve terminal), the AD group displayed even greater cortical increases and again, the putamen samples were not affected. The correlation between the transporter activity measured in the first study with fresh tissue[8] to the expression of [^{3}H]-hemicholinium binding sites in the frozen tissue in the second study[7] from patients with corresponding samples in both studies was highly significant ($r = 0.7$, $p < 0.0001$).

DISCUSSION

These data indicate that, when samples are obtained within two hours of death, the expression and activity of the choline transporter protein is greatly

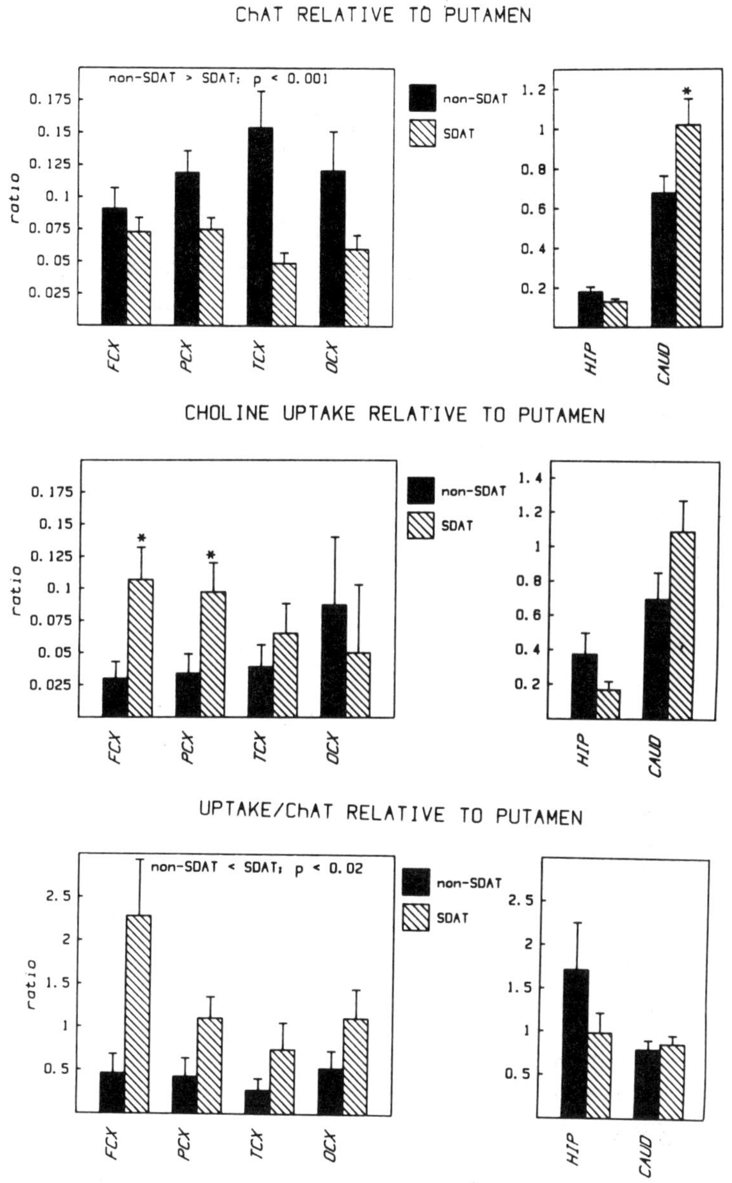

FIGURE 1. Choline acetyl-transferase activity and high-affinity [^{3}H]choline uptake into synaptosomes, assessed as a ratio to values in the putamen. Statistical differences are indicated for choline acetyltransferase and uptake/choline acetyltransferase and reflect main effects attributable to diagnosis (two-way ANOVA). Asterisks indicate individual regions where the non-AD and AD groups differ (Duncans multiple range test), assessed *post hoc* only where ANOVA indicated a significant interaction of region and diagnosis: significance was $p < 0.01$ for caudate (CAUD) choline acetyltransferase relative to putamen, $p < 0.01$ for frontal cortex (FCX) choline uptake relative to putamen, and $p < 0.02$ for parietal cortex (PCX) choline uptake relative to putamen. Other abbreviations: choline acetyltransferase (ChAT), hippocampus (HIP), temporal cortex (TCX), and occipital cortex (OCX).

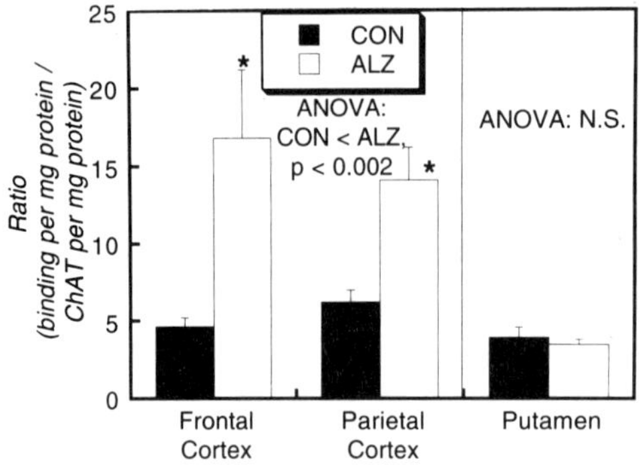

FIGURE 2. Ratio of [H]hemicholinium-3 binding capacity to choline acetyltransferase activity in rapid autopsy tissue obtained from controls (CON) and patients with Alzheimer's disease (ALZ). Two-factor ANOVA across all three regions indicates a significant main effect of diagnosis ($p < 0.0001$) and an interaction of diagnosis X region ($p < 0.0008$). ANOVAs conducted separately for the two cortical regions and for the putamen are shown within the panels, and asterisks denote significant differences from corresponding control values. In addition, the effect in the frontal cortex is distinguishable from that in the putamen (interaction of diagnosis X region, $p < 0.0003$), as is the effect in the parietal cortex ($p < 0.004$). Effects in frontal cortex are equivalent to those in the parietal cortex (no interaction of diagnosis X region).

increased in AD cortical regions undergoing loss of choline acetyltransferase activity. These results run counter to the view that the loss of cholinergic function in AD involves neuronal death as the primary cause of dysfunction, since if that were true, all cholinergic presynaptic markers would decline similarly. The absence of any such changes in transporter expression or function in the putamen, a region unaffected by the disease, indicates the abnormality in choline transport is not an inherent genetic defect that would be expressed throughout the brain, but rather is specifically associated with regions undergoing cholinergic degeneration.

The rise in choline transport is far too large to be accounted for simply by compensatory adjustments to the decline in the number of terminals. Compensatory changes would restore uptake to control levels, but in fact the cortical AD regions show elevations well above normal. Indeed, animal studies with cholinergic lesions do not indicate any net upregulation of choline transporter activity in the surviving neurons; thus, it is unlikely that the increase is secondary to neurodegeneration, but rather that the change in transporter expression may be a primary component of the disease. Choline is utilized for membrane phospholipid formation in every cell, but only in cholinergic neurons is it also needed for neurotransmitter synthesis. Our results thus support the idea[11] that a defect in choline utilization within the cell, in the presence of increased neurotransmitter turnover, leads to neuronal "autocannibalization" of membrane phospholipids containing choline, and eventually to degeneration. In this scenario, inadequate incorporation of choline into membrane components and/or neurotransmitter leads to a positive feedback loop in which depletion of transmitter stores contributes to failure of neurotransmission, producing compensatory increases in impulse activity and a further increase in demand for choline. Increased choline transport, while enhancing availability of choline, would not correct the underlying defect. Furthermore, because choline transport is responsive to impulse activity, the degree of upregulation seen in AD cortical regions may represent a pathologic state of neuronal "overdrive." Convulsant doses of stimulants in laboratory animals, treatments that clearly damage neurons, raise transporter activity only 30%, whereas we found increases of up to 400%.

Our results represent an apparent contradistinction from several reports of transporter activity in standard autopsy material, which show a general decrease in binding in cortical regions and hippocampus in AD. These differences are most likely attributable to short vs. prolonged postmortem delay in obtaining the tissues, as it is precisely the upregulation associated with elevated impulse activity that decays the most rapidly after death. Using tissue from rapid autopsy, we have shown that choline uptake is elevated in cortical regions of Alzheimer's disease patients,[8] whereas after a lengthy postmortem wait, the elevation is lost.[12] The results obtained from rapid autopsy are thus more likely to correspond to the actual conditions present in the intact presynaptic terminal, because they exhibit activity-related changes whereas the samples obtained after a delay do not.

An underlying defect in choline utilization in AD has important consequences for potential therapeutic approaches to restoring cholinergic function. Attempts to increase cholinergic nerve impulse activity, in the presence of already pathologic upregulation, would worsen the metabolic consequences of the defect, and might be expected to produce short-term improvement in synaptic function at the expense of accelerating neurodegeneration. Similarly, without correcting the underlying metabolic defect, choline loading would be ineffective. The proper strategy is to break the positive feedback loop between aberrant choline utilization and upregulation of cholinergic activity. Based on our findings, administration of drugs that *reduce* cholinergic nerve activity and demand for choline would be appropriate; because they would worsen cholinergic components of cognitive function, these effects could be corrected with agents that act on postsynaptic cholinergic function, such as cholinergic receptor agonists and cholinesterase inhibitors.

REFERENCES

1. DAVIES, P. & A. J. F. MALONEY. 1976. Selective loss of central cholinergic neurons in Alzheimer's Disease. Lancet **2:**1403.
2. WHITEHOUSE, P. J. 1993. Cholinergic therapy in dementia. Acta Neurol. Scand. **88:**42–45.

3. SIMON, J. R., S. ATWEH & M. J. KUHAR. 1976. Sodium-dependent high affinity choline uptake, a regulatory step in the synthesis of acetylcholine. J. Neurochem. **26:**909–922.
4. SIMS, N. R., D. M. BOWEN, S. J. ALLEN, C. C. T. SMITH, D. NEARY, D. J. THOMAS & A. N. DAVISON. 1983. Presynaptic cholinergic dysfunction in patients with dementia. J. Neurochem. **40:**503–509.
5. SHELTON, D. L., J. V. NADLER & C. W. COTMAN. 1979. Development of high affinity choline uptake and associated acetylcholine synthesis in the rat fascia dentata. Brain Res. **163:**263–275.
6. LAU, C., F. J. SEIDLER, A. M. CAMERON, H. A. NAVARRO, J. M. BELL, J. BARTOLOME & T. A. SLOTKIN. 1988. Nutritional influences on adrenal chromaffin cell development, comparison with central neurons. Pediatr. Res. **24:** 583–587.
7. SLOTKIN, T. A., C. B. NEMEROFF, G. BISSETTE & F. J. SEIDLER. 1994. Overexpression of the high affinity choline transporter in cortical regions affected by Alzheimer's Disease, evidence from rapid autopsy studies. J. Clin. Invest. **94:** 696–702.
8. SLOTKIN, T. A., F. J. SEIDLER, B. J. CRAIN, J. M. BELL, G. BISSETTE & C. B. NEMEROFF. 1990. Regulatory changes in presynaptic cholinergic function assessed in rapid autopsy material from patients with Alzheimer disease: Implications for etiology and therapy. Proc. Natl. Acad. Sci. USA **87:**2452–2455.
9. VICKROY, T., W. ROESKE & H. YAMAMURA. 1984. Sodium-dependent high-affinity binding of [^{3}H]hemicholinium-3 in the rat brain: A potentially selective marker for presynaptic cholinergic sites. Life Sci. **35:**2335–2343.
10. KHACHATURIAN, Z. S. 1985. Diagnosis of Alzheimer's Disease. Arch. Neurol. **42:**1097–1105.
11. WURTMAN, R. J. 1992. Choline metabolism as a basis for the selective vulnerability of cholinergic neurons. Trends Neurosci. **15:**117–122.
12. RYLETT, R. J., M. J. BALL & E. H. COLHOUN. 1983. Evidence for high affinity choline transport in synaptosomes prepared from hippocampus and neocortex of patients with Alzheimer's disease. Brain Res. **289:**169–175.

Occurrence of Interleukin-6 in Cortical Plaques of Alzheimer's Disease Patients May Precede Transformation of Diffuse into Neuritic Plaques[a]

M. HÜLL,[b] M. BERGER,[b] B. VOLK,[c] AND J. BAUER[b,d]

[b] *Department of Psychiatry and* [c]*Department of Neuropathology, Freiburg University Medical School, D-79104 Freiburg, Germany*

ABSTRACT: Interleukin-6 (IL-6) immunoreactivity has previously been shown in plaques in Alzheimer's disease (AD), and elevated IL-6 concentrations have been measured biochemically in brains of AD patients. In this report, we present data on the appearance of IL-6 immunoreactivity in AD plaques according to the stage of plaque formation. Diffuse plaques are found in the early stages of plaque formation, whereas primitive and classic plaques are thought to represent later stages of plaque pathology. We classified plaques using the Bielschowsky silver stain method in serial sections of paraffin-embedded cortices of clinically diagnosed and histopathologically confirmed AD patients and patients with no clinical history of dementia. In the brains of nondemented and demented persons, we found plaques using the silver staining method or immunohistochemistry with antibodies against the amyloid precursor protein. In the nondemented group, diffuse plaques were the predominant plaque type, whereas primitive plaques formed the larger proportion of lesions in the group of AD brains.

IL-6 was only detectable in plaques of demented patients. In AD cases, IL-6 was found in a significantly higher ratio in diffuse plaques as would have been expected from a random distribution of IL-6 in all plaque types.

We conclude that the presence of IL-6 immunoreactivity correlates with clinically detectable dementia. In addition to the ubiquitous presence of amyloid in nondemented and demented brains, an IL-6-related immunological mechanism may be involved both in the transformation from diffuse to primitive plaques in AD and in the development of dementia.

INTRODUCTION

The cytokine interleukin-6 (IL-6) plays an important role in a number of chronic inflammatory diseases or degenerative processes (for reviews see ref.

[a] This study was supported by DFG grant Ba883/5-1, SFB364/A5 and Thyssen Stiftung grant 1993/1/69.

[d] *Send correspondence to:* Dr. Joachim Bauer, Prof. of Psychoneuroimmunology, Psychiatrische Universitätsklinik, Hauptstr. 5, D-79104 Freiburg, Germany; Fax: 0761-270-6917.

1). In the central nervous system, IL-6 is involved in diseases such as meningitis, encephalitis and HIV infection.[2] In Alzheimer's disease (AD), IL-6 immunoreactivity has previously been found in plaques,[3] and high concentrations of IL-6 have been detected biochemically in AD brains.[4] IL-6 has recently been shown to cause neurodegeneration in a transgenic mice model.[5]

After silver staining of histological sections, four different plaque types can be classified: diffuse ("pre-primitive"), primitive, classic ("mature") and compact plaques ("burned-out").[6–9] The terms in brackets are sometimes used to emphasize the developmental aspects of plaque formation from a pre-primitive to a primitive and finally to a mature stage.[9] In diffuse plaques, neuritic changes are absent, whereas marked neuritic pathology prevails in primitive and classic plaques.

To determine the time point of IL-6 involvement during plaque formation, we identified the plaque type by the Bielschowski silver staining and used immunohistochemistry with antibodies against IL-6 on adjacent sections. We were interested in whether IL-6 immunoreactivity correlates with a certain plaque type.

MATERIALS

Tissues from ten autopsy cases of patients between the ages of 53 and 85 years (73.8 ± 10.9) with clinically diagnosed and neuropathologically conformed Alzheimer's disease and ten age-matched control cases without any clinical history of dementia were used for this histological study. There was no history of any other neurological disease in both groups.

From the AD cases as well as from the control cases, serial sections of 5 μm were obtained from formalin-fixed and paraffin-embedded tissues of isocortical and hippocampal areas. Consecutive sections were stained with either a modified Bielschowski silver stain[10] or immunohistochemically with a polyclonal antibody against human IL-6 (Genzyme, 1:50), or a monoclonal antibody against amyloid-precursor-protein (Boehringer, 1:50) directed against amino acids 60–100 of APP. Before immunohistochemistry, sections were heated in a tissue-antigen unmasking fluid (TUF, Dianova, Germany), which selectively enhanced primary antibody binding. Treatment with primary antibody was followed by standard avidin-biotin techniques using the Vectastain Elite ABC kit (Vector Laboratories) and diaminobenzidine (DAB, Serva). Omission of the primary antibody or preincubation of IL-6 antibodies with recombinant IL-6 (Boehringer) abolished staining. Sections were counterstained with haematoxylin, dehydrated and then mounted with Vitro-Clud (Langenbrinck, Germany).

After silver staining, plaques were classified according to morphology as diffuse, primitive, classic or compact as described previously.[7,9] Silver stained sections of each region of each brain were used to determine the proportion of each plaque type. The prevalence of each plaque type was expressed as a percentage of the total of all classified plaques for each case and region. The plaque type of the IL-6 positive plaques was determined by comparing IL-6-immunoreactive spots and silver stained plaques in serial sections. The distribution of plaque morphology of IL-6 positive plaques was expressed in percent of all IL-6 positive plaques.

RESULTS

We stained serial sections using the Bielschowsky method and with antibodies against APP. In all cases nearly every lesion identified by the silver staining was also APP positive and vice versa (FIG. 1), similar to other studies comparing immunohistochemistry and silver staining.[11] However, we were unable to distinguish diffuse or primitive plaques solely by immunohistochemical methods.

In clinically nondemented cases, the distribution of the different plaque types did not vary much between different cases or brain regions. Most plaques were diffuse (75.7% ± 9%) while the remaining were primitive (24.3% ± 9%). In this nondemented group, nearly no classic or compact plaques were detectable. In the AD group, the distribution of plaque type showed a wide variation between different cases, and even between different cortical areas of one case; 44.2% (± 17.4%) of all lesions were diffuse, 52.3% (± 16.1%) were primitive, 2.8% (± 4.5%) were classic and 0.6% (± 1%) were compact plaques.

No IL-6-immunoreactivity could be detected in plaques of nondemented cases. In contrast, all 10 AD brains displayed IL-6 positive plaques. We used 6 AD cases for detailed analyse of how IL-6 immunoreactivity was distributed among the different plaque types (FIG. 2).

IL-6 immunoreactivity was rarely seen in classical plaques and completely absent in compact plaques. Most of the IL-6 positive plaques were diffuse (70.8% ± 11.2%) and the remaining IL-6 positive plaques were primitive (27.7% ± 11.7%).

We calculated for each case and cortical area the ratio of [percentage of diffuse plaques in IL-6 positive plaques] divided by [percentage of diffuse plaques in all plaques]. If IL-6 is preferentially present in the early plaque types then the ratio should be more than 1 (FIG. 3). Among the various AD cases, this ratio showed a range from a slight overrepresentation of IL-6 in diffuse plaques (ratio: 1.08) to a strong overrepresentation of IL-6 in diffuse plaques (ratio: 2.55). The mean of this ratio for all six AD cases was 1.67 (± 0.55). IL-6 is therefore found 1.6 times more often in diffuse early plaques than expected from a random distribution.

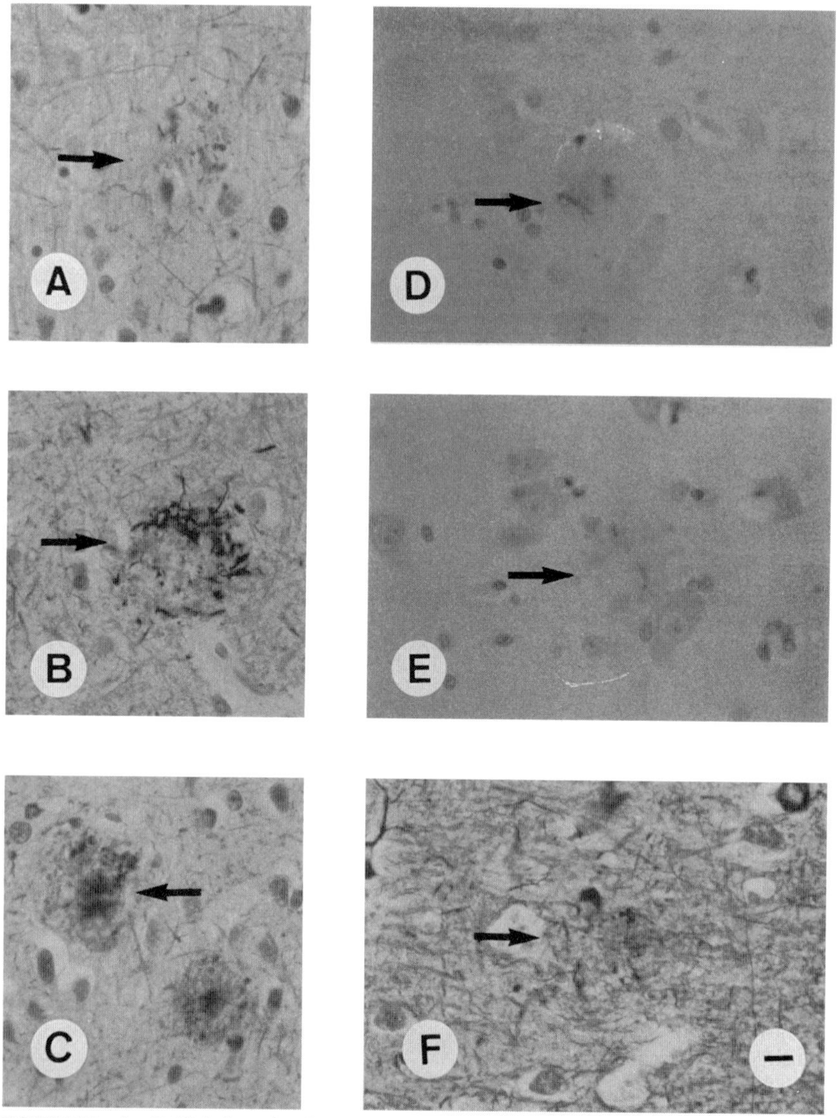

FIGURE 1. **A–C:** Bielschowsky silverstaining of different plaque types: diffuse **(A)**, primitive **(B)** and classic **(C)** plaque. Spacebar in **(F)** 10 mm. **D–F:** Serial sections from the cortex of an AD patient: **D:** Immunohistochemical staining with an anti-interleukin-6 antibody. **E:** Immunohistochemical staining with an anti-APP antybody of a section adjacent to **D. F:** Bielschowsky silverstaining of a section adjacent to **E,** revealing a diffuse plaque morphology.

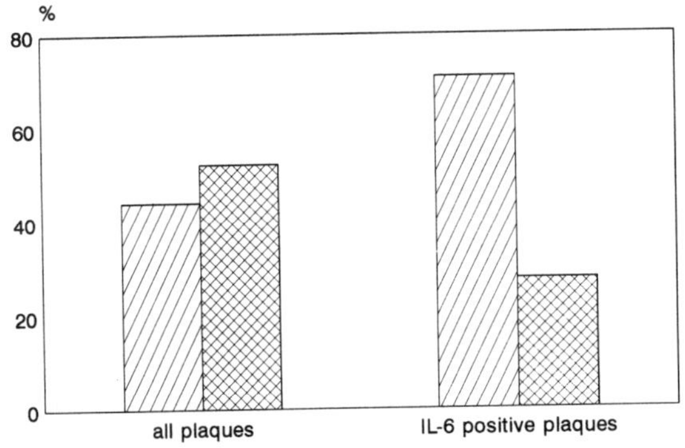

FIGURE 2. Histogram showing the mean proportion of diffuse and primitive plaques for 6 AD cases and the mean proportion of diffuse and IL-6 immunoreactive or primitive and IL-6 immunoreactive plaques of these cases. ▨ = diffuse; ⊠ = primitive.

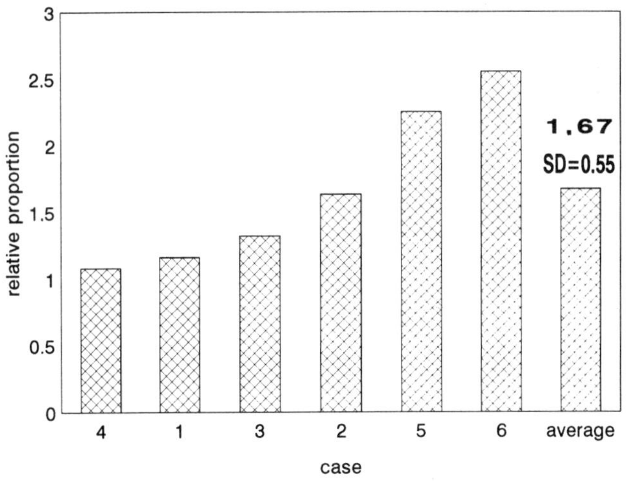

FIGURE 3. Histogram showing relative proportion of diffuse IL-6-immunoreactive plaques for 6 AD cases.

DISCUSSION

An involvement of the immune system in the pathogenesis of dementia has been suggested by recent studies (for review, see ref. 12). IL-6 is the most potent inducer of acute phase proteins like α_1-antichymotrypsin or c-reactive protein which are present in elevated amounts in AD brains.[3,13] IL-6 can be synthesized by microglial cells and astrocytes,[14] and neuroblastoma cells react to IL-6 by synthesizing metallothionein.[15] Both IL-6 and complement proteins[3,16] are present in plaques in AD. Therefore, a local "brain-specific acute phase response" may be involved in the initiation or progression of dementia in AD.

Our study focused on the correlation of plaque type and the presence of IL-6. Four different plaque types can be identified by silver staining procedures: diffuse, primitive, classic and compact plaques.[7,9] Most investigators believe in a temporal progression of neuropathology from diffuse (early) to primitive and classic plaques.[11,17]

In the nondemented control group we classified approximately 75% of the plaques as diffuse while neuritic plaques were predominant (52%) in the AD cases. These results are in agreement with previous studies which show that diffuse plaques represent the vast majority of plaques found in nondemented cases and less than half of the plaques in AD subjects.[6,7]

Our results demonstrate that IL-6 is present only in AD cases.[3] Consistent with this observation elevated concentrations of IL-6 were found only in brains of AD patients.[4] We found IL-6 immunoreactivity mainly in plaques, however in some cases IL-6 immunoreactivity was not accompanied by the presence of a silverstainable plaque on adjacent sections. Diffuse plaques were more than 1.5 fold overrepresented among IL-6 positive plaques. This clearly shows that IL-6 is associated with the early stages of plaque formation in AD brains,[11] indicating that the occurrence of IL-6 does precede neuritic degeneration.

In addition to IL-6, α_1-antichymotrypsin and activated complement are present already in diffuse plaques of AD cases.[17] Complement factors might be able to form membrane attack complexes and destroy neurites.[16]

Most studies have found only a weak correlation between the number of all plaques and dementia.[6] The amount of diffuse plaques, representing deposits of βA4 protein, showed no correlation with dementia.[6,8,18] Neuritic changes mark the transformation from diffuse to primitive plaques. Since a high proportion of primitive plaques seems to correlate better with dementia, neuritic changes may be critical for intellectual decline.

Two theories are discussed to explain the neuritic changes in AD. One theory focuses on a cascade driven by pathological changes inside the plaque.[9] A reverse cascade theory favors a general pathological process throughout the neuropil.[19] Independent from these theories, our data suggest that an IL-6-mediated process could be involved in neuritic pathology. This transformation might be an important step in the development of dementia.[9,18]

The possible importance of IL-6 as part of the neurodegenerative cascade in AD was recently demonstrated in a transgenic animal model in which IL-6-is overexpression in astrocytes.[5]

IL-6 synthesis is classically induced by viral or bacterial infection which is not the case in chronic inflammatory diseases or AD. Changes of cortical neurotransmitters may induce IL-6 synthesis.[14] In addition, psychological stress has been shown to cause elevated peripheral IL-6 concentrations.[20]

During aging in humans, baseline IL-6 levels gradually rise, suggesting a decrease in the control of IL-6 synthesis. Therefore, neurons and glial cells might also be exposed to higher concentrations of IL-6 during aging which might, along with other factors, induce pathological changes.

In conclusion, our results show an early and specific involvement of IL-6 in the progression of morphological changes in AD. Therefore, inflammatory processes may influence the development of dementia in AD.

ACKNOWLEDGMENT

We thank Dr. K. Tamai for valuable discussion of the manuscript.

REFERENCES

1. ERSHLER, W. B. 1993. Interleukin-6: A cytokine for gerontologists. J. Am. Geriatr. Soc. **41**:176–181.
2. LAURENZI, M. A., A. SIDEN, M. A. PERSSON, G. NORKRANS, L. HAGEBERG & F. CHIODI. 1990. Cerebrospinal fluid interleukin-6 activity in HIV infection and inflammatory and non-inflammatory diseases of the nervous system. Clin. Immunol. Immunopathol. **57**:233–241.
3. BAUER, J., S. STRAUSS, U. SCHREITER-GASSER, U. GANTER, P. SCHLEGEL, I. WITT, B. VOLK & M. BERGER. 1991. Interleukin-6 and α2-macroglobulin indicate an acute phase state in Alzheimer's disease cortices. FEBS Lett. **285**: 111–114.
4. WOOD, J. A., P. L. WOOD, R. RYAN, N. R. GRAFF-RADFORD, C. PILAPIL, Y. ROBITAILLE & R. QUIRION. 1993. Cytokine indices in Alzheimer's temporal cortex: No changes in mature IL-1 beta or IL-1 RA but increases in the associated acute phase proteins IL-6, alpha-2-macroglobulin and C-reactive protein. Brain Res. **629**:245–252.
5. CAMPBELL, I. L., C. R. ABRAHAM, E. MASLIAH, P. KEMPER, J. D. INGLIS, M. B. A. OLDSTONE & L. MUCKE. 1993. Neurologic disease induced in transgenic mice by cerebral overexpression of interleukin 6. Proc. Natl. Acad. Sci. USA **90**:10061–10065.
6. DELAERE, P., C. DUYCKAERTS, Y. HE, F. PIETTE & J. J. HAUW. 1991. Subtypes and differential laminar distributions of βA4 deposits in Alzheimer's disease: Relationship with the intellectual status of 26 cases. Acta Neuropathol. Berl. **81**:328–335.
7. HALLIDAY, G., D. FLOWERS & L. BAUM. 1994. Analysis of staining methods for different cortical plaques in Alzheimer's disease. Acta Neuropathol. Berl. **87**:174–188.
8. SPARKS, D. L., H. LIU, S. W. SCHEFF, C. M. COYNE & J. C. HUNSAKER. 1993. Temporal sequence of plaque formation in the cerebral cortex of non-demented individuals. J. Neuropathol. Exp. Neurol. **52**:135–142.
9. WISNIEWSKI, H. M. & R. D. TERRY. 1973. Reexamination of the pathogenesis of the senile plaque. Prog. Neuropathol. **2**:1–26.

10. YAMAMOTO, T. & A. HIRANO. 1986. A comparative study by modified Bielschowsky, Bodian and thioflavin S stains on Alzheimer's neurofibrillary tangles. Neuropathol. Apl. Neurobiol. **12:**3–9.
11. IKEDA, S., N. YANAGISAWA, D. ALLSOP & G. G. GLENNER. 1990. Early senile plaques in Alzheimer's disease demonstrated by histochemistry, immunocytochemistry and electron microscopy. Hum. Pathol. **21:**1221–1226.
12. AISEN, P. S. & K. L. DAVIS. 1994. Inflammatory mechanisms in Alzheimer's disease: Implications for therapy. Am. J. Psychiatry **151:**1105–1113.
13. ABRAHAM, C. R., D. J. SELKOE & H. POTTER. 1988. Immunochemical identification of the serine protease inhibitor alpha 1-antichymotrypsin in the brain amyloid deposits of Alzheimer's disease. Cell **52:**487–501.
14. NORRIS, J. G. & E. N. BENVENISTE. 1993. Interleukin-6 production by astrocytes: induction by the neurotransmitter norepinephrine. J. Neurol. Immunol. **45:**137–146.
15. BAUER, J., U. GANTER, J. ABEL, S. STRAUSS, U. JONAS, R. WEISS, P. GEBICKE-HAERTER, B. VOLK & M. BERGER. 1993. Effects of interleukin 1 and interleukin 6 on metallothionin and amyloid precursor protein expression in human neuroblastoma cells. J. Neurol. Immunol. **45:**163–173.
16. MCGEER, P. L., H. AKIYAMA, S. ITAGAKI & E. G. MCGEER. 1989. Activation of the classical complement pathway in brain tissue of Alzheimer patients. Neurosci. Lett. **107:**341–346.
17. ROZEMULLER, J. M., P. EIKELENBOOM, F. C. STAM, K. BEYREUTHER & C. L. MASTERS. 1989. βA4 protein in Alzheimer's disease: Primary and secondary cellular events in extracellular amyloid depositin. J. Neuropathol. Exp. Neurol. **48:**674–691.
18. CRYSTAL, H. A., D. W. DICKSON, M. J. SLIWINSKI, R. B. LIPTON, E. GROBER, H. MARKS-NELSON & P. ANTIS. 1993. Pathological markers associated with normal aging and dementia in the elderly. Ann. Neurol. **34:**1682–1687.
19. PERRY, G. 1993. Neuritic plaques in Alzheimer's disease originate from neurofibrillary tangles. Med. Hypotheses **40:**257–258.
20. ZHOU, D., A. W. KUSNECOV, M. R. SHURIN, M. DEPAOLI & B. S. RABIN. 1993. Exposure to physical and psychological stressors elevates plasma interleukin-6, relationship to the activation of hypothalamic-pituitary-adrenal axis. Endocrinology **133:**2523–2530.

Anti-Inflammatory Drugs in the Fight against Alzheimer's Disease[a]

P. L. McGEER AND E. G. McGEER
Kinsmen Laboratory of Neurological Research, University of British Columbia, Vancouver, B.C., Canada

ABSTRACT: Lesions in Alzheimer disease are characterized by the assembly of a variety of cells and proteins associated with the immune system. Activated microglia express high levels of MHC glycoproteins and receptors for complement. Small numbers of T-lymphocytes infiltrate tissue. Proteins of the classical complement pathway are closely connected with β-amyloid deposits. Several materials associated with senile plaques, including β-amyloid protein itself, bind C1q *in vitro* and activate the pathway. The membrane attack complex of complement, as well as proteins which defend against that complex, colocalize with dystrophic neurites. These data imply that an autodestructive process is occurring in Alzheimer's disease, and that anti-inflammatory drugs might be an effective form of therapy. Some epidemiological evidence and results of a pilot clinical trial support this hypothesis.

INTRODUCTION

It has often been suggested that the brain is immunologically privileged. The existence of the blood brain barrier, the absence of conventional lymphatic drainage, and the unusual tolerance of the brain to transplanted tissue have reinforced the idea that brain is somehow isolated from immunological processes. However, recent evidence, derived from the modern techniques of immunohistochemistry and molecular biology, indicates that the brain is neither immunologically privileged nor immunologically isolated. In fact, it appears to be immunologically vulnerable, owing to the fragility of neurons and their complex interconnections. The immune system is intended to be injurious to invading organisms. Any response to an immunological challenge places host tissue at risk. Neurons are post-mitotic cells, with no ability to divide, and little ability to recover from injury. Therefore, processes that would have little effect on many organs of the body could have serious consequences for brain.

[a] This research was supported by grants from the Alzheimer Society of B.C. and the Jack Brown and Family A.D. Research Fund, as well as donations from individual British Columbians.
[b] *Send correspondence to:* Dr. Patrick L. McGeer, Kinsmen Laboratory of Neurological Research, University of British Columbia, 2255 Wesbrook Mall, Vancouver, B.C. V6T 1Z3 Canada; Tel: (604) 822-7377; Fax: (604) 822-7086.

In this chapter we will describe briefly evidence that there is a prominent immune response in Alzheimer's disease (AD). Particular attention will be paid to the autodestructive forces that are inherent to such a response and the potential harm they can cause. The possibility of ameliorating such damage through the use of anti-inflammatory drugs will be discussed. Detailed references to much of the original literature may be found in recent reviews.[1,2] Although the emphasis here is on AD, there is some evidence that similar immune response may play a role in a number of other chronic neurodegenerative disorders.[3]

THE IMMUNE RESPONSE OF BRAIN

A strong immune reponse is essential to protect the body against infection, but if the underlying pathology is not resolved, a chronic inflammatory state occurs as, for example, in the joints in rheumatoid arthritis. There are two forms of immune response—innate and adaptive. The innate response is immediate and requires no previous exposure to an offending agent. The adaptive response is different. This is where B cells recognize an antigen and clone antibodies to it, and where T cells identify a peptide and clone cells responding specifically to it. The response is slow, but gains power over days to weeks and has a lingering memory. But the important point to note is, whether the reponse is innate or adaptive, the efferent arm of the immune system swings into action. The complement system acts as the main method of amplification. Offending agents are marked for phagocytosis by opsonizing components of complement. They are then phagocytosed by professional phagocytes. In the case of brain, these are activated microglia. Offending cells are also directly attacked by the membrane attack complex (the MAC) of complement and by toxic oxygen species produced by microglia. The only problem is that neurons can get caught in the cross-fire.

An immune response may be local or systemic. A local response does not require many of the components that are routinely considered to be hallmarks of an autoimmune disease. Antibodies, T cells, B cells, neutrophils, peripheral monocytes, and serum proteins are all manifestations of a systemic response, but will be absent if the inflammatory response is primarily local.

Contemporary research is establishing the fact that brain can generate a local immune response. There is a growing list of immune system proteins that have been found to be produced by brain cells. For example, they can make complement protiens, inflammatory cytokines, coagulation proteins, and such toxic products as the MAC of complement and superoxide radicals.

A variety of inflammatory cytokines, including interleukin-1α, interleukin-1β, interleukin-3, interleukin-6, tumor necrosis factor-α and colony-stimulatory factor-1, are made by both microglia and astrocytes. This list will un-

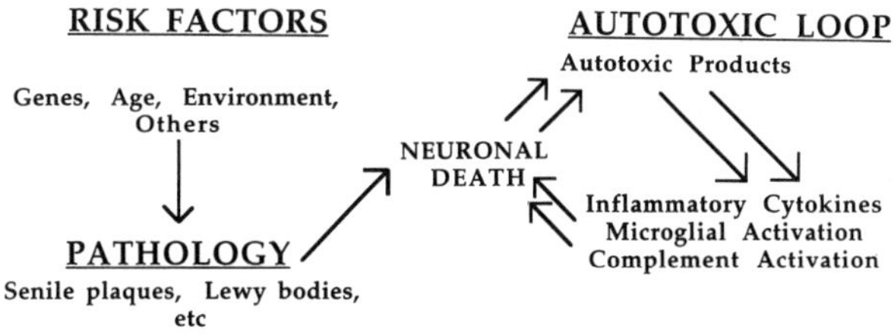

FIGURE 1. Autotoxic loop hypothesized as responsible for much of the neuronal death in AD.

doubtedly be greatly expanded in the next few years. There is a remarkable parallelism in the inflammatory cytokines made by microglia and astrocytes. T cells make some different inflammatory cytokines, namely γ-interferon, interleukin-2, and tumor necrosis factor-β. T cells seen in the vicinity of a lesion in AD[2] may well be contributing inflammatory cytokines, quite apart from any specific action by a cloned subset recognizing a distinct peptide.

THE COMPLEMENT CASCADE

The complement cascade is central to the immune response. Culture experiments indicate that complement proteins are produced by a surprising variety of brain cells. Astrocytic and neuroblastoma cell lines have been shown to make C3, C4 and C9, while microglia are known to produce C1q and C3. Many people think that the complement system is a series of serum enzymes that rush into a site of injury. This is not so at all, and complement may hold the key to many disorders where destruction of neurons occurs.

FIGURE 1 illustrates this point, with a pathology flow chart. There are certain risk factors we cannot do much about with regard to diseases like AD. We are stuck with age and our genetic make-up. These and other factors can lead to a primary pathology which results in neuronal death. But if an immune response is generated, as the autotoxic loop shows, and if this generation leads to significant autodestruction of neurons, then a potential method of intervention appears. That method is to dampen the immune response.

There are two complement pathways—the classical and the alternative—but is is only the classical which is involved in AD. It is activated when the C1q component of C1 binds to a target. Most people think this must be an antibody, but that is only one of many ways in which this cascade can be started. The β-amyloid protein, which is a prominent component of senile plaques, can activate complement—as can the pentraxins amyloid P and C-reactive protein which are also present. Binding of C1q dissociates C1 and starts the proteolytic cascade. C1r acts upon C1s which cleaves C2 and C4.

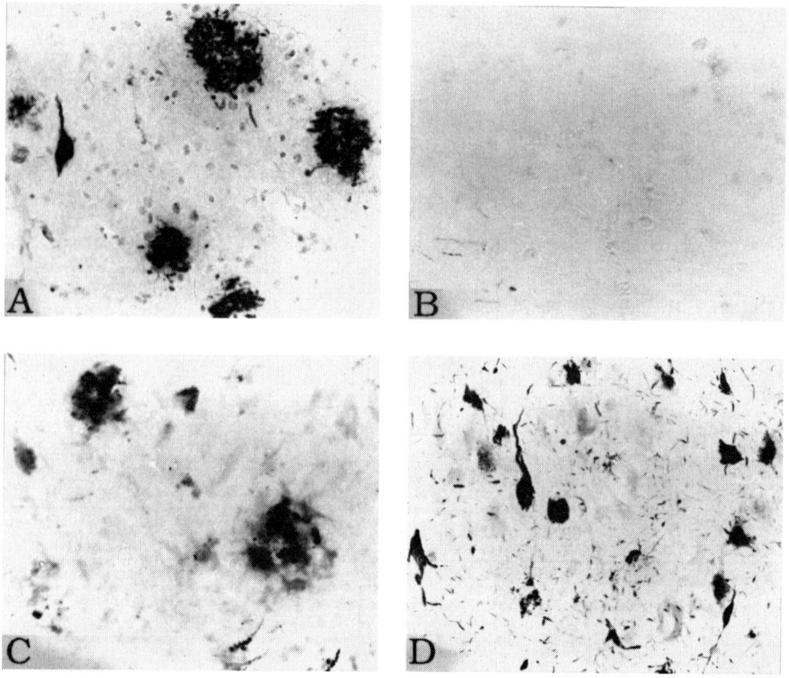

FIGURE 2. (**A**,**B**) staining for C4d in the entorhinal cortex of a case of AD (**A**) and an aged control (**B**). Amyloid deposits and neurofibrillary tangles are intensely stained. (**C**) Dark staining (purple on slide) of aggregates of HLA-DR positive microglia on senile plaque stained by an antibody to β-amyloid peptide (light staining brown on original slide). (**D**) Staining for C5b-9 (the MAC) in AD hippocampus. Strongly positive dendritic neurons and intracellular tangles can be seen; amyloid deposits are not stained.

The C4b and C2a fragments combine and form an ester or amide bond with the target tissue near the site of activation. This really nails the target, because, not only is there amplification, but, when the chemical bond forms, there is no getting away of the target. The fixed portion then cleaves C3 and the C3b portion again forms a chemical attachment to the target tissue. This completes the opsonization phase. But C5 is cleaved, starting the formation of MAC. C5b joins with C6, C7 and C8. This amphiphilic complex seeks a viable membrane in which to bury itself. If successful, multiple molecules of C9 then join the complex to create a lytic fraction.

The pathological entities of AD, senile plaques and neurofibrillary tangles, are richly decorated by C1q, C4d and C3d (FIG. 2A). The interpretation is that they are assisting with phagocytosis and acting as ligands for complement receptors. Such receptors are exclusively present on microglia and are sharply upregulated when the microglia are converted into the reactive state. Reactive microglia are embedded in senile plaques and gather near extracellular tangles

(FIG. 2C). Small numbers of T cells infiltrate the tissue.[2] Levels of inflammatory cytokines are raised.[2] However, the most direct evidence that autodestruction of neurons is occurring in AD comes from immunohistochemistry for the MAC. This is found on distorted neurites in or near senile plaques and in tangle-containing neurons in the vicinity (FIG. 2D). It is not found in association with the senile plaques, because it only attaches to viable cell membranes. The interpretation is that the MAC has formed near the opsonizing sites and is attacking neuronal membranes in the vicinity, causing damage to them and ultimately destroying them. This interpretation is strengthened by the finding that three proteins which defend host cells from the MAC are also prominently associated with these lesions. The three proteins, protectin, clusterin, and vitronectin, like the MAC itself, are not seen in normal tissue under the same immunohistochemical conditions. The interpretation is that they have been upregulated in a last ditch effort to try to fend off the MAC.

When the classical complement pathway is activated, the smaller fragments C4a, C3a and C5a diffuse away from the site to stimulate more inflammation, and especially the respiratory burst apparatus of the microglia. So this turns on yet another attack system.

THE RESPIRATORY BURST

The respiratory burst system is possessed only by professional phagocytes. Its objective is to generate toxic oxygen molecules to destroy unwanted cells. Again, the purpose is to ward off invaders, but host cells may be damaged. Appropriate tools have not yet been developed to detect evidence of respiratory burst activity in postmortem brain tissues. However, microglial cells can be studied in culture, and the capability of this system assessed. There are many different ways in which respiratory burst can be activated, including treatment of microglia with the β-amyloid peptide.[4] The end result is the production in the extracellular space of superoxide radicals, which can form hydrogen peroxide and other toxic oxygen species. When professional phagocytes like microglia are appropriately stimulated, the amount of oxygen converted into superoxide anions is many-fold higher than the amount the cells use for respiration. The amounts of toxic oxygen species generated by these cells can be orders of magnitude higher than you would find for neurons or any cell that is not a professional phagocyte. One powerful way to stimulate microglial cells to generate superoxide anions *in vitro* is to expose them to yeast particles coated with complement proteins. An effective way to prevent this is to treat the microglial cells with antibodies to complement receptors.[5] Thus, complement proteins are important to the generation of toxic oxygen species.

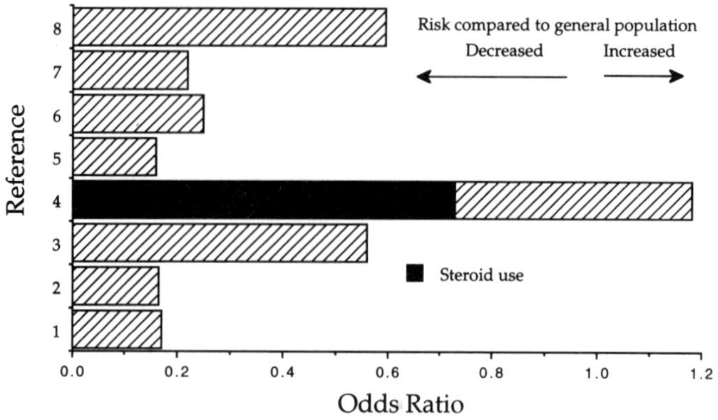

FIGURE 3. Published epidemiological studies on the frequency of clinically diagnosed AD in persons with arthritis 3 (columns 2–5,7) or those taking antiinflammatory drugs (6,8) divided by the frequency in age-matched control groups. Column 1 shows the frequency of arthritis in an AD population divided by that in controls. A frequency of AD of 2.5% in the general population over the age of 65 is assumed for the calcuation of the odds ratio for column 2. The data are taken from: (1) Jenkinson *et al.*,[10] (2) McGeer *et al.*,[11] (3) Broe *et al.*,[12] (4) Graves *et al.*,[13] (5) Li et al.,[14] (6) Breitner *et al.*,[7] (7) Myllykangas-Luosujarvi & Isomaki,[15] and (8) Beard *et al.*[16]

EPIDEMIOLOGICAL STUDIES AND PILOT CLINICAL TRIALS

The accumulating evidence of a chronic innate immune reaction in AD brain, and particularly the abundance of the MAC, led to the hypothesis that immune system attack is responsible for much of the neuronal destruction in AD. Based on this hypothesis, it was suggested that chronic treatment with antiinflammatory agents might slow the progress of AD or retard its onset.[6]

One group that routinely take such anti-inflammatory drugs, often of the non-steroidal type (NSAIDs) are arthritics, and studies of the incidence of dementia in elderly persons with arthritis have generally shown a much lower prevalence than found in the age-matched general population (FIG. 3). Perhaps the most interesting study was that of John Breitner and colleagues of Duke University.[7] They identified 50 elderly twin pairs, one of whom had AD. In going over all the factors that could account for the difference, the only one that was significant was the taking of anti-inflammatory drugs by the spared twin.

Another group that chronically takes drugs of a different type, also having anti-inflammatory actions, are sufferers from leprosy. Dapsone is the classic anti-leprosy drug. It has anti-inflammatory properties and also appears to inhibit the deposition of amyloid, even in the brain. A study of the incidence of dementia in all Japanese leprosy cases over the age of 65 indicated that those remaining on dapsone, or its close relative promin, had a significantly lower incidence of dementia than those who had been taken off treatment for at least 5 years.[8]

Proof of the hypothesis must, however, await large scale clinical trials. A pilot study has been done using indomethacin, an NSAID that crosses the blood/brain barrier.[9] In this small, but double-blind and placebo-controlled trial, the patients on 100–150 mg/day of indomethacin held their own over the six month period on a battery of cognitive tests, while the placebo-treated group showed the expected significant decline. Again, this study is only suggestive and many more trials will need to be conducted before we will know whether the anti-inflammatory therapeutic approach will work. Agents which cross the blood brain barrier will need to be selected, and different therapeutic regimens tried. It seems possible, however, that searches for new types of drugs capable of selective modulation of brain immune responses may lead to important new therapies for AD.

REFERENCES

1. AISEN, P. S. & K. L. DAVIS. 1994. Inflammatory mechanisms in Alzheimer's disease: Implications for therapy. Am. J. Psychiatry, **151**:1105–1113.
2. McGEER, P. L., J. ROGERS & E. G. McGEER. 1994. Neuroimmune mechanisms in Alzheimer disease pathogenesis. Alzheimer Dis. Assoc. Disord. **8**:149–165.
3. McGEER, E. G. & P. L. McGEER. 1994. Neurodegeneration and the immune system. *In* Neurodegenerative Disorders. D. B. Calne, Ed.: 277–300. Philadelphia: WB Saunders.
4. KLEGERIS, A., D. G. Walker & P. L. McGEER. 1994. Activation of macrophages by Alzheimer β-amyloid peptide. Biochem. Biophys. Res. Commun. **199**: 984–991.
5. KLEGERIS, A. & P. L. McGEER. 1994. Inhibition of respiratory burst in macrophages by complement receptor blockade. Eur. J. Pharmacol. **260**:273–277.
6. McGEER P. L. & J. ROGERS. 1992. Anti-inflammatory agents as a therapeutic approach to Alzheimer's disease. Neurology **42**:447–449.
7. BREITNER, J. C. S., B. A. GAU, K. A. WELSH, B. L. PLASSMAN W. M. McDONALD, M. J. HELMAS & J. C. ANTHONY 1994. Inverse association of anti-inflammatory treatments and Alzheimer's disease: Initial results of a co-twin control study. Neurology **44**:227–232.
8. McGEER, P. L., N. HARADA, H. KIMURA, E. G. McGEER & M. SCHULZER. 1992. Prevalence of dementia amongst elderly Japanese with leprosy: Apparent effect of chronic drug therapy. Dementia **3**:146–149.
9. ROGERS, J., L. C. KIRBY, S. R. HEMPELMAN, D. L. BERRY, P. L. McGEER, A. W. KASZNIAK, J. ZALINSKI, M. COFIELD, L. MANSUKHANI, P. WILLSON & F. KOGAN 1993. Clinical trial of indomethacin in Alzheimer's disease. Neurology **43**:1609–1611.
10. JENKINSON, M. I., M. R. BLISS, A. T. BRAIN & D. L. SCOTT. 1989. Rheumatoid

arthritis and senile dementia of the Alzheimer's type. *Brit. J. Rheumatol.* **28:** 86–87.

11. McGeer, P. L., E. G. McGeer, J. Rogers & J. Sibley. 1990. Anti-inflammatory drugs and Alzheimer disease. Lancet **335:**1037.
12. Broe, G. A., Henderson, H. Creasey, E. McCusker, H. E. Korten, A. F. Jorm, W. Longley & J. C. Anthony 1990. A case-control study of Alzheimer's disease in Australia, *Neurology* **40:**1698–1707.
13. Graves, A. B., E. White, T. D. Koepsell, B. V. Reifler, G. van Belle, E. B. Larson & M. Raskind 1990. A case-control study of Alzheimer's disease. *Ann. Neurol.* **28:**766–774.
14. Li, G., Y. C., Shen, C. H., Chen Y. W., Zhau & J. M., Silverman 1992. A case-control study of Alzheimer's disease in China. Neurology **42:**1481–1482.
15. Myllykangas-Luosujarvi, R. & H., Isomaki 1994. Alzheimer's disease and rheumatoid arthritis, *Brit. J. Rheumatol.*, **33:**501–502.
16. Beard, C. M., E. P. C., Kokmen, O'Brien & L. T. Kurland. 1994. Non-steroidal anti-inflammatory drug use and Alzheimer's disease in Rochester, Minnesota, *Neurobiol. Aging* **15:**S45 (abstract).

Disorders of Visual and Spatial Perception in the Early Stage of Alzheimer's Disease

G. BINETTI,[a,c] S. F. CAPPA,[b] E. MAGNI,[a] A. PADOVANI,[b]
A. BIANCHETTI,[a] AND M. TRABUCCHI[a]

[a] *Alzheimer' Disease Unit, Istituto S. Cuore Fatebenefratelli, Brescia, Italy*
[b] *Department of Neurology, University of Brescia, Italy*

ABSTRACT: A battery of visual and spatial perception tests (VOSP) has been administered to a sample of 25 mild Alzheimer's disease (AD) patients and 25 age- and education-matched controls in order to assess visuospatial skills in the early phase of the disease. Among visual object perception tests, AD patients were impaired only in the task where they had to name the stimulus (Silhouettes task). The score in this task significantly correlated with the Boston Naming Test score ($r = .73$; $p < 0.0001$). The performance on the Object decision task (which does not require naming), figure-ground analysis, and identification of degraded stimuli was unimpaired. Similarly there were no significant differences for any of the spatial perception tests. The results showing that the performances on most measures of visual and spatial perception is unimpaired, indicate that the "early" perceptual processes, as well as the access to structural knowledge, are not defective in early AD. On the other hand, AD patients are impaired at the level of semantic access and name retrieval. The finding of an highly significant correlation among tests requiring visual access to structural and semantic knowledge suggests that a rigid separation between perceptual and semantic mechanisms may be unwarranted.

INTRODUCTION

An impairment of visual and spatial perceptual abilities, or agnosia, is included among the diagnostic criteria for dementia.[1,2] The pattern of visual dysfunction present in Alzheimer's disease (AD) is considered to reflect the pathology in primary visual and association areas, without evidence of retinocalcarine (precortical) involvement.[3,4] The assessment of "higher order" perceptual abilities in AD is difficult, as most tests (for example, the naming of objects and pictures) necessarily involve semantic memory,[5] while drawing

[c] *Send correspondence to:* Dr. Giuliano Binetti, Alzheimer's Disease Unit, S.Cuore Fatebenefratelli Institute, Via Pilastroni, 29068 Brescia, Italy; Tel: 003930 3501368; Fax: 0039 30 3501366.

TABLE 1. Demographic and Clinical Data

		Controls	AD
		(n. 25)	(n.25)
Age (years)		72.1 (8.1)	72.6 (8.7)
Education (years)		8.0 (4.0)	6.9 (3.6)
Gender	M	n. 7	n. 3
	F	n. 18	n. 22
Duration of disease (mos)		—	32.4 (20.1)
MMSE (score)	***	27.2 (2.4)	21.5 (2.2)
CDR	0	n. 25	—
	0.5	—	n. 7
	1	—	n. 18
Geriatric Depression Scale (score)		10.4 (7.1)	11.6 (7.3)

*** t-test, sig $p < 0.001$.

tasks rely heavily on executive functions.[6] Figure-ground analysis has been reported to be severely affected in AD patients.[7,8]

In the present study a battery of tests of visual and spatial perception[9] has been administered to a homogeneous sample of patients with mild AD. The main objectives of the study were:

1. to assess the presence of visuospatial disorders in the early phase of disease and to investigate the profile of impairment
2. to study the correlations between impaired visuospatial abilities and other cognitive areas (in particular, language and executive functions).

MATERIALS AND METHODS

Patients and Controls

Twenty-five patients, enrolled in a large scale longitudinal study of neuropsychological function in early AD at the Alzheimer's Disease Unit at S. Cuore Fatebenefratelli Hospital (Brescia) were included. Written, informed consent was obtained from all subjects or the caregivers when appropriate. All included patients met NINCDS-ADRDA criteria for probable Alzheimer's disease.[2] Cognitive status was assessed using the Mini Mental State Examination (MMSE-10) and an extensive neuropsychological test battery.

A closely matched control group (n = 25), recruited among spouses and relatives of the patients, was submitted to the same evaluation. None had a history of neurological illness or mental decline, and all had a normal neurological examination. Demographic and clinical data are summarized in TABLE 1.

TABLE 2. VOSP Scores

VOSP Subtests	Controls	AD	Significance
• Visual perception			
Screening	19.5 (0.8)	19.4 (1.4)	ns
Incomplete letters	16.0 (3.5)	13.6 (5.1)	ns
Silhouettes	13.1 (4.5)	8.7 (4.6)	.001
Object decision	10.5 (4.7)	8.3 (3.4)	ns
Progressive silhouettes	13.2 (2.5)	13.2 (4.6)	ns
• Spatial perception			
Dot counting	9.7 (0.7)	9.5 (0.7)	ns
Position discrimination	18.1 (3.9)	18.0 (2.6)	ns
Number location	7.6 (2.4)	6.4 (2.5)	ns
Cube analysis	7.8 (2.2)	6.2 (1.9)	ns

Visual Object and Space Perception Battery (VOSP)[9]

This battery consists of 4 visual-object and 4 space perception tests. The majority of these tests require very simple responses. Each was devised to focus on one component of visual perception, while minimizing the involvement of other cognitive skills.

RESULTS

The results are summarized in TABLE 2. The performances of AD and control subjects were compared using the t-test, with Bonferroni correction for multiple comparisons. There were no significant differences for any of the spatial perception tests. Among visual object perception tests, figure-ground analysis (screening task) and identification of degraded stimuli was unimpaired. An impairment was found in the Silhouettes task, where the patients were required to name the stimulus (silhouette drawings of animals and inanimate objects, rotated through varying degrees from the lateral axis), whereas performance on the Object decision task, which does not require naming, was comparable in the two groups. Also the Progressive silhouettes task did not differentiate AD patients from controls.

The patients' scores in the Silhouettes were correlated with the following variables: age, education, Boston Naming test (BNT), Token test, WAIS Vocabulary, Wisconsin Card Sorting test (categories, errors, perseverations), Raven Coloured Progressive Matrices, Attentional matrices. The only significant correlation (with Bonferroni correction) was with the BNT score ($r = .73$; $p < 0.0001$).

DISCUSSION

The present results indicate that, in the early stages of AD, the performance on most measures of visual and spatial perception is unimpaired. A comparison of the control data from our study with a younger normative sample[9] suggests that after age 69 a considerable decrease of the performance can be expected also in the "normal" population. While great attention was paid in order to exclude subjects with visual complaints and to ensure adequate correction in normal controls, the influence of peripheral visual impairment in both the AD and the control population cannot be excluded in the absence of an in-depth neuro-ophthalmologic evaluation.

According to current models of object recognition,[11] the present study indicates that the "early" perceptual processes, as well as the access to structural knowledge, are not defective in early AD. On the other hand, AD patients are clearly impaired at the level of semantic access and name retrieval. Parallel findings have been reported in a recent study assessing famous faces recognition in AD.[12]

The impairment in tasks requiring semantic access from visual information, as well as lexical retrieval, is a well-known finding in AD neuropsychological literature.[5] The present study indicates that it can be considered one of the early features of AD. The precise locus of impairment within these different stages of information processing has been object of a lively debate. Most authors favor the central, "semantic" stage, but the role of lexical retrieval disorders cannot be discounted. The finding of a highly significant intercorrelation among tasks requiring visual access to structural and semantic knowledge suggests that a rigid separation between perceptual and semantic mechanisms may be unwarranted.[5]

REFERENCES

1. AMERICAN PSYCHIATRIC ASSOCIATION. 1987. Diagnostic and Statistical manual. 3rd edition, revised. Washington, DC: APA.
2. MCKHANN, G., D. DRACHMAN, M. FOLSTEIN, R. KATZMAN, D. PRICE & E. M. STADLAN. 1984. Clinical diagnosis of Alzheimer's disease: Report of the NINCDS-ADRDA Work Group under the auspices of Department of Health and Human Services Task Force on Alzheimer's disease. Neurology **34**: 939–944.
3. CRONIN-GOLOMB, A., S. CORKIN, J. F. RIZZO, J. COHEN, J. H. GROWDON & K. S. BANNKS. 1991. Visual dysfunction in Alzheimer's disease: Relation to normal aging. Ann. Neurol. **29**:41–52.
4. RIZZO, J. F., A. CRONIN-GOLOMB, J. H. GROWDON, S. CORKIN, T. J. ROSEN, M. A. SANDBERG, K. A. CHIAPPA & S. LESSELL. 1992. Retrocalcarine dysfunc-

tion in Alzheimer's disease. A clinical and electrophysiological study. Arch. Neurol. **49**:93–101.

5. Chertkow, H., D. Bub & D. Caplan. 1992. Constraining theories of semantic memory processing: Evidence from dementia. Cog. Neuropsychol. **9**:327–365.
6. Kirk, A. & A. Kertesz. 1991. On drawing impairment in Alzheimer's disease. Arch. Neurol. **48**:73–77.
7. Mendez, M. F., M. A. Mendez, R. Martin, K. A. Smyth & P. J. Whitehouse. 1990. Complex visual disturbances in Alzheimer's disease. Neurology **40**:439–443.
8. Trobe, J. D. & C. M. Butter. 1993. A screening test for integrative visual dysfunction in Alzheimer's disease. Arch. Ophthalmol. **111**:815–818.
9. Warrington, E. K. & M. James. The Visual Object and Space Perception Battery. Thames Valley Test Company, Bury S. Edmunds, UK, 1991.
10. Folstein, M. F., S. E. Folstein & P. R. McHugh. 1975. The Mini Mental State Examination. J. Psychiatric Res. **12**:189–198.
11. Humphreys, G. W. & M. J. Riddoch. 1987. The fractionation of visual agnosia. *In* Visual Object Processing: A Cognitive Neuropsychological Approach. G. W. Humphreys & M. J. Riddoch, Eds. London: Lawrence Erlbaum.
12. Hodges, J. R., D. P. Salmon & N. Butters. 1993. Recognition and naming of famous faces in Alzheimer's disease: A cognitive analysis. Neuropsychologia **31**:775–788.

Atrophy of the Hippocampal Formation in Early Familial Alzheimer's Disease

A Longitudinal MRI Study of at-Risk Members of a Family with an Amyloid Precursor Protein 717$_{\text{VAL-GLY}}$ Mutation[a]

N. C. FOX,[b–d] E. K. WARRINGTON,[b] J. M. STEVENS,[b,c] AND M. N. ROSSOR[b,c]

[b] *The National Hospital for Neurology and Neurosurgery, London, United Kingdom*

[c] *St Mary's Hospital, London, United Kingdom*

ABSTRACT: The hippocampal formation (HF) is known from pathological and magnetic resonance imaging (MRI) studies to become severely atrophied in established Alzheimer's disease (AD). This study examined whether changes in the HF could also be detected in very early AD by scanning subjects at risk of developing familial AD (FAD). Five at risk members of a pedigree with the amyloid precursor protein (APP) 717 valine to glycine mutation underwent serial MRI scanning with volumetric measurement of the HF as well as neurological and neuropsychological assessments. Over a period of two years two subjects became clinically affected, a loss of up to 20% of the volume of the HF occurred in the two years over which symptoms first appeared. Asymmetrical HF atrophy was shown to have been present before the development of overt symptoms. This may have important implications for early diagnosis in AD more generally.

INTRODUCTION

It is in the earliest stages of AD that clinical diagnosis is most difficult. Making an accurate early diagnosis is however important to research and to clinical practice. With the development of therapies that may slow the

[a] N. C. Fox is an Alzheimer's Disease Society Research Fellow and this research was funded by this fellowship.

[d] *Send correspondence to*: Dr. Nick Fox, Dementia Research Group, Department of Neurology, St Mary's Hospital, London W2 1NY, United Kingdom; Tel: 44-171-725-1264; Fax: 44-171-725-1422; Email: n.fox@sm.ic.ac.uk.

progression of AD it will become even more important. At present a definitive diagnosis of AD relies on histopathological confirmation and is not usually available in life. Neuroimaging has the potential to detect early changes *in vivo* in AD and follow them longitudinally. This study used magnetic resonance imaging (MRI) to quantify the earliest changes in the hippocampal formation (HF) in AD.

The acquisition of thin contiguous slice MR images allows accurate quantification of the atrophy of cerebral structures. Reproducible MR-based volumetric measurement of the HF is well established in AD.[1] Hippocampal volumes have been shown to be reduced by as much as 40% in patients with clinically diagnosed AD patients of moderate severity.[1,2] The majority of MRI studies of the HF in AD have the disadvantage that patients are scanned when they are already moderately affected. Even patients described as mildly affected (*e.g.* MMSE > 21) have had symptoms for some months or years. Studies comparing such mildly affected patients with normal controls have shown that the HF has already lost 25% of its volume at this stage of AD.[3,4] Scanning of subjects at risk of familial (FAD) provides a way of identifying the early presymptomatic cerebral changes. Longitudinal scanning allows smaller volume changes to be detected since for each subject, the initial scan can be used as a within-subject control. Mutations in the amyloid precursor protein (APP) gene are associated with familial AD with an early and relatively predictable age at onset (45–55 years) in those family members possessing the mutation.[5,7] Such families are ideal for imaging studies since the compounding effects of age-related illnesses are avoided and subjects may be scanned before symptoms appear.

METHODS

At-risk members of a large FAD APP $717_{\text{Val-Gly}}$ pedigree who were within five years of the earliest age at onset of the disease within the family were asked to participate in a longitudinal study. Details of this pedigree have previously been described.[6] Fully informed, written consent was obtained from five (all male) at-risk family members of the appropriate age (range 43–53 years). The study was approved by the ethics committees of the National Hospital for Neurology and Neurosurgery and St Mary's Hospital, London.

Annual assessments included medical and neurological examinations, neuropsychometry and an MRI volumetric scan. On each visit a complete history was taken from the subject and from the subject's spouse or other close family member. Neuropsychological tests performed at each visit comprised the WAIS-R and detailed assessments of memory (verbal and visual), language and visuospatial functions. In addition, neurological examination included the

Mini Mental State Examination (MMSE) and the Clinical Dementia Rating (CDR).

Magnetic resonance imaging was performed on a GE 1.5 Tesla Signa unit (GE Medical Systems, Milwaukee, USA). In addition to routine scanning, volumetric imaging was performed in the coronal plane, using a spoiled gradient echo technique (SPGR) with a 24 cm field of view yielding 124 contiguous 1.5 mm thick slices through the head on a 256·128 image matrix with acquisition parameters (35/5/1/35 - TR/TE/NEX/FLIP).

Images were transferred to a GE IC workstation (GE Medical Systems, Milwaukee, USA), where measurements were performed in a randomized and blinded fashion. All patient details were removed from the images and the images were chosen at random from a larger group including normal controls. In addition the images were displayed either normally or flipped left to right so that it was not possible for the observer to know which subject or which HF was being measured. Each scan was measured on three separate occasions spread over a three week period. The hippocampus was manually outlined in every slice starting from the longest length of the fornix and proceeding anteriorly. A standard anatomical atlas was used to define the structures involved.[7] The subiculum and alveus were included in the areas, but the fornix was excluded as were CSF containing spaces. Oblique sections through the associated clefts and fissures can give the appearance of an enclosed CSF space when in fact these areas are in continuity with the lateral ventricles or ambient cistern. This method has been previously described and validated in this unit and by others using normal controls and phantom studies.[8] The volume of the hippocampal formation (VHF) so defined was then calculated by summing the areas outlined and multiplying by the slice thickness (Cavalieri's principle). In addition the hippocampal cross-sectional area outlined in each slice was plotted against slice number to allow slice by slice comparison with previous scans. This allows the identification of specific regions of hippocampal volume change with time.

RESULTS

At the initial assessments, all five at-risk family members were asymptomatic in that neither they, nor their families, had any complaints relating to their cognitive function. Two years later two subjects had become symptomatic and showed significant changes in their neuropsychometry. The results here relate to the assessments and hippocampal measurements performed at the beginning and at the end of this period.

At entry into the study all subjects were working full-time. All had normal neurological assessments and all scored 29/30 or 30/30 on the MMSE and 0 on the CDR. One subject (4.1), however, commented that he was worried

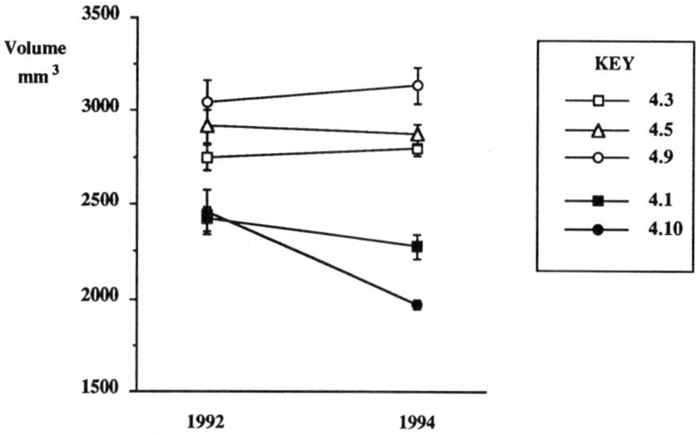

FIGURE 1. Left HF volumes of the at-risk subjects, showing dramatic atrophy over two years in the two subjects who became affected (*solid symbols*).

about his memory. He had at that stage a visual memory score which was greater than the 50th percentile and a verbal memory score between the 5th and 25th percentile. By contrast Subject 4.10 scored beneath the 1st and 5th percentile for visual memory and below the 1st percentile for verbal memory. However, neither he nor his wife considered that there was any problem and he thus fulfilled the criteria for entry into the study as an asymptomatic at-risk individual.

Over the following two years there was no significant change either clinically or in the test scores of three of the subjects. Two subjects, 4.1 and 4.10, however, became clinically affected and sought medical advice because of memory impairment at or just after the date of the second MRI scan. At this stage Subject 4.1 was still working and scored 27/30 on the MMSE and 0.5 on the CDR. Subject 4.10 became worried about his memory in the year following his first scan and began to have difficulties route finding and coping at work during the next year. He went on sick leave four months before his second scan. At the time of his second scan he was still able to score 28/30 on the MMSE and had a CDR of 0.5–1.0.

The three clinically unaffected subjects had less than 5% volume difference between left and right hippocampi and no more than 5% change in volume measurements over the two years. However, for the two subjects who became symptomatic, the volumes of their HFs changed markedly over the same period with volume losses of 6% and 13% for the left and right hippocampi of Subject 4.1 and 20% and 18% loss for the left and right hippocampi of Subject 4.10 (see FIGS. 1 and 2). Furthermore there was already significant asymmetry in these subjects hippocampi at the time of their initial scan: the volume of Subject 4.1's left HF was 8% smaller than his right when

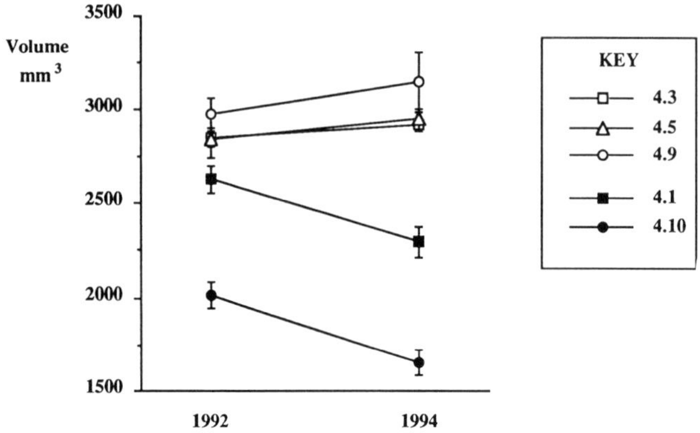

FIGURE 2. Right HF volumes of the at-risk subjects again showing a decline in hippocampal size in the affected subjects. Comparing with FIGURE 1, asymmetry of the HF is very obvious for subject 4.10 (*solid circles*) in 1992, the right being 18% smaller than the left.

asymptomatic while Subject 4.10's right HF was 18% smaller than his left at his initial assessment (TABLE 1).

DISCUSSION

This study shows that in familial AD significant hippocampal asymmetry is detectable at a very early stage in the disease, and at a time when patients may be unaware of any deficit. This suggests that atrophy of the hippocampal has occurred over some period before symptoms appear. The two subjects

TABLE 1. HF Volumes: The Lower Half Shows the at-Risk Subjects Who Became Affected

Subject (age at scans)	Hippocampus	VHF/mm^3 ($\pm$95%CI) 1992	VHF/mm^3 ($\pm$95%CI) 1994	%VHF 1992/1994
4.3 (39 & 41 yrs)	Left	2746 (±68)	2796 (±42)	102%
	Right	2852 (±110)	2918 (±32)	102%
4.5 (50 & 52 yrs)	Left	2918 (±91)	2873 (±57)	98%
	Right	2841 (±44)	2952 (±56)	104%
4.9 (53 & 55 yrs)	Left	3044 (±121)	3139 (±99)	103%
	Right	2980 (±82)	3148 (±158)	105%
4.10 (46 & 48 yrs)	Left	2456 (±120)	1971 (±25)	80%
	Right	2014 (±69)	1657 (±69)	82%
4.1 (43 & 45 yrs)	Left	2421 (±64)	2274 (±64)	94%
	Right	2625 (±71)	2293 (±83)	87%

who became affected during this study had had abnormal memory test results at the time of their first scan. It remains to be established by continued longitudinal follow up of other at-risk subjects whether changes are detectable with structural imaging even before neuropsychological abnormalities appear. The rate of atrophy of the HF is clearly considerable at up to 10% per year even at this early stage. These results also suggest that HF volume loss associated with AD may be asymmetric initially but progress to affect both sides equally. From the results given here no conclusions are possible about selective hippocampal changes relative to other cerebral structures, but further work is in progress to determine this and to identify those areas within the HF that are affected first. The consistency of the findings described in these family members needs to be confirmed by longitudinal scanning of further subjects in this and other families. Analysis of future scans may provide information on how the rate of HF change varies with time and between sides. The confirmation of these results in a larger sample may have important implications for the early diagnosis of AD and the development of reliable measures of early disease progression.

ACKNOWLEDGMENTS

We thank the staff of the MRI unit at St Mary's Hospital and the family members who helped with the study.

REFERENCES

1. Jack, C. R., Jr., R. C. Petersen & P. C. O'Brien. 1992. MR-based hippocampal volumetry in the diagnosis of Alzheimer's disease. Neurology **42**:183–188.
2. Seab, J. B., W. J. Jagust & S. T. S. Wong. 1988. Quantitative NMR Measurements of hippocampal atrophy in Alzheimer's disease. Magn. Reson. Med. **8**: 200–208.
3. Lehericy, S., M. Baulac & J. Chiras. 1994. Amydalohippocampal MR volume measurements in the early stages of Alzheimer disease. AJNR **15**:927–937.
4. Killiany, R. J., M. B. Moss & M. S. Albert. 1993. Temporal lobe regions on magnetic resonance imaging identify patients with early Alzheimer's disease. Arch. Neurol. **50**:949–954.
5. Rossor, M. N., S. Newman & R. S. J. Frackowiak. 1993. Alzheimers-disease families with amyloid precursor protein mutations. Ann. NY Acad. Sci. **695**: 198–202.
6. Kennedy, A. M., S. Newman & A. McCaddon. 1993. Familial alzheimers-disease—A pedigree with a mis-sense mutation in the amyloid precursor protein gene (amyloid precursor protein 717 valine - glycine). Brain **116**:309–324.

7. DUVERNOY, H. M. 1988. The Human Hippocampus: An Atlas of Applied Anatomy. Berlin: Springer-Verlag.
8. COOK, M. J., D. R. FISH & S. D. SHORVON. 1992. Hippocampal volumetric and morphometric studies in frontal and temporal lobe epilepsy. Brain **115**: 1001–1015.

Visual versus Auditory Memory Stimulation in Patients with Probable Alzheimer's Disease

A PET Study with 18 FDG[a]

J. KESSLER, M. GHAEMI, R. MIELKE, K. HERHOLZ, AND W. D. HEISS[b]

Max-Planck Institut für neurologische Forschung and Neurologische Universitätsklinik, Köln, Germany

ABSTRACT: Patients with probable Alzheimer's disease were either stimulated with a continuous visual recognition task (n = 18) or with a continuous auditory recognition task during PET measurement with 18 FDG. The PET scanner was a Siemens CTI, ECAT EXACT with 5 mm transaxial and 6 mm axial resolution. The global stimulation effects were nearly identical in both groups and increased of 4.17 ± 8.14% in the auditory stimulated group and of 4.03 ± 11.78% in the visual stimulated group. Beyond regional stimulation effects in both groups a common cerebellar stimulation effect was measured. It is concluded that the cerebellar stimulation effect reflects a modality independent cognitive process and that the only small enhancement of global glucose metabolism indicates disturbed stimuli processes and finally explained the patient's failure in memory tasks.

INTRODUCTION

In a previous study patients with probable Alzheimer's disease (AD) were stimulated by a continuous visual recognition task during 18F-2-fluoro-2-deoxy-D-glucose (FDG) measurement and PET.[1] The rate of glucose metabo FDG. The PeT scanner walism was compared with their resting PET and with that of a healthy control group (CG). Compared to the control group the AD patients showed only a reduced enhancement of glucose metabolism in selected brain areas (5.7 ± 11.1% vs 21 ± 18% of a CG). This lack of stimulation was interpreted as a reduced reserve capacity of their brains. However, it is not clear as to how far modality-specific factors of the memory

[a] This work was supported in part by BMFT grant No. 070 1763 A/8.

[b] *Send correspondence to:* Dr. W. D. Heiss, Max-Planck-Institut für neurologische Forschung, Gleueler Straße 50, D-50931 Köln, Germany; Tel: 49-221-4726-220; Fax: 49-221-4726-298.

task—the mode of presentation—influenced the glucose pattern. Furthermore, it is also not determinable as to how far processes that are independent of modality, like sustained attention, motivation, and recognition process, influenced metabolic activity. Therefore, we now applied a continuous auditory recognition task with identical task structure and difficulty as the continuous visual recognition task to patients with AD and compared the stimulation effect as measured with PET and FDG of this two stimulation paradigms.

PATIENTS

Thirty-five patients with the clinical diagnosis of probable Alzheimer's disease according to research criteria of the NINCDS-ADRDA work group criteria[2] and the DSM-III-R criteria[3] of primary degenerative dementia were studied. CT or MRI was obtained in each case to rule out ischemic infarct or brain tumors. The modified Hachinski score was 2 or less. The degree of mental deterioration was quantified on Mini-Mental-State-Test (MMST) by Folstein *et al.*[4] The degree of dementia was comparable in both groups and expressed in MMST-scores: 20.33 (3.81) in the auditory stimulation group (N = 17) versus 21.72 (5.14) in the visual stimulation group (N = 18). These groups were comparable in education, profession, and severity of dementia. Mean age was 65.11 (SD = 7.86 range 48–76) years in the visual stimulation group and 65.23 (SD = 7.53 range 54–78) years in the auditory stimulation group.

PET

Positron emission tomography 18F-2-fluoro-2-deoxy-D-glucose (FDG) was used to measure $rCMR_{glu}$. The PET scanner was a Siemens CTI, ECAT EXACT with 5 mm transaxial and 6 mm axial resolution. 370 MBq FDG was injected i.v. and arterialized venous blood samples were withdrawn at predetermined intervals as described previously[5] and used together with the measured tissue activity to calculate $rCMR_{glu}$ with a fixed lumped constant of 0.42. The quantitative regional analysis of $rCMR_{glu}$ for resting and stimulation PET is described elsewhere.[6] Each subject was studied at rest and during stimulation, usually on successive days. In the resting condition the patients were supine with eyes closed, ears unplugged, in a darkened room with low ambient noise. For visual stimulation, a continuous visual recognition task was used. It consisted of 200 items, *e.g.* black and white pictures of concrete objects, numbers, words, meaningless syllables, combinations of letters and numbers, figures from the recurring figure test of Kimura[7] and the random shapes of Vanderplas and Garvin.[8] Verbal and nonverbal items were balanced in order to obtain bihemispheric cortical stimulation. The subjects were instructed to identify any previously presented picture among the new ones with spoken yes/no responses. Each picture was presented for at least 5 seconds, the intertrial interval was usually 3 seconds. The task was performed for 30 minutes and was started immediately after FDG injection. During the auditory stimulation condition, spoken words and numbers were used rather than printed stimuli. The words were taken from a pool of 800 substantives, rated for concreteness, imagery, meaningfulness and frequency.[9] The numbers ranged from one to four place decimal. The instruction, stimuli presentation, intertrial interval and scoring procedure were identical as with the visual stimulation condition. Performance in both tasks was described with the d′ from the signal detection theory.

RESULTS

Task performance in the two recognition tasks expressed in d′ indicated that auditory memory task was a little bit more difficult for the AD patients. Task performance was unrelated in both groups to rate of glucose metabolism. The main PET results are summarized in TABLE 1. In the resting condition, the rate of glucose metabolism expressed in μmol/100g/min in selected brain areas was not different between the two groups. Also was the global stimulation effect nearly identical in both groups and increased of 4.17% (SD = 8.14) in the auditory group and 4.03% (SD = 11.78) in the visual group.

The high standard deviation reflected a great range of individual stimulation. The amount of stimulation in different brain regions is demonstrated in FIGURE 1.

Within comparison of the groups (resting vs stimulation) showed a significant increase in the rate of glucose metabolism only in thalamus ($p = 0.0096$) and cerebellum ($p = 0.0004$) in the auditorily stimulated group. The only significant increase in the visually stimulated group was measured in occipital lobe ($p = 0.0256$), especially in the primary visual area ($p = 0.0018$). Glucose enhancement in cerebellum was significant at 5 to 10% level. Beyond occipital lobe and cerebral activation next highest stimulation was observed in thalamus. No age effect on stimulation capacity was found for both groups ($p > .10$). Asymmetric glucose pattern at rest was preserved in both groups also in the stimulation condition. Analysis of variance with age, brain side, region, and degree of dementia as covariate showed no main effect or significant interactions. The only significant correlation was found in the auditory stimulated group with MMST-scores and left temporal cortex ($r = 0.53$ $p < 0.05$) and left frontal cortex ($r = 0.48$ $p < 0.05$).

DISCUSSION

In spite of the different kinds of stimuli presentation, the global stimulation effect was comparable in both groups, even though in both groups the in-

TABLE 1. Glucose Metabolism (μmol/100g/min) in Both Stimulation Conditions

	Auditory				Visual			
	Rest		Stimulation		Rest		Stimulation	
	Left	Right	Left	Right	Left	Right	Left	Right
Temporal	28.37 (4.13)	29.03 (5.03)	29.32 (4.04)	29.89 (5.11)	28.49 (2.97)	30.28 (3.11)	30.05 (5.45)	31.35 (5.18)
Frontal	33.89 (4.34)	34.12 (5.02)	35.17 (4.06)	35.45 (4.89)	33.11 (3.84)	34.30 (4.17)	33.93 (5.93)	34.88 (6.32)
Parietal	29.19 (4.32)	29.96 (5.42)	30.41 (4.38)	31.01 (5.72)	29.36 (3.37)	31.53 (3.10)	30.62 (5.48)	32.09 (5.47)
Occipital	30.88 (4.14)	30.84 (5.73)	31.82 (4.19)	31.76 (5.64)	30.94 (3.16)	31.44 (2.84)	33.58 (6.07)	34.12 (6.12)
Visual	36.54 (4.56)	36.41 (5.63)	37.86 (4.23)	37.91 (5.45)	33.18 (3.68)	34.14 (2.69)	36.87 (6.98)	38.12 (5.49)
Thalamus	36.38 (4.31)	36.32 (4.85)	38.42 (4.13)	38.30 (4.11)	32.40 (3.88)	33.15 (3.29)	33.58 (4.49)	34.75 (4.08)
Cerebellum	32.68 (4.01)	33.01 (4.12)	34.99 (4.12)	35.28 (4.50)	32.61 (4.22)	32.37 (4.24)	35.16 (6.06)	34.28 (6.20)
Global	31.64 (4.22)		32.82 (4.15)		31.54 (2.67)		32.87 (5.15)	

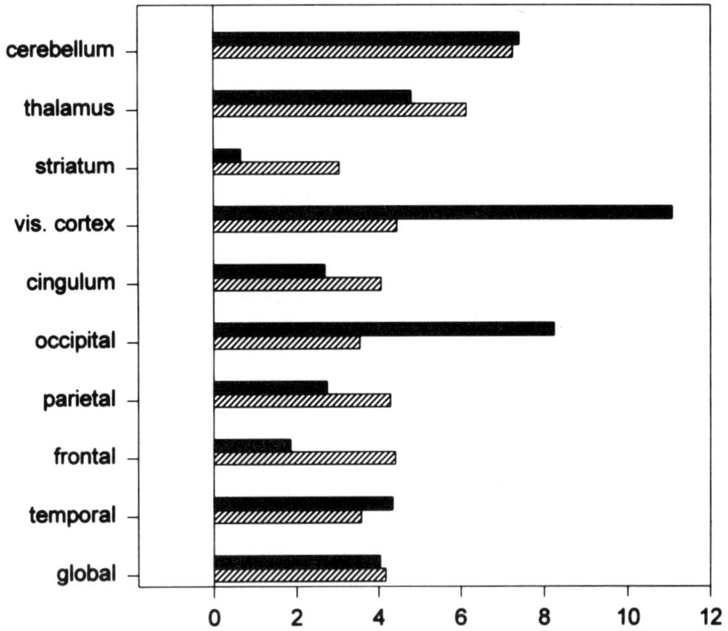

FIGURE 1. Percent of stimulation induced by visual and auditory memory task: visual ■ auditory ▨.

crease of glucose metabolism was effected by activation of different brain regions. Whereas in visually stimulated AD patients the predominant stimulation effect could be demonstrated in primary visual cortex, in auditorily stimulated AD patients major enhancement of glucose uptake in primary and secondary auditory cortex could not be detected. The resting state was eyes closed but ears unplugged in a room with diminished light and low ambient noise. Thus, the primary sensory contrast to eyes open during visual stimulation was probably larger than to ears still unplugged during auditory stimulation. This difference in relative primary stimulus intensity might explain the difference with regard to primary sensory cortex stimulation. In addition, detection of small regions like the auditory cortex is very sensitive to moving artifacts during PET measurement which could not be completely avoided in all patients. For that reason only large, clearly defined structures were used as ROIs (*e.g.* temporal, parietal, frontal, occipital) and significant differences between rest and stimulation can then be regarded as very reliable and valid. In both groups a significant cerebellar stimulation effect could be proved. Independent on motoric functions there were indications that the cerebellum takes part not only in motoric aspects of tasks, but also in cognitive processes. In both tasks stimuli were presented in sequences and probably this sequenced aspect of the tasks lead to the demonstrated increase of glucose in cerebellum in both groups. But also in the thalamus the average glucose metabolism was increased significantly in the auditory stimulated group and tendentiously in the visual stimulated group (FIG. 1). It is well known that thalamus structures are involved in memory processes and the increase of glucose metabolism in this structure reflects the stimuli processing independent its modality of presentation. As there did not exist a control group of the auditory stimulated AD patients, the continuous visual recognition task was carried out also in a group of healthy controls. In this group a significantly higher stimulation effect than in the group of AD patients was detected. The comparably low increase of glucose metabolism in both stimulation paradigms proved the AD patients to be impaired independently on the stimulation paradigm. In the beginning, the stimulus presentation—visual or auditory—definitely lead to a glucose enhancement dependent on modality, but lack of glucose activation in other brain regions referred to a disturbed stimuli process and finally explained the patients' failure in memory tasks.

REFERENCES

1. KESSLER, J., K. HERHOLZ, M. GROND & W. D. HEISS. 1991. Impaired metabolic activation in Alzheimer's disease: A PET study during continuous visual recognition. Neuropsychologia **29**:229–243.
2. MCKHANN, G., D. DRACHMAN, M. F. FOLSTEIN, R. KATZMAN, D. PRICE & E. M. STADLAN. 1984. Clinical diagnosis of Alzheimer's disease: Report of the NINCDS-ADRDA work group under the auspices of department of health and human services task force on Alzheimer's disease. Neurology **35**:939–944.
3. DSM-III-R. Diagnostisches und Statistisches Manual Psychischer Störungen. Weinheim Beltz Verlag, 1987.
4. FOLSTEIN, M. F., S. E. FOLSTEIN & P. R. MCHUGH. 1975. Mini Mental State: A practical method for grading the cognitive state of patients for the clinician. J. Psychiat. Res. **12**:189–198.
5. HEISS, W. D., G. PAWLIK, K. HERHOLZ, R. WAGNER, H. GÖLDNER & K. WIENHARD. 1984. Regional kinetic constants and cerebral metabolic rate of glucose in normal human volunteers determined by dynamic positron-emission-tomography of 18F-2-fluoro-2-deoxy-D-glucose. J. Cereb. Blood Flow Metab. **4**: 212–223.
6. HEISS, W. D., J. KESSLER, R. MIELKE, B. SZELIES & K. HERHOLZ. 1994. Long-term effects of phosphatidylserine, pryritinol and cognitive training in Alzheimer's disease. Dementia **5**:88–98.
7. KIMURA, D. 1963. Right temporal lobe damage. Arch. Neurol. **8**:264–271.
8. VANDERPLAS, J. M. & E. A. GARVIN 1959. The association value of random shapes. Exp. Psychol. **57**:147–154.
9. BASCHEK, I. L., J. BREDENKAMP, B. OEHRLE & W. WIPPICH. 1977. Bestimmung der Bildhaftigkeit (I), Konkretheit (C) und der Bedeutungshaltigkeit (m') von 800 Substantiven. Z. Exp. Angew. Psychol. **24**:353–396.

Alterations in Functional Neuroanatomical Connectivity in Alzheimer's Disease

Positron Emission Tomography of Auditory Verbal Short-Term Memory[a]

J. T. BECKER,[b,c,e] M. A. MINTUN,[d] K. ALEVA,[b] M. B. WISEMAN,[b] T. NICHOLS,[d] AND S. T. DEKOSKY[b,c]

Alzheimer's Disease Research Center and Center for Functional Brain Imaging, [b]Department of Psychiatry, [c]Department of Neurology, and [d]Department of Radiology, University of Pittsburgh Medical Center, Pittsburgh, Pennsylvania 15213 USA

ABSTRACT: Conscious recall of past events which have specific temporal and spatial contexts, termed episodic memory, is mediated by a system of interrelated brain regions. In Alzheimer's disease (AD) this system breaks down, resulting in an inability to recall events from the immediate past. Studies of normal human auditory-verbal short-term memory suggest that the brain system underlying these processes has distinct components, and the present study utilized the methods of functional brain mapping to determine the nature and extent of the breakdown that occurs in AD. Using subtraction techniques of PET-acquired images of regional cerebral blood flow we demonstrate that AD patients show a compensatory hyperactivation of various regions of cerebral cortex normally involved in these tasks, as well as activation of cortical areas not activated by normal elderly subjects. These results provide clear evidence of functional plasticity in the AD patient's brain even if those changes do not result in normal memory function.

INTRODUCTION

Functional neuro-imaging studies using regional cerebral blood flow and cerebral glucose metabolism techniques of Alzheimer's disease (AD) patients have demonstrated decreased functioning of the parietal and temporal cortices

[a] This research was supported by grants from the National Institute of Mental Health (MH49815, K02-MH0177) and the National Institute on Aging (AG05133).

[e] *Send correspondence to:* Dr. James T. Becker, Neuropsychology Research Program, 502 Iroquois Building, 3600 Forbes Avenue, Pittsburgh, PA 15213; Tel: (412) 624-0142; Fax: (412) 383-1755; Email: JTBECKER@VMS.CIS.PITT.EDU.

when patients are at rest, consistent with the pattern of neurodegenerative changes in these regions.[1] Recent studies of young, healthy normal volunteers have revealed distinct patterns of functional activity as the subjects engage in tasks related to auditory verbal short-term memory.[2–4] These studies are of particular relevance because the procedures have the potential of revealing specific functional defects related to specific information processing deficits, prior to any observable structural abnormalities. The purpose of this report is to describe alterations in cerebral blood flow of AD patients during the performance of verbal memory tasks that place differential demands on cognition, and presumably different functional processing systems.

METHODS

Five women and two men meeting the NINCDS/ADRDA criteria for Probable AD participated in the study. Their overall level of cognition was mildly impaired (mean MMS score = 21.7 ± 3.4). These patients were compared with seven age-, education-, and sex-matched controls who all completed the same PET procedures.

Each subject was scanned eight times measuring regional cerebral blood flow (rCBF) using ^{15}O-water with standard laboratory procedures. During the first and last scans, the subjects were instructed to lie still and fix their gaze on a cross-hair target on a video screen over perpendicular to their line of vision ("Rest" condition). During the remaining six scans, the subjects were tested on single word repetition (1WR), three word repetition (3WR), and eight word free recall (8WR).

RESULTS

Contrasting 3WR with Rest among the control subjects revealed those brain regions more active during auditory processing, working memory, and speech output than at rest. The healthy elderly subjects showed significant increases in rCBF in brain regions responsible for audition, phonological storage, and articulatory rehearsal and articulated speech. The AD patients, like the normal control subjects, showed significant activations in brain regions associated with speech and language function, and with phonological storage. Of particular note, however, is that in all regions examined, the field of activity observed in the AD patients was greater than the activation in the same region in the normal controls.

By comparing the pattern of rCBF seen in the 8WR condition with that seen in the 3WR condition, it is possible to isolate those brain mechanisms responsible for the encoding and retrieval of information from secondary episodic memory. The control subjects demonstrated an area of significant activation in BA10 of the middle frontal gyrus bilaterally.

By contrast, the AD patients showed significant activity in the dorsolateral prefrontal cortex across the convexity, bilaterally. Posteriorly, the AD patients had significant regional activation in the supramarginal gyrus on the left (BA40), and superior temporal gyrus (BA39). Along the midline posteriorly, significant activity was observed in the cuneate (BA31).

DISCUSSION

The finding that the spatial extent of activation seen in the AD patients is greater than that seen in the control subjects suggests that there is a broadening of the cortical field in response to the altered cortical connections due to the neuropathological changes. Although the specific locus of activity remains the same (*e.g.*, supramarginal gyrus) the area around these foci has changed, consistent with the observations that areas of cortical representation of information can be altered by changes in the input to those brain regions.[5]

The differences in patterns of activation in the 8WR condition relative to 3WR may be related to specific frontal neuropathological changes. Among the normal elderly control subjects, it was lateral orbital frontal cortex, bilaterally, that was specifically activated during the 8WR task relative to 3WR. The finding that this brain region was not activated by the AD patients, and that the adjacent dorsolateral cortex *was* activated, would be consistent with the hypothesis that the orbital region was dysfunctional, and that related areas of the frontal cortex were attempting to compensate for this functional loss. An abnormality in the region of frontal cortex would occur in Braak's Stage B of amyloid deposition, and Stages V-VI of neurofibrillary changes. Although there is little other evidence consistent with the notion that *extracellular* amyloid can affect neuronal function, the early stage of the dementia in these patients leads one away from focusing exclusively on the neurofibrillary changes. The finding that there was consistent and significant activation in the region of the supramarginal gyrus among the AD patients is less clear. This brain region is most often associated with phonological storage and was activated by both patients and controls when the 3WR was contrasted to Rest. The fact that significant activation was seen in the 8WR condition, over and above that seen in the 3WR, suggests that the AD patients may have attempted to alter their cognitive strategy to deal with the demands of the 8WR task.

REFERENCES

1. Haxby, J. V., R. Duara & C. L. Grady. 1985. Relations between neuropsychological and cerebral metabolic asymmetries in early Alzheimer's disease. J. Cereb. Blood Flow Metab. **5**:193–200.

2. BECKER, J. T., M. A. MINTUN, D. J. DIEHL, J. DOBKIN, A. MARTIDIS, D. C. MADOFF & S. T. DEKOSKY. 1994. Functional neuroanatomy of verbal free recall: A replication study. Human Brain Mapping **1**:284–292.
3. GRASBY, P. M., C. D. FRITH, K. J. FRISTON, C. BENCH, R. S. J. FRACKOWIAK & R. J. DOLAN. 1993. Functional mapping of brain areas implicated in auditory-verbal memory function. Brain **116**(1):1–20.
4. PAULESCU, E., C. D. FRITH & R. S. J. FRACKOWIAK. 1993. The neural correlates of the verbal component of Working Memory. Nature **362**:342–345.
5. SHALLICE, T. & E. K. WARRINGTON. 1970. Independent functioning of verbal memory stores: A neuropsychological study. Quart. J. Exp. Psychol. **22**: 261–273.

Varieties of Progressive Non-fluent Aphasia[a]

S. F. CAPPA,[b,d] D. PERANI,[c] C. MESSA,[c] A. MIOZZO,[b] AND F. FAZIO[c]

[b] *Laboratorio di Neuropsicologia, Clinica Neurologica, University of Brescia, Brescia, Italy*

[c] *Dipartimento di Neuroscienze e Bioimmagini, Ospedale S. Raffaele, University of Milano, Milan, Italy*

ABSTRACT: We report four patients with progressive aphasia of the non-fluent type as the presenting clinical manifestation. The patients were included in a longitudinal study of focal progressive neuropsychological syndromes, and were periodically submitted to neuropsychological evaluations and neuroimaging studies (TC, MRI, SPET or PET). The pattern of neuropsychological impairment was in good agreement with the results of functional imaging studies, which indicated involvement of the anterior regions of the left hemisphere. The evolution of the clinical picture was extremely heterogeneous in the four patients, ranging from a relatively stable picture of transcortical motor aphasia to a severe progressive frontal lobe syndrome. Progressive non-fluent aphasia appears to be a reliable clinical marker of the localization of the pathological process; whether this is related to specific neuropathological conditions, such as Pick's disease, remains for the moment a matter of speculation.

INTRODUCTION

The distinction between fluent and non-fluent aphasia is a traditional one in clinical asphasiology[1]: non-fluent aphasias are characterized by reduced speech production with shortened phrase length. Other characteristics which may be present are hesitant speech with impairment of articulation, reduction of syntactic complexity with omission of function words, anomia and altered melodic line. In the case of focal (usually vascular) lesions, non-fluent aphasias are typically associated with damage including the anterior regions of the left hemisphere.[2,3] Non-fluent aphasia can also be observed as a progressive neuropsychological syndrome. Caselli and Jack[4] included in the "asymmetric cortical degeneration syndromes" two forms of progressive aphasia (dysfluent

[a] Supported by a National Research Council (CNR) grant to S.F.C.

[d] *Send correspondence to:* Dr. Stefano F. Cappa, Clinica Neurologica-II Neurologia Spedali Civili, P.le Ospedale, 25125 Brescia, Italy; Tel: +30-3995489; Fax: +30-3384086; Email: cappa@master.cci.unibs.it.

and fluent). Weintraub and Mesulam[5] distinguished a "primary progressive aphasia" among several other forms of progressive dementia. The same authors have suggested that a non-fluent aphasia profile is useful in differentiating "primary progressive aphasia" from probable Alzheimer's disease.[6]

In the course of a longitudinal study of patients presenting with focal neuropsychological dysfunction, which were assessed periodically with neuropsychological tests and functional imaging studies, we have had the opportunity to observe four cases presenting as progressive non-fluent aphasia, with extremely heterogeneous patterns of progression. Neuroimaging studies were in good agreement with the clinical presentation, showing a functional involvement of the anterior regions of the left hemisphere. Follow-up studies indicated a close parallelism between clinical progression and cerebral functional involvement.

CASE REPORTS

Case 1

IM was a 55-year-old right-handed woman with 5 years of schooling, who was referred for difficulty in finding words. Neurological and language examination were within normal limits (TABLE 1); only performance on the Wisconsin Card Sorting test was impaired. Brain MRI was normal. PET with [18F]FDG showed a hypometabolic area in the left prefrontal region. In the following months there was a progressive deterioration of language, memory, and reasoning abilities (data about language are summarized in TABLE 1). Three years later the patient had echolalia and a severe impairment of naming and comprehension. Initiative was severely reduced. A follow-up PET study with [^{18}F]FDG showed an extension of hypometabolism in both hemispheres.

Case 2

MV was a 46-year-old right-handed man with 5 years of education, who was referred for the evaluation of slurred speech and auditory comprehension impairment.[7] The neurological examination was unremarkable. Spontaneous speech was severely dysarthric and limited to single content words; auditory comprehension and written language were severely impaired. The most striking feature of the neuropsychological evaluation was the severe oro-facial apraxia, associated with flawless performance on an extensive test of limb praxis. MRI showed diffuse atrophy, most severe in the left temporo-parietal region; SPET indicated a marked hypoperfusion in the left frontal lobe and temporal pole. In the following months the patient deteriorated at a rapid

TABLE 1. Case 1. Summary of Language Evaluations (Western Aphasia Battery)

	Oct. '91	Oct. '92	Apr. '93	Max
Spontaneous speech				
• Information content	7	7	4	10
• Fluency	9	8	4	10
Comprehension				
• Yes/no questions	57	36	12	60
• Auditory word recognition	57	54	51	60
• Sequential commands	64	28	50	80
Repetition	98	79	51	100
Naming				
• Object naming	60	58	37	60
• Word fluency	17	9	4	20
• Sentence completion	10	10	8	10
• Responsive speech	10	10	6	10
Reading and writing				
• Reading	81	44	29	100
• Writing	59.5	50.5	17.5	100
Praxis	46	30	18	60
Construction				
• Drawing	17	16	14.5	30
• Block design	8	3	0	9
• Calculation	24	20	0	24

rate: severe behavioral disturbances (agitation, violence outbursts) required hospital admission. The patient died one year later: pathological verification was denied.

Case 3

MF was a 44-year-old right-handed man with 8 years of education, who was referred for difficulties in finding words. At the time of the first evaluation, the patient had a mild aphasia, which could be classified as transcortical motor: the remaining neuropsychological examination was within normal limits. During the following two years, there was a progressive worsening of language abilities, without any other manifestation: the patient was still employed as a skilled laborer in a factory and went to work alone driving his car. The results of language assessment are reported in TABLE 2. The copy and recall of Rey's figure were now impaired. The first SPET study showed a left hemispheric hypoperfusion, particularly severe in the dorsolateral frontal cortex; on the follow-up study two years later the right dorsolateral frontal cortex was also hypoperfused.

TABLE 2. Case 2. Summary of Language Evaluations[a]

	Nov. '91	Oct. '93
Description of an event	30	0
Visual naming	95	65
Auditory comprehension		
• Words	100	95
• Semantically related words	100	75
• Sentences	100	10
• Token test	72	42
Repetition		
• Words	100	95
• Nonwords	80	100
• Sentences	80	0
Written naming	80	5
Reading comprehension		
• Words	100	100
• Semantically related words	100	85
• Sentences	100	0
Reading aloud		
• Words	100	100
• Nonwords	80	20
• Sentences	90	20
Dictation		
• Words	60	0
• Nonwords	90	0
• Sentences	30	0
Copying		
• Words	90	90
• Transposition	100	80

[a] Scores are expressed as percent correct.

Case 4

MM was a 66-year-old right handed man with 5 years of schooling, who was referred for evaluation of progressive impairment of oral expression and auditory comprehension. The aphasia was severe and characterized by reduced spontaneous speech, preserved repetition with echolalia and impaired comprehension. There was a mild bucco-facial and limb apraxia. There was indication of a memory impairment which extended to nonverbal material (Rey's figure). At re-evaluation, three years later, the patient was mute, with no evidence of auditory comprehension and repetition limited to single words. Perseverations were now evident in praxic and constructional tasks. The patient, who lived in a rural setting, was still working full time in the fields and required no assistance in the activities of daily life. A SPET study showed a severe hypoperfusion of the frontal lobes, more extensive on the left.

DISCUSSION

These four patients show different varieties of "anterior" progressive neuropsychological dysfunction. Case 1 had a clinical picture characterized by severe "frontal lobe" signs, associated with a progressive transcortical aphasia, and fulfills the criteria for dementia of the "frontal lobe" type.[8] On the other hand, Case 2 was an instance of progressive loss of speech output and oro-facial dyspraxia,[9,10] associated with a rapidly progressive dementia. Cases 3 and 4 had a progressive non-fluent, transcortical-type aphasia,[11] with a moderate and slowly progressive impairment in other cognitive functions. Functional imaging studies (SPET, PET) provide clinico-anatomical correlations which are in agreement with the data from studies in patients with vascular lesions. These four patients provide further evidence that a careful analysis of the clinical picture may lead to predictions about specific lesion localization. It is also noteworthy that the clinical progression towards widespread neuropsychological impairment correlates with the extension of perfusional and metabolic cerebral involvement shown by SPET and PET. Nonfluent aphasia can thus be considered a clinical marker of pathological conditions involving the anterior language areas. Whether this finding indicates specific nosological correlations (such as Pick's disease) must await confirmation from clinico-pathological studies.

REFERENCES

1. GOODGLASS, H., F. A. QUADFASEL & W. H. TIMBERLAKE. 1964. Phrase length and the type and severity of aphasia. Cortex **2**:74–89.
2. BENSON, D. F. 1967. Fluency in aphasia: Correlation with radioactive scan localisation. Cortex **3**:373–394.
3. DAMASIO, H. & A. R. DAMASIO. 1989. Lesion Analysis in Neuropsychology. New York: Oxford University Press.
4. CASELLI, R. J. & C. R. JACK, JR. 1992. Asymmetric cortical degeneration syndromes. A proposed clinical classification. Arch. Neurol. **49**:770–780.
5. WEINTRAUB, S. & M.-M. MESULAM. 1993. Four neuropsychological profiles in dementia. *In* Handbook of Neuropsychology, Vol. 8. F. Boller & J. Grafman, Eds. 253–289. Amsterdam: Elsevier.
6. WEINTRAUB, S., N. P. RUBIN & M.-M. MESULAM. 1990. Primary progressive aphasia. Longitudinal course, neuropsychological profile and language feaures. Arch. Neurol. **47**:1329–1335.
7. CAPPA, S. F., C. A. DE FANTI, R. DE MARCO, E. MAGNI, C. MESSA & F. FAZIO. 1993. Progressive dysphasic dementia with bucco-facial apraxia: A case report. Behav. Neurol. **6**:159–163.
8. NEARY, D., J. S. SNOWDEN, B. NORTHEN & P. GOULDING. 1988. Dementia of frontal lobe type. J. Neurol. Neurosurg. Psychiat. **51**:353–361.

9. TYRREL, P. J., E. K. WARRINGTON, R. S. J. FRACKOWIAK & M. N. ROSSOR. 1990. Heterogeneity in progressive aphasia due to focal cortical atrophy: A clinical and PET scan study. Brain **113**:1321–1326.
10. TYRREL, P. J., L. D. KARTSOUNIS, R. S. J. FRACKOWIAK, L. J. FINDLEY & M. N. ROSSOR. 1991. Progressive loss of speech output and orofacial dyspraxia associated with frontal lobe hypometabolism. J. Neurol. Neurosurg. Psychiat. **541**:351–357.
11. KARTSOUNIS, L. D., R. F. CRELLIN, H. CREWES & B. K. TOONE. 1991. Primary progressive non-fluent aphasia: A case study. Cortex **27**:121–129.

Computer-based Cognitive Training in Alzheimer's Disease Patients

M. HOFMANN,[a] C. HOCK, AND F. MÜLLER-SPAHN

Department of Psychiatry, University of Basel, Basel Switzerland

ABSTRACT: Memory training programs for cognitively impaired patients have often been criticized for their lack of relevance to everyday activities. We therefore report our experience with four patients suffering from probable Alzheimer's disease, who were trained with a new computer-based program recently developed by our research group. An everyday task of personal relevance to the patient is simulated and trained on a PC-touch-screen using personal photographs of the patient's surroundings and biography. According to the degree of cognitive impairment, training has three major aspects 1) social competence in patients with beginning deficiencies, 2) orientation in patients with moderate disease, and 3) emotional aspects in patients with advanced deficiencies. The patient's training performance improved substantially. While psychopathometric tests showed no significant effects with regard to general cognitive performance, levels of motivation were high and there was a positive acceptance of the training and signs of emotional activation.

INTRODUCTION

To date, there are no safe and efficient treatments for the increasing number of Alzheimer's disease patients to improve cognition and memory.[1] In addition to pharmacological, psychosocial, and psychotherapeutic strategies, cognitive training programs have been chosen as a possible therapeutic approach. Only a few case reports exist for Alzheimer's disease patients, showing only few and short-term effects of cognitive training as far as specifically trained cognitive subfunctions (*e.g.*, short term memory) are concerned.[2] Since conventional training is mainly based on abstract tests, the lack of relevance to everyday life has been a target of constant criticism. We therefore developed a computer-based cognitive training program, specifically tailored for the individual patient, employing personal and biographic material (*e.g.*, pictures of the patient's surroundings, biographic photographs) which relates directly

[a] *Send correspondence to:* Dr. M. Hofmann, Department of Psychiatry, University of Basel, Wilhelm Klein-Strasse 27, CH.4025 Basel, Switzerland; Tel.: 41-61-325 5111; Fax.: 41-61-325 5512.

TABLE 1. Patient Characteristics and Training Programs

Patient	Sex	Age (y)	AD-Type	AD-(y)	CDR	MMS	Training Contents
1	m	82	LO	5	1	19	Orientation, finding the way to the bakery, social competence, shopping
2	m	85	LO	6	2	23	Orientation, finding the way to the caregiver's flat in the neighborhood
3	f	66	EO	7	2	18	Initiating and coordinating household tasks
4	m	50	EO	5	2	12	Biography, pictures from early childhood to the present time

Legend: AD(y) = Duration of disease; CDR = Clinical Dementia Rating (8); EO = early onset AD; LO = Late onset AD; MMS = Mini-Mental-State (6).

to the patient's local and social environment. Depending upon the stage of impairment, there are three major aspects to the training: 1) social competence in patients with beginning deficiencies, simulating—for instance—a shopping-situation on the PC-screen; 2) orientation in patients with moderate disease; and 3) emotional aspects, in severely demented patients, using—for instance—biographic material.

SUBJECTS AND METHODS

Four patients, each with a diagnosis of probable Alzheimer's disease according to DSM-III-R[3] and NINCDS-ADRDA criteria,[4] were selected for the training program. The severity of the disease ranged from mild to middle according to the CDR-Staging.[5] In cooperation with the patients, who lived in their own homes and each had a close caring relative or a caregiver, a task to be trained was specified. The patient's characteristics and the training programs are shown in TABLE 1. Fifty to 100 photographs were taken, illustrating successive steps of the task, scanned into the memory of an IBM-compatible 486-PC, and integrated into a presentation-software program that displays the pictures on a special large size-color monitor and allows the interactive operation of the program via touch-screen-functions. The patients were instructed to "move through" the simulated scenes step-by-step, by touching correct screen areas (*e.g.*, pictures) and to complete the task (*e.g.*, shopping), with as little help from the instructor as possible and to try to avoid wrong stimuli in order to be as fast as possible. The time needed, the number of mistakes, and the number of times the test instructor gave advice were registered. Training frequency was 3–4 times per week. Two training phases, each lasting 4 weeks, were separated by a 4-week interval without training to allow intra-individual comparison with and without training, and to see whether training effects persisted. A set of psychometric tests and scales including SIDAM,[6] MMS,[7] SKT (Syndrom Kurztest,[8]) AAT,[9] IDDD,[10] and MADRS[4] were rated to document cognitive performance, state of mood and competence in daily living activities before and after each training period (week 0, 4, 8, 12). In a questionnaire (scale from −9 to +9), which we developed (unpublished) to evaluate subjective effects of the training, the patients and caregivers rated three questions (from −3 = "not at all" to +3 = "very much") regarding the effect of the training on cognition and memory, the possibility of transferring trained skills into real situations and how well the training was liked.

RESULTS

As shown in FIGURE 1, the patients achieved a marked reduction of the quantified parameters (time, advice, mistakes) during the two training periods. Each column in the histograms represents the mean of the parameter of either the first two, or the last two, training sessions within a training period. The figures at the top of the columns (week 4, 8, 12) signify the percentage in relation to the mean performance of the initial two training session (week 1) which is set at 100%. Three patients performed markedly better in the first two sessions of the second training period compared to the first two sessions of the first period. Psychopathometric tests and scales revealed no significant changes over the course of time (see bottom parts of the graphs). Our questionnaire was rated by the patients (caregivers): +7 (−1), +8 (+5), +6 (+2) and +3 (+1).

DISCUSSION

Our results, similar to those of Cameron and Stevens,[2] provide evidence that the performance of Alzheimer's disease patients can be improved through specific training, even in advanced stages of the disease. Differing from Powell-Procter and Miller,[6] who found only very short-lasting effects, our data suggests that at least three patients were able to maintain parts of the trained skills over a period of 4 weeks, as indicated in the better results at the beginning of the second training period compared to those of the initial sessions. This may be due to the fact that operating the programs via touch-screen draws on motoric-automatized operations involving implicit memory processes to a certain degree. It has been suggested[6,11] that motoric and implicit

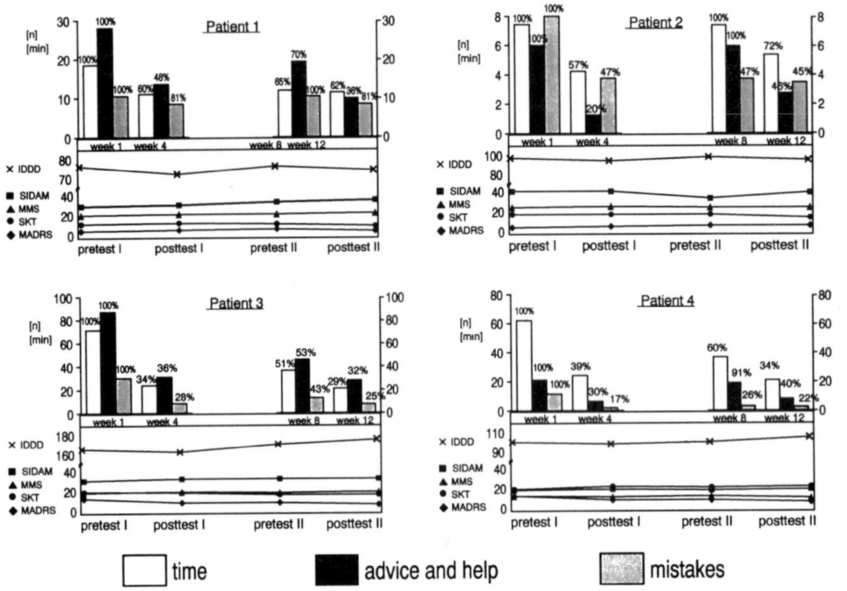

FIGURE 1. The patients training performance and results of psychometric tests. LEGEND: IDDD = Interview for the Deterioration in Daily Living Activities in Dementia[13] (33–231 points*); MADRS = Montgomery and Asberg Depression Rating Scale[14] (0–60 points**); MMS = Mini-Mental-State[15] (0–30 points*), SIDAM = Structured Interview for the Diagnosis of Dementia of the Alzheimer Type, Multi-infarct Dementias and Dementias of other Etiology according to ICD-10 and DSM-III-R[15] (0–55 points**); SKT = Syndrom Kurztest[8] (0–27 p.**); *: higher scores = getting worse; **: higher = scores getting better.

(procedural) memory processes may be relatively spared in AD. Similar to other investigators,[2,10] we found no evidence for a general improvement of cognition or explicit memory as measured in the psychopathometric tests. There were no measurable significant effects with regard to daily living activities, which may be due in part to the difficulty in measuring these unspecific effects and a poor validation of the IDDD-scale we used. Apart from anecdotal reports about a transfer of the trained task into real situations, according to caregiver's reports and also partly reflected in the patient's positive evaluation of the training, the patients experienced the occupation with familiar issues as motivating and emotionally rewarding. They seemed to profit from a sense of achievement resulting in signs of an improved self-esteem and improved self-conception.

In accordance with Bäckman's[12] criteria describing features yielding positive findings of memory training in AD, our approach integrates the patient's personal (social- and local-) environment and is adjustable to the degree of impairment. Thus we hope to have emphasized the issue of this training approach, which is to train tasks that are of relevance to the patient's personality and self-esteem and relate to everyday activities and quality of living. The aim is to stimulate the patient mentally, to improve self-conception and self-confidence via feelings of success, in order to help to preserve personal autonomy and a feeling of personal identity as long as possible.

ACKNOWLEDGMENT

We would like to thank Mr. Markus Fischer for his support in all questions of computer hard- and software and for his help in programming.

REFERENCES

1. GROWDON, J. H. 1992. Treatment for Alzheimer's Disease? New Engl. J. Med. **327**(No 18):1306–1308.
2. CAMERON, J. C. & A. B. STEVENS. 1990. Spaced-retrieval: A memory intervention for dementia of the Alzheimer's type, Clin. Gerontol. **10**(1):58–61.
3. AMERICAN PSYCHIATRIC ASSOCIATION. 1987. Diagnostic and Statistical Manual of Mental Disorders, 3rd revised edition. Washington DC: American Psychiatric Association.
4. MCKHANN, G. *et al.* 1984. Clinical diagnosis of Alzheimer's Disease: Report of the NINCDS-ADRDA work group under auspices of the Department of Health and Human Service Task Force in Alzheimer's Disease. Neurology **34**: 939–944.
5. HUGES, C. P., L. BERG, W. L. DANZINGER, L. A. COBEN & R. L. MARTIN. 1982. A new clinical scale for the staging of dementia. Br. J. Psychiatry **140**: 566–572.
6. OBER, B. A. & G. K. SHENAUT. 1988. Lexical decision and priming in Alzheimer's disease. Neuropsychologica **26**:273–286.
7. FOLSTEIN, M. F., S. E. FOLSTEIN & P. R. MCHUGH. 1975. Mini-mental State. A practical method for grading the cognitive state of patients for the clinician. J. Psych. Res. **12**:189–198.
8. ERZIGKEIT, H. 1986. Der SKT zur Beurteilung theraprutischer Effekte nootroper Substanzen. Proceedings of the 3rd Symposium on Nootropics 3–5 March 1986, Dresden.
9. HUBER, W., K. POECK & D. WENIGER. 1973. Aachener Aphasie Test. Hogrefe Göttingen.
10. POWELL-PROCTER, L. & E. MILLER. 1982. Reality orientation. A critical appraisal. Br. J. Psychiatry **140**:457–463.
11. ESLINGER, P. J. & A. R. DAMASIO. 1986. Preserved motor learning in Alzheimer's disease: Implications for anatomy and behavior. J. Neurosci. **6**: 3006–3009.

12. BÄCKMANN, L. 1992. Memory training and memory improvement in Alzheimer's disease: Rules and exceptions. Acta Neurol. Scand. **139**(Suppl.):84–89.
13. TEUNISSE, S. & M. M. DERIX. 1992. Measurement of activities of daily living in patients with dementia living at home: Development of a questionnaire. Tijdschr-Gerontol-Geriatr. **22**(2):53–59.
14. MONTGOMERY, S. A. & M. ASBERG. 1979. A new depression scale designed to be sensitive to change. Br. J. Psychiatry **134**:382–389.
15. ZAUDIG, M., J. MITTERHAMMER & W. HILLER. 1989. Structured interview for the diagnosis of dementia of the Alzheimer type, multi-infarct dementia and dementias of other etiology according to ICD-10 and DSM-R-III-SIDAM. Logomed: Munich.

ApoE2 Allele, Down's Syndrome, and Dementia[a]

M. C. ROYSTON,[b,i] D. MANN,[c] S. PICKERING-BROWN,[c] F. OWEN,[d] R. PERRY,[e] R. RAGBAVAN,[e] C. KHIN-NU,[f] S. TYNER,[f] K. DAY,[f] R. CROOK,[g] J. HARDY,[g] AND G. W. ROBERTS[h]

[b] *Departments of Psychiatry and Anatomy, Charing Cross and Westminster Medical School, London, United Kingdom*

[c] *Department of Pathological Sciences, University of Manchester, Manchester, United Kingdom*

[d] *Division of Neurosciences, University of Manchester, Manchester, United Kingdom*

[e] *Department of Neuropathology, Newcastle General Hospital, Newcastle-Upon-Tyne, United Kingdom*

[f] *Department of Psychiatry, Northgate and Prudhoe NHS Trust Hospital, Morpeth, Northumberland, United Kingdom*

[g] *Suncoast Alzheimer's Disease Laboratories, Departments of Psychiatry, Pharmacology, Neurology and Biochemistry, University of South Florida, Tampa, Florida, USA*

[h] *Department of Molecular Neuropathology, SmithKline Beecham Pharmaceuticals, Harlow, United Kingdom*

ABSTRACT: All individuals with Down's syndrome (trisomy 21-DS) develop the pathogenic hallmarks of Alzheimer's disease in old age (+40 years).[1–4] The extent of pathology is variable, but it has been shown that the amount of β-amyloid pathology is variable and related to age and the degree of dementia.[3] Thus, in DS, growing old is associated with a progressive pathological process which results in cognitive decline. However, neuropsychological studies of older DS subjects have identified a clinical dementia in only a proportion of cases.[4–7] These contradictory observations could be reconciled if some factor existed which modulated the rate and amount of β-amyloid pathology. Recent studies demonstrate an association between the apolipoprotein E4 (ApoE4) allele and the earlier age of onset in both sporadic[8–9] and familial[10] AD. Increased amounts of β-amyloid pathology can also be related to the E4 allele. However, at present there are no data documenting the effects of ApoE genotype on the expression or degree of clinical symptoms of the disease. We have examined the ApoE genotype in a cohort of clinically evaluated elderly patients with DS in order to examine the effects of ApoE genotype on the clinical symptoms of dementia. We report

[a] MCR was supported by The Mental Health Foundation and the Wellcome Trust, JH by NIA grants (R01 AG11871) and private donations.

[i] *Send correspondence to:* Dr. M. Claire Royston, Departments of Psychiatry and Anatomy, Charing Cross and Westminster Medical School, London, W6 8RP, UK; Tel: 44-181-846-7680; Fax: 44-181-846-7025.

here that, despite the presence of an active disease process, the ApoE E2 allele is associated with longevity and preservation of cognitive functioning.

Central to this study is the difficulty of making an accurate diagnosis of dementia of the AD-type in the DS population. In DS there is a wide variation in the premorbid intellectual level (verbal and performance) and the level of social/functional skills attained.[4–7] This variable baseline makes it difficult to identify and quantify specific cognitive changes associated with the emergence of dementia. Furthermore, reliable markers identifying the early changes of dementia in DS have not been established, since the symptoms of memory dysfunction and disorientation which characterize the early clinical picture of Alzheimer's disease (AD) are not reliable markers of dementia in DS individuals as they also occur in association with their developmental disability. One approach to solving these problems has been to base a diagnosis of dementia in DS on a change in the level of functioning of an individual patient.[3–11] Using this approach dementia can be characterized by changes in the following criteria; reduced speech and comprehension, loss of self-help skills, deterioration in social and vocational skills, and personality changes.[3,12,15] The occurrence of AD-type dementia in DS is usually ascribed to the effects of overproduction of β-amyloid protein, caused by a gene dosage effect on the amyloid precursor protein (APP).[16] Mutations in this gene, some of which lead to the overproduction of βAP from APP, are another cause of AD. However, it has recently become clear that apolipoprotein E (ApoE) genotype plays an important role both as a risk factor in the etiology and as a modulator of the age of onset of AD. Studies have shown that the E4 allele predisposes to disease and the E2 allele, in some circumstances, delays onset.[8,9] Furthermore, ApoE appears to modulate the age of onset of disease in a family in which the primary defect is a mutation in the gene which causes increased βAPP production.[10] In this family, E4 homozygotes had an earlier age of onset of dementia than E3 homozygotes and E2 heterozygotes had a later age of onset. Preliminary studies have also indicated that the E2 allele is associated with longevity in the general population[17] and in DS.[18]

This series of observations led us to examine the ApoE genotype in clinically assessed elderly patients with DS. Cases of DS were available from 2 centers: Manchester and Newcastle-Upon-Tyne (UK). A retrospective assessment of the case records was made (including medical and nursing notes) by clinicians blind to the neuropathological data and genotyping results (see TABLE 1). Patients for this study met the following specific inclusion criteria: adequate quality and quantity of clinical information, long-stay residence in institutions with good level of training and educational opportunities, no recent changes in institution, trisomy 21 karyotype, screening for and docu-

TABLE 1. ApoE Genotype of Patients with Down's Syndrome and Presence of Dementia

Case	Age	ApoE Genotype	Dementia
A	52	34	Yes
B	54	34	Yes
C	56	34	Yes
D	56	33	Yes
E	57	33	Yes
F	57	34	Yes
G	58	33	Yes
H	58	33	Yes
I	58	33	Yes
J	59	34	Yes
K	59	34	Yes
L	59	33	Yes
M	59	34	Yes
N	60	34	Yes
P	60	33	Yes
Q	61	32	No
R	64	32	No
S	65	33	Yes
T	67	33	No
U	68	32	Yes
V	68	22	No
W	71	32	No

mented treatment of hypothyroidism, and no concurrent significant medical illness which could have contributed to a deterioration in neuropsychological functioning. For subjects from both centers, a diagnosis of dementia was made only if there was clear evidence of a deterioration in self-care skills, continence, sleep pattern, occupational skills, or the emergence of new dysfunctional behaviors or psychological symptoms. For the purpose of this study a simple dichotomy of the patients into a demented and nondemented group was made. ApoE genotyping was carried out using our modification of a standard procedure. We observed no patients with the E4/E4 genotype. Patients with one E4 allele had the youngest age of death (mean age 57 years), E3/E3 genotypes were intermediate (mean age at death 60 years), and those with at least one E2 were the longest surviving individuals in the study (mean age at death 66). The one patient who was homozygous for E2 died at age 68. The effect of a single E2 allele on longevity was highly significant (t-test, $p < 0.01$, TABLE 2). The presence of an E2 allele had a marked relationship to the cognitive state of the patients. Despite the expectation that longevity would be associated with more marked cognitive decline, those patients with an E2 allele were less likely to develop dementia (chi square, $p < 0.01$, TABLE 2).

TABLE 2. Analysis of the Effects of ApoE2 on Age at Death and Dementia

	Age at Death		
No E2 allele	mean age at death	58.5 years + 3.4	n = 18
E2 allele	mean age at death	66.4 years + 3.9	n = 5
Significant difference $p < 0.007$			

	Presence of Dementia		
		Dementia	
		No	Yes
E2 allele	No	1	17
	Yes	4	1
Significant difference $p < 0.002$.			

In the general population the E2 allele is associated with longevity. The increased life span of DS patients who carry an E2 allele shown in a previous study[18] is therefore unsurprising. Aging is the most important risk factor associated with the development of AD, and studies on DS patients have clearly demonstrated a correlation between age and the amount of β-amyloid protein neuropathology.[3] Thus, one might predict a significant trend for DS subjects with an E2 allele (by virtue of their increased life expectancy) to exhibit the highest accumulation of β-amyloid protein neuropathology and therefore be at greatest risk of developing AD-type dementia. Paradoxically, our study does not support this hypothesis. Subjects with at least one E2 allele were significantly less likely to develop an AD-type dementia. Our data also demonstrate that the trisomy 21 phenotype can be modified by genes on other chromosomes. It is likely that other characteristics of the DS phenotype are similarly variable.

The mechanism underlying the association between ApoE genotype and the development of AD-type neuropathology is uncertain. However, both DS and AD patients with E2 alleles can become demented and, as we have noted previously,[18] develop both β-amyloid- and tau-related pathologies. We interpret this to indicate that the role of the ApoE genotype is probably one of modulating the basic disease process rather than playing a causal role.

Our data indicate that, in patients at risk of developing, or in the early stages of an Alzheimer-type disease, determination of the ApoE genotype and in particular the possession of an E2 allele may well offer a considerable advantage in determining the likely prognosis of a patient. Such prognostications will have considerable utility in the design planning and implementation of clinical care of DS and AD patients. In addition because the clinical expression of dementia is related to the progression and degree of pathology, determination of the ApoE genotype will provide guidance in the selection and implementation of modes of therapeutic intervention both in clinical trials and clinical practice.

REFERENCES

1. MANN, D. M. A. 1988. Mech. Ageing Dev. **43**:99–136.
2. OLIVER, C. & A. J. HOLLAND. 1986. Psychol. Med. **16**:307–322.
3. ROYSTON, M. C., *et al.* 1994. Neurodegeneration **3**:43–51.
4. THASE, M., *et al.* 1984. Biol. Psychiat. **19**:571–585.
5. CALTAGIRONE, C., U. NOCENTINI. 1990. Int. J. Neurosci. **54**:221–230.
6. JOHANSON, A., *et al.* 1991. Dementia **2**:159–168.
7. SCHAPIRO, M. B., *et al.* 1989. Neurology **39**:1349–1353.
8. SAUNDERS, A. M., *et al.* 1993. Neurology **43**:1467–1472.
9. CORDER, E. M., *et al.* 1994. Nature Genetics **7**:180–183.
10. HOULDEN, H., *et al.* 1993. Lancet **342**:737–738.
11. WARING, S. C., *et al.* 1994. Lancet **343**:611.
12. LAI, F. & R. S. WILLIAMS. 1989. Arch. Neurol. **57**:230–236.
13. EVENHUIS, H. M. 1990. Arch. Neurol. **47**:263–267.
14. SCHAPIRO, M. B., *et al.* 1989. Neurology **39**:1349–1353.
15. MANN, D. M. A., M. C. ROYSTON & C. R. I. RAVINDA. 1990. Neurol. Sci. **99**:153–164.
16. HARDY, J. & D. A. ALLSOP. 1991. Trends in Pharmacological Sci. **12**:383–388.
17. RUMBLE, B., *et al.* 1989. N. Engl. J. Med. **320**:1446–1462.
18. CAI, X. D., T. E. GOLDE & S. G. YOUNKIN. 1993. Science **259**:514–516.
19. CITRON, M., *et al.* 1992. Nature **360**:672–674.
20. SCHACHTER, F., *et al.* 1994. Nature Genetics **6**:29–32.
21. HARDY J., *et al.* 1994. Lancet **343**:979–980.
22. ROBERTS, G. W., *et al.* 1993. NeuroReport **4**:7–9.

Alzheimer's Disease and Apolipoprotein E in Italy[a]

S. SORBI,[b] B. NACMIAS, P. FORLEO, S. PIACENTINI, AND L. AMADUCCI

Department of Neurological Sciences, University of Florence, Florence Italy

ABSTRACT: Recent studies have provided evidence of association of apolipoprotein E (ApoE) ϵ4 allele and late onset familial and sporadic Alzheimer's disease (AD). Epidemiological studies have established allelic variation at the ApoE locus. We have analyzed the ApoE gene polymorphism in a sample of 416 Italian subjects. Our data confirm a significant association between ϵ4 allele and sporadic AD. The frequency of ϵ4 allele in early onset familial AD patients was comparable to control values suggesting that ϵ4 allele does not represent a risk factor for early onset familial AD (EOFAD). Moreover, we found a not-previously reported association between ApoE ϵ2 allele and sporadic AD and EOFAD. We included in this study two EOFAD families with the APP717 Val → Ile mutation in the Amyloid Precursor Protein (APP) gene on chromosome 21.

In any of the EOFAD families there was a significant effect of the ApoE genotype on the age of onset with the exception of one of the two mutated EOFAD families in which the 2 allele delays the age of onset.

INTRODUCTION

Alzheimer's disease (AD) is a neurodegenerative disorder characterized by the presence of senile plaques, neurofibrillary tangles and congophilic angiopathy. Although the pathogenesis of amyloid deposition is not completely clarified, recent studies suggest the involvement of ApoE in this process. The ApoE gene is located in a cluster with the genes for ApoC-I and ApoC-I′ pseudogene and the ApoCII gene on chromosome 19. It contains four exons encoding a 299 aminoacid 34kD plasma protein involved in cholesterol transport produced and secreted in the nervous system by astrocytes. The three most common alleles of the ApoE gene have been identified: the ϵ3 allele is probably the wild type and codes for the isoform of ApoE called E3, the ϵ2

[a] Supported by the Consiglio Nazionale delle Ricerche (C.N.R., Rome, Italy), Progetto finalizzato Invecchiamento, Publication code 952552. B.N. is partially supported by Telethon-Italy (grant n. E096).

[b] *Send correspondence to:* Dr. Sandro Sorbi, Department of Neurological Sciences, University of Florence, Viale Morgagni, 85, 50134 Florence, Italy; Tel: 0039 55 412634; Fax: 0039 55 418718.

codes for an Arg 158 → Cys substituted product called E2 and the ϵ4 codes for a Cys 112 → Arg substituted product called E4. The three alleles are combined in the general population in six different genotypes, either heterozygous (ApoE 2/3, 4/3 and 4/2) or homozygous (ApoE 2/2, 3/3 and 4/4).

Genetic studies have provided evidence of association of apolipoprotein E (ApoE) ϵ4 allele and late onset familial and sporadic Alzheimer's disease.[1] It has been established that inheritance of the ϵ4 allele of the ApoE gene is a major risk factor for late onset AD, increasing the incidence of AD and decreasing the age of onset as a function of the inherited dose of this allele. Immunohistochemical investigations demonstrated that AD affected individuals homozygous for ApoE4 show increased β-amyloid peptide immunoreactivity in neuritic plaques and congophilic angiopathy[2]: *in vitro* studies found that ApoE binds to β-amyloid with high affinity.[3]

These combined data account for the vast majority of Alzheimer's disease cases that are of late onset: most of these are sporadic, although some are familial. Familial forms with an early onset of the disease comprise 5 to 10 percent of all AD cases. Chromosomal localization has been reported to be on chromosome 14q24.3.[4] Recent studies found that APOE genotype does not modify age of onset in the chromosome 14 linked AD families.[5] Furthermore, a small number of early onset families is associated to point mutations in the Amyloid Precursor Protein (APP) gene on chromosome 21,[6] suggesting that in at least some cases, APP is mainly involved. A role of the APOE genotype could be demonstrated in these mutated families by the fact that it modulates the age of onset of AD families with APP717 and APP670 mutations although it does not influence the clinical expression of the APP gene codon 693 and 692 mutations.[7] Hyslop *et al.*[8] recently suggested that the genotypes of APOE and APP genes may interact to determine the final pathologic phenotype of AD. Moreover, there is even more convincing evidence that ApoE ϵ2 allele is a protective factor for risk of developing AD in both familial and sporadic late onset AD.[9]

In order to further analyze the polymorphism of the ApoE gene, we studied the ApoE genotype distribution and the corresponding allele frequencies in a sample of Italian AD patients, including those with sporadic AD, early onset familial AD (EOFAD) with family offsprings of affected members younger than the age at onset and escapees. We also included two extended Italian families with EOFAD carrying the APP 717 Val → Ile mutation, recently described by our group.[10]

METHODS

The study has included 416 Italian individuals. AD patients were recruited in the Department of Neurological Sciences of Florence, Italy. A careful clinical and eventually neuropathological evaluation has been made according to NINCDS-ADRDA criteria. We divided these subjects as following: 64 patients with sporadic AD (mean age at onset 60.2 ± 5.5), 104 patients with early onset familial Alzheimer's disease, including APP717 mutated families (mean age at onset, defined as the onset of first cognitive changes, 54.6 ± 3.4), 52 subjects at risk to develop AD (mean age 32.2 ± 9.1), being offspring of an affected member of EOFAD families and younger than the age at onset in that family, 7 escapees (offspring of affected members who did not have the disease and are two standard deviations older than the mean age at onset in their family) from the same EOFAD families (mean age 70.5 ± 4.3), 84 apparently normal controls (mean age 59.3 ± 6.5), 105 subjects with other neurological diseases (mean age 43.5 ± 7.6). The pathological controls were patients with various neurological diseases (olivo-ponto-cerebellar atrophy, Creutzfeldt-Jakob disease, Huntington disease, Familial Amyloidotic Polyneuropathy, Freidreich ataxia, Bipolar-unipolar disorder, Down's syndrome and Hereditary Cerebral Hemorrhage). DNA from affected and unaffected individuals was extracted with the phenolchloroform procedure. Amplification from 200ng of genomic DNA was performed using a Thermal Cycler 9600 (Perkin-Elmer). The primers were used are complementary to the region of the allelic variation in exon IV of the ApoE gene: a 332bp region of DNA that spans all the ApoE polymorphic sites was amplified. In all samples PCR amplification resulted in a single product of the predicted size which was ascertained on minigel using as a molecular weight standard the 123bp ladder HhaI (Biolabs) digests were carried out at 37°C for 18h, then resolved through 2% Metaphor (FMC) Agarose plus 1% standard Agarose and stained with ethidium bromide. The size of the resultant fragments was estimated by comparison with a known molecular size marker (plasmid PBR322 digested with HinfI). The frequencies of $\epsilon 2$-, $\epsilon 3$- and $\epsilon 4$-alleles were estimated by gene counting.

RESULTS

The frequencies of the three common alleles of the ApoE gene in the sample of the apparently normal controls resulted as follows: $\epsilon 2 = 0.047$; $\epsilon 3 = 0.869$; $\epsilon 4 = 0.084$. Besides, there were no differences in allele frequencies between normal and pathological control subjects, suggesting that the Italian general non-Alzheimer population is strongly characterized by a higher prevalence of the $\epsilon 3$ allele. The frequency of the $\epsilon 2$ allele was significantly increased in AD and EOFAD patients (0.125 and 0.115 versus 0.045 in total controls) and in subjects at risk to develop FAD (0.105), suggesting a high association between the $\epsilon 2$ allele and Alzheimer's disease.

In any of the EOFAD families there was a significant effect of the ApoE genotype on the age of onset with the exception of one of the two mutated EOFAD families in which the ϵ2 allele delay the age of onset. In the first APP mutated family, Flo12, (mean age at onset of cognitive symptoms 58 ± 4 y) all (n = 3) living clinically affected members and the two younger offsprings carrying the mutation had genotype ϵ3/ϵ3. In the second family we observed a possible relationship between delay of age at onset and the ϵ2 allele. In this family, Flo13, (mean age at onset 50 ± 3 y) two affected members whose age at onset was 48 and 46 years respectively, had genotype ϵ3/ϵ3, while a third affected member with later age at onset (52 years) had genotype ϵ2/ϵ3. In this family we have also recently identified a person, 55 years old, who, at the age of 1.8 standard deviation (SD) above the mean age of onset in this pedigree, carries the APP717 Val → Ile mutation but does not show any sign of clinical disease on neurologic or neuropsycologic tests or on computerized axial tomography or magnetic resonance imaging scans and he also has the ϵ2/ϵ3 genotype.

In the offspring of this family, 6 members, with age below the mean age at onset and carrying the mutation, had either the ϵ2/ϵ3 or the ϵ3/ϵ3 genotype. In none of the subjects belonging to these two families did we find the ϵ4 allele.

DISCUSSION

The results of our study on the Italian population confirm previous observations that there is a highly significant association between ϵ4 allele and sporadic AD. We found a 5-fold increase of ApoE ϵ4 allele frequency in Italian sporadic AD patients (0.406 versus 0.084 in control subjects) confirming earlier reports.[1,11] The frequency of ϵ4 allele in early onset familial AD patients was comparable to control values, suggesting that ϵ4 allele does not represent a risk factor for early onset familial AD. Moreover, in our Italian AD patients we found a not–previously reported association between ApoE ϵ2 allele and Alzheimer's disease. The increased ϵ2 frequency in AD and EOFAD patients and at-risk subjects provide evidence that, in at least Italian Alzheimer's disease patients, ϵ2 is not a rare allele. Furthermore, our results are consistent with the recent report by Gerdes *et al.*[12] who compared 46 different populations around the world and found that the Italian population showed the highest prevalence of ϵ3 allele (0.857).[13] The allele frequencies we found in the general Italian population are within the range of those observed in other Southern regions of Europe but different from those observed in Northern Europe.[12]

In none of the patients of the chromosome 14–linked AD family is the onset age modified by the ApoE genotype, in accordance with the result by Van Broechkoven *et al.*[14] Furthermore in none of the patients of the other

11 EOFAD families was there any relationship between age at onset and APOE genotypes. Our findings demonstrate that the ApoE $\epsilon 2$ allele does not have a protective effect in EOFAD patients and that the $\epsilon 3$ allele, which is the most common allele in the human population, remains the most frequent allele even in the affected Alzheimer's population.

In contrast with our results on Italian EOFAD families, the observations on our APP mutated families suggests: 1) the $\epsilon 4$ allele is not sufficient for the phenotypic appearance of AD in APP717 Val → Ile mutated subjects; 2) the presence of the $\epsilon 2$ allele may probably interact with the APP717 Val → Ile mutation and delay the age at onset.

A significant support for the protective effect of $\epsilon 2$ allele comes from the observations of Hardy *et al.*[14] who noted a later age of onset for two $\epsilon 2/\epsilon 3$ heterozygotes in EOFAD segregating an APP mutation. This hypothesis is supported also by the study from Corder *et al.*,[9] who recently reported a protective effect of $\epsilon 2$ allele in LOFAD and by the data from Hyslop *et al.*,[8] who reported that in their family too, the older, asymptomatic carriers of the APP717 Val → Ile mutation had the $\epsilon 2/\epsilon 3$ genotype. These results suggest that the possible interaction between ApoE alleles and AD clinical expression is not a common mechanism shared by all forms of familial AD but that is probably specific of APP717 Val → Ile mutated families.

REFERENCES

1. SAUNDERS, A. M., W. J. STRITTMATTER, D. SCHMECHEL, *et al.* 1993. Association of apolipoprotein E allele $\epsilon 4$ with late onset familial and sporadic Alzheimer's disease. Neurology **43**:1476–1472.
2. NAMBA, Y., M. TOMONAGA, H. KAWASAKI, *et al.* 1991. Apolipoprotein E immunoreactivity in cerebral amyloid deposits and in neurofibrillary tangles in Alzheimer's disease and kuru plaque amyloid in Creutzfeldt-Jacob disease. Brain Res. **541**:163–166.
3. STRITTMATTER, W. J., K. H. WEISGRABER, D. Y. HUANG, *et al.* 1993. Binding of human apolipoprotein E to synthetic amyloid beta peptide: Isoform-specific effects and implications for late-onset Alzheimer disease. Proc. Natl. Acad. Sci USA **90**:8098–8102.
4. ST. GEORGE HYSLOP, P. H., J. HAINES, E. ROGAEV, *et al.* 1992. Genetic evidence for a novel familial Alzheimer's disease locus on chromosome 14. Nature Genetics **2**:330–334.
5. VAN BROECKHOVEN, C., H. BACKHOVENS, M. CRUTS, *et al.* 1994. APOE geneotype does not modulate age of onset in families with chromosome 14 encoded Alzheimer's disease. Neurosci. Lett. **169**:179–180.
6. GOATE, A., M.-C. CHARTIER-HARLIN, M. MULLAN, *et al.* 1991. Segregation of a missense mutation in the amyloid precursor protein gene with familial Alzheimer's disease. Nature **349**:704–706.
7. HAAN, J., C. VAN BROECKHOVEN, C. M. VAN DUIJN, *et al.* 1994. The apolipo-

protein E ϵ4 allele does not influence the clinical expression of the amyloid precursor protein gene codon 693 and 692 mutation. Ann. Neurol. **36**: 434–437.

8. St. George Hyslop, P. H., D. Crapper McLachlan, T. Tuda, *et al.* 1994. Alzheimer's disease and possible gene interaction. Science **263**:537.
9. Corder, E. H., A. M. Saunders, N. J. Risch, *et al.* 1994. Protective effect of apolipoprotein E type 2 allele decreases risk of late onset Alzheimer disease. Nature Genet. 7:180–183.
10. Sorbi, S., B. Nacmias, P. Forleo, *et al.* 1993. APP717 and Alzheimer's disease in Italy. Nature Genetics **4**:10.
11. Corder, E. H., A. M. Saunders, W. J. Strittmatter, *et al.* 1993. Gene dose of apolipoprotein E type 4 allele and the role of Alzheimer's disease in late onset families. Science **261**:921–923.
12. Gerdes, L. U., I. C. Klausen, I. Sihm, *et al.* 1992. Apolipoprotein E polymorphism in a Danish population compared to findings in 45 other study populations around the world. Genet. Epidemiol. **9**:155–167.
13. Hallman, D. M., E. Boerwinkle, N. Saha, *et al.* 1992. The apolipoprotein E polymorphism: A comparison of allele frequencies and effects in nine populations. Am. J. Hum. Genet. **49**:338–349.
14. Houlden, H., J. Collinge, A. Kennedy, *et al.* 1993. ApoE genotype and Alzheimer's disease. Lancet **342**:737–738.

Apolipoprotein E Alleles and Brain Vascular Pathology in Alzheimer's Disease[a]

R. N. KALARIA,[b] D. L. COHEN, AND D.R.D. PREMKUMAR

Departments of Neurology, Neurosciences and Pathology, Case Western Reserve University, Cleveland, Ohio 44106 USA

ABSTRACT: The presence of apolipoprotein E (ApoE)-E4 allele has been implicated as a risk factor for Alzheimer's disease (AD). We examined the occurrence of ApoE 4 alleles in AD associated with cerebral amyloid angiopathy and other vascular lesions. We found significantly high frequency of the ApoE 4 allele in AD with moderate to severe CAA. The frequency of the allele was also high in AD cases with other vascular lesions such as multiple infarcts and lacunes. As previously reported, we confirm a greater frequency of the ApoE 4 allele in the diffuse Lewy body variant of AD. Our results suggest ApoE 4 allele to be a significant factor in the development of CAA in AD. While this may be related to increased brain amyloid load as a consequence of ApoE genotype, the possibility exists that ApoE may be a specific factor in vascular abnormalities associated with AD.

INTRODUCTION

It is now clear that a number of proteins or factors are associated with amyloid deposits and other lesions in AD. Some of these include the heparan sulfate glycosaminoglycans, amyloid P component, complement proteins, serine protease inhibitors, and most recently the apolipoproteins.[1] The significance of the associated factors is unknown but among these apolipoprotein E (ApoE) is being intensively studied. ApoE is a polymorphic protein that plays a central part in the metabolism of cholesterol and the triglycerides. In the human body, the synthesis of ApoE is controlled by a single gene locus on chromosome 19 which determines three codominant alleles E2, E3 and E4. The polymorphism results in six ApoE phenotypes, E2/2, E3/3 and E4/4 as homozygotes and E2/3, E3/4 and E2/4 as heterozygotes. Specific ApoE

[a] The research was supported by NIH Grants RO1AG10030, POAG08012B2 and R5AG08992A2.

[b] *Send correspondence to:* Dr. R. N. Kalaria, Department of Neurology (BRB5), Case Western Reserve University, 10900 Euclid Avenue, Cleveland, Ohio 44106, USA; Tel: (216)368-1113; FAX: (216)368-1144.

alleles were first implicated in atherosclerosis.[2] Most recent studies showed the high frequency of ApoE 4 alleles to be associated with late-onset familial and sporadic AD.[3] This implicates an important functional role for ApoE isoforms in the pathophysiology of AD. However, related studies have also reported an increased inheritance of ApoE 4 allele in patients with multi-infarct dementia,[4] vascular dementia,[5] diffuse Lewy body disease[6] and recently Creutzfeldt-Jakob disease.[7] Notwithstanding that AD is likely a group of heterogenous disorders and to delineate the vascular pathology in AD, we assessed ApoE genotypes in AD with and without vascular lesions. We specifically examined the association of ApoE 4 allele in AD subjects exhibiting remarkable cerebral amyloid angiopathy and other vascular lesions including infarcts.

MATERIAL AND METHODS

Tissues

Human brain tissue from a total of 184 subjects with either AD, no pathological diagnosis, or other diseases was obtained at autopsy immediately upon removal within 2 to 7 hours after death (age range 55–90 years).

Immunocytochemical Analyses

Patterns of vascular amyloid deposition and degree of CAA in neocortical tissue sections and in some cases in isolated vessel fractions were examined with antibodies to Aβ protein.[8,9]

Isolation of DNA and PCR

DNA was extracted by standard methods from the cortical tissues, digested and used for the PCR to assess ApoE genotype essentially as described previously.[10] The ApoE 4 allele frequency was defined as the fraction of the number of samples with at least one ApoE 4 allele from the total alleles per group.

RESULTS AND DISCUSSION

We examined the occurrence of ApoE alleles in 184 samples from subjects who were clinically diagnosed with possible AD (FIG. 1). Upon histopathological analyses after autopsy, 11 of these turned out to be various non-AD

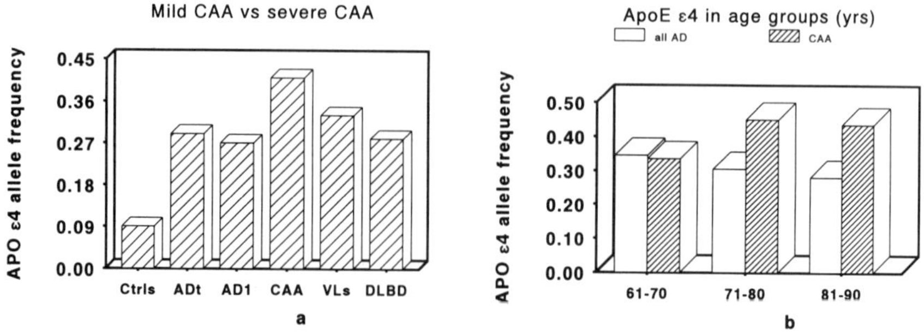

FIGURE 1. ApoE genotypes in AD in relation to CAA and other vascular lesions. **a** (ApoE genotypes in AD) shows distribution of at least one ApoE 4 allele in various groups of subjects: Controls (Ctrls) without AD, all AD subjects (ADt), AD with minimal CAA (AD1), AD with moderate to severe CAA (CAA), AD with vascular lesions (VLs) and AD with diffuse Lewy body disease (DLBD). These results were derived from analyses of a total of 184 individual samples with 15 or more (except Ctrls) samples in each group. The frequencies in the CAA and VLs groups were significantly different from the AD1 group. **b** (ApoE genotypes by age) shows distribution of ApoE4 allele in the ADt and CAA groups by age. Note the apparent increased frequency with age in CAA.

controls. The presence of at least one ApoE 4 allele in these samples was low (FIG. 1a), comparable to previously reported frequencies.[3,11] The two subjects with E4 allele exhibited the E2/4 and E3/4 genotypes. The rest were either E3/3 or E2/3 diagnosed with a variety of disorders, including hereditary dysphasic dementia and ALS. In contrast to the non-AD controls and consistent with previous reports,[12,13] the frequency of ApoE 4 allele in the total AD pool was expectedly high (FIG. 1a). However, when we divided the AD pool into subgroups, each characterized by the presence of a significant feature in the pathology, we noted that the ApoE 4 frequency in AD with moderate to severe CAA[14] was much higher than those with only focal or negligible CAA (FIG. 1b). This frequency seemed to increase with age and was apparently higher in the 7th and 8th decades compared to all the AD subjects (FIG. 1a). We also noted that among these three subjects who suffered intracerebral hemorrhage and were predominantly characterized with severe CAA also carried the ApoE 4 allele. Interestingly, a group of subjects with a pathological diagnosis of AD, but who exhibited some vascular pathology including infarcts, lacunes, Binswanger lesions and petechial hemorrhages had a higher frequency of the ApoE genotype with E4 allele than those without such lesions (FIG. 1a). These results are not inconsistent with the greater frequency of ApoE 4 allele in vascular conditions associated with dementia (see above). We further observed that the percent frequency of ApoE 4 allele in the diffuse Lewy body variant of AD[6] was increased but not as high as that in AD subjects exhibiting CAA (FIG. 1a).

In summary, our analysis of a relatively small sample suggests the development of CAA in AD is increased by inheritance of at least one ApoE 4 allele. These observations also suggest that ApoE 4 allele is associated with vascular lesions in AD. It is possible that specified ApoE gene products may be factors in promoting pathogenetic processes within the cerebral vasculature in concert with aging.

REFERENCES

1. NAMABA, Y., M. TOMONAGA, H. KAWASAKI, E. OTOMO & K. IKEDA. 1991. Apolipoprotein E immunoreactivity in cerebral amyloid deposits and neurofibrillary tangles in Alzheimer's disease and kuru plaque amyloid in Creutzfeldt-Jakob disease. Brain Res. **541**:163–166.
2. DAVIGNON, J., R. E. GREGG & C. F. SING. 1988. Apolipoprotein E polymorphism and atherosclerosis. Atrerioscler. Thromb. **8**:1–21.
3. ROSES, A. D. 1994. Apolipoprotein E affects the rate of Alzheimer disease expression: β-amyloid burden is a secondary consequence dependent on ApoE genotype and duration of disease. J. Neuropathol. Exp. Neurol. **53**(5): 429–437.
4. NOGUCHI, S., K. MURAKAMI & N. YAMADA. 1993. Apolipoprotein E geneotype and Alzheimer's disease. Lancet **342**:737.
5. FRISONI, G. B., O. CALABRESI, C. GEROLDI, A. BIANCHETTI, A. L. D'ACQUARICA, S. GOVONI, C. R. SIRTORI, M. TRABUCCHI & G. FRANCESCHINI. 1994. Apolipoprotein E + 4 allele in Alzheimer's disease and vascular dementia. Dementia **5**:240–242.
6. ARAI, H., S. HIGUCHI, T. MURAMTSU, T. IWATSUBO, H. SASAKI & J. Q. TROJANOWSKI. 1994. Apolipoprotein E gene in diffuse Lewy body disease with or without co-existing Alzheimer's disease. Lancet **344**:1307.
7. AMOUYEL, P., O. VIDAL, J. M. LAUNAY & J. L. LAPLANCHE for the FRENCH GROUP. 1994. The apolipoprotein E alleles as major susceptibility factors for Creutzfeldt-Jakob disease. Lancet **344**:1315–1318.
8. KALARIA, R. N., T. E. GOLDE, S. N. KROON & G. PERRY. 1993. The serine protease inhibitor antithrombin III and its mRNA in the pathogenesis of Alzheimer's disease. Am. J. Pathol. **143**:886–893.
9. KAWAI, M., R. N. KALARIA, P. CRAS, S. L. SIEDLAK, M. E. VELASCO, E. R. SHELTON, H. W. CHAN, B. D. GREENBERG & G. PERRY. 1993. Degeneration of amyloid precursor protein-containing smooth muscle cells in cerebral amyloid angiopathy. Brain Res. **623**:142–146.
10. WENHAM, P. R., W. H. PRICE & D. BLUNDELL. 1991. Apolipoprotein of genotyping by one stage PCR. Lancet **337**:1158–1159.
11. REBECK, G. W., J. S. REITER, D. K. STRICKLAND & B. T. HYMAN. 1993. Apolipoprotein E in sporadic Alzheimer's disease: Allelic variation and receptor interactions. Neuron **11**:575–580.
12. POIRIER, J., J. DAVIGNON, D. BOUTHILLIER, S. KOGAN, P. BERTRAND & S. GAUTHIER. 1993. Apolipoprotein E polymorphism and Alzheimer's disease. Lancet **342**:697–699.

13. SAUNDERS, A. M., K. SCHMADER, J. C. S. BREITNER, D. BENSON, W. T. BROWN, L. GOLDFARB, D. GOLDGABER, M. G. MANWARING, M. H. SZYMANSKI, N. MCCOWN, K. C. DOLE, D. E. SCHMECHEL, W. J. STRITTMATTER, M. A. PERICAK-VANCE & A. D. ROSES. 1993. Apolipoprotein E e4 allele distribution in late-onset Alzheimer's disease and in other amyloid-forming diseases. Lancet **342**:710–711.

14. SCHMECHEL, D. E., A. M. SAUNDERS, W. J. STRITTMATTER, B. J. CRAIN, C. M. HULETTE, S. H. JOO, M. A. PERICIAK-VANCE, D. GOLDGABER & A. D. ROSES. 1993. Increased amyloid β-peptide deposition in cerebral cortex as a consequence of apolipoprotein E genotype in late-onset Alzheimer disease. Proc. Natl. Acad. Sci. USA **90**:9649–9653.

Amyloid β-Protein, APOE Genotype and Head Injury

J. A. R. NICOLL,[a,c] G. W. ROBERTS,[b] AND D. I. GRAHAM[a]

[a] *Department of Neuropathology, University of Glasgow, Glasgow, United Kingdom*
[b] *Molecular Neuropathology Department, SmithKline Beecham Pharmaceuticals, United Kingdom*

ABSTRACT: Deposition of amyloid β-protein (Aβ) in the brain plays a key role in the pathogenesis of Alzheimer's disease. Head injury is an epidemiological risk factor for Alzheimer's disease, and deposition of Aβ occurs in approximately one third of individuals dying shortly after a severe head injury. Of the three common apolipoprotein E alleles (APOE-ϵ2, ϵ3, and ϵ4) APOE-ϵ4 allele is a strong risk factor for both sporadic and some familial cases of Alzheimer's disease and there is *in vitro* evidence that apolipoprotein E is directly involved in Aβ deposition. We have examined the frequency of APOE-ϵ4 in those individuals with Aβ deposition following head injury and found that the APOE-ϵ4 frequency (0.52) is higher than in most studies of sporadic Alzheimer's disease. In those head-injured individuals without amyloid deposition the APOE-ϵ4 frequency (0.16) is similar to that in non-Alzheimer's disease controls ($p < 0.00001$). Our data indicate an interaction between known environmental and genetic risk factors for Alzheimer's disease and underlines the importance of convergence of data around the common mechanism of Aβ deposition. Furthermore, it indicates a genetic susceptibility to the effects of a head injury which may be of significance both to those who have recently sustained such an injury and to those whose activities put them at risk of trauma.

The evidence linking head injury and AD includes epidemiological studies[1,2] and the Aβ deposition described previously.[3–5] We show here Aβ deposition following head injury is more likely in those with a genetic susceptibility conferred by apolipoprotein (APOE) ϵ4. Amyloid β-protein (Aβ) deposits resembling diffuse plaques were present in the cerebral cortex in 23 out of 90 (26%) individuals who died within 2 weeks of a head injury (TABLE 1). The proportion of patients with Aβ deposition for each of the 6 APOE genotypes is shown in TABLE 2. Every one of the 6 patients who were homozygous for the APOE-ϵ4 allele had Aβ deposits. APOE-ϵ4 allele frequency in the 23 individuals with deposition of Aβ (Aβ+) was 0.52 compared with 0.16 for those individuals without Aβ deposition (Aβ−)($\chi^2 = 23.013$, 1

[c] *Send correspondence to:* Dr. J. Nicoll, Department of Neuropathology, Institute of Neurological Sciences, Southern General Hospital, Glasgow G51 4TF, United Kingdom; Tel: 041-201-2046; Fax: 041-201-2998.

TABLE 1. Details of Head-injured Patients with and without Amyloid β-Protein Deposition (Aβ+, Aβ−)

		Aβ+	Aβ−
Number		23	67
Age (years)	mean + SD	52 ± 19	28 ± 18
	range	14–75	0.15–79
Survival following head injury (days)			
	mean + SD	3.3 ± 4	2.9 ± 3
	range	<1–13	<1–13

df, $p < 0.00001$; see TABLE 3). APOE-ε4 allele frequency in the Aβ+ head-injured group was higher than most previously published frequencies in sporadic Alzheimer's disease (AD), whereas the APOE-ε4 frequency in the Aβ− group is similar to non-AD controls.[6] A gene dose effect was also apparent with the proportion of head-injured patients with Aβ deposition increasing in line with the number of APOE-ε4 alleles (from 10% for those without an ε4 allele, to 35% for those with one ε4 allele, to 100% for the ε4 homozygotes (χ^2-for trend = 22.85, 1 df, $p < 0.001$; see TABLE 4). APOE-ε2 occurs with a relatively low frequency in patients with AD and it has been suggested that it may exert a protective effect.[7] The ε2 frequency was lower in the Aβ+ head-injured patients (0.02; representing only one ε2 allele) than the Aβ− group (0.1) although this finding was not significant (TABLE 3).

Patients in the Aβ+ group (mean 52 years) were older than those of the Aβ− group (mean 28 years) and it could be suggested that the Aβ deposits are a reflection of early AD that was present prior to the head injury and not related to trauma.

However, none of the 23 Aβ+ patients was cognitively impaired prior to the head injury, had a family history of AD or had Down's syndrome (retrospective analysis of clinical notes) and it should be noted that the youn-

TABLE 2. Head-injured Patients, APOE Genotype, and Aβ Deposition

APOE Genotype	Percentage with Aβ Deposition
2/2	0% (0/2)
2/3	0% (0/7)
3/3	12% (5/41)
2/4	25% (1/4)
3/4	37% (11/30)
4/4	100% (6/6)

Note: APOE genotyping was performed on formalin-fixed paraffin-embedded tissue by standard methods.

TABLE 3. APOE Allele Frequences in Head-injured Patients

A. All head-injured patients		
APOE allele	Aβ+	Aβ−
ε2	1/46 (0.02)	14/134 (0.1)
ε3	21/46 (0.46)	98/134 (0.73)
ε4[a]	24/46 (0.52)	22/134 (0.16)
B. Head-injured patients Under 60 Years of Age		
APOE allele	Aβ+	Aβ−
ε4[b]	14/28 (0.5)	22/126 (0.17)

[a] $\chi^2 = 23.013$, 1 df, $p < 0.00001$.
[b] $\chi^2 = 13.542$, 1 df, $p < 0.001$.

gest patient with Aβ deposits was only 14 years of age, far younger than the age at which AD pathology is to be expected. Our previous study of 45 non-head injury controls (age range 51 to 80 years) detected no Aβ deposits in those patients under 60 years of age.[4] In this study a separate analysis of the head-injured patients under the age of 60 years showed that the APOE-ε4 frequency for the 14 Aβ+ cases is 0.5 compared with 0.17 for the controls under 60 years ($\chi^2 = 13.542$, 1 df, $p < 0.001$; see TABLE 3B). In addition, although the numbers are small, the relationship also holds when the 9 Aβ+ head-injured patients under the age of 50 years are considered as a separate group; the APOE = ε4 frequency is 0.39 compared with 0.18 for the controls under 50 years ($\chi^2 - 4.250$, 1 df, $p = 0.039$).

We believe that our analysis demonstrates an association, at least as strong as that in AD, between possession of APOE-ε4 and deposition in the brain of Aβ following head injury. This has a number of implications. The demonstration of this association in a novel situation supports the contention that apolipoprotein E isoform E4, or alternatively absence of isoforms E2 and E3, influences Aβ metabolism or that of its precursor (APP) *in vivo* in such a way as to augment the formation of Aβ. Furthermore, this effect can be seen in young patients indicating that the APOE genotype is of significance in influencing metabolic processes in the brain much earlier in life than the age at which AD usually becomes manifest (over 65 years). When patients

TABLE 4. Proportion of Head-injured Patients with Deposition of Aβ (Aβ+) According to APOE-ε4 Gene Dose

APOE-ε4 Gene Dose	Proportion of Head-injured Patients Aβ+	Percentage
0	5/50	10%
1	12/34	35%
2	6/6	100%

χ^2 for trend = 22.985, 1 df, $p < 0.001$.

of all ages are considered together, an age difference between Aβ+ an Aβ− groups is to be predicted from previous work indicating that head injury is a greater risk factor for AD in older patients.[2]

Apolipoprotein E, and especially isoform E4, promotes Aβ formation *in vitro* within a few hours.[8] APP is over-expressed in all individuals following head injury,[4] but this over-expression is considerably more likely to result in deposition of Aβ in those individuals possessing an APOE-ϵ4 allele: 45% of those with one or more ϵ4 alleles are Aβ+ compared with 10% of those with no ϵ4 allele.

We have argued that head injury–related Aβ deposition could trigger the development of the full spectrum of AD pathology later in life in long-term survivors. Recent epidemiological evidence supports the existence of a synergistic interaction between head injury and APOE-ϵ4.[9] If the development of AD is a potential long-term consequence of trauma-related Aβ deposition, then the data presented here indicate that APOE genotyping will allow definition of the risk of subsequent neurodegeneration in individuals who are in jeopardy of sustaining a head injury through social or professional activities (*e.g.*, amateur and professional boxers, and other sportsmen such as mountain climbers).

In addition, the association between APOE genotype and CNS vulnerability might generalize to encompass other types and causes of CNS disease/damage. In this latter case APOE genotyping may prove to be a universal marker for predicting the prognosis of patients with CNS disorders.

REFERENCES

1. Mortimer, J. A., C. M. Van Juijn, V. Chandra, L. Fratiglioni, A. B. Graves, A. Heyman, A. F. Jorm, E. Kokmen, K. Kondo, W. A. Rocca, S. L. Shalat, H. Soininen & A. Hofman. 1991. Int. J. Epidemiol. **20**(Suppl. 2), S28–S35.
2. Mayeux, R., R. Ottman, M.-X. Tang, L. Noboa-Bauza, K. Marder, B. Gurland & Y. Stern. 1993. Ann. Neurol. **33**:494–501.
3. Roberts, G. W., S. M. Gentleman, A. Lynch & D. I. Graham. 1991. Lancet **ii**:1422–1423.
4. Roberts, G. W., S. M. Gentleman, A. Lynch, L. Murray, M. Landon & D. I. Graham. 1994. J. Neurol. Neurosurg. & Psychiatry **57**:419–425.
5. Roberts, G. W., D. Allsop & C. J. Bruton. 1990. J. Neurol. Neurosurg. & Psychiatry **53**:373–378.
6. Saunders, A. M., W. J. Strittmatter, D. Schmechel, P. H. St. George-Hyslop, M. A. Pericak-Vance, S. H. Joo, B. L. Rosi, J. F. Gusella, D. R. Crapper-MacLachlan, M. J. Alberts, C. Hulette, B. Crain, D. Goldgaber & A. D. Roses. 1993. Neurology **43**:1467–1472.

7. CORDER, E. H., A. M. SAUNDERS, N. J. RISCH, W. J. STRITTMATTER, D. E. SCHMECHEL, P. C. O. GASKELL, J. B. RIMMLER, P. A. LOCKE, P. M. CONNEALLY, K. E. SCHMADER, G. W. SMALL, A. D. ROSES, J. L. HAINES & J. N. PERICAK-VANCE. 1994. Nature Genetics 7:180–184.
8. MA, J., A. YEE, B. BREWER, S. DAS & H. POTTER. 1994. Nature **372**:92–94.
9. MAYEUX, R., R. OTTMAN, G. MAESTRE, C. NGAL, M.-X. TANG, H. GINSBERG, M. CHUN, B. TYCKO & M. SHELANSKI. 1995. Neurology **45**:555–557.

Apolipoprotein E Polymorphism Is Associated with Both Senile Plaque Load and Alzheimer-Type Neurofibrillary Tangle Formation[a]

W. MÄRZ,[b,f] H. SCHARNAGL,[b] M. KIRÇA,[c] J. BOHL,[d] W. GROß,[c] AND T. G. OHM[e]

[b] *Division of Clinical Chemistry, Department of Medicine, Albert Ludwigs-University, Freiburg, Germany*

[c] *Gustav Embden-Center of Biological Chemistry, Johann Wolfgang Goethe-University, Frankfurt am Main, Germany*

[d] *Department of Neuropathology, Johannes Gutenberg-University, Mainz, Germany*

[e] *Institute of Anatomy, Charité, Berlin, Germany*

ABSTRACT: Recent work provided evidence that the apolipoprotein (apo) E polymorphism is associated with late-onset sporadic Alzheimer's disease. The major histological hallmarks of Alzheimer's disease are the extraneuronal deposition of A4/β-amyloid and the intraneuronal formation of neurofibrillary tangles, the latter correlating strongly with the psychometric status. We examined the relationship between the apo E polymorphism and Alzheimer's disease-related histological changes using a staging system which accounts for the progression of the disease over time and correlates well with the cognitive decline ante mortem. We observed a significant positive correlation between both neurofibrillary changes and A4/β-amyloid deposits and the ε4 gene dose. We estimated that the presence of one apo E4 allele leads to an earlier onset of the histopathological process of about one decade. The association of both types of Alzheimer's disease-related changes with the prevalence of the ε4-allele suggests that the apo E polymorphism causally contributes to the development of Alzheimer's disease.

INTRODUCTION

Alzheimer's disease (AD) is the major cause of dementia in the elderly. Only recently, evidence was provided that the apolipoprotein (apo) E polymorphism is associated with late onset sporadic AD.[1–3] Histologically, AD is characterized by intraneuronal neurofibrillary tangles (NFT) and extraneuronal A4/β-amyloid (A4/β) deposition. The NFT-load correlates with the

[a] This research was sponsored by the Hedwig and Herbert Eckelmann-Stiftung and the Dr. Adam Stiftung (to W.M.) and by the Deutsche Forschungsgemeinschaft (to T.G.).

[f] *Send correspondence to:* Dr. Winfried März, Division of Clinical Chemistry, Department of Medicine Albert Ludwigs-University, Hugstetter Straße 55, 79106 Freiburg, Germany; Tel: +49 (761)270-3516; Fax: +49 (761)270-3444.

ante mortem psychometrical scores much better than the A4/β burden.[4] In the autopsy-based studies conducted so far, carriers of the ε4 allele were found having a higher cerebral load of A4/β, but not of neurofibrillary changes when compared to ε3 homozygotes.[2,3,5] In these studies, cases and controls were compared. Provided that NFT had to reach a threshold amount before causing death, then autopsy analysis of NFT in full-blown AD would detect a similar NFT burden, irrespective of the apo E genotype. To avoid this problem, we studied a random sample of consecutive autopsies comprising a broad, but clearly defined spectrum of AD-related morphological changes using a staging system[6] which takes into account the gradual development of AD.[7]

METHODS

The 191 brains used in this study came from routine autopsies carried out at the University Hospitals of Mainz or Frankfurt. To assure that the individuals were at risk for AD, age threshold was set at 60 years. Neurofibrillary changes and A4/β depositions were analyzed in polyethylene glycol embedded sections by means of silver staining techniques[8–11] and staged.[6] Apo E genotypes were determined in DNA isolated from deep-frozen brain tissue according to Hixson and Vernier[12] with minor modifications.[13]

RESULTS

The overall relative frequencies of apo E genotypes in our study (ε2/2, 0.01; ε3/3, 0.607; ε4/4, 0.031; ε2/3, 0.136; ε2/4 0.031; ε3/4 0.183) were close to data published for Caucasians.[14] We examined the relationship between the apo E polymorphism and NFT (0, I to VI) or A4/β (0, A to C) stages by contingency table analysis. Both kinds of histopathological changes were significantly associated with the apo E genotype (NFT: $chi^2 = 50.9$, $p = 0.01$; A4/β: $chi^2 = 30.1$, $p = 0.012$). The mean NFT and A4/β stages were significantly different with the apo E genotype, even after adjusting for the risk factor age as a covariable (TABLE 1). The mean stages were higher in all of the samples with at least one ε4 allele (ε4/2, ε4/3, ε4/4), compared to those samples without the ε4 allele (ε2/2, ε3/2, ε3/3). In addition, mean stages were higher in the ε4/4 homozygotes than in ε4/3 and ε4/2 heterozygotes,indicating that the ε4 gene dose is related to the risk of AD. We analyzed the relationship between age and NFT or β/A4 stages separately for those samples with at least one ε4 allele ($n = 47$) and those lacking an ε4 allele ($n = 144$) by exponential regression analyses (FIG. 1). A slightly accelerated development of NFT, but not of A4/β deposition was seen in the carriers of at least one ε4 allele. At a given age, the presence of an ε4-allele shifted the mean NFT and A4/β stages by approximately one.

TABLE 1. Apo E Genotypes and Mean Stages of M. Alzheimer-associated Histological Changes in 191 Brain Samples

Apo E		Observed Means		Adjusted Means[a]	
Genotype	n	NFT	A4/β	NFT	A4/β
ε2/2	2	1.00	0.00	2.23	0.80
ε3/2	26	2.89	1.54	2.33	1.18
ε3/3	116	2.25	1.42	2.07	1.30
ε4/2	6	3.00	2.50	2.44	2.14
ε4/3	35	3.37	2.09	3.05	1.88
ε4/4	6	4.17	3.00	4.56	3.25
univariate F		4.26	4.92	4.98	5.78
p		<0.001	<0.001	<0.001	<0.001

[a] Adjusted for the effect of age as a covariable using the SPSS-X MANOVA procedure. NFT: stage of neurofibrillary tangles according to Braak and Braak.[6] A4/β: stage of A4/β deposition according to Braak and Braak; however, to allow for numerical treatment of the data, stage A was converted to 1, B to 2, and C to 3.

DISCUSSION

In AD, the cerebral NFT load correlates with *ante mortem* psychometrical assessments much better than the A4/β burden.[4] We, therefore, regard as an important aspect of this study that the mean stage of NFT increased in parallel with the ε4 gene dose, a finding which has not been reported in autopsy studies so far.[2,3,5] In the earlier studies, cases were classified dichotomously, either affected or not. Thus, only the NFT load of full-blown AD was quantified. It is conceivable that NFT formation has to reach a threshold level before causing an individual's death. Thus, fully developed AD cases may show only minor differences in NFT between carriers of different apo E alleles. We circumvented this problem by analyzing consecutive brain samples using a staging system which accounts for the gradual development of AD related changes over time[6] and which correlates with the degree of dementia.[7] This approach allowed to assess the NFT or the A4/β formation in individuals who died before the full-blown picture of AD had developed. The analysis revealed that, at a given age, the mean NFT and A4/β stages were shifted upwards by one in ε4 carriers. We previously determined the duration of a single stage to equal one decade.[15] Taken together, the data suggest that the presence of an ε4 allele leads to an earlier onset of AD related changes of approximately ten years. This estimate is in excellent agreement with a report by Poirier *et al.*[1] who analyzed the apo E polymorphism in relation to the age at clinical onset of sporadic AD and with recent findings in late onset AD families.[16] The fact that both intraneuronal NFT and extraneuronal amyloid formation correlate positively with the gene dose of ε4 suggests that apo E is causally linked to the pathogenesis of AD.

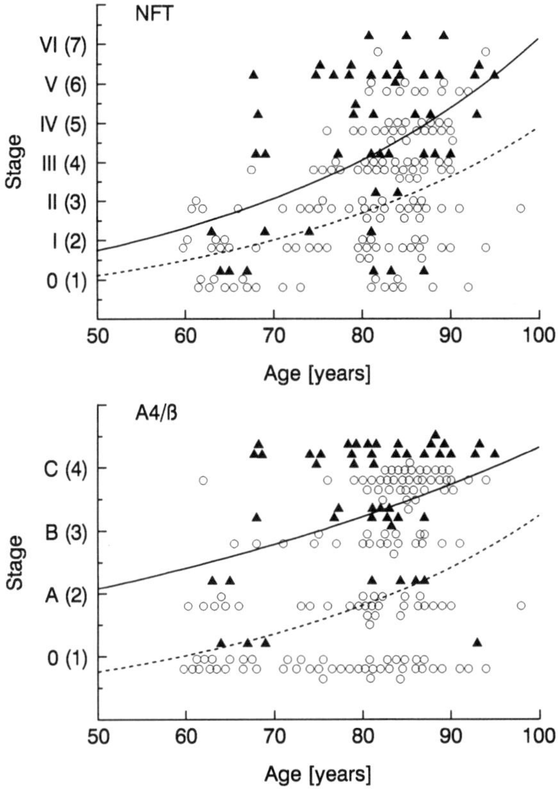

FIGURE 1. Relationship between age and AD-associated histopathological changes. Triangles and solid lines: $\epsilon 4$ allele present; circles and dashed lines: $\epsilon 4$ allele absent. *Top panel*: Stages of neurofibrillary changes versus age (non-linear correlation coefficients: ϵ allele present, $r = 0.316$; $\epsilon 4$ allele absent, $r = 0.383$). *Bottom panel*: Stages of A4/β deposition versus age (non-linear correlation coefficients: $\epsilon 4$ allele present, $r = 0.413$; $\epsilon 4$ allele absent, $r = 0.426$). The arabic numbering of the stages (in parentheses) was used to calculate the correlation coefficients.

REFERENCES

1. POIRIER, J., J. DAVIGNON, D. BOUTHILLIER, S. KOGAN, P. BERTRAND & S. GAUTHIER. 1993. Apolipoprotein E polymorphism and Alzheimer's disease. Lancet **342**:697–699.
2. REBECK, G. W., J. S. REITER, D. K. STRICKLAND & B. T. HYMAN. 1993. Apolipoprotein E in sporadic Alzheimer's disease, allelic variation and receptor interactions. Neuron **11**:575–580.
3. SAUNDERS, A. M., W. J. STRITTMATTER, D. SCHMECHEL, P. H. ST. GEORGE-HYSLOP, M. A. PERICAK-VANCE, S. H. JOO, B. L. ROSI, J. F. GUSELLA, D. R. CRAPPER-MACLACHLAN, M. J. ALBERTS, C. HULETTE, B. CRAIN, D. GOLD-

GABER & A. D. ROSES. 1993. Association of apolipoprotein E allele e4 with late-onset familial and sporadic Alzheimer's disease. Neurology **43**:1467–1472.
4. ARRIAGADA, P. V., J. H. GROWDON, E. T. HEDLEY-WHITE & B. T. HYMAN. 1992. Neurofibrillary tangles but not senile plaques parallel duration and severity of Alzheimer's disease. Neurology **42**:631–639.
5. SCHMECHEL, D. E., A. M. SAUNDERS, W. J. STRITTMATTER, B. J. CRAIN, C. M. HULETTE, S. H. JOO, M. A. PERICAK-VANCE, D. GOLDGABER & A. D. ROSES. 1993. Increased amyloid β-peptide deposition in cerebral cortex as a consequence of apolipoprotein E genotype in late-onset Alzheimer's disease. Proc. Natl. Acad. Sci. USA **90**:9649–9653.
6. BRAAK, H. & E. BRAAK. 1991. Neuropathological staging of Alzheimer related changes. Acta Neuropathol. **82**:239–259.
7. BRAAK, H., C. DUYCKAERTS, E. BRAAK & F. PIETTE. 1992. *In* Alzheimer's disease: Advances in Clinical and basic research. B. Corain, K. Iqbal, M. Nicolini, B. Winblad, H. Wisniewski & P. Zatta, Eds. 131–137. Chichester: John Wiley & Sons.
8. BRAAK, H., E. BRAAK, J. BOHL & T. G. OHM. 1989. Alzheimer's disease, mismatch between amyloid plaques and neuritic plaques. Neurosci. Lett. **103**: 24–28.
9. CAMPBELL, S., R. C. SWITZER & T. L. MARTIN. 1989. Alzheimer's plaques and tangles: A controlled and enhanced silver staining method. Soc. Neurosci. Abstr. **13**:678.
10. GALLYAS, F. 1971. Silver staining of Alzheimer's neurofibrillary changes by means of physical development. Acta Morphol. Acad. Sci. Hung. **19**:1–8.
11. OHM, T. G., E. BRAAK & H. BRAAK. 1989. A competition between silver stain and immunocytochemistry. Neural Transm. [P-D Sect] **1**:111.
12. HIXSON, J. E. & D. T. VERNIER. 1990. Restriction isotyping of human apolipoprotein E by gene amplification and cleavage with HhaI. J. Lipid Res. **31**: 545–548.
13. RUZICKA, V., W. MÄRZ, A. RUSS, W. MONDORF & W. GROß. 1993. Characterization of the gene for apolipoprotein E5-Frankfurt ($Gln^{81} \rightarrow Lys$, $Cys^{112} \rightarrow Arg$) by polymerase chain reaction, restriction isotyping, and temperature gradient gel electrophoresis. Electrophoresis **14**:1032–1037.
14. DAVIGNON, J., R. E. GREGG & C. F. SING. 1988. Apolipoprotein E polymorphism and atherosclerosis. Arteriosclerosis **8**:1–21.
15. OHM, T. G., MÜLLER, H. BRAAK & J. BOHL. 1995. Close-meshed prevalence rates as a tool to uncover the pace of Alzheimer's disease-related changes Neuroscience **64**:209–217.
16. CORDER, E. H., A. M. SAUNDERS, W. J. STRITTMATTER, D. E. SCHMECHEL, P. C. GASKELL, G. W. SMALL, A. D. ROSES, J. L. HAINES & M. A. PERICAK-VANCE. 1993. Gene dose of apolipoprotein E type 4 allele and the risk of Alzheimer's disease in late onset families. Science **261**:921–923.

APP Gene Family

Alternative Splicing Generates Functionally Related Isoforms[a]

R. SANDBRINK,[b,d] C. L. MASTERS,[c] AND K. BEYREUTHER[b]

[b] *Zentrum für Molekulare Biologie Heidelberg (ZMBH), University of Heidelberg, Heidelberg, Germany*

[c] *Department of Pathology, University of Melbourne, Parkville, Australia*

ABSTRACT: The Alzheimer's βA4-amyloid protein precursor (APP) and the APP-like proteins (APLPs) are transmembrane glycoproteins with a similar modular domain structure. APP exists in 8 isoforms generated by alternative splicing of exons 7, 8, and 15, of which the L-APP mRNAs lacking exon 15 are ubiquitously expressed in rat tissues but not in neurons. Rat APLP2, the nearest relative of APP, is similarly expressed in 4 different isoforms due to alternative splicing of inserts encoding a Kunitz protease inhibitor domain (KPI, homologous to exon 7 of APP) and a divergent region of 12 amino acids on the NH2-terminal side of the transmembrane domain (12 aa exon). KPI-APLP2 transcripts are highly expressed in neurons, in contrast to KPI-APPs, while L-APLP2 mRNA isoforms lacking the 12 aa exon are predominantly expressed in non-neuronal rat tissues, similar to L-APPs. Further examination of the divergent domains in APP and APLP2 harboring the similarly alternatively spliced APP exon 15 and the 12 aa exon of APLP2 revealed some structural similarities of the amino acid sequences and the predicted secondary structures. In both L-APLP2 and L-APP, a putative xylosyl-transferase recognition site for chondroitin sulfate glycosaminoglycan attachment is present that is interrupted in APP and APLP2 isoforms expressing APP exon 15 or the 12 aa exon of APLP2. Thus, a related function of the divergent domains and the corresponding alternatively spliced APP and APLP2 isoforms in regulation of the binding properties of the ectodomain is suggested. Additionally, β-secretase cleavage of APP might be sterically hindered selectively in proteoglycan L-APP but not in APP lacking the proteoglycan attachment site. Neurons which have a uniquely low portion of L-APP and high content of APP might therefore be especially susceptible to βA4-protein liberation. This could explain the selective vulnerability of neurons that is observed in Alzheimer's disease.

[a] This research was sponsored by Grants SFB317 and 258 from the Deutsche Forschungsgemeinschaft, the Bundesminister für Forschung und Technologie of Germany, the Metropolitan Life Foundation, the Fonds der Chemischen Industrie, the Forschungsschwerpunkt Baden-Württemberg (to K.B.), the National Health and Medical Research Council of Australia, the Victorian Health Promotion Foundation, and the Aluminum Development Corporation of Australia (to C.L.M.).

[d] *Send correspondence to:* Dr. Rupert Sandbrink, Zentrum für Molekulare Biologie Heidelberg (ZMBH), University of Heidelberg, Im Neuenheimer Feld 282, D-69120 Heidelberg, Germany; Tel: +49-6221-566848; Fax: +49-6221-565891.

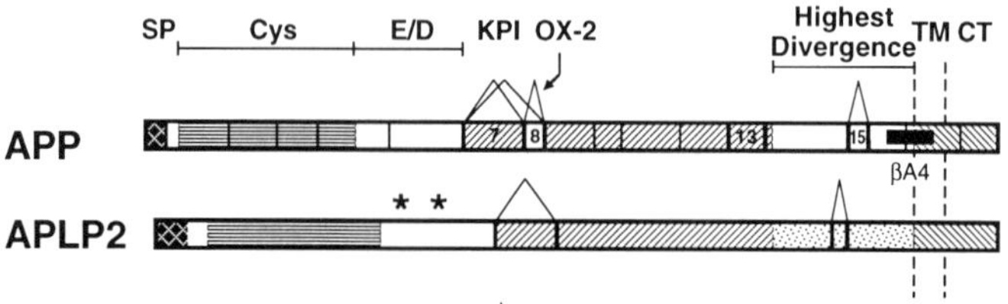

FIGURE 1. APP and APLP2 structure. Full length rat APP (rAPP770) and rat APLP2 (rAPLP2-765) and alternatively spliced exons (exons 7, 8 and 15 in APP, two exons in APLP2) are shown. Regions of highest homology are hatched, SP, signal peptide; Cys, cysteine-rich region; E/D, acidic domain; KPI, Kunitz protease inhibitor domain; OX-2, the MRC OX-2 domain of APP, βA4, the βA4-amyloid region of APP *(black bar)*; TM, transmembrane domain; CT, cytoplasmatic domain. For APP, boundaries of regions encoded by individual exons are shown. The region with the highest divergence is indicated; white in APP and dotted in APLP2. Two asterisks mark the positions where rat APLP2 differs from its human and murine homologues by a deletion of 2 amino acids *(left)* and an insertion of 4 amino acids *(right)*.

INTRODUCTION

Extracellular amyloid deposits characteristic for Alzheimer's disease (AD) predominantly contain aggregates of the βA4 or Aβ protein, derived from larger transmembrane glycoproteins denoted as βA4 amyloid protein precursors (APP) (FIG. 1).[1,2] Several APP isoforms exist, due to alternative splicing of three exons (7, 8, 15) (TABLE 1): Exon 7 is homologous to the Kunitz family of protease inhibitors (KPI),[3] and exon 8[3] is related to the MRC OX-2 antigen in thymocytes (OX-2).[2] Transcripts excluding exon 15 were first

TABLE 1. APP and APLP2 Isoforms Generated by Alternative Splicing

Species h = human m = murine r = rat		Non-L-APP/APLP2 Isoforms +Exon 15/12 aa Exon				L-APP/APLP2 Isoforms −Exon 15/12 aa Exon			
		+KPI		−KPI		+KPI		−KPI	
		+OX2	−OX2	+OX2	−OX2	+OX2	−OX2	+OX2	−OX2
h, m, r	APP	770	751	714	695	752	733	696	677
h, m	APLP2	—	763[a]	—	707[a]	—	751	—	695
r	APLP2	—	765	—	709	—	753	—	697

Owing to alternative splicing of 3 (APP) or 2 (APLP2) exons, 8 different APP isoforms and 4 APLP2 isoforms are generated (see text), denoted according to their lengths in amino acids as given in the table. Rat APLP2 isoforms are two amino acids longer than their human and murine homologues.

[a] Murine APLP2 transcripts including the 12 aa exon have not yet been described but are postulated in analogy to human and rat APLP2.

discovered in lymphocytes and microglial cells and thus denoted as leukocyte-derived APP (L-APP) mRNAs.[4] For rat, it was then shown that L-APP mRNA is significantly expressed in all non-neuronal tissues comprising between 25% and 70% of total APP transcripts.[5] The low expression of L-APP mRNA in the adult rat brain (approx. 4%) is due to a characteristic neuronal APP mRNA isoform expression pattern: Among several non-neuronal brain tissues and primary cultured brain cells analyzed, only primary cultured neurons did not express detectable levels of L-APP transcripts.[5]

Recently, APP-like proteins were described. The nearest APP-relative, now termed amyloid precursor-like protein (APLP2), was identified in human,[6,7] mouse,[8–11] and rat.[12,13] APLP2 domain structure is similar to APP and includes 3 regions of approx. 70% amino acid identity to APP including the KPI domain (FIG. 1). However, 125 amino acids of the ectodomain directly adjacent to the NH2-terminus of the transmembrane domain are highly divergent to APP. Consequently, a βA4-protein-like sequence is also not present within APLP2.

To characterize APLP2 alternative splicing, we performed a RT-PCR analysis of APLP2 mRNA isoform expression in perfused rat tissues. By comparison to APP, common features of regulation of alternative splicing were identified. Sustained by structural similarities of amino acid sequences and predicted secondary structures, related functions of the divergent domains and the corresponding alternatively spliced isoforms of APP and APLP2 are suggested.

METHODS

Total RNA of perfused rat tissues was extracted by using the acid phenol method, and cDNA was prepared using oligo$(dT)_{17}$ as primer as previously described.[5] Primers used for APLP2-PCR amplification were: sml 838: 5′-GGAGACGATTACAATGAGGAGAATCC-3: and αrl1877: 5′-CCATCTTGCTCTGCCATTCCAG-3′.[9] Usually, 28 PCR cycles were performed using standard cycle parameters.[5] 5′-Endlabeling of the antisense primers was performed with polynucleotide kinase and [γ-33P]ATP. PCR products were analyzed by denaturing polyacrylamide gel electrophoresis and quantitated by phosphorimaging.[5]

RESULTS AND DISCUSSION

For transmembrane APP, alternative splicing of three exons has been described, involving exon 7 and 8,[3] and 15.[4,5] Thus, eight APP isoforms can be generated (TABLE 1). By analysis of perfused rat tissues and primary cultured brain cells, all eight APP transcripts were detected in non-neuronal cells, while expression of L-APP 752, 733, 696 and 677 mRNAs was not observed in neurons.[5]

To analyze alternative splicing of APLP2 in perfused rat tissues, we performed a series of PCR analysis covering almost all of the APLP2 sequence. In accordance with the heterogenicity of cDNA clones isolated from human and murine tissues,[6–8,10,11] four APLP2 isoform were detected due to alternative splicing of two exons.[9] The two alternatively spliced inserts involve the KPI domain homologous to exon 7 of the APP gene and a region of 12 amino acids located on the NH_2-terminal side of the transmembrane domain (12 aa exon). Interestingly, the latter insert is part of the region of highest divergence between APP and APLP2 (FIG. 1). Isoforms lacking this insert were denoted as L-APLP2 isoforms in analogy to alternative splicing of APP exon 15. No APLP2 equivalent to exon 8 of APP as detected.[9]

As with APP, the corresponding 4 APLP2 mRNA isoforms are denoted according to their size in amino acids. Full-length human APLP2 including the KPI domain and the region encoded by the 12 aa exon consists of 763 amino acids as predicted by its cDNA sequence.[6] Thus, APLP2-763, L-APLP2-751, APLP2-707 and L-APLP2-695 isoforms are expected. For murine APLP2 isoforms, similar lengths are postulated, although the presence of the 12 aa exon has not yet been shown in the cDNA clones described.[8,10,11] Because of a deletion of two amino acids and an insertion of 4 amino acids within the acidic domain, rat APLP2 isoforms are two amino acids longer than the corresponding human and murine APLP2 isoforms.[12] By using the prefix 'r' to indicate the rat species, these isoforms are denoted as rAPLP2-765, rL-APLP2-753, rAPLP2-709 and rL-APLP2-697 isoforms. A comparison of APP and APLP2 isoforms in human, mouse, and rat is given in TABLE 1.

To determine the relative amount of each of the 4 APLP2 isoforms in rat tissues, we employed a radiolabeled antisense primer in our RT-PCR analysis and quantitated the products by phosphorimaging.[9] Typically, in non-nervous tissues, most APLP2 transcripts include the KPI exon (61–94%) but not the 12 aa exon (63–92% L-APLP2). A lower L-APLP2 portion was detected in cardiac and skeletal muscle (43 and 35%, respectively), strikingly similar to the low L-APP expression that we had previously observed in these issues.[2] In adult rat cerebral cortex, the full length APLP2 transcript is predominantly expressed (73%); only small amounts of L-APLP2 (13%) and KPI-minus isoforms (81% KPI-APLP2) were detected. A very similar expression pattern was observed in other brain regions as well as for primary cultured neurons (7% L-APLP2, 92% KPI-APLP2), while non-neuronal brain tissues like meninges as well as microglia and astrocytes in culture predominantly express L-APLP2 mRNA isoforms (70–80%).[9]

High expression of L-APLP2 isoforms in non-neuronal cells and weak expression in primary cultured neurons resemble the tissue specific alternative splicing of exon 15 of APP. In contrast to this, expression of the KPI encoding exon of APLP2 is abundant in both neuronal and non-neuronal tissues and thus differs from APP.

Although L-APP and L-APLP2 isoforms are similarly expressed predominantly in non-neuronal cells, several differences exist. While L-APP expression is virtually not detectable in neurons *in vitro* (<0.5%), L-APLP2 mRNA is well observed. Generally, significantly higher portions of L-APLP2 than L-APP were detected in non-neuronal tissues. During rat brain development and aging, expression of APP exon 15 and the 12 aa exon of APLP2 is differently regulated.[14]

Because of the similarities in non-neuronal L-APP and L-APLP2 expression, we analyzed the amino acid sequences of the divergent regions in more detail. In both APP and APLP2, these regions are of similar length (130 amino acids in APP, 125 amino acids in APLP2) and hydrophilicity. Although the two alternatively spliced inserts differ in length (18 amino acids in APP, 12 amino acids in APLP2) and are located at a different distance from the transmembrane domain, there are some parallels further supporting a relationship of the divergent domains and the two alternatively spliced inserts. Firstly, in both APP and APLP2, these inserts represent the carboxy-terminal end of a proline-rich region, 9 prolines in APP, 8 in APLP2, presumably modifying secondary structures in a similar manner. Secondly, predictions for the corresponding secondary structures support a related structure of the divergent domains: *e.g.*, using the Chou-Fasman algorithm, for both proteins an extended α-helical domain between the transmembrane regions and the alternatively spliced inserts is expected. In APLP2, this putative α-helical region is shifted aminoterminally, in parallel with the alternatively spliced insert. Thirdly, both alternatively spliced inserts contain two acidic residues in their aminoterminal part and two basic residues at their carboxyterminal end. Fourthly, the amino acids immediately preceding and following the alternatively spliced inserts {in brackets} are highly related: TENE-{18 aa, exon 15}-GSGLT in rat APP, and SENE-{12 amino acids}-GSGMA in APLP2. For L-APP and L-APLP2 isoforms lacking the inserts, potential xylosyl-transferase recognition sites for chondroitin sulfate glycosaminoglycan attachment are generated by fusing the preceding acidic amino acids to the GSG sequence containing the serine residue potentially modified. Although there is some deviation from the proposed typical xylosyl-transferase recognition sequence (acidic amino acids followed by SGXG, where X is any amino acid[15]; however, this sequences is not well defined as discussed in ref. 16), chondroitin sulfate glycosaminoglycan attachment has been demonstrated to occur only at this serine residue in murine L-APLP2-751.[17] Chondroitin sulfate proteoglycan-APP has been isolated from the rat glioma cell line C6,[18] which is expected to produce high amounts of L-APPs,[5] and recently also from primary cultured astrocytes.[19] In rat neurons lacking L-

APPs,[5] however, chondroitin sulfate proteoglycan-APP was not detected.[13] This is explained by interruption of the xylosyl-transferase recognition sequence due to the presence of positively charged amino acids encoded by APP exon 15, and indeed, full length human APP770 is not modified by xylosyl-transferase.[17]

In summary, predominant expression of L-APP isoforms lacking exon 15 and L-APLP2 isoforms lacking the 12 aa exon in non-neuronal cells and the described structural similarities of the amino acid sequences and the predicted secondary structures suggest a related function of the divergent domains and the corresponding alternatively spliced isoforms. If there is a predominant chondroitin sulfate glycosaminoglycan attachment to L-APP and L-APLP2 isoforms converting them into proteoglycans, the consequences of the absence or low abundance of this modification in neurons have to be considered. Notably, β-secretase cleavage might be sterically hindered in proteoglycan L-APP, favoring the non-amyloidogenic α-secretase cleavage. Neuronal APP with its uniquely low portion of L-APP, however, would be especially susceptible to the amyloidogenic pathway and βA4-protein liberation. Since the highest amounts of APP are produced by neurons *in vivo,* this might well explain the selective vulnerability of neurons observed in Alzheimer's disease.

The notion of a related influence of the divergent domains on APP and APLP2 processing and metabolism has recently been supported by demonstration of α-secretase-like proteolytic cleavage of murine APLP2.[10] Whether there is also a β-secretase-like processing of APLP2, however, remains to be shown. The latter would be expected to be different for proteoglycan L-APLP2 and chondroitin sulfate-free APLP2 in much the same way as postulated for L-APP and chondroitin sulfate-free APP.

REFERENCES

1. KANG, J., H. G. LEMAIRE, A. UNTERBECK, J. M. SALBAUM, C. L. MASTERS, K. H. GRZESCHIK, G. MULTHAUP, K. BEYREUTHER & B. MUELLER-HILL. 1987. The precursor of Alzheimer's disease amyloid A4 protein resembles a cell-surface receptor. Nature **325:**733–736.
2. MÜLLER-HILL, B. & K. BEYREUTHER. 1989. Molecular biology of Alzheimer'sdisease. Annu. Rev. Biochem. **58:**287–307.
3. KITAGUCHI, N., Y. TAKAHASHI, Y. TOKUSHIMA, S. SHIOJIRI & H. ITO. 1988. Novel precursor of Alzheimer's disease amyloid protein shows protease inhibitor activity. Nature **331:**530 532.
4. KÖNIG, G., U. MÖNNING, C. CZECH, R. PRIOR, R. B. BANATI, U. SCHREITER-GASSER. J. BAUER, C. L. MASTERS & K. BEYREUTHER. 1992. Identification and differential expression of a novel alternative splice isoform of the βA4 amyloid precursor protein (APP) mRNA in leukocytes and brain microglial cells. J. Biol. Chem. **267:**10804–10809.
5. SANDBRINK, R., C. L. MASTERS & K. BEYREUTHER. 1994. βA4-amyloid protein precursor mRNA isoforms without exon 15 are ubiquitously expressed in rat tissues including brain, but not in neurons. J. Biol. Chem. **269:**1510–1517.

6. SPRECHER, C. A., F. J. GRANT, G. GRIMM, P. J. O'HARA, F. NORRIS, K. NORRIS & D. C. FOSTER. 1993. Molecular cloning of the cDNA for a human amyloid precursor protein homolog: Evidence for a multigene family. Biochemistry **32:**4481–4486.
7. WASCO, W., S. GURUBHAGAVATULA, M. D. PARADIS, D. M. ROMANO, S. S. SISODIA, B. T. HYMAN, R. L. NEVE & R. E. TANZI. 1993. Isolation and characterization of APLP2 encoding a homologue of the Alzheimer's associated amyloid β protein precursor. Nature Genet. **5:**95–100.
8. HANES, J., H. VON DER KAMMER, G. K. KRISTJANSSON & K. H. SCHEIT. 1993. The complete cDNA coding sequence for the mouse CDEI binding protein. Biochim. Biophys. Acta **1216:**154–156.
9. SANDBRINK, R., C. L. MASTERS & K. BEYREUTHER. 1994. Similar alternative splicing of a non-homologous domain in βA4-amyloid protein precursor-like proteins. J. Biol. Chem. **269:**14227–14234
10. SLUNT, H. H., G. THINAKARAN, C. VON KOCH, A. C. Y. LO, R. E. TANZI & S. S. SISODIA. 1994 Expression of a ubiquitous, cross-reactive homologue of the mouse β-amyloid precursor protein (APP). J. Biol. Chem. **269:**2637–2644.
11. VIDAL, F., A. BLANGY, M. RASSOULZADEGAN & F. CUZIN. 1992. A murine sequence-specific DNA binding protein shows extensive local similarities to the amyloid precursor protein. Biochem. Biophys. Res. Commun. **189:** 1336–1341.
12. SANDBRINK, R., C. L. MASTERS & K. BEYREUTHER. 1994. The complete nucleotide and deduced amino acid sequence of rat amyloid precursor-like protein 2: Two amino acids length difference to human and murine homologues. Biochim. Biophys. Acta **1219:**167–170.
13. YAN, Y. C., Y. BAI, L. WANG, S. MIAO & S. S. KOIDE. 1990. Characterization of cDNA encoding a human sperm membrane protein related to A4 amyloid protein. Proc. Natl. Acad. Sci. USA **87:**2405–2408.
14. SANDBRINK, R., C. L. MASTERS & K. BEYREUTHER. 1994. APP gene family: Unique age-associated changes in splicing of Alzheimer's βA4-amyloid protein precursor. Neurobiol. Dis. **1:**13–24.
15. BOURDON, M. A., T. KRUSIUS, S. CAMPBELL, N. B. SCHWARTZ & E. RUOSLAHTI. 1987. Identification and synthesis of a recognition signal for the attachment of glycosaminoglycans to proteins. Proc. Natl. Acad. Sci. USA **84:** 3194–3198.
16. KJELLEN, L. & U. LINDAHL. 1991. Proteoglycans: structures and interactions. Annu. Rev. Biochem. **60:**443–475.
17. THINAKARAN, G. & S. S. SISODIA. 1994. Amyloid precursor-like protein 2 (APLP2) is modified by the addition of chondroitin sulfate glycosaminoglycan at a single site. J. Biol. Chem. **269:**22099–22104.
18. SHIOI, J., J. P. ANDERSON, J. A. RIPELLINO & N. K. ROBAKIS. 1992. Chondroitin sulfate proteoglycan form of the Alzheimer's β-amyloid precursor. J. Biol. Chem. **267:**13819–13822.
19. SHIOI, J., M. N. PANGALOS, C. MYTILINEOU & N. K. ROBAKIS. 1994. Chondroitin sulfate proteoglycan form of Alzheimer's amyloid precursor protein is produced mostly in astrocytes but not in neurons. Soc. Neurosci. Abstracts **20:** 1642.

Altered Expression of Amyloid β Precursor mRNAs in Cerebral Vessels, Meninges, and Choroid Plexus in Alzheimer's Disease[a]

D. R. D. PREMKUMAR AND R. N. KALARIA[b]

Departments of Neurology, Neurosciences and Pathology, Case Western Reserve University, Cleveland, Ohio 44106 USA

ABSTRACT: Altered tissue-specific processing or production of the amyloid precursor protein (APP) is thought to be central to amyloid deposition in cerebrovascular and the neocortical tissues in Alzheimer's disease (AD). We demonstrate that Aβ peptide(s) is readily detectable and increased in cerebral vessels, meninges and choroid plexus obtained at autopsy from AD subjects compared to age-matched controls. Using the reverse transcription (RT)-polymerase chain reaction (PCR), we further found that Aβ transcripts encoding the Aβ sequence in all forms of APP containing exons 16 and 17 (of APP770) were significantly increased in vessel samples in AD subjects. This was also evident in the neocortical samples and not related to pre-mortem factors or postmortem interval. It is possible that the increased Aβ mRNAs reflect enhanced expression of the L-APP isoform (APP770 without exon 15) expressed in leukocytes and glia alike. We also found evidence for changed proportions of APP 770, 756 and 695 mRNAs in cerebral vessel samples from AD subjects compared to controls. Whereas APP770 and APP751, the predominant forms, were significantly decreased, APP695 transcript was increased in vessel samples from AD subjects. Such changes were not evident in neocortical samples from the same subjects. These observations suggest tissue-specific changes in expression of APP isoforms implicating one of the mechanisms for increased accumulation of Aβ in cerebrovascular tissues in AD.

INTRODUCTION

Amyloid β (Aβ) protein is cleaved from a membrane-bound glycoprotein termed the amyloid precursor protein(s) (APP) that is common to both cerebral amyloid angiopathy (CAA) and parenchymal amyloid deposits of Alzheimer's disease (AD).[1] The released Aβ consisting of 39–43 residues may subse-

[a] These research efforts were supported by NIH Grants RO1AG10030, POAG08012B2 and R5AG08992A2.

[b] *Send correspondence to:* Dr. R. N. Kalaria, Department of Neurology (BRB5), Case Western Reserve University, 10900 Euclid Avenue, Cleveland, Ohio 44106, USA; Tel: (216) 368-1113; Fax: (216) 368-1144.

quently undergo aggregation to form amyloid fibrils under the influence of various amyloid associated factors.[2] Recent studies in cells transfected with APP genes suggest that Aβ is produced by normal cellular metabolism and that cells transfected with mutant forms of APP secrete substantially increased amount of Aβ or modified carboxyl (C)-terminal fragments of APP.[2] It is not certain whether these modifications occur in sporadic AD.

Several studies have also addressed these issues by examination of the expression of the APP gene and its alternatively spliced forms in AD. In general, these studies do not suggest a clear pattern of altered APP gene expression in AD.[2] While we[3–5] and others (see refs. in Glenner[1] and Selkoe[2]) have described localization of APP and its fragments in the cerebral vasculature, tissue-specific APP gene expression in cerebrovascular tissues in AD cases with significant CAA has not been elucidated.

MATERIAL AND METHODS

Tissues

Human brain tissue from 20 subjects with AD and 8 age-matched controls was sampled at autopsy immediately upon removal within 2 to 7 hours after death (age range 55–80 years; AD subjects 73 $\pm$ 3 and controls 69 $\pm$ 4 years). Cerebral vessels, meningeal vessels and choroid plexus were isolated and prepared as described previously.[6]

Immunoblotting Analyses

Vascular tissue samples were immunoblotted and Aβ peptides were detected by ECL method, and quantified by densitometry.[7]

RNA Extraction and PCR

Total RNA was isolated from the tissues, reverse transcribed (RT) and used for the PCR with quantification as described previously.[7] The cDNA templates, isotope label and number of PCR cycles were maintained to be equal for all samples. Primer sequences for the Aβ domain were as follows: Aβ domain exon 17 primer 1872 sense GGT GCA ATC ATT GGA CTC ATG GTG and 2147 antisense CGA TGG GTA GTG AAG CAA TGG TTT (APP695) and others according to Golde *et al.*[8]

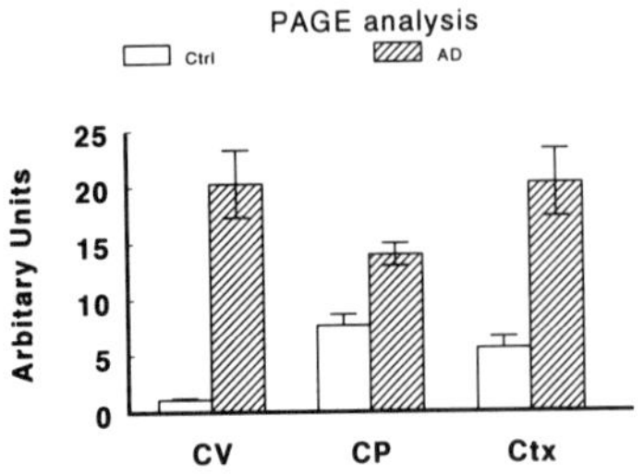

FIGURE 1. Semi-quantification of Aβ in cerebral vessels (CV) and choroidal (CP) samples. Immunoblots of tissue samples were probed with antibody 6E10 or 4G8 and the detected Aβ peptide was quantified by densitometry. Results show mean ± SEM *(bars)* for 3–6 samples. Significant differences by two tailed t-test, $p < 0.005$. There was general lack of Aβ in controls and high mw aggregates were often apparent in AD samples.

RESULTS AND DISCUSSION

We assessed soluble Aβ protein in cerebral vessel homogenates from AD and control subjects. Immunoblotting experiments showed that soluble Aβ was markedly increased in vessel and choroidal preparations from subjects with neocortical amyloidotic pathology and undetectable in subjects lacking sufficient pathology or in the cerebellum (Fig. 1). Most of these peptides were of the Aβ 39–40 form (Kalaria *et al.* unpublished results).

While it is known that cerebral vessels express APP mRNAs,[8] the relative alterations in their expression are not clear. Figure 2a shows the proportional changes in transcripts of three isoforms, APP770, APP751 and APP695. These results demonstrate alterations in APP transcripts in cerebral vessel fractions from AD subjects compared to age-matched controls. We observed significantly decreased APP770 and APP751 transcripts but increased

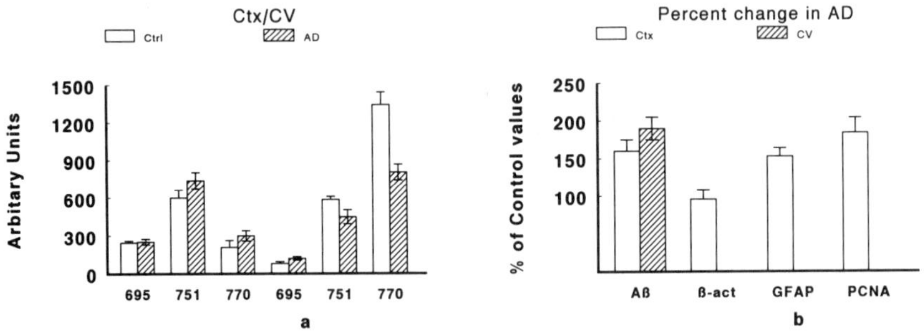

FIGURE 2. RT-PCR amplified products from control (Ctrl) and AD cDNA samples. The results represent mean ± SEM *(bars)* for 5–8 samples. Quantification of the autoradiograms achieved by calibration with phosphor image analyses. **a:** Changed proportions of APP mRNAs in cerebral vessels (CV). APP770 and APP756 were decreased but APP695 was increased in vessels fractions from AD compared to controls. Similar results were evident in meningeal and choroidal samples. **b:** Increased mRNAs for Aβ in both cerebral vessels and neocortex (Ctx). The mRNAs for GFAP and PCNA (proliferating cell nuclear antigen) were also increased in cortex. The specificity of these results was shown by the unaltered β-actin mRNA between controls and AD.

APP695 in vascular preparations from AD subjects. The changes were not related to pre-mortem factors and postmortem intervals, which were similar for both groups.[9] In addition, we observed that the proportions of APP transcripts were different in cortex from those in cerebral vessels (FIG. 2a). This is consistent with the results of Golde *et al.*[8]

We also found that the total mRNA transcripts encoding the Aβ sequence were significantly increased in both cerebral vessels and neocortex from AD compared to age-matched controls (FIG. 2b). This suggests that the enhanced concentrations of Aβ peptide could be derived through altered APP mRNAs containing the Aβ sequence although other factors and possibilities exist. We propose that at least in neocortex, Aβ may be produced by increased L-APPs, perhaps microglia[10,11] rather than from the conventional APP isoforms since these were not proportionally altered between controls and AD. Our observations offer an explanation of the suggested differences in APP processing and resultant size of Aβ peptide(s) in cerebrovascular and neocortical tissues.

In summary, we found increased Aβ peptides associated with cerebral vessels in AD. We further observed increased expression of APP mRNAs containing the Aβ sequence. In contrast to the neocortex, expression of APP770 and APP751 mRNA transcripts were decreased whereas APP695 was increased in cerebral vessels in AD. These alterations may explain the proposed differences in APP processing and Aβ deposition in cerebrovascular tissues and the brain parenchyma.

REFERENCES

1. GLENNER, G. G. 1992. Converging pathways in Alzheimer's disease. Lab. Invest. **67**:271–273.
2. SELKOE, D. J. 1994. Normal and abnormal biology of the β-amyloid precursor protein. Annu. Rev. Neurosci. **17**:489–517.
3. KALARIA, R. N., S. N. KROON & I. LIEBERBURG. 1992. Amyloid protein precursor and A4/beta protein in cerebral microvessels from Alzheimer's disease subjects. Soc. Neurosci. Abstr. **18**:776.
4. KAWAI, M., R. N. KALARIA, P. CRAS, S. L. SIEDLAK, M. E. VELASCO, E. R. SHELTON, H. W. CHAN, B. D. GREENBERG & G. PERRY. 1993. Degeneration of amyloid precursor protein-containing smooth muscle cells in cerebral amyloid angiopathy. Brain Res. **623**:142–146.
5. TAMAOKA, A., R. N. KALARIA, I. LIEBERBURG & D. J. SELKOE. 1992. Identification of a stable fragment of the Alzheimer amyloid precursor containing the β-protein in brain microvessels. Proc. Natl. Acad. Sci. USA **89**:1345–1349.
6. KALARIA, R. N., S. SROMEK, I. G. GRAHOVAC & S. I. HARIK. 1992. Transferrin receptors of human and rat cerebral microvessels and brain: assessment in Alzheimer's Disease. Brain Res. **585**:87–93.
7. KALARIA, R. N., T. E. GOLDE, S. N. KROON & G. PERRY. 1993. The serine

protease inhibitor antithrombin III and its mRNA in the pathogenesis of Alzheimer's disease. Am. J. Pathol. **143:**886–893.
8. GOLDE, T. E., S. ESTUS, M. USIAK, L. H. YOUNKIN & S. YOUNKIN. 1990. Expression of β amyloid protein precursor mRNAs: Recognition of a novel alternatively spliced form and quantitation in Alzheimer's disease using PCR. Neuron **4:**253–267.
9. HARRISON, P. J., A. J. L. BARTON, A. W. PROCTER, D. M. BOWEN & R. C. A. PEARSON. 1994. The effects of Alzheimer's disease, other dementias, and premortem course on β-amyloid precursor protein messenger RNA in frontal cortex. J. Neurochem. **62:**635–644.
10. KONIG, G., U. MONNING, C. CZECH, R. PRIOR, R. BANATI, U. SCHREITER-GASSER, J. BAUER, C. L. MASTERS & K. BEYREUTHER. 1992. Identification and differential expression of a novel alternative splice isoform of the βA4 amyloid precursor protein (APP) mRNA in leukocytes and brain microglial cells. J. Biol. Chem. **267:**10804–10809.
11. MONNING, U., G. KONIG, R. BANATI, H. MECHLER, C. CZECH, J. GEHRMANN, U. SCHREITER-GASSER, C. L. MASTERS & K. BEYREUTHER. 1992. Alzheimer βA4-amyloid protein precursor in immunocompetent cells. J. Biol. Chem. **267:** 23950–23956.

β-APP Cognitive Function versus β-Amyloid–induced Cell Death

M. GSCHWIND, J. R. MARTIN, J.-L. MOREAU, AND G. HUBER[a]

Pharma Division, Preclinical CNS Research, F. Hoffmann-La Roche Ltd., CH-4002 Basel, Switzerland

ABSTRACT: The β-amyloid precursor protein (β-APP) has been hypothesized to play an important role in the establishment of synaptic connections. Icv injections of anti-βAPP antibodies into rat brains produced no appreciable effect on subsequent learning of a passive avoidance task whereas memory assessed 1 day later in a retention test was impaired in anti-β-APP– but not control-IgG–injected animals. This suggests a possible involvement of β-APPs in cognitive functions. In order to evaluate the properties of the proteolytic Aβ-fragment accumulating in Alzheimer's disease brains, four different neuronal cell types were exposed to $A\beta_{1-42}$ for 24 hours. All cells degenerated in response to Aβ, yet chromosomal condensation and internucleosomal DNA fragmentation, typical for apoptosis, occurred in only three of the cell types tested. These findings suggest that β-APPs may play an important role in cognitive processes and additionally, that their alternative proteolytic product Aβ may be differentially toxic to neuronal cell types, inducing cell death either by necrosis or by apoptosis.

INTRODUCTION

Deposition and accumulation of β-amyloid peptide in the brain of Alzheimer's disease (AD) patients is an early event in the disease process.[1] This involves proteolytic processing of β-APPs, a family of integral membrane glycoproteins derived by alternative mRNA splicing from a gene located on chromosome 21.[2] Little is known about the physiological role of β-APPs. The most abundant mRNA species in brain is β-APP695, which lacks a Kunitz protease inhibitor (KPI) domain that is found in other β-APPs more broadly expressed in various tissues. During brain maturation very high levels of β-APPs are expressed during synaptogenesis[3] and mature isoforms can be localized at synaptic sites.[4] Proteolytic processing of β-APPs can result in extracellularly secreted molecules or alternatively β-APPs can undergo an amyloidogenic processing yielding the Aβ-peptide which is deposited in AD senile plaques.[1] The accumulation of β-amyloid is hypothesized to trigger

[a] *Send correspondence to:* Dr. Gerda Huber, Pharma Division, Preclinical, CNS Research, Bldg. 69/452, 4002 Basel, Switzerland; Tel: 61/688 3721; Fax: 61/688 1720.

neuronal degeneration leading to subsequent clinical symptoms in AD. In this study, the potential involvement of β-APP in cognitive processes as well as the potential neurotoxicity of Aβ for different neuronal cell types were investigated.

RESULTS AND DISCUSSION

In order to evaluate the role of β-APPs in brain function 20 μl of anti-β-APP-specific antibody (affinity purified; 1.3 mg/ml; anti-N-terminal epitope aa 16–32) or control rabbit IgG (1.5 mg/ml) were injected intracerebroventricularly into rats. The animals were then submitted to 3 training sessions to learn a passive avoidance task (*i.e.* not to step down from a safe area onto an electrified grid during a 2-min test period). All animals were able to learn this task adequately (FIG. 1). However, on the following day, the anti-β-APP–injected animals showed significantly shorter step-down latencies than the control-IgG-injected animals in the same test situation, suggesting that shock exposure was forgotten (FIG. 1). This significant decrease in performance in the retention test induced by blockade of β-APP function with an antibody suggests a role for β-APPs in cognitive processes. This might occur by synaptic β-APPs connecting to extracellular matrix components or the postsynaptic cell thereby stabilizing synapses during consolidation or retrieval. β-APP functions in cognitive processes may be relevant in AD brain where β-APPs are misprocessed to give rise to the accumulating Aβ peptide.

The potential neurotoxic effects of the Aβ peptide were investigated in different neuronal cell cultures in parallel experiments. After 24-hour exposure to 0.5 mg/ml of $A\beta_{1-42}$ all cell types tested showed amyloid-induced degeneration. Lactate dehydrogenase (LDH) release became evident as early

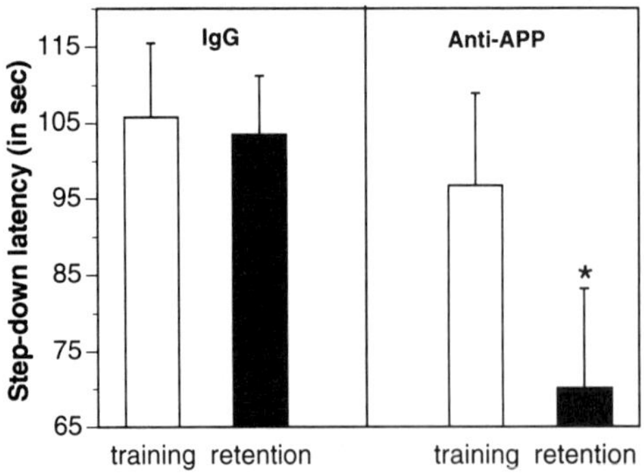

FIGURE 1. Passive avoidance performances. Step-down latencies of control IgG- or anti-β-APP-injected rats in the third training session and 1 day later (retention). Group size N = 14. Values are mean ± SEM. $p < 0.05$, Student's *t*-test.

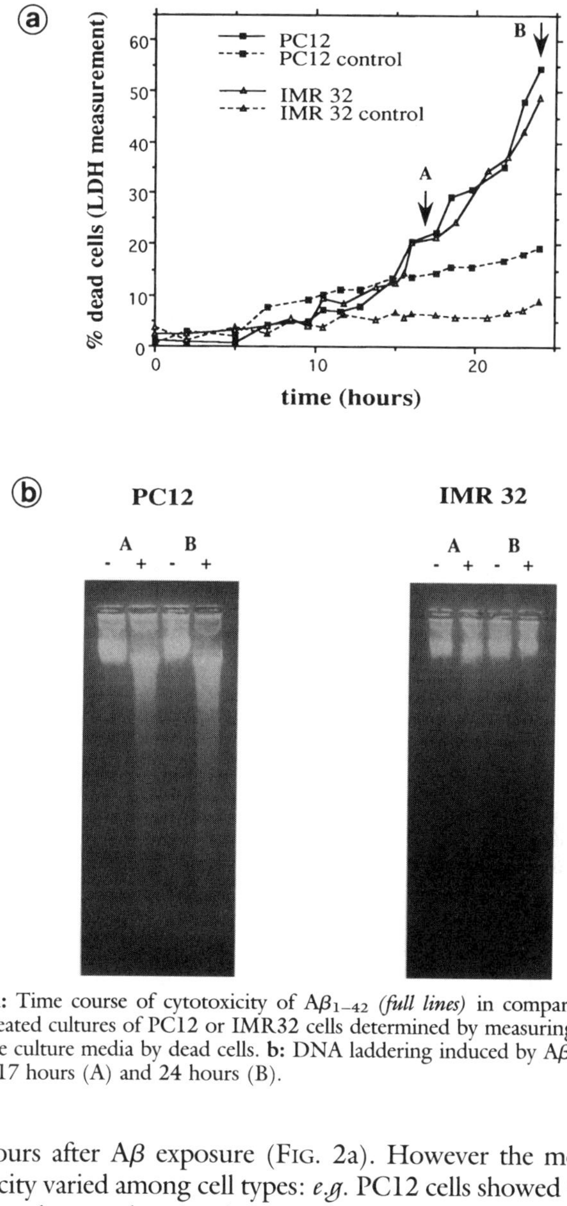

FIGURE 2. **a:** Time course of cytotoxicity of $A\beta_{1-42}$ *(full lines)* in comparison to vehicle *(broken lines)* treated cultures of PC12 or IMR32 cells determined by measuring LDH activity released into the culture media by dead cells. **b:** DNA laddering induced by $A\beta_{1-42}$ in parallel cell cultures at 17 hours (A) and 24 hours (B).

as 10–15 hours after Aβ exposure (FIG. 2a). However the mechanism of amyloid toxicity varied among cell types: *e.g.* PC12 cells showed typical signs of apoptosis such as nuclear condensation and DNA fragmentation whereas IMR32 neuroblastoma cells died without exhibiting these typical apoptotic features (FIG. 2b). These data showed that amyloid can induce cell death either by an apoptotic or by a necrotic mechanism depending on the cell type.

A combination of changes in β-APP synaptic functions together with amyloid-induced cell death might be important for neurodegeneration in AD brains.

REFERENCES

1. SELKOE, J. D. 1993. Physiological production of the β-amyloid protein and the mechanism of Alzheimer's Disease. TINS **16:**403–409.
2. GLENNER, G. G. 1988. Alzheimer's Disease: Its proteins and genes. Cell **52:** 307–308.
3. LÖFFLER, J. & G. HUBER. 1992. β-Amyloid precursor protein isoforms in various rat brain regions and during brain development. J. Neurochemistry **59:** 1316–1324.
4. SHIGEMATSU, K., P. L. MCGEER & E. G. MCGEER. 1992. Localization of amyloid precursor protein in selective postsynaptic densities of rat cortical neurons. Brain Res. **592:**353–357.

Amyloid β-Protein Induces the Cerebrovascular Cellular Pathology of Alzheimer's Disease and Related Disorders[a]

W. E. VAN NOSTRAND,[b] J. DAVIS-SALINAS, AND S. M. SAPORITO-IRWIN

Department of Microbiology and Molecular Genetics, College of Medicine, University of California, Irvine, California 92717-4025 USA

ABSTRACT: One of the hallmark pathologic characteristics of Alzheimer's disease (AD) and related disorders is deposition of the 39–42 amino acid amyloid β-protein (Aβ) in the walls of cerebral blood vessels. The cerebrovascular Aβ deposits in these disorders are associated with degenerating smooth muscle cells in the vessel wall which have been implicated in the expression of the amyloid β-protein precursor (AβPP) and formation of Aβ. We have established primary cultures of human cerebrovascular smooth muscle cells as a model for investigating the cellular pathologic processes involved in the cerebral amyloid angiopathy of AD and related disorders. Recently, we have shown that $A\beta_{1-42}$, the predominant pathologic cerebrovascular form of Aβ, causes extensive cellular degeneration that is accompanied by a striking increase in the levels of cellular AβPP, potentially amyloidogenic carboxyl terminal AβPP fragments, and soluble Aβ peptide in the cultured human cerebrovascular smooth muscle cells. Together, these studies provide evidence that Aβ contributes to the onset and progression of the cerebrovascular pathology associated with AD and related disorders and suggests the mechanism involves a molecular cascade with a novel product-precursor relationship that results in the adverse production and accumulation of Aβ.

INTRODUCTION

Cerebral amyloid angiopathy is characterized pathologically by deposition of amyloid in the media of primarily small and medium-sized arteries and

[a] This work was supported by NIH Grants AG00538, HL49566 and Grant-in-Aid Award 94006240 from the American Heart Association. S. M. Saporito-Irwin was supported by National Institute on Aging National Research Service Award AG0096-11 from the University of California, Irvine.

[b] *Send correspondence to:* Dr. William E. Van Nostrand, Department of Microbiology and Molecular Genetics, University of California, Irvine, CA 92717-4025; Tel: (714)824-2556; Fax: (714)824-8598; Email: wevannos@uci.edu.

arterioles of the cerebral cortex and leptomeninges.[1] Cerebral amyloid angiopathy is a common pathology of Alzheimer's disease (AD) and related disorders including Down's syndrome and hereditary cerebral hemorrhage with amyloidosis-Dutch type.[1] The cerebral amyloid angiopathy in these disorders share a common amyloid subunit known as the amyloid β-protein (Aβ).[2–4] Aβ is an insoluble, self-aggregating peptide of 39–42 amino acids.[3,5] A recent study showed that cerebrovascular Aβ is predominantly 42 amino acids, the same isoform found in senile plaques in patients with AD.[6] Aβ is proteolytically derived from a set of alternatively spliced precursor proteins, termed the amyloid β-protein precursor (AβPP), which are encoded by a single gene located on chromosome 21.[7,8]

Several findings have implicated smooth muscle cells in the cerebrovascular pathology and production of AβPP and Aβ in cerebral amyloid angiopathy. For example, Aβ deposition in the walls of cerebral blood vessels is accompanied by degeneration and eventual disappearance of smooth muscle cells, suggesting that the amyloid is toxic to these cells *in vivo*.[2,9] Immunohistochemical studies have localized AβPP to smooth muscle cells in cortical and leptomeningeal blood vessels.[10] Furthermore, ultrastructure and immunohistochemical studies have suggested smooth muscle cells of cerebral blood vessels are involved in the production of AβPP and the formation of Aβ.[9,11]

Previously, we demonstrated that cultured human leptomeningeal smooth muscle (HLSM) cells synthesize and secrete AβPP proteins.[12] Cultured HLSM cells provide a useful paradigm to investigate certain aspects of the cellular pathology associated with cerebral amyloid angiopathy.

$A\beta_{1-42}$ CAUSES DEGENERATION OF CULTURED HLSM CELLS

In the cerebral amyloid angiopathy of AD and related disorders vascular smooth muscle cells show pronounced signs of degeneration suggesting that the Aβ deposits are toxic to these cells *in vivo*.[2,9] Similarly, degenerating neurons are associated with parenchymal Aβ deposits in senile plaques in AD.[13] Previous studies have shown that cultured neurons degenerate in the presence of the different Aβ isoforms including $A\beta_{1-39}$, $A\beta_{1-40}$, $A\beta_{1-42}$ and the biologically active Aβ fragment $A\beta_{25-35}$.[14] $A\beta_{1-42}$ has been identified as the predominant form of Aβ found in cerebral blood vessels of AD patients with cerebral amyloid angiopathy;[16] however the shorter $A\beta_{1-39}$ and $A\beta_{1-40}$ have also been reported.[14,15] Therefore, we investigated the effects of the different length synthetic Aβ peptides on the degeneration of cultured human leptomeningeal smooth muscle (HLSM) cells to determine if these cells are sensitive to one or more forms of the peptide. HLSM cells incubated for six days with 25 μM $A\beta_{1-42}$ exhibited signs of extensive degeneration including loss of their characteristic morphology and evidence of severe cellular atrophy.

Similar findings on HLSM cellular degeneration were observed with higher concentrations (100 μM) of the biologically active fragment $A\beta_{25-35}$. In contrast, HLSM cells incubated for six days in the absence of peptide or with 25 μM $A\beta_{1-39}$, $A\beta_{1-40}$ or a scrambled sequence $A\beta_{1-42}$ peptide showed no apparent signs of degeneration and retained their characteristic morphology.

The time course of the cellular degeneration and cell death was also examined. When HLSM cells were continuously incubated with 25 μM $A\beta_{1-42}$ cell death was first observed at ≈6 days, gradually increased through day 12, and sharply increased by day 18 with ≈60% cell death. In contrast, HLSM cells incubated in the absence of peptide or with 25 μM $A\beta_{1-39}$, $A\beta_{1-40}$ or scrambled sequence $A\beta_{1-42}$ showed neither signs of cellular degeneration nor loss of cell viability even after >18 days of incubation. Together, these findings suggested that $A\beta_{1-42}$ is selectively toxic to cultured HLSM cells.

$A\beta_{1-42}$ INDUCES INCREASED LEVELS OF CELLULAR AβPP IN CULTURED HLSM CELLS

Since degenerating vascular smooth muscle cells in cerebral amyloid angiopathy have been implicated in the production of AβPP,[9,11] we performed immunoblotting studies to examine the effects of the different length Aβ peptides on the secreted and cellular levels of AβPP in the cultured HLSM cells. Incubation of the HLSM cells with 25 μM $A\beta_{1-42}$ for six days caused a striking >15-fold increase in the amount of cellular AβPP; however, no change in the levels of secreted AβPP in the culture medium was detected. In contrast, incubation of the HLSM cells with 25 μM of $A\beta_{1-39}$, $A\beta_{1-40}$, or scrampled sequence $A\beta_{1-42}$ had no appreciable effect on the levels of secreted or cellular AβPP. Dose response studies showed that elevated levels of cellular AβPP were observed at ≈10 μM $A\beta_{1-42}$ and the effect was maximal between 25–50 μM peptide. No effect on HLSM cellular AβPP levels were observed over the concentration ranges used in these studies (10–100 μM) with the shorter Aβ peptides. The selective effects of $A\beta_{1-42}$ on the HLSM cells may arise from differences in aggregation, solubility or cellular turnover compared to the shorter $A\beta_{1-39}$ and $A\beta_{1-40}$ peptides. This finding suggests a dynamic product-precursor relationship where $A\beta_{1-42}$ stimulates a large increase in HLSM cellular AβPP.

$A\beta_{1-42}$ AND $A\beta_{25-35}$ INDUCE INCREASED LEVELS OF CELLULAR CARBOXYL TERMINAL AβPP FRAGMENTS AND SOLUBLE Aβ PEPTIDE IN CULTURED CEREBROVASCULAR SMOOTH MUSCLE CELLS

Cells normally produce soluble, extracellular Aβ,[16,17] possibly through an endosomal/lysosomal mediated event involving carboxyl terminal AβPP fragments.[18,19] In parallel to the markedly elevated levels of cellular $A\beta PP$, HLSM cells incubated with $A\beta_{1-42}$ exhibited an $\approx$14-fold increase in the levels of carboxyl terminal $A\beta PP$ fragments compared to untreated cells or cells incubated with the scrambled sequence $A\beta_{1-42}$ peptide. Presumably, this may lead to higher levels of soluble, extracellular $A\beta$ being formed by the HLSM cells. However, this point was very difficult to address since high concentrations of exogenous $A\beta_{1-42}$ (25 μM) were used to induce the increased levels of cellular $A\beta PP$ and carboxyl terminal $A\beta PP$ fragments which compromised our ability to quantitate the production of $A\beta$ with $A\beta$-specific antibodies. To circumvent this problem we conducted parallel studies using the biologically active fragment $A\beta_{25-35}$. $A\beta_{25-35}$ at 100 μM caused a similar, but less robust, on cellular degeneration and increased levels of cellular $A\beta PP$ in the HLSM cells compared to $A\beta_{1-42}$. Since our anti-$A\beta$ antibodies recognized epitopes within the first 17 amino acids of $A\beta$ using $A\beta_{25-35}$ allowed us to conduct immunoprecipitation studies to determine the effects of this biologically active $A\beta$ fragment on the production of soluble $A\beta$. Incubation of the HLSM cells with 100 μM $A\beta_{25-35}$ resulted in an $\approx$3-fold increase in the levels of carboxyl terminal $A\beta PP$ fragments and soluble $A\beta$ peptide. The close relationship between the levels of carboxyl terminal $A\beta PP$ fragments and soluble $A\beta$ peptides in the HLSM cells incubated with $A\beta$25–35 suggests that even higher levels of soluble $A\beta$ peptides are likely generated in the HLSM cells incubated with the more robust acting $A\beta_{1-42}$.

PROPOSED MODEL FOR AN $A\beta$ CASCADE MECHANISM IN CEREBRAL AMYLOID ANGIOPATHY

Taken together, our findings support the critical role of $A\beta$ in AD and suggest a possible mechanism for the accumulation and spread of the cerebrovascular pathology of cerebral amyloid angiopathy. We propose that vascular smooth muscle cells in the walls of cerebral blood vessels normally produce $A\beta PP$ and presumably small amounts of soluble $A\beta$. Alternatively, the initial $A\beta$ may originate from another source such as the neuropil or circulation. At some point there is a pathologic interaction between the $A\beta$ and the cerebrovascular smooth muscle cells. This may involve insoluble aggregated forms of $A\beta$ that arise from increased levels of soluble $A\beta$ and/or expression of aggregation promoting factors in the microenvironment of cell surfaces and/or the cerebrovascular wall. Once initiated this process can lead to degeneration of the cerebrovascular smooth muscle cells with a concomitant increase in the levels of cellular $A\beta PP$, carboxyl terminal $A\beta PP$ fragments and additional $A\beta$ peptide. Our findings suggest that the process could initiate with a single or small group of smooth muscle cells producing a positive feedback cascade that with time spreads the pathology through distinct regions in the cerebrovasculature and, possibly, into the neuropil.

REFERENCES

1. VINTERS, H. V. 1987. Cerebral amyloid angiopathy: A critical review. Stroke **18:**311–324.
2. CORIA, F., M. LARRONDO-LILLO & B. FRANGIONE. 1989. Degeneration of smooth muscle cells in beta-amyloid angiopathies. J. Neuropathol. Exp. Neurol. **48:**368–375.
3. GLENNER, G. G. & C. W. WONG. 1984. Alzheimer's disease: Initial report of the purification and characterization of a novel cerebrovascular amyloid protein. Biochem. Biophys. Res. Commun. **122:**885–890.
4. VAN DUINEN, S. G., E. M. CASTANO, F. PRELLI, G. T. A. M. BOTS, W. LUYENDIJK & B. FRANGIONE. 1987. Hereditary cerebral hemorrhage with amyloidosis in patients of Dutch origin is related to Alzheimer's disease. Proc. Natl. Acad. Sci. USA **84:**5991–5994.
5. MASTERS, C. L., G. SIMMS, N. A. WEINMAN, G. MULTHAUP, B. L. MCDONALD & K. BEYREUTHER. 1985. Amyloid plaque core protein in Alzheimer disease and Down syndrome. Proc. Natl. Acad. Sci. USA **82:**4245–4269.
6. ROHER, A. E., J. D. LOWENSON, S. CLARKE, A. S. WOODS, R. J. COTTER, E. GOWING & M. J. BALL. 1993. β-amyloid-(1–42) is a major component of cerebrovascular amyloid deposits: Implications for the pathology of Alzheimer's disease. Proc. Natl. Acad. Sci. USA **90:**10836–10840.
7. GOLDGABER, D., M. I. LERMAN, O. W. MCBRIDE, U. SAFFIOTI & D. C. GAJDUSEK. 1987. Characterization and chromosomal localization of a cDNA encoding brain amyloid of Alzheimer's disease. Science **235:**877–880.
8. KANG, J., H. G. LEMAIRE, A. UNTERBECK, J. M. SALBAUM, C. L. MASTERS, K.-H. GRZESCHIK, G. MULTHAUP, K. BEYREUTHER & B. MULLER-HILL. 1987. The precursor of Alzheimer's disease amyloid A4 protein resembles a cell-surface receptor. Nature **325:**733–736.
9. KAWAI, M., R. N. KALARIA, P. CRAS, S. L. SIEDLAK, M. E. VELASCO, E. R. SHELTON, H. W. CHAN, B. D. GREENBERG & G. PERRY. 1993. Degeneration of vascular muscle cells in cerebral amyloid angiopathy of Alzheimer's disease. Brain Res. **623:**142–146.
10. SHOJI, M., S. HIRAI, Y. HARIGAYA, T. KAWARABAYASHI & H. YAMAGUCHI. 1990. The amyloid β-protein precursor is localized in smooth muscle cells of leptomeningeal vessels. Brain Res. **530:**113–116.
11. WISNIEWSKI, H. M., J. FRACKOWIAK, A. ZOLTOWSKA & K. S. KIM. 1994. Vascular β-amyloid in Alzheimer's disease angiopathy is produced by proliferating and degenerating smooth muscle cells. Amyloid **1:**8–16.
12. VAN NOSTRAND, W. E., A. J. M. ROZEMULLER, R. CHUNG, C. W. COTMAN & S. M. SAPORITO-IRWIN. 1994. Amyloid β-protein precursor in cultured leptomeningeal smooth muscle cells. Amyloid **1:**1–7.
13. CUMMINGS, B. J., J. H. SU, J. W. GEDDES, W. E. VAN NOSTRAND, S. L.

Wagner, D. D. Cunningham & C. W. Cotman. 1992. Aggregation of the amyloid precursor protein (APP) within degenerating neurons and dystrophic neurites in Alzheimer's disease. Neuroscience **48:**763–777.

14. Prelli, F., E. Castano, G. G. Glenner & B. Frangione. 1988. Differences between vascular and plaque core amyloid in Alzheimer's disease. J. Neurochem. **51:**648–651.
15. Joachim, C. L., L. K. Duffy, J. H. Morris & D. J. Selkoe. 1988. Protein chemical and immunocytochemical studies of meningovascular beta-amyloid protein in Alzheimer's disease and normal aging. Brain Res. **474:**100–111.
16. Haass, C., M. G. Schlossmacher, A. Y. Hung, C. Vigo-Pelfrey, A. Mellon, B. L. Ostaszewski, I. Lieberburg, E. H. Koo, D. B. Schenk, D. B. Teplow & D. J. Selkoe. 1992. Amyloid β-peptide is produced by cultured cells during normal metabolism. Nature **359:**322–325.
17. Shoji, M., T. E. Golde, J. Ghiso, T. T. Cheung, S. Estus, L. M. Shaffer, X.-D. Cai, D. M. McKay, R. Tintner, B. Frangione & S. G. Younkin. 1992. Production of the Alzheimer amyloid β protein by normal proteolytic processing. Science **258:**126–129.
18. Estus, S., T. E. Golde, T. Kunishita, D. Blades, D. Lowery, M. Eisen, M. Usiak, X. Qu, T. Tabira, B. D. Greenberg & S. G. Younkin. 1992. Potentially amyloidogenic, carboxyl-terminal derivatives of the amyloid protein precursor. Science **255:**726–728.
19. Golde, T. E., S. Estus, L. H. Younkin, D. J. Selkoe & S. G. Younkin. 1992. Processing of the amyloid protein precursor to potentially amyloidogenic derivatives. Science **255:**728–730.

Participation of Gene Expression in the Protection against Amyloid β-Peptide Toxicity by the β-Amyloid Precursor Protein[a]

S. W. BARGER[b] AND M. P. MATTSON[b–d]

[b] *Sanders-Brown Research Center on Aging and* [c]*Department of Anatomy and Neurobiology, University of Kentucky, Lexington, Kentucky 40536 USA*

ABSTRACT: The amyloid β-peptide (Aβ) is a toxic derivative of the β-amyloid precursor protein. Alternative processing of this precursor also yields large soluble forms (APP^Ss) which are secreted from many cell types. These APP^Ss have neuritogenic and neuroprotective activities; indeed, APP^Ss can protect primary neurons from the toxicity of Aβ itself. To begin to explore the regulation of gene expression by APP^S, we have focused on the NF-κB transcription factor family. NF-κB is induced by conditions of stress, including cellular oxidation. We report that NF-κB can also be induced by APP^S. Furthermore, we effected direct activation of NF-κB through disinhibition using antisense oligonucleotide technology. This means of activating NF-κB resulted in protection of neuroblastoma cells from the toxicity of a calcium ionophore and protection of primary hippocampal neurons from the toxicity of Aβ. Together, these data suggest that NF-κB may exist as a common agent inducing a neuroprotective pattern of gene expression in response to either trophic cytokines or stress itself.

INTRODUCTION

The amyloid β-peptide (Aβ), which forms the characteristic plaques of Alzheimer's disease (AD), can be neurotoxic both in cultured cells and *in vivo*. This toxicity apparently involves adverse effects on cellular calcium homeostasis.[1] More recent evidence suggests that this effect involves the ability of Aβ to reduce the activity of ion-motive pumps in the neuronal plasma membrane.[2] These ATPases are exquisitely sensitive to oxidation, consistent with recent observations that Aβ toxicity involves reactive oxygen radicals.[3–5]

Aβ is proteolytically derived from the larger β-amyloid precursor protein (βAPP). Alternative processing can also liberate most of the extracellular

[a] This research was sponsored by funds from the NIH (M.P.M. and S.W.B.), the French Foundation (S.W.B.) and the Alzheimer's Association (M.P.M.).

[d] *Send correspondence to:* Dr. Mark P. Mattson, 211 Sanders-Brown Building, University of Kentucky, Lexington, KY 40536-0230; Tel: (606)257-6040; Fax: (606)323-2866

portion of this precursor as a secreted protein (APP^S) with several cytokinetic activities. We have previously reported that APP^S protects CNS neurons from various toxic conditions, including the toxicity of Aβ itself.[4,6] Neuroprotection by APP^S is maximal after a pretreatment of 12–24 hours, suggesting the involvement of gene expression. We have begun to explore the molecular responses to APP^S treatment by analyzing the NF-κB family of transcription factors. The prototypical member of this family is a heterodimer of p50 and p65 subunits. NF-κB is constitutively present in the cytosol held in an inactive form by the binding of a third class of subunits generally termed I-κB. Activation of transcription factor activity results from dissociation of I-κB from the p50/p65 dimer.

Aβ was recently shown to activate NF-κB[3] consistent with the demonstrated responsiveness of this transcription factor to oxidation.[7] However, it is unclear whether NF-κB activation participates in Aβ toxicity or whether it is a compensatory response involved in an attempt to alleviate damage. We have examined the involvement of NF-κB in neuronal viability with two meaningful results: first, we have found that APP^S, a neuroprotective stimulus, also activates NF-κB; second, specific induction of NF-κB is neuroprotective. Together, these data implicate NF-κB as an agent orchestrating neuroprotective patterns of gene expression in response to both neurotrophic factors and oxidative insults.

METHODS

Materials

APP^S was purified from the culture medium of transfected HEK293 cells as described.[8] Aβ 25–35 was from Bachem (Torrance, CA). Oligonucleotides were synthesized by the Macromolecular Synthesis and Analysis Facility (U. Kentucky) or Integrated DNA Technologies (Coralville, IA).

Cell Culture

Mouse hippocampal cultures were established from E17 C57BL/6 mice by methods previously described for rat hippocampal cultures.[9] On culture day 4, the medium was exchanged for Neurobasal (Gibco) with defined medium supplements described previously.[9] Survival was determined by visual examination of defined morphological criteria as described.[9]

The SH-SY5Y human neuroblastoma line was generously provided by Dr. J. Biedler (Sloan-Kettering Cancer Inst.) and was maintained in minimal essential medium (MEM) supplemented with 10% FBS. For survival experiments, SH-SY5Y cells were plated overnight at 24,000 cells/cm^2 in 96-well plates. For differentiated cultures, the medium was then replaced with MEM containing 0.5% FBS and 10 μM retinoic acid for five days; undifferentiated cultures were switched to MEM with 0.5% FBS and used within three days of plating. The fluorescent vital dye alamar Blue (Alamar) was used as an index of survival as previously described.[9] The neuro-2A mouse neuroblastoma line was obtained from American Type Culture Collection and maintained in MEM with 10% FBS.

Oligonucleotides

Antisense oligonucleotide against the human I-κB clone MAD3.[10] was synthesized as follows: 5′-GCCGCCTGGAACATG-3′; its mismatch control was 5′-GCCGGCTCGAATTTG-3′. Antisense against mouse MAD3[11] was 5′-TGGCTGAAACATGGC-3′; its mismatch control was 5′-TAGTTGGAA-CACGGC-3′. Oligonucleotide sequences were compared to all vertebrate Genbank sequences and no significant matches were found other than the targeted genes.

Transfections and CAT Assays

A chloramphenicol acetyltransferase (CAT) reporter vector driven by an NF-κB responsive promoter (pκB/TK5[12]) was used as an index of NF-κB activity. Transfections were performed with Lipofectamine (Gibco), essentially per manufacturer's instructions. Lysates were assayed for CAT activity essentially as described.[13]

RESULTS

Initial experiments using the electrophoretic mobility shift assay (EMSA) demonstrated that a factor binding to the NF-κB consensus DNA motif was elevated in primary hippocampal neurons treated with APP^S. Consistent with our previous demonstration of a role for cyclic GMP in APPs actions,[8] this induction was reduced by the cGMP-dependent kinase inhibitor Rp-8-Br-cGMPS (RGS). Densitometric scans of the EMSA results indicated that this factor was elevated 15-fold in cells treated with APP^S for 45 min, but this was reduced to only an 8-fold elevation by coaddition of 30 μM RGS. To determine whether this APP^S-inducible factor was a transactivator, we transfected neuroblastoma cells with an NF-κB-responsive CAT reporter construct. APP^S elevated CAT activity approximately 3-fold over a 20-h treat-

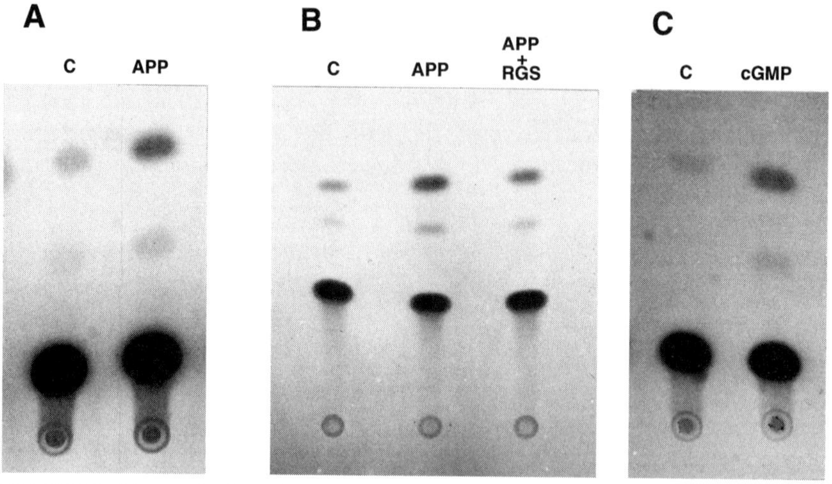

FIGURE 1. Transactivation of an NF-κB-dependent promoter by APPs. Neuroblastoma cells were transfected with pκB/TK5 and exposed to indicated treatments for 20 h. Lysates were assayed for CAT activity, and reaction products separated by thin-layer chromatography and visualized by autoradiography. The most abundant species represents unreacted [^{14}C]chloramphenicol; higher species are mono- and diacetylated [^{14}C]chloramphenicol resulting from CAT activity. **A:** Neuro-2A cells were exposed to control vehicle (C) or 1 nM APP^{S}695 for 20 h. **B:** Differentiated SH-SY5Y cells were exposed to control vehicle, 0.5 nM APPs695, or 0.5 nM APP^{S}695 combined with 30 μM RGS. **C:** Differentiated SH-SY5Y cells were treated with vehicle or 0.3 mM 8-Br-cGMP.

ment (FIG. 1). This induction was also inhibited by RGS, and a similar induction resulted from application of a cell-permeant cGMP analog.

To determine whether the induction of NF-κB activity by APP^S was involved in neuroprotection, we sought to specifically activate NF-κB by diminishing the levels of one of its inhibitory subunits, termed MAD3.[10] An antisense oligonucleotide directed against human MAD3 was applied to SH-SY5Y cells. Differentiated cultures showed elevation of NF-κB DNA-binding activity by EMSA. Viability was then compared in differentiated or undifferentiated SH-SY5Y cultures after a 24-h exposure to calcium ionophore (FIG. 2A). Differentiated cultures displayed substantial calcium-dependent cell death, which was alleviated by treatment with MAD3 antisense oligonucleotide.

APP^S has been shown to protect against Aβ toxicity;[4] therefore, we also assayed the effects of MAD3 antisense on this parameter. Hippocampal cultures from embryonic mice were treated with antisense oligonucleotides directed against mouse MAD3. When exposed to the 25–35 amino acid fragment of Aβ, neurons in these cultures died at significantly lower rates than controls (FIG. 2B).

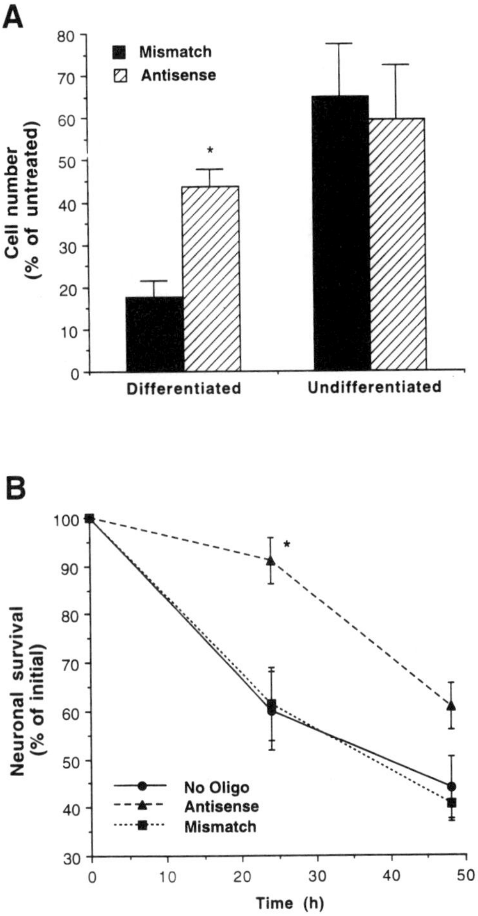

FIGURE 2. Neuroprotection by induction of NF-κB activity. **A:** SH-SY5Y cells were treated with 35 μM MAD3 antisense oligonucleotide or mismatch control oligonucleotide with or without differentiation in retinoic acid. After 48 h, some cultures of each treatment were exposed to 600 nM A23187 for 24 h. Viability was determined as described in METHODS and is expressed as percentage of the value from the corresponding differentiation and oligonucleotide treatment without ionophore. Values reflect the mean of 4 cultures + SEM ($*p < 0.05$ vs. differentiated cells treated with mismatch control oligonucleotide and ionophore) **B:** Primary mouse hippocampal cultures were treated for 24 h with 20 μM MAD3 antisense oligonucleotide, mismatch control oligonucleotide, or no oligonucleotide. Photomicrographs were then taken of identical fields before 24 h after, and 48 h after addition of 20 μM Aβ 25–35 fragment. Viable cells were counted and are expressed as percentage of the initial cell number in the corresponding field. Values reflect the mean of 3 cultures + SEM ($*p < 0.05$ vs. mismatch control oligonucleotide or no oligonucleotide).

DISCUSSION

We have demonstrated that APP^S induces an NF-κB-like transcription factor in both neuroblastoma cell lines and primary neuronal cultures. Independent induction of NF-κB activity by antisense-mediated disinhibition resulted in protection against calcium-dependent cell death and Aβ toxicity. These data suggest that one component of APP^S neuroprotection involves enhanced expression of genes controlled by NF-κB. Furthermore, the results suggest that induction of NF-κB by conditions of stress is an attempt by the cells to alleviate damage. Others have noted an induction of NF-κB by induced seizures *in vivo*.[14] While oxidation may induce NF-κB in a cell-autonomous manner, several paracrine cytokines which induce NF-κB are also elevated by brain injury (reviewed in ref. 15). Tumor necrosis factors are also neuroprotective[16] and induce NF-κB in primary neurons.[17]

It is of interest that NFκB induction protected cells against both calcium ionophore and Aβ. Loss of calcium regulation plays a role in Aβ toxicity. Future studies will focus on whether NF-κB induces genes which enhance calcium homeostasis and/or genes which alleviate problems downstream of $[Ca^{2+}]_i$ increases. For instance, elevated $[Ca^{2+}]_i$ can enhance the generation of reactive oxygen species,[9] and mitochondrial superoxide dismutase has an NF-κB-responsive promoter.[18]

REFERENCES

1. MATTSON, M. P., B. CHENG, D. DAVIS, K. BRYANT, I. LIEBERBURG & R. E. RYDEL. 1992. β-Amyloid peptides destabilize calcium homeostasis and render human cortical neurons vulnerable to excitotoxicity. J. Neurosci. **12:**379–389.
2. MARK, R. J., K. HENSLEY, D. A. BUTTERFIELD & M. P. MATTSON. 1995. Amyloid β-peptide impairs ion-motive ATPase activities: Evidence for a role in loss of neuronal Ca^{2+} homeostasis and cell death. J Neurosci **15:**6239–6249.
3. BEHL, C., J. DAVIS, R. LESLEY & D. SCHUBERT. 1994. Hydrogen peroxide mediates amyloid β protein toxicity. Cell 77:817–827.
4. GOODMAN, Y. & M. P. MATTSON. 1994. Secreted forms of β amyloid precursor protein protect hippocampal neurons against amyloid β-peptide-induced oxidative injury. Exp. Neurol. **128:**1–12.
5. HENSLEY, K., J. CARNEY, M. MATTSON, M. AKSENOVA, M. HARRIS, J. WU, R. FLOYD & A. BUTTERFIELD. 1994. A model for β-amyloid aggregation and neurotoxicity based on free radical generation by the peptide: Relevance to Alzheimer's disease. Proc. Natl. Acad. Sci. USA **91:**3270–3274.
6. MATTSON, M. P., B. CHENG, A. R. CULWELL, F. S. ESCH, I. LIEBERBURG & R. E. RYDEL. 1993. Evidence for excitoprotective and intraneuronal calcium-regulating roles for secreted forms of the β-amyloid precursor protein. Neuron **10:**243–254.
7. SCHRECK, R., P. RIEBER & P. A. BAEUERLE. 1991. Reactive oxygen intermediates as apparently widely used messengers in the activation of the NF-κB transcription factor and HIV-1. EMBO J. **10:**2247–2258.
8. BARGER, S. W., R. R. FISCUS, P. RUTH, F. HOFMANN & M. P. MATTSON. 1995. Role of cyclic GMP in the regulation of neuronal calcium and survival by secreted forms of β amyloid precursor. J. Neurochem. **64:**2087–2096.
9. MATTSON, M. P., S. W. BARGER, J. BEGLEY & R. J. MARK. 1994. Calcium, free radicals, and excitotoxic neuronal death in primary cell culture. Meth. Cell Biol. **46:**187–216.
10. HASKILL, S., A. A. BEG, S. M. TOMPKINS, J. S. MORRIS, A. D. YUROCHKO, A. SAMPSON-JOHANNES, K. MONDAL, R. RALPH & A. S. BALDWIN, JR. 1991. Characterization of an immediate-early gene induced in adherent monocytes which encodes IκB-like activity. Cell **65:**1281–1289.

11. CHIAO, P. J., S. MIYAMOTO & I. M. VERMA. 1994. Autoregulation of IκBα activity. Proc. Natl. Acad. Sci. USA **91**:28–32.
12. SHAKHOV, A. N., M. A. COLLART, P. VASSALLI, S. A. NEDOSPASOV & C. V. JONGENEEL. 1990. κB-Type enhancers are involved in lipopolysaccharide-mediated transcriptional activation of the tumor necrosis factor α gene in primary macrophages. J. Exp. Med. **171**:35–47.
13. GORMAN, C. M., L. F. MOFFAT & B. H. HOWARD. 1982. Recombinant genomes which express chloramphenicol acetyltransferase in mammalian cells. Mol. Cell Biol. **2**:1044–1051.
14. PRASAD, A. V., W. H. PILCHER & S. A. JOSEPH. 1994. Nuclear factor-κB in rat brain: Enhanced DNA-binding activity following convulsant-induced seizures. Neurosci. Lett. **170**:145–148.
15. MATTSON, M. P. & S. W. SCHEFF. 1994. Endogenous neuroprotection factors and traumatic brain injury: Mechanisms of action and implications for therapies. J. Neurotrauma **11**:3–33.
16. CHENG, B., S. CHRISTAKOS & M. P. MATTSON. 1994. Tumor necrosis factors protect neurons against metabolic-excitotoxic insults and promote maintenance of calcium homeostasis. Neuron. **12**:139–153.
17. BARGER, S. W., D. HORSTER, K. FURUKAWA, Y. GOODMAN, J. KRIEGLSTEIN & M. P. MATTSON. 1995. Tumor necrosis factors α and β protect neurons against amyloid β-peptide toxicity: Evidence for involvement of a κβ-binding factor and attenuation of peroxide and Ca^{2+} accumulation. Proc. Natl. Acad Sci USA **92**:9328–9332.
18. PANG, X.-P., N. S. ROSS, M. PARK, G. J. F. JUILLARD, T. M. STANLEY & J. M. HERSHMANN. 1992. Tumor necrosis factor-α activates nuclear factor κB and induces manganous superoxide dismutase and phosphodiesterase mRNA in human papillary thyroid carcinoma cells. J. Biol. Chem. **267**:12826–12830.

The Role of Arachidonic Acid in the Secretion of the Amyloid Precursor Protein (APP)[a]

M. R. EMMERLING, D. T. DUDLEY, R. D. DYER, R. T. CARROLL, P. D. DOYLE, AND R. E. DAVIS

Parke-Davis Pharmaceutical Research, Division of Warner-Lambert Co., 2800 Plymouth Road, Ann Arbor, Michigan 48106 USA

ABSTRACT: We have studied the activation of human ml-muscarinic receptors in a genetically engineered Chinese hamster ovary cell line (CHO-m1) to determine which second messenger systems affect the secretion of APP via the non-amyloidogenic route. Carbachol activation of the signaling pathways in CHO-ml cells promotes APP secretion by activation of both protein kinase C (PKC)-dependent or Ca^{++}-dependent second messenger pathways. Both pathways converge to increase the enzyme activity of phospholipase A_2 (PLA_2), the enzyme that releases arachidonic acid from cellular stores. Directly activating PLA_2 with melittin, a peptide from bee venom, or by adding arachidonic acid directly to cultured cells increases the secretion of APP. Thus, our results indicate that arachidonic acid is yet another cellular second messenger involved in regulating the metabolism of APP in addition to PKC and cytoplasmic Ca^{++}. Moreover, activation of PLA_2 appears to be an obligatory event in increasing the secretion of APP from CHO-ml cells by the various methods of activation that we have tried thus far.

INTRODUCTION

The loss of cholinergic neurons and innervation in the hippocampus and cortical areas is one of the pathological hallmarks of Alzheimer's Disease (AD). This cholinergic deficit is implicated in the cognitive loss seen in AD and has prompted efforts to develop cholinergic replacement therapies to improve the cholinergic deficit. The cholinomimetic approach has yielded positive effects on the cognitive functions of individuals with AD.[1] However, recent research suggests that the cholinomimetic strategy may ameliorate

[a] *Send correspondence to:* Dr. Mark R. Emmerling, Parke-Davis Pharmaceutical Research, Division of Warner-Lambert Co., 2800 Plymouth Road, Ann Arbor, Michigan 48106, USA; Tel: (313)998-5917; Fax: (313)996-1193; Email: EMMERLM@AA.WL.COM.

more than cognitive difficulties. The seminal work of Nitsch and his colleagues[2] revealed that cholinomimetics influence the metabolic processing of APP in cultured cells. Activation of cells with muscarinic receptors linked to phospholipase C (PLC) increases the secretion of APP by a non-amyloidogenic route while decreasing the production of the β-amyloid peptide (Aβ).[3] This effect may have the net benefit of decreasing Aβ-associated pathology, while promoting the secretion of APP and its associated trophic activity.[4]

The signal transduction pathways that mediate the change in APP metabolism are under intense investigation, since understanding these pathways may provide insights into the pathophysiology of AD as well as means to render therapeutic assistance. Thus far, studies indicate that protein phosphorylation resulting from m1-receptor activation of PKC increases the non-amyloidogenic secretion of APP.[5] However, work by Felder and others[6] shows that activation of m1-muscarinic receptors, in addition to activating PKC, also initiates a variety of other signal transduction events, involving PLA_2, phospholipase D, adenylate cyclase, receptor-operated calcium channels, and tyrosine kinase activity. The role(s) of these other second messenger systems in the regulation of APP metabolism is largely unexplored.

Circumstantial evidence already implicates PLA_2 in APP secretion. The levels of APP in brain tissue increase extracellularly upon trauma, suggesting that nerve cell injury promotes the secretion of APP. One consequence of brain trauma is the activation of the enzyme PLA_2,[7] which can be activated by a variety of G-protein linked receptors. Because the activation of PLA_2 influences secretion of a variety of substances from various cells, we sought to determine if PLA_2 activation also affects APP secretion in CHO-m1 cells treated with the muscarinic agonist carbachol.

MATERIALS AND METHODS

The methods for the following experiments have been described elsewhere.[8] Briefly, CHO-m1 cells were treated with media containing the agents to be studied, after one hour media was harvested, concentrated and prepared for gel electrophoresis and immunoblotting. APP was detected by 22C11 using either an alkaline phosphatase or chemiluminescence detection system. Band density was determined as integrated optical density (IOD).

RESULTS

To determine if activation of PLA_2 is involved in carbachol-induced secretion of APP, we treated CHO-m1 cells with 1 mM carbachol in the presence and absence of the irreversible PLA_2 inhibitors, manoalide and sclaradial.

Both PLA_2 inhibitors decrease carbachol-stimulated release of APP at drug concentration of 1 μM to 10 μM. Manoalide inhibition of PLA_2 also decreases the increased APP secretion by CHO-ml cells produced after PKC activation by phorbol 13-myristate 12-acetate (PMA) and from increasing cytoplasmic Ca^{++} with the ionophore A23187. Our studies show that the increase in APP secretion by A23187 is not sensitive to down regulation of PKC by an overnight treatment with 10 μM PMA, indicating that the effect of Ca^{++} is independent of PMA-sensitive PKC.

Activating PLA_2 directly, bypassing muscarinic receptors, also increases APP secretion from CHO-m1 cells. Bee venom melittin directly activates PLA_2, stimulating the production of arachidonic acid.[9] Mellittin increases the release of arachidonic acid and of APP from CHO-m1 cells. The melittin-induced increase in APP secretion is dependent on the dose, 1μg/mL being optimal, and is inhibited by manoalide, but not by down regulation of PKC by overnight pretreatment of cells with 10 μM PMA. Treatment with staurosporine and down regulation of muscarinic receptors by carbachol also have no discernible effect on melittin-stimulated APP secretion. The inability of the latter two treatments to change the melittin response implies that melittin works independently of PKC and activation of muscarinic receptors. Synthetic melittin causes CHO-m1 cells to increase their release of APP. However, phospholipase activating protein (PLAP), a protein reported to have melittin-like effects on cells, produces no increase in the secretion of APP or in the production of arachidonic acid by CHO-ml cells. Arachidonic acid added directly to culture media dose-dependently increases the secretion of APP from CHO-m1 cells. The optimal concentration of arachidonic acid for increasing APP secretion is between 0.6 μg/mL to 2.0 μg/mL. Thus, arachidonic acid treatment alone is sufficient to promote the release of APP from CHO-m1 cells.

DISCUSSION

Agonist occupancy of the muscarinic receptors of our CHO-m1 cells activate multiple second messenger systems involving, but perhaps not limited to, PKC, cytoplasmic Ca^{++} and ultimately PLA_2. The activation of PLA_2 occurs through two distinct pathways, one that involves PKC and the other that is independent of PKC activity. The PKC-dependent pathway can be activated by agonist occupation of the m1 receptor or directly by the activation of PKC with phorbol esters. Agonist activation of the PKC-dependent pathway is known to work through a heterotrimeric G-protein that activates PLC. PLC digests membrane lipids to generate diacylglycerol that in turn stimulates PKC activity. Based on work in other laboratories, we infer that

PKC leads to the activation of mitogen-activated-protein (MAP) kinase that subsequently phosphorylates PLA_2 in the cytoplasm. The phosphorylated PLA_2 is translocated from the cytoplasm to membranes by a Ca^{++}-dependent process and digests membrane lipids to produce arachidonic acid. The PKC-independent pathway associated with m1-muscarinic receptor activation increases PLA_2 activity through a Ca^{++}-dependent process. The increase in cytoplasmic Ca^{++} results from production of inositol triphosphate (IP_3) and activation of a receptor-operated Ca^{++} channel. Our results show that the Ca^{++} mediated increase in APP secretion involves PLA_2 activity. The intervening steps that lead to PLA_2 activation after increasing cytoplasmic Ca^{++} are unclear, but our results indicate that extended treatments with carbachol, but not PMA, can down regulate or desensitize the process(es). The ineffectiveness of PKC-down regulation by PMA to block the effects of Ca^{++} indicates that PMA-sensitive PKC is not involved in the Ca^{++}-dependent process.

Independently activating these second messenger systems reveals that each is capable of increasing the secretion of APP and shows that PLA_2 activity is required to obtain stimulated secretion of APP regardless of the mode of activation. This implies that increased formation of arachidonic acid, the product of PLA_2 enzyme activity, occurs downstream from the other signal transduction events. It also implies that arachidonic acid is a necessary part of the final common pathway mediating stimulated secretion of APP. Our results show that addition of arachidonic acid alone is sufficient to increase the secretion of APP by our CHO-m1 cells. This suggests that any receptor-ligand interaction that ultimately increases arachidonic acid formation will also potentiate the secretion of APP. Conversely, a decrease in PLA_2 activity should diminish APP secretion. This relationship between PLA_2 and APP secretion may be relevant to understanding the pathophysiology of AD since both the level of cholinergic activity and of PLA_2 activity[10] is decreased in AD brains. One would predict that hypo-functioning PLA_2 would contribute to reduced APP secretion and possibly changes in the production of $A\beta$.

The involvement of PLA_2 and arachidonic acid in secretory processes is not unusual. Both are implicated in the secretion of neurotransmitters,[11,12] amylase,[13] insulin,[14] and parathyroid hormone.[15] The effects of PLA_2 enzyme activity on secretory processes are thought to be mediated by the effects of arachidonic acid on the fusion of intracellular vesicles with one another and with plasma membrane.[16] PLA_2 activity also controls the transport of proteins thought the Golgi network and post-Golgi, secretory pathways of cells.[17] This effect of PLA_2 may influence both APP and $A\beta$ synthesis since the secretory pathways of cells are implicated in their post-translational processing.[18]

ACKNOWLEDGMENTS

The authors wish to thank Ms. Amy Smith and Ms. Jennifer Rodriguez for their invaluable technical support.

REFERENCES

1. KNAPP, M. J., S. I. GRACON, C. S. DAVIS, P. R. SOLOMON, W. W. PENDLEBURY & D. S. KNOPMAN. 1994. Efficacy and safety of high-dose tacrine—A 30-week evaluation. Alzheimer Disease & Associated Disorders **8:**S22–S31.
2. NITSCH, R. M., B. E. SLACK, R. J. WURTMAN & J. H. GROWDON. 1992. Release of Alzheimer amyloid precursor derivatives stimulated by activation of muscarinic acetylcholine receptors. Science **258:**304–307.
3. HUNG, A. Y., C. HAASS, R. M. NITSCH, W. Q. QIU, M. CITRON, R. J. WURTMAN, J. H. GROWDON & D. J. SELKOE. 1993. Activation of protein kinase C inhibits cellular production of the amyloid β-protein. J. Biol. Chem. **268:** 22959–22962.
4. ROCH, J. M., L. W. JIN, H. NINOMIYA, D. SCHUBERT & T. SAITOH. 1993. Biologically active domain of the secreted form of the amyloid β/A4 protein precursor. Ann. N. Y. Acad. Sci. **695:**149–157.
5. CAPORASO, G. L., S. E. GANDY, J. D. BUXBAUM, T. V. RAMABHADRAN & P. GREENGARD. 1992. Protein phosphorylation regulates secretion of Alzheimer β/A4 amyloid precursor protein. Proc. Natl. Acad. Sci. USA **89:**3055–3059.
6. FELDER, C. C., A. L. MA, I. M. BRILEY & J. AXELROD. 1993. Muscarinic acetylcholine receptor subtypes associated with release of Alzheimer amyloid precursor derivatives activate multiple signal transduction pathways. Ann. N. Y. Acad. Sci. **695:**15–18.
7. SHOHAMI, E., Y. SHAPIRA, G. YADID, N. REISFELD & S. YEDGAR. 1989. Brain phospholipase A_2 is activated after experimental closed head injury in the rat. J. Neurochem. **53:**1541–1546.
8. EMMERLING, M. R., C. J. MOORE, P. D. DOYLE, R. T. CARROLL & R. E. DAVIS. 1993. Phospholipase A_2 activation influences the processing and secretion of the amyloid precursor protein. Biochem. Biophys. Res. Commun. **197:** 292–297.
9. SHIER, W. T. 1979. Activation of high levels of endogenous phospholipase A_2 in cultured cells. Proc. Natl. Acad. Sci. USA **76:**195–199.
10. GATTAZ, W. F., A. MARAS, N. J. CAIRNS, R. LEVY & H. FÖRSTL. 1994. Decreased Phospholipase A_2 activity in Alzheimer's brains. Biol. Psychiatry. In press.
11. NIGISHI M., S. ITO & O. HAYAISHI. 1990. Arachidonic acid stimulates phosphoinositide metabolism and catecholamine release from bovine adrenal chromaffin cells. Biochem. Biophys. Res. Commun. **169:**773–779.
12. PRABHATI, R., J. D. BERMAN, W. MIDDLETON & J. BRENDLE. 1993. Botulinum toxin inhibits arachidonic acid release associated with acetylcholine from PC12 cells. J. Biol. Chem. **268:**11057–11064.

13. GROSFILS, K., F. GOMEZ & J. P. DEHAYE. 1992. Inhibition by mepacrine and amylase secretion from intact and permeabilized rat pancreatic acini. Biochem. Biophys. Res. Commun. **184**:408–413.
14. BAND, A. M., P. M. JONES & S. L. HOWELL. Arachidonic acid-induced insulin secretion from rat islets of Langerhans. Biochim. Biophys. Acta. **1176**:64–68.
15. BOURDEAU A., J. C. SOUBERBIELLE, P. BONNET, P. HERVIAUX, C. SACHS & M. LIEBERHERR. 1992. Phospholipase-A_2 action and arachidonic acid metabolism in calcium-mediated parathyroid hormone secretion. Endocrinology **130**: 1339–1344.
16. CREUTZ, C. E. 1981. Cis-unsaturated fatty acids induce the fusion of chromaffin granules aggregated by synexin. J. Cell Biol. **91**:247–256.
17. TAGAYA, M., N. HENOMATSU, T. YOSHIMORI, A. YAMAMOTO, Y. TASHIRO & T. FUKUI. 1993. Correlation between phospholipase A_2 activity and intra-golgi protein transport reconstituted in a cell-free system. Fed. Eur. Biochem. Soc. **324**:201–204.
18. HAASS, C. & D. J. SELKOE. 1993. Cellular processing of beta-amyloid precursor protein and the genesis of amyloid beta-peptide. Cell **75**:1039–1042.

The Role of Heparan Sulfate Proteoglycans in the Pathogenesis of Alzheimer's Disease[a]

D.H. SMALL,[b,e] T. WILLIAMSON,[b] G. REED,[b] H. CLARRIS,[c] K. BEYREUTHER, C. L. MASTERS,[b] AND V. NURCOMBE[d]

[b] *Department of Pathology and* [d]*Department of Anatomy and Cell Biology, The University of Melbourne, and Mental Health Resource Institute, Parkville, Victoria 3052, Australia*

[c] *Centre for Molecular Biology, University of Heidelberg, Heidelberg, Germany*

ABSTRACT: The hallmark of Alzheimer's disease (AD) is the deposition of amyloid plaques and neurofibrillary tangles in the brain. The relationship between amyloid deposition and the cognitive deficit is still unclear. The amyloid βA4 protein is produced by proteolytic cleavage of the amyloid protein precursor (APP). Very little is known about the normal function of APP and the role the protein may play in pathogenesis. Several studies have shown that APP is important for the regulation of neurite outgrowth. Our studies support these findings and indicate that the neurite outgrowth-promoting effects of APP are stimulated by an interaction between APP and specific proteoglycans. Using site-directed mutagenesis, a heparan sulfate binding site which mediates this effect has been mapped to the N-terminus of APP (residues 96-110, HBD-1). A peptide homologous to HBD-1 blocks the trophic effects of APP in cell culture. To purify specific proteoglycans which stimulate the action of APP, an affinity column was constructed using a biotinylated peptide homologous to HBD-1 coupled to streptavidin-agarose. Two proteoglycans were isolated from a crude brain cell conditioned medium by affinity chromatography. The purified proteoglycans bound APP saturably with high affinity and stimulated the action of APP on neurite outgrowth from chick sympathetic neurons. Digestion of the proteoglycan fraction with heparitinase I or chondroitinase ABC demonstrated the presence of two major proteins, a heparan sulfate proteoglycan with a core protein of 63-67 kD molecular mass and a chondroitin sulfate proteoglycan with a core protein of 100-110 kD molecular mass. The results demonstrate that APP binds to at least two proteoglycans and that this interaction may regulate the trophic effects of the protein. The interaction of specific APP-binding proteoglycans with amyloid plaques may disturb the normal function of APP and contribute to the neuritic degeneration that is commonly seen around the amyloid plaque cores.

[a] This work was supported by grants from the National Health and Medical Research Council of Australia. K. Beyreuther is supported by grants from the Bundesministerium für Forschung und Technologie and from the Deutsche Forschungsgemeinschaft.

[e] *Send correspondence to:* Dr. David H. Small, Department of Pathology, University of Melbourne, Parkville, Victoria 3052, Australia; Tel: 61-3-344-4205, Fax: 61-3-344-4004, Email: david_small @ muwayf.unimelb.edu.au.

INTRODUCTION

Studies which have identified a locus for familial Alzheimer's disease (AD) within the amyloid protein precursor (APP) gene on chromosome 21 have demonstrated the importance of APP in the pathogenesis of AD.[1] However, the relationship between the production of βA4 and the pathogenic mechanism which causes AD has yet to be elucidated. It is unknown whether the production of amyloid plaques is the direct cause of AD or whether a defect in the normal function of APP is responsible. Our approach has been to study the normal function of APP with a view to elucidating potential pathogenic mechanisms.

THE FUNCTION OF APP

Several studies suggest that APP may play a role in cell growth, repair or adhesion. APP expression is increased in response to focal ischemia.[2] The expression of APP also correlates with the extent of cellular differentiation of SY5Y neuroblastoma cells.[3] There is strong evidence linking APP to the regulation of neurite outgrowth. In the developing chick brain, APP expression increases during the developmental period when neurite outgrowth is maximal.[4] In PC12 cells[5] and in primary hippocampal explant cultures,[6] NGF concomitantly stimulates neurite outgrowth and APP expression. In addition, recombinant APP may mediate the actions of NGF on neurite outgrowth from PC12 cells.[7] Studies by Schubert *et al.*,[8] Whitson *et al.*,[9] and Breen *et al.*[10] have implicated APP in neuronal adhesion or neurite outgrowth mechanisms. The domain responsible for the neurite outgrowth promoting effects of APP is still unknown. Saitoh and coworkers[11] have reported that a region in the middle of APP, containing the amino acid sequence RERMS, binds to a specific cell surface receptor which mediates neurite outgrowth. Other studies place a receptor binding domain closer to the C-terminus of APP.[12]

APP BINDS TO HEPARAN SULFATE PROTEOGLYCANS IN THE EXTRACELLULAR MATRIX

There is increasing evidence that the physiological effects of secreted forms of APP are mediated through an interaction with the components of the extracellular matrix (ECM), specifically the heparan sulfate proteoglycans (HSPG). APP can bind saturably and with high affinity to purified ECM[4] and to heparin,[13] an analog of heparan sulfate. When purified ECM is treated with agents which selectively decrease the level of heparan sulfate proteoglycans, there is a corresponding decrease in the number of high affinity binding sites for APP, demonstrating that APP binds predominantly to proteoglycans in the ECM.[4] Multhaup[14] has reported that a heparin-binding domain (HBD-2) is located between residues 316-337 of the APP695 sequence. We have identified a heparin-binding domain (HBD-1) close to the N-terminus of APP which proves to be necessary for the neurite outgrowth promoting effect of APP. Mutagenesis of three basic residues within this region decreases the ability of APP to bind to heparin.[13]

To study the function of APP-ECM interactions, we have examined the effects of substratum-bound APP on neurite outgrowth from primary cultures of chick sympathetic and mouse hippocampal neurons.[6,13] The presentation of APP to cells in substratum-bound form to some extent mimics the way in which APP is presented to cells *in vivo* bound to the ECM. We found that substratum-bound APP, in combination with a preparation of HSPG purified from postnatal day 3 mouse brain, stimulated neurite outgrowth in culture. Both the binding of APP to HSPG and the stimulation of neurite outgrowth was blocked by a peptide homologous to HBD-1, demonstrating that the binding of APP to HSPG is an important step in the activation of the growth-promoting effect of substratum-bound APP.

IDENTIFICATION OF SPECIFIC PROTEOGLYCANS WHICH BIND APP

Although HSPG purified from a culture of postnatal day 3 mouse brain cells stimulated the neurite outgrowth effect of APP, a crude HSPG fraction from cultures of embryonic day 10 (E10) cells (a period prior to the major phase of neurite outgrowth in the mouse brain) did not stimulate the effect. The proteoglycan composition of the rodent brain becomes increasingly complex during development.[15] Thus a specific developmentally regulated proteoglycan may be important for the neurite outgrowth effect of APP.

To examine this hypothesis, we have purified proteoglycans from P3 mouse brain cultures which bind selectively and with high affinity to APP (Williamson *et al.*, submitted for publication). Conditioned medium from cultures of mouse brain cells was applied to an affinity column in which a biotinylated peptide homologous to HBD-1 domain of APP was coupled to streptavidin-agarose. The bound eluate from the affinity column was 17-fold enriched in proteoglycans compared with the crude conditioned medium. The purified proteoglycan fraction stimulated the action of APP on neurite outgrowth from chick sympathetic neurons and was found to bind APP saturably. Both high affinity ($K_d = 1.4 \times 10^{-11}$ M) and low affinity ($K_d = 6.5 \times 10^{-10}$ M) binding sites for APP were identified. Digestion of the affinity-purified proteoglycan fraction with heparitinase I or chondroitinase ABC demonstrated the presence of two major proteoglycans in the purified fraction, an HSPG with a core protein molecular mass of 63-69 kD and a chrondoitin proteoglycan (CSPG) with a core protein molecular mass of 100-110 kD.

The results suggest that the expression of specific proteoglycans may be an important step in the regulation of the neurite outgrowth-promoting activity of APP. Furthermore, our data suggest that secreted APP (released from growing neurites) may act in an autocrine manner, *i.e.* by binding to proteoglycans in the ECM and to a specific cell-surface receptor to regulate neurite outgrowth.

PROTEOGLYCANS AND THE PATHOGENESIS OF AD

Attempts to explain the neurotoxic effects of amyloid plaques must take in account the fact that both cellular and cognitive changes following the initial appearance of amyloid plaques are very slow. It has been estimated that the cognitive changes may follow 35 years after the appearance of amyloid plaques in the brain.[16] This slow time course of disease progression is in contrast to the rapid neurotoxic changes that aggregated forms of βA4 induce in cell culture.[17]

Although the precise role proteoglycans play in the pathogenesis of AD is unknown, it seems quite probable that there could be a defect in the normal interaction of proteoglycans with APP around the amyloid plaque. HSPGs are known to be constituents of amyloid plaques.[18] The binding of HSPGs to amyloid plaques could interfere with the function of APP in the region of the plaque, which could, in turn, result in increased expression of APP and increased amyloidogenic processing. Increased production of βA4 would result in further sequestration of HSPGs and further loss of the normal function of APP. Thus a cascade of biochemical events involving amyloid deposition would occur, the ultimate consequence of which would be the loss of normal trophic support leading to synaptic loss and dementia. Interference with this cascade of events could have therapeutic consequences. For this reason, further studies on the normal function of APP are warranted.

REFERENCES

1. GOATE, A., M. C. CHARTIER-HARLIN, M. MULLAN, J. BROWN, F. CRAWFORD, L. FIDANI, L. GIUFFRA, A. HAYNES, N. IRVING, L. JAMES, R. MANT, P. NEWTON, K. ROOKE, P. ROQUES, C. TALBOT, M. PERICAK-VANCE, A. ROSES, R. WILLIAMSON, M. ROSSOR, M. OWEN & J. HARDY. 1991. Segregation of a missense mutation in the amyloid precursor protein gene with familial Alzheimer's disease. Nature **349:**704–706.

2. ABE, K., R. E. TANAI & K. KOGURE. 1991. Selective induction of Kunitz-type protease inhibitor domain-containing amyloid precursor protein mRNA after persistent focal ischemia in rat cerebral cortex. Neurosci. Lett. **125:**172–174.
3. KÖNIG, G., C. L. MASTERS & K. BEYREUTHER. 1990. Retinoic acid induced differentiated neuroblastoma cells show increased expression of the βA4 amyloid gene of Alzheimer's disease and an altered splicing pattern. FEBS Lett. **269:** 305–310.
4. SMALL, D. H., V. NURCOMBE, R. MOIR, S. MICHAELSON, D. MONARD, K. BEYREUTHER & C. L. MASTERS. 1992. Association and release of the amyloid protein precursor of Alzheimer's disease from chick brain extracellular matrix. J. Neurosci. **12:**4143–4150.
5. REFOLO, L. M., S. R. J. SALTON, J. P. ANDERSON, P. MEHTA & N. K. ROBAKIS. 1989. Nerve and epidermal growth factors induce the release of the Alzheimer amyloid precursor from PC12 cell cultures. Biochem. Biophys. Res. Commun. **164:**664–670.
6. CLARRIS, H., V. NURCOMBE, D. H. SMALL, K. BEYREUTHER & C. L. MASTERS. 1994. Secretion of nerve growth factor from septum stimulates neurite outgrowth and release of the amyloid protein precursor of Alzheimer's disease from hippocampal explants. J. Neurosci. Res. **38:**248–258.
7. MILWARD, E., R. PAPADOPOULOS, S. J. FULLER, R. D. MOIR, D. SMALL, K. BEYREUTHER & C. L. MASTERS. 1991. The amyloid protein precursor of Alzheimer's disease is a mediator of the effects of nerve growth factor on neurite outgrowth. Neuron. **9:**129–137.
8. SCHUBERT, D., L. W. JIN, T. SAITOH & G. COLE. 1989. The regulation of amyloid β protein precursor and its modulatory role in cell adhesion. Neuron **3:**689–694.
9. WHITSON, J. S., C. G. GLABE, E. SHINTANI, A. ABCAR & C. W. COTMAN. 1990. β-Amyloid protein promotes neuritic branching in hippocampal cultures. Neurosci. Lett. **110:**319–324.
10. BREEN, K. C., M. BRUCE & B. H. ANDERTON. 1991. Beta amyloid precursor protein mediates neuronal cell-cell and cell surface adhesion. J. Neurosci. Res. **28:**90–100.
11. JIN, L. W., H. HINOMIYA, J.-M. ROCH, D. SCHUBERT, E. MASLIAH, D. A. C. OTERO & T. SAITOH. 1994. Peptides containing the RERMS sequence of amyloid β/A4 protein precursor bind cell surface and promote neurite extension. J. Neurosci. **14:**5461–5470.
12. KOO, E. L., L. PARK & D. J. SELKOE. 1994. Amyloid beta protein as a substrate interacts with extracellular matrix to promote neurite outgrowth. Proc. Natl. Acad. Sci. USA **90:**4748–4752.
13. SMALL, D. H., V. NURCOMBE, G. REED, H. CLARRIS, R. MOIR, K. BEYREUTHER & C. L. MASTERS. 1994. A heparin-binding domain in the amyloid protein precursor of Alzheimer's disease is involved in the regulation of neurite outgrowth. J. Neurosci. **14:**2117–2127.
14. MULTHAUP, G. 1994. Studies of the amyloid precursor protein (APP) in brain: Regulation of APP-ligand binding. *In* Amyloid Protein Precursor in Development, Aging and Alzheimer's Disease. C. L. Masters *et al.*, Eds.: 76–89. Springer-Verlag.

15. HERNDON, M. E. & A. D. LANDER. 1990. A diverse set of developmentally regulated proteoglycans is expressed in rat central nervous system. Neuron **4:** 949–961.
16. RUMBLE, B., R. RETALLACK, C. HILBICH, G. SIMMS, G. MULTHAUP, R. MARTINS, A. HOCKEY, P. MONTGOMERY, K. BEYREUTHER & C. L. MASTERS. 1989. Amyloid A4 protein and its precursor in Down's syndrome and Alzheimer's disease. N. Eng. J. Med. **320:**1446–1452.
17. YANKNER, B. A., L. R. DAWES, S. FISHER, L. VILLA-KOMAROFF, M. L. OSTER-GRANITE & R. L. NEVE. 1989. Neurotoxicity of a fragment of the amyloid precursor associated with Alzheimer's disease. Science **245:**417–420.
18. SNOW, A. D., H. MAR, D. NOCHLIN, K. KIMATA, M. KATO, S. SUZUKI, J. HASSELL & T. N. WIGHT. 1988. The presence of heparan sulphate proteoglycans in the neuritic plaques and congophilic angiopathy in Alzheimer's disease. Am. J. Pathol. **133:**456–463.

Differential Regulation of APP Secretion by Apolipoprotein E3 and E4[a]

B. L. WOLOZIN,[b,d] J. BASARIC-KEYS,[b] R. CANTER,[b] Y. LI,[b] D. VANDERPUTTEN,[c] AND T. SUNDERLAND[b]

[b] *Section on Geriatric Psychiatry and* [c]*Laboratory of Biochemical Genetics, NIMH, Bethesda, Maryland 20892 USA*

ABSTRACT: The apolipoprotein E isozyme, apolipoprotein E4, has been implicated as a risk factor for Alzheimer's disease. One reason for the increased risk may be that apolipoprotein E binds to the Aβ peptide, but there may be other factors as well. We show that apolipoprotein E is a potent regulator of the secretion of amyloid precursor protein. In cultures of PC12 cells, nanomolar levels of apolipoprotein E3 induce a rapid decrease in the secretion of APP, being observable in 30 min. and stable over 24 hours. Apolipoprotein E4, in contrast, increases secretion of APP over a similar time course. Reciprocal changes occur in cellular amyloid precursor protein. Differential characteristics are also seen in apo E binding to the cells, where apo E4 binds over a slower time course than apo E3. These results suggest a novel mechanism by which apolipoprotein E may be influencing the metabolism of amyloid precursor protein.

INTRODUCTION

Recent studies indicate that the apolipoprotein E allele, ϵ4, is an important risk factor for Alzheimer's disease.[1] In some families, the risk for Alzheimer's Disease (AD) increases from 20% to 90% for individuals with increasing numbers of apo ϵ4 alleles.[2] Two lines of evidence suggest that the increased risk of AD derives from the association of the Aβ peptide with apolipoprotein E4 (apo E4). Studies of Alzheimer pathology show that apo E4 accumulates with Aβ in neuritic plaques.[1] *In vitro* studies show that, in high micromolar range, apo E and Aβ can bind to each other.[3,4] This association may promote

[a] Y. Basaric-Keys is supported through a cooperative research and development agreement with Molecular Geriatrics (Libertyville, IL).

[d] *Send correspondence to:* Dr. Benjamin Labe Wolozin, Section on Geriatric Psychiatry, NIMH, Bldg. 10 Rm 3D41, 10 Center Dr. MSC 1264, Bethesda, MD 20892-1264; Tel: (301) 496-3421; Fax: (301) 402-0188; Email: ben @ codon.nih.gov.

the aggregation of Aβ, which is more toxic than monomeric Aβ. Hence the term "pathologic chaperone" has been used to describe one possible role of apo E in Alzheimer's pathophysiology.[5] From a biophysical perspective, the ability of apo E to promote Aβ aggregation may reflect a higher affinity of apo E for aggregated as opposed to monomeric Aβ. Consequently, it is currently unknown whether the higher affinity of apo E for aggregated Aβ actually increases neuritic plaque formation or, rather, leads to increased binding of apo E to pre-existing neuritic plaques. If the latter scheme is true, then other mechanisms may contribute to the role of apo E4 in increasing the risk of AD. Hence, we have studied the regulation of amyloid precursor protein (APP) by apo E4 in order to determine whether differential actions of apo E4 account for its role as a risk factor for AD.

METHODS

Cell Culture

PC12 cells were serum starved and then treated with apo E (Pan Vera, Madison, WI) as described.

Analysis of Secreted APP

Following treatment of the cells, the medium was removed and spun at 14,000 rpm for 15 min. APP was then isolated from the supernatant using agarose-coupled heparin (BioRad) and immunoblotted on 10% tris/glycine gels using antibodies, 22C11 (Boehringer-Mannheim) or anti-9 (generously donated by Ivan Leiberburg, Athena Neurosciences).

Analysis of Cellular APP

Cells were washed twice with PBS, scraped and immunoblotted using 10% tris/glycine gels, 30 μg protein per lane and the anti-APP antibodies 22C11 or anti-9 or an anti-apolipoprotein E antibody (Calbiochem).

RESULTS

The regulation of APP secretion by apo E3 and apo E4 was examined first. PC12 cells (300,000 cells per 60 mm plate) were incubated overnight in serum-free medium and then treated with apo E3 or apo E4 (0, 1, 2, 4 or 10 μg/ml). After 30 min. of incubation, the medium was removed and immunoblotted as described in METHODS. Both apo E3 and apo E4 induced dose-dependent changes in the secretion of APP from the PC12 cell. Apo E3 produced a dose-dependent decrease in APP secretion inducing maximal decreases in APP secretion of 74% at 10 μg/ml apo E3. In contrast, apo E4 had the opposite effect, producing a dose-dependent increase in APP secretion of 28%. Interestingly, at the higher doses of apo E4, an additional 140 kD form of APP was observed in the medium. Whether this is APP alone or a complex of APP and apo E4 is being investigated; however, stripping and reprobing of the immunoblots with an antibody to apo E, have not identified any bands in this size range as would be expected for an APP/apo E complex (data not shown). Incubation of apo E3 or E4 for longer periods of time, up to 24 h, produced similar changes in APP levels, indicating the changes in APP induced by apo E3 and E4 were constant (data not shown).

Previous studies on the regulation of APP secretion by muscarinic agonists show that changes in APP secretion produce the opposite affects on cellular APP; so, for instance, an increase in APP secretion is manifest by a corresponding decrease in cellular APP.[6] Similar phenomena were observed upon stimulation of the PC12 cells with apo E3 and E4, where treatment with apo E3 for 30 min produced a dose-dependent increase in the expression of cellular APP, with a dose response curve that paralleled that of the secreted APP. The same experiment performed with apo E4, produced a corresponding decrease in cellular APP, manifest predominantly in the 116 kD form of APP. The changes in cellular APP, however, were transient. Increases in cellular APP in response to addition of exogenous apo E3 were evident within 30 min of addition of apo E (10 μg/ml) and constant over 6 h, but by 24 h the affects were absent. Thus, the apo E induces changes in secretion of APP that are matched by reciprocal changes in cellular APP.

Further evidence of the differential activities of Apo E3 and E4 was seen in binding studies. Serum-starved PC12 cells were incubated for varying times with serum-free medium containing 5 μg/ml of apo E3 or E4. The medium was then removed, the cells were washed twice with cold PBS and immunoblots of the cellular homogenates were generated as described above. While no endogenous apo E was detected in the PC12 cells addition of the apo E to the cells resulted detectable binding of apo E to the cells. Interestingly, the two isozymes showed differential kinetics. Apo E3 binding maximized rapidly, within 1 h, and then decayed over the next 24 hours. In contrast, apo E4 binding steadily increased over a 6-h time course, after which it began to decay. The transient binding of apo E3 to the cells may account for the transient affects of apo E on cellular APP levels. These differential responses of the apo E3 and E4 may also reflect differential mechanisms of actions of the two isozymes, which could contribute to the contrasting effects of the two isozymes on APP processing.

DISCUSSION

These results indicate that apo E3 and E4 differentially regulate APP secretion, acting in a range of 1–10 μg/ml, which corresponds to 22–220 nM. The rapid response of secreted APP as well as the reciprocal response of cellular APP both support the suggestion that the acute actions of apo E primarily affect post-translational rather than transcriptional processing. The chronic affects of apo E, as judged by the 24 h time points, are less clear. Decreased levels of secreted APP remain evident; however, this could reflect residual affects of acute regulation by apo E. In contrast to the changes seen in APP secretion, a single dose of apo E has no effect on cellular APP levels 24 h after application. This either reflects transcriptional changes that compensate for decreased APP levels or the decreased binding of apo E at 24 h, which is evident in the time courses of apo E binding. Further studies of APP protein and RNA need to be done in order to determine whether chronic application of apo E alters APP transcriptional regulation at all.

A critical issue that is currently being investigated is the affects of apo E on Aβ production. One mechanism by which apo E4 could increase the risk of AD is by increasing production of Aβ, which would lead to increased plaque formation and neuronal loss in AD. The results presented above are actually not consistent with such a mechanism, though. The increased secretion of APP seen with apo E4 most likely correlates with decreased Aβ production, which is opposite to that expected for an Alzheimer's risk factor. However, PC12 cells are unusual in that they do not make much Aβ, and other cells that do produce Aβ (such as 293 cells) may show relevant changes in Aβ production. In addition, the system may have added complexities giving rise to unexpected results.

Finally, the mechanism of action of apo E remains to be investigated. Apo E3 and E4 could be binding differentially to a receptor such as the LDL receptor, the LDL receptor-related protein or APP itself. Binding of apo E to APP could alter its exposure to secretases and affect secretion in the process. Alternatively, the differential effects with apo E3 and apo E4 could result from changes in signal transduction, such as those that may be operating in dorsal root ganglion neurons, where apo E3 and E4 act differentially.[7]

[NOTE ADDED IN PROOF: The antibodies used to analyze APPs, 22C11 and anti-9, crossreact with both APP and amyloid like precursor protein 2, APLP2. We have now analyzed these results using an antibody specific to APP, 10D5 (generously donated by I. Leiberburg, Athena Neurosciences) and an antibody specific to APLP2, APLP2-D2 (generously donated by S. Sisodia, Johns Hopkins). The results using these specific antibodies show that the changes in secretion induced by apo E result solely from changes in secretion of APLP2.]

REFERENCES

1. STRITTMATTER, W., A. SAUNDERS, D. SCHMECHEL, M. PERICAK-VANCE, J. ENGHILD, *et al.* 1993. Apolipoprotein E: High-avidity binding to β-amyloid and increased frequency of type 4 allele in late-onset familial Alzheimer's disease. Proc. Natl. Acad. Sci. USA **90:**1977–1981.
2. CORDER, E., A. SAUNDERS, W. STRITTMATTER, D. SCHMECHEL, P. GASKELL, *et al.* 1993. Gene dose of apolipoprotein E type 4 allele and the risk of Alzheimer's disease in late onset families. Science. **261:**921–923.
3. SCHMECHEL, D., A. SAUNDERS, W. STRITTMATTER, B. CRAIN, C. HULETTE, *et al.* 1993. Increased amyloid β-peptide deposition in cerebral cortex as a consequence of apolipoprotein E genotype in late-onset Alzheimer disease. Proc. Natl. Acad. Sci. USA **90:**9649–9653.
4. STRITTMATTER, W., K. WEISGRABER, D. HUANG, L. DONG, G. SALVESON, *et al.* 1993. Binding of human apolipoprotein E to synthetic amyloid beta peptide: Isoform-specific effects and implications for late-onset Alzheimer disease. Proc. Natl. Acad. Sci. USA **90:**8098–8102.
5. MA, J., A. YEE, H. BREWER, S. DAS & H. POTTER. 1994. Amyloid-associated proteins alpha(1)-antichymotrypsin and apolipoprotein E promote assembly of Alzheimer beta-protein into filaments. Nature **372:**92–94.
6. NITSCH, R., B. SLACK, R. WURTMAN & J. GROWDON. 1992. Release of Alzheimer amyloid precursor derivatives stimulated by activation of muscarinic acetylcholine receptors. Science **258:**304–307.
7. NATHAN, B., S. BELLOSTA, D. SANAN, K. WEISGRABER, R. MAHLEY, *et al.* 1994. Differential effects of apolipoprotein E3 and E4 on neuronal growth in vitro. Science **264:**850–852.

Regulation of APP Processing by Intra- and Intercellular Signals[a]

J.D. BUXBAUM AND P. GREENGARD

Laboratory of Molecular and Cellular Neuroscience, The Rockefeller University, New York, New York, USA 10021

ABSTRACT: APP processing appears to be under complex regulation. This regulation is apparently important under both normal and pathological conditions. Of direct clinical interest is the observation that Aβ formation can be regulated by various means. This raises the possibility that altered APP processing may cause an increase in Aβ formation in AD, and suggests that it may be possible to regulate the production of Aβ as a therapeutic approach in AD. As an example of the utility of the latter approach, consider a patient carrying the Swedish APP mutation. If it is true that the cause of AD in such a patient is due to increased Aβ production, then decreasing Aβ production should delay the onset of the disease. Even in individuals where increased Aβ formation is not the cause of AD but there is some other causes, such as the presence of an allele of apolipoprotein E which causes Aβ accumulation and hence synaptic loss, decreasing Aβ formation may be beneficial. It is of course a very long way from *in vitro* experiments to therapy. The current emphasis on studying APP processing *in vivo* represents the next step towards this goal.

PROTEIN KINASE C REGULATES APP PROCESSING

It had been shown that APP undergoes cleavage and secretion[1] phosphorylation. To study this question, specific antibodies against the various parts of the APP molecule were used. The first experiments directed at studying the regulation of APP processing made use of a carboxyl-terminal antibody which precipitated full-length APP, as well as the carboxyl-terminal of APP which remains behind after the cleavage and secretion of the secreted portion of APP (APP_S). It was observed that when cells were treated with either phorbol dibutyrate (PBt_2), an agent which activates protein kinase C, and/or okadaic acid, an agent which inhibits protein phosphatases 1 and 2A, the levels of

[a] This research was supported by NIH Grants AG-09464 and AG-10491. J.D.B. is the recipient of an Alzheimer Association Faculty Scholars Award.

[b] *Send correspondence to:* Dr. Joseph D. Buxbaum, Laboratory of Molecular and Cellular Neuroscience, The Rockefeller University, 1230 York Avenue, New York, NY 10021.

mature (fully glycosylated and sulfated) APP diminished significantly, while the levels of carboxyl-terminal derivatives of APP increased.[2] Because APP_S production from full-length APP involves the loss of full-length APP and the generation of the carboxyl-terminal fragment, it was tempting to speculate that these results could be explained by arguing that APP_S production was regulated by protein phosphorylation.

Subsequent to the first study of phosphorylation on APP processing, the effects of protein phosphorylation on APP_S production were studied by several groups. Using antibodies against the amino-terminal of APP and studying the APP_S released into the extracellular space, it could be convincingly shown that phorbol esters and/or okadaic acid stimulate the production of APP_S.[3,4] The regulation of APP processing and secretion by phorbol esters and okadaic acid was observed in many cell types and appears to be a ubiquitous means of regulating APP cleavage and secretion. These cell types include various cell types that might model the cells important in producing $A\beta$ in the brain, such as human endothelial, neuroblastoma and glioma cells.[5,6]

After it was shown that $A\beta$ was produced normally by cultured cells, the role of protein kinase C and protein phosphatases 1 and 2B in the regulation of APP processing was extended to include a role in regulating the production of $A\beta$ as well as p3. p3 is a $A\beta$ fragment which is assumed to be derived after normal ($\alpha-$) cleavage of APP.[7] Several laboratories have recently shown that activation of protein kinase C and/or inhibition of protein phosphatases 1 and 2A lead to dramatically decreased production of $A\beta$ with increased production of p3.[8–10] If p3 is in fact derived after α-cleavage of APP, then it is not surprising that increasing α-cleaved APP_S production is associated with increased p3 production. The decrease in $A\beta$ production may be in part due to the amounts of full-length APP being limiting: increasing α-cleavage would then decrease the amounts of APP available for α-cleavage.[8] As before, the effects of protein kinase C activation on $A\beta$ production have been observed in a variety of cell types including primary human astrocytes.[10] Cells expressing any of several mutations in APP which cosegregate with AD, including the Swedish mutation, still respond to protein kinase C activation with decreased $A\beta$ production,[8,9] suggesting that regulating the production of $A\beta$ as a therapeutic approach may be possible even in individuals with such mutations.

REGULATION OF APP PROCESSING BY PHOSPHOLIPASE C-LINKED FIRST MESSENGERS

With the demonstration of a role for protein kinase C in the regulation of APP processing, it was predicted that various first messengers which activate the phospholipase C/protein kinase C cascade would also be capable of regulating APP processing. This prediction was confirmed in several subsequent studies. Cholinergic agonists were shown to regulate APP_S formation by several groups: the effects could be mediated by muscarinic receptors, particularly muscarinic receptors known to be coupled to the phospholipase C/protein kinase C cascade.[5,11] Acetylcholine is altered in Alzheimer disease brain, making these studies particularly interesting. Interleukin 1 is also altered in Alzheimer disease brain and cerebrospinal fluid[12,13] and can activate the phospholipase C/protein kinase C cascade. In these early studies interleukin 1 was also shown to be able to regulate APP_S formation.[5,6]

With the observation that $A\beta$ is normally produced by cells, the effects of activation of the phospholipase C/protein kinase C cascade on $A\beta$ formation were studied. Direct activation of this cascade by mastoparan and mastoparan X increased the formation of APP_S while decreasing the formation of $A\beta$.[8] Similarly, muscarinic agonists could decrease $A\beta$ production in cells overexpressing the M1 or M3 muscarinic receptors and in human neuroglioma cells.[6,9] The surprise came when the protein kinase C in cells overexpressing muscarinic receptors was down-regulated. In such cells it was predicted that cholinergic agonists should have little effect; however, no difference was observed in cells lacking phorbol ester stimulated protein kinase C when compared to control cells, when examined for muscarinic agonist regulation of APP processing.[6] This was interpreted as suggesting that the effects of the muscarinic agonists could be mediated by either the phospholipase C/protein kinase C cascade or by the phospholipase C/calcium cascade.

A recent report raises the possibility that the effects of phospholipase C-linked first messengers on APP processing may involve the activation of phospholipase A_2.[14] Comparing the relative efficiencies of different first messengers for their ability to regulate APP_S and $A\beta$ production indicates that for a given cell compounds which are better able to stimulate APP_S production are generally better able to inhibit $A\beta$ formation,[6] consistent with APP being rate-limiting in the formation of these metabolites.[8]

REGULATION OF APP PROCESSING BY CALCIUM

A potential role for intracellular calcium in the regulation of APP processing was first demonstrated using the calcium ionophore A23187. Treating B-104 neuroblastoma cells or differentiated PC12 cells with A23187 led to increased production of APP_S.[15] These data suggest that voltage or ligand-gated calcium channels could regulate APP processing. Calcium released from intracellular stores has also been implicated in the regulation of APP processing. Treating various cells with thapsigargin or cyclopiazonic acid, compounds which inhibit the endoplasmic reticulum calcium ATPase, leading to the release of calcium into the cytoplasm, caused increased APP_S formation.[6]

Significantly, the effects of these compounds on APP_S formation were still observed in cells which had been treated for 24 h with phorbol esters to down-regulate protein kinase C, suggesting that the effects of these compounds are protein kinase C-independent. Under most conditions increased APP_S formation induced with thapsigargin and cyclopiazonic acid was accompanied by decreased Aβ formation. However, in the presence of low concentrations of thapsigargin, increases in Aβ were observed. In HK_{293} cells calcium ionophore treatment dramatically stimulated Aβ formation without affecting APP_S levels.[16]

Evidence for a physiological role of calcium in the regulation of APP_S and Aβ formation was mentioned above. In cells transfected with the M3 receptor and treated for 24 h with phorbol esters to down-regulate protein kinase C, carbachol was able to stimulate APP_s formation and inhibit Aβ production. This suggests that the phospholipase C/calcium cascade is able to regulate APP processing in a protein kinase C-independent manner.[6] Similar results were observed with cells treated for 24 hr with phorbol esters to down-regulate protein kinase C and subsequently treated with ATP.

It is interesting to consider the analogies between pro-TGFa and APP: both proteins can undergo secretory processing which is stimulated by protein kinase C or calcium[17] and for neither protein is phosphorylation of the holoprotein required to mediate the effects of the stimulatory agent(s).

REFERENCES

1. WEIDEMANN A., G. KONIG, D. BUNKE, P. FISCHER, J. M. SALBAUM, C. L. MASTERS & K. BEYREUTHER. 1989. Identification, biogenesis, and localization of precursors of Alzheimer's disease A4 amyloid protein. Cell **57:**115–126.
2. BUXBAUM J. D., S. E. GANDY, P. CICCHETTI, M. E. EHRLICH, A. J. CZERNIK, R. P. FRACASSO, T. V. RAMABHADRAN, A. J. UNTERBECK & P. GREENGARD. 1990. Processing of Alzheimer β/A4 amyloid precursor protein: modulation by agents that regulate protein phosphorylation. Proc. Natl. Acad. Sci. USA **87:** 6003–6006.
3. CAPORASO, G. L., S. E. GANDY, J. D. BUXBAUM, T. V. RAMABHADRAN & P. GREENGARD. 1992. Protein phosphorylation regulates secretion of Alzheimer β/A4 amyloid precursor protein. Proc. Natl. Acad. Sci. USA **89:**3055–3059.
4. Secretory processing of the Alzheimer amyloid β/A4 protein precursor is increased by protein phosphorylation. Biochem. Biophys. Res. Commun. **187:** 1285–1290.
5. BUXBAUM, J. D., M. OISHI, H. I. CHEN, R. PINKAS-KRAMARSKI, E. A. JAFFE, S. E. GANDY & P. GREENGARD. 1992. Cholinergic agonists and interleukin 1 regulate processing and secretion of the Alzheimer β/A4 amyloid protein precursor. Proc. Natl. Acad. Sci. USA **89:**10075–10078.
6. BUXBAUM, J. D., A. A. RUEFLI, C. A. PARKER, A. M. CYPESS & P. GREENGARD.

1994. Calcium regulates processing of the Alzheimer amyloid protein precursor in a protein kinase C-independent manner. Proc. Natl. Acad. Sci. USA **91:** 4489–4493.

7. HAASS, C., A. T. HUNG, M. G. SCHLOSSMACHER, D. B. TEPLOW & D. J. SELKOE. 1993. β-Amyloid peptide and a 3-kDa fragment are derived by distinct cellular mechanisms. J. Biol. Chem. **268:**3021–3024.
8. BUXBAUM, J. D., E. H. KOO & P. GREENGARD. 1993. Protein phosphorylation inhibits production of Alzheimer amyloid β/A4 peptide. Proc. Natl. Acad. Sci. USA **90:**9195–9198.
9. HUNG, A. Y., C. HAASS, R. M. NITSCH, W. Q. QIU, M. CITRON, R. J. WURTMAN, J. H. GROWDON & D. J. SELKOE. 1993. Activation of protein kinase C inhibits cellular production of the amyloid β-protein. J. Biol. Chem. **268:** 22959–22962.
10. GABUZDA, D., J. BUSCIGLIO & B. A. YANKNER. 1993. Inhibition of β-amyloid production by activation of protein kinase C. J. Neurochem. **61:**2326–2329.
11. NITSCH, R. M., B. E. SLACK, R. J. WURTMAN & J. H. GROWDON. 1992. Release of Alzheimer amyloid precursor derivatives stimulated by activation of muscarinic acetylcholine receptors. Science **258:**304–307.
12. GRIFFIN, W. S., L. C. STANLEY, C. LING, L. WHITE, V. MACLEOD, L. J. PERROT, C. L. WHITE & C. ARAOZ. 1989. Brain interleukin 1 and S-100 immunoreactivity are elevated in Down syndrome and Alzheimer disease. Proc. Natl. Acad. Sci. USA **86:**7611–6715.
13. CACABELOS, R., M. BARQUERO, P. GARCIA, X. A. ALVAREZ & E. VARELA DE SEIJAS. 1991. Cerebrospinal fluid interleukin-1 β (IL-1 β) in Alzheimer's disease and neurological disorders. Methods Find. Exp. Clin. Pharmacol. **13:**455–458.
14. EMMERLING, M. R., C. J. MOORE, P. D. DOYLE, R. T. CARROLL & R. E. DAVIS. 1993. Phospholipase A_2 activation influences the processing and secretion of the amyloid precursor protein. Biochem. Biophys. Res. Commun. **197:** 292–297.
15. LOFFLER J. & G. HUBER. 1993. Modulation of β-amyloid precursor protein secretion in differentiated and nondifferentiated cells. Biochem. Biophys. Res. Commun. **195:**97–103.
16. QUERFURTH, H. W. & D. J. SELKOE. 1994. Calcium ionophore increases amyloid β peptide production by cultured cells. Biochemistry **33:**4550–4561.
17. PANDIELLA, A. & J. MASSAGUE. 1991. Multiple signals activate cleavage of the membrane transforming growth factor-α precursor. J. Biol. Chem. **266:** 5769–5773.

Defective Protein Kinase C α Leads to Impaired Secretion of Soluble β-Amyloid Precursor Protein from Alzheimer's Disease Fibroblasts[a]

S. GOVONI,[b] M. RACCHI,[c,e,g] S. BERGAMASCHI,[c]
M. TRABUCCHI,[f] F. BATTAINI,[f] A. BIANCHETTI,[d]
AND G. BINETTI[d]

[b] *Institute of Pharmacology, University of Pavia, Pavia, Italy*
[c] *Institute of Pharmacological Sciences, University of Milano, Milano, Italy*
[d] *Alzheimer's Disease Research Unit and* [e] *Laboratory of Molecular and Cellular Neurobiology, Ospedale "S. Cuore di Gesu"' FBF Brescia, Italy*
[f] *Department of Experimental Medicine and Biochemical Sciences, University of Roma "Tor Vergata", Italy*

ABSTRACT: The present study shows that cultured fibroblasts from sporadic AD patients present: a) reduced (−30%) cytosolic protein kinase C (PKC) activity; b) increased K_D of phorbol ester binding (+94%) in cytosolic fractions; c) reduced (−30%) soluble protein kinase Cα immunoreactivity; d) lower (−27.5%) basal soluble APP secretion and e) reduced soluble APP secretion in response to low phorbol ester concentrations (over threefold difference using 9 nM phorbol-12,13-dibutyrate-PdBu). Since the PKC-stimulated secretion of APP leads to the cleavage of the precursor within the amyloidogenic β-A4 fragment, the reduced PKC activity in AD patients may lead to accumulation of potentially amyloidogenic or toxic APP fragments. A defect in the secretion of soluble amyloid β-protein precursor is indeed suggested by literature data on familial AD fibroblasts as well as by the reported results.

INTRODUCTION

Much research supports speculation that Alzheimer's disease is not limited to the brain, although the expression is prominent in neuronal tissue. In

[a] This research was partially supported by grants from Regione Lombardia and from the Ministry of Health.

[g] *Send correspondence to:* Dr. M. Racchi, Institute of Pharmacological Sciences, University of Milano, Via Balzaretti 9, 20133 Milano, Italy; Tel: + +39-2-20488333; Fax: + +39-2-29404961; Email: Racchi@isfunix.farma.unimi.it.

particular, abnormalities occur in a number of accessible non-neuronal tissues including skin fibroblasts.[1] A number of metabolic/biochemical processes have been found abnormal in cultured skin fibroblasts derived from AD patients including alterations in calcium content, potassium channels, mitochondrial functions, choline transport, adhesiveness, expression of Alzheimer antigens, and protein kinase activity.[2–5] Although the prominent clinical symptoms in AD relate to brain damage, abnormalities in clinically unaffected tissue may lead to improved understanding of the cellular pathophysiology as well as to useful diagnostic tests.

A variety of observations have pointed out the involvement of protein kinase C (PKC) in AD. The enzyme levels (measured as phorbol ester binding) and activity (assessed as phosphorylation of exogenous (histone) and endogenous (protein P86, possibly MARCKS) substrates are decreased in frontal cortex of AD patients.[6] The alteration of PKC dependent phosphorylation could represent a biochemical defect linked to the mechanism involved in the development of the degenerative pathology. Indeed this view is sustained by numerous *in vitro* studies that demonstrate a direct involvement of PKC in the regulation of APP metabolism. PKC increases the rate of secretion of soluble APP following cleavage of membrane bound APP within the sequence of βA4[7–9] thus preventing the formation of amyloidogenic peptides.[10,11]

Alterations concerning PKC have also been described in peripheral tissues such as skin fibroblasts from AD patients suggesting that alterations in this kinase system at central level are not merely the consequence of neuronal degeneration and death, but rather represent causative events. The observation of a PKC deficit is indeed one of the most reproducible biochemical defects associated with AD in peripheral tissues.[2,4,5]

The similar changes of PKC in fibroblasts and in the brain of AD patients underscore the possibility that the study of this family of enzymes may represent a useful approach to investigate the cellular pathophysiology of AD. Within this context we have taken the effort to build a fibroblast cell line bank with the purpose to study the molecular defects connected to PKC and AD.

PATIENTS AND METHODS

All subjects included in this study were evaluated in the Alzheimer's Disease Unit of the S. Heart Hospital, Brescia. The protocol of the study was approved by the Ethical Committee of the Hospital. Written informed consent was obtained from all subjects or, where appropriate, the caregivers. Biopsies from non-AD, age matched, controls and AD patients were obtained. The diagnosis of probable AD was made by a senior neurologist according to the criteria developed by National Institute of Neurological and Communicative Disorders and Stroke (NINCDS) and the Alzheimer's Disease and Related Disorders Association (ADRDA). All AD patients presented a 2–7 years history of progressive cognitive impairment predominantly affecting memory. Cognitive status was quantified using the Mini Mental State Examination (MMSE). The scores were as follows: AD, 8.8 ± 7.0; non-AD, 19.5 ± 10.8. Seven of the control subjects were diagnosed as having, respectively, cortico-basal degeneration, late onset schizophrenia, normotensive hydrocephalus, and delirium in a patient with breast cancer. The mean duration of illness in non-AD patients with 46.0 ± 55.3 months.

The methodology for obtaining the explants of shoulder skin (epidermis plus dermis), establishment of fibroblast cell lines, growth and storage conditions of the fibroblasts has been previously described in detail.[4]

Preparation of cytosolic and membrane fractions, measurement of PdBu binding and PKC histone-directed activity, evaluation of PKC isoform by Western blotting have been performed as described in detail.[4,12]

Proteins precipitated from the conditioned media were gel electrophoresed transferred by Western blotting and immunodetected using GID antiserum (kindly provided by Dr. Tsunao Saitoh) at a 1:2000 dilution. To detect immunoreactive bands the blots were subsequently incubated with goat anti-rabbit IgG-alkaline phosphatase conjugate. After incubation with the secondary antibody the blots were developed with substrates for alkaline phosphatase.

For quantification of the immunoreactive bands on western blots, the relative abundance was calculated from the area of the peak corresponding to the selected band following acquisition of the blot image through a Nikon CCD video camera module and analysis by means of the Image 1.44 program using the subroutine for gel plot analysis.

Data of AD and non-AD groups were compared using the one way analysis of variance followed, when significant, by two-tailed Student's t-test: a p value <0.05 was considered significant. The values reported in the text are means ± SD.

RESULTS AND DISCUSSION

TABLE 1 summarizes the observed differences in PKC activity and immunoreactivity in fibroblasts from AD patients as compared to age matched controls. Phosphorylating activity in the soluble fraction was reduced when using histone as substrate and an *in vitro* condition allowing full stimulation of the enzyme by addition of all the co-factors and calcium. The lower phosphorylating activity appeared to be due to a reduced (−27%) cytosolic content of PKCα, the only calcium-dependent PKC isoform we detected in

TABLE 1. Cytosolic Protein Kinase C Changes in Fibroblasts of AD Patients

Parameter	Unit	Non-AD	AD	% Change
PKC activity	pmol/min/mg prot.	442 ± 87	309 ± 73[a]	−30
Phorbol ester binding (Kd)	nM	3.26 ± 0.52	6.33 ± 2.82[a]	+94
PKC α immunoreactivity	Arbitrary units	1.0 ± 0.26	0.7 ± 0.1[a]	−30

[a] $p < 0.05$ in comparison with non-AD.

human fibroblasts.[12] The significant reduction of PKCα in the cytosol of AD fibroblasts may also explain the finding of a significant reduction of affinity for PdBu in such subcellular fraction. In fact the affinities for phorbol esters binding are different for the different isoforms; in particular the calcium-dependent isoforms show a binding affinity that is on average fourfold higher than the calcium independent PKCs.[13] The reduced levels of PKCα in AD may allow an increased contribution to the binding kinetic by PKCδ (the other prelavent cytosolic PKC isoform).

The impaired PKC activities appear to be associated with an altered ability of AD fibroblasts to release soluble APP in the medium both in basal conditions and following stimulation with low concentrations of PdBu. As shown in FIGURE 1A the soluble APP released into the media of AD cells was 27.5% less than non-AD fibroblasts indicating a constitutive deficiency in APP pro-

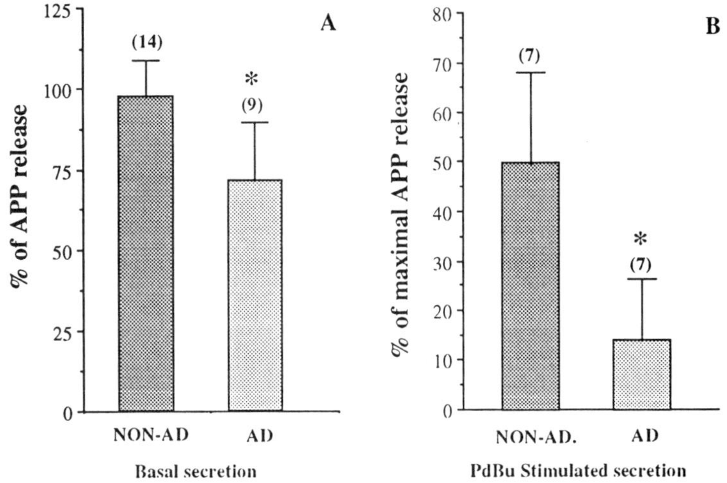

FIGURE 1. Basal and stimulated APP secretion from fibroblasts of AD and non-AD patients. **A:** Basal secretion of APP was measured in AD and control fibroblasts as described in the METHODS section. **B:** Release of APP from AD and control fibroblasts after treatment with 9 nM phorbol-12,13-dibutyrate. Proteins secreted in the conditioned media in 2 hours were analyzed as described in the METHODS section. The number in parentheses indicates the number of cell lines used for the experiment. * $p < 0.05$.

cessing. Moreover when non-AD fibroblasts were exposed to a low concentration (9nM) of PdBu secreted APP at 50% of their maximal release, while in these conditions the AD fibroblasts reached only 14% of the maximal release (FIG. 1B). High concentrations of phorbol esters (typically between 75–300 nM, as frequently used in the literature) induced equivalent maximal release of APP from both groups of cells (not shown). The concentrations of PdBu needed to elicit half maximal stimulation of APP release in AD fibroblasts was almost twice that needed in control fibroblasts a ratio which is consistent with the fact that the affinity for the ligand in AD cells is reduced by an average of twofold (TABLE 1[4]).

These data suggest that a subtle defect in PKC regulated processing of APP may, over time, become of pathogenetic significance in AD assuming that the same defect could be present in the brain of AD patients. Moreover the consistency of the data obtained from cultured skin fibroblasts underscores the importance of this model in the study of the pathogenesis of AD.

REFERENCES

1. BLASS J. P., A. C. BAKER, R. K. F. SHEU, R. S. BLACH & A. SMITH. 1989. Use of cultured skin fibroblasts in studies of Alzheimer's disease. *In* Boller F et al eds. Biological markers of Alzheimer's Disease. F. Boller *et al.,* Eds. New York: Springer Verlag. 153–162.
2. BRUEL, A., G. CHERQUI, J. L. COLUMELLI, D. MARGELIN, M. ROUDIER, P. M. SINET, M. PRIEUR, J.-L. PERIGNON & J. DELABAR. 1991. Reduced protein kinase C activity in sporadic Alzheimer's disease fibroblasts. Neurosci. Lett. **133:**89–92.
3. ETCHEBERRIGARAY, R., E. ITO, K. OKA, B. TOFEL-GREHL, G. E. GIBSON & D. L. ALKON. 1993. Distinguishing features of potassium channels from Alzheimer, aged and young donors. Proc. Natl. Acad. Sci. **90:**8209–8213.
4. GOVONI, S., S. BERGAMASCHI, M. RACCHI, F. BATTAINI, G. BINETTI, A. BIANCHETTI & M. TRABUCCHI. 1993. Cytosol protein kinase C down regulation in fibroblasts from Alzheimer's disease patients. Neurology **43:**2581–2586.
5. VAN HUYNH, T., G. COLE, R. KATZMAN, K. P. HUANG & T. SAITOH. 1989. Reduced PKC immunoreactivity and altered protein phosphorylation in Alzheimer's disease fibroblasts. Arch. Neurol. **46:**1195–1199.
6. COLE, G., K. R. DOBKINS, L. A. HANSEN, R. D. TERRY & T. SAITOH. 1988. Decreased levels of protein kinase C in Alzheimer brain. Brain Res. **452:** 165–174.
7. ESCH, F. S., P. S. KEIM, K. E. BEATTIE, R. W. BLACHER, A. R. CULWELL, T. OLTERSDORF, D. MCCLURE & P. J. WARD. 1990. Cleavage of amyloid β-peptide during constitutive processing of its precursor. Science **248:** 1122–1124.
8. SLACK, B. E., R. M. NITSCH, E. LIVNEH, G. M. KUNZ JR., J. BREU, A. ELDAR & R. J. WURTMAN. 1993. Regulation by phorbol ester of amyloid precursor

protein release from Swiss 3T3 fibroblasts overexpressing protein kinase Cα J. Biol. Chem. **268:**21097–21101.

9. SISODIA, S. S., E. H. KOO, K. BEYREUTHER, A. UNTERBECK & D. L. PRICE. 1990. Evidence that β-amyloid protein in Alzheimer's disease is not derived by normal processing. Science **248:**492–495.
10. BUXBAUM, J. D., E. H. KOO & P. GREENGARD. 1993. Protein phosphorylation inhibits production of Alzheimer amyloid β/A4 peptide. Proc. Natl. Acad. Sci. USA **90:**9195–9198.
11. HUNG, A. Y., C. HAAS, R. M., NITSCH, W. Q. QIU, M. CITRON, R. J. WURTMAN, J. H. GROWDON & D. SELKOE. 1993. Activation of PKC inhibits cellular production of the amyloid b-protein. J. Biol. Chem. **268:**22959–22962.
12. RACCHI, M., S. BERGAMASCHI, S. GOVONI, W. C. WETSEL, A. BIANCHETTI, G. BINETTI, F. BATTAINI & M. TRABUCCHI. 1994. Characterization and distribution of protein kinase C isoforms in human skin fibroblasts. Arch. Biochem. Biophys. **314:**107–111.
13. KAZANIETZ M. G., L. B. ARECES, A. BAHADOR, H. MISCHAK, J. GOODNIGHT, F. MUSHINSKI & P. M. BLUMBERG. 1993. Characterization of ligand and substrate specificity for the calcium-dependent and calcium-independent protein kinase C isozymes. Mol. Pharmacol. **44:**298–307.

Metabotropic Glutamate Receptors Regulate APP Processing in Hippocampal Neurons and Cortical Astrocytes Derived from Fetal Rats[a]

R. K. K. LEE,[b] J. JIMENEZ, A. J. COX, AND R. J. WURTMAN

Department of Brain and Cognitive Sciences, Massachusetts Institute of Technology, Cambridge, Massachusetts 02139 USA

ABSTRACT: It has previously been shown that stimulation of muscarinic m1 or m3 receptors can, by generating diacylglycerol (DAG) and activating protein kinase C (PKC), accelerate the breakdown of the amyloid precursor protein (APP) to form soluble, non-amyloidogenic peptides (APPs). This relationship has been demonstrated in human glioma and neuroblastoma cells as well as in transfected human embryonic kidney (HEK) cells and PC12 cells.[1,2]

We now provide evidence that stimulation of metabotropic glutamate receptors (mGluRs), which also are coupled to DAG and PKC, similarly accelerates processing of APP into non-amyloidogenic APPs in hippocampal neurons and cortical astrocytes derived from normal fetal rats. The mGluR antagonist, L(+)-2-amino-3-phosphonopropionic acid (L-AP3), and GF 109203X, an inhibitor of PKC, both blocked the release of APPs from hippocampal neurons and astrocytes evoked by glutamate receptor stimulation. Inasmuch as glutamatergic neurons in cortex and hippocampus are known to be damaged in Alzheimer's disease, our findings suggest that amyloid formation may be enhanced by the resulting glutamate deficiency and that selective mGluR agonists may be useful in facilitating synaptic efficacy and treating the disease.

INTRODUCTION

The amyloid protein found in senile plaques of patients with Alzheimer's disease (AD) is derived from proteolytic processing of a much larger amyloid precursor protein (APP). Like some growth factor precursors, APP is anchored to the cell membrane and is cleaved to form a soluble peptide, APPs, which is released into the extracellular space. Our laboratory previously showed that the neurotransmitter acetylcholine, acting via m1 or m3 muscar-

[a] This research was supported by the Center for Brain Sciences and Metabolism Trust Fund.

[b] *Send correspondence to:* Dr. Robert K. K. Lee, Department of Brain and Cognitive Sciences, E25-604, M.I.T., Cambridge, MA 02118 USA; Tel: (617) 253 6733; Fax: (617) 253 6882.

inic receptors coupled to phospholipases, DAG and PKC, accelerates this process.[1] Another consequence of the secretory cleavage of APP is disruption of its amyloidogenic βA4 domain.

Alternative pathways of APP metabolism leave the amyloidogenic region intact to form a 39-42 amino acid peptide, βA4, which can potentially aggregate into β-pleated sheets to form amyloid. The aberrant processing of APP via disruption of the secretory pathway could initiate a cascade of events that result in neurodegeneration and dementia. Biochemical mechanisms that prevent the formation of intact βA4 or promote the release of APPs could retard the progression of AD.

Our present studies were designed to determine if the modulation of APP breakdown by neurotransmitters can be demonstrated in neurons and astrocytes. We also asked whether glutamate could act through its metabotropic receptors, some of which also enhance DAG formation, to mimic acetylcholine's stimulation of APPs formation.

METHODS

Astrocytes cultures were prepared from neonatal rat cortex according to protocols similar to those of McCarthy and DeVellis.[3] Hippocampal neurons were obtained from 18/19 day rat embryos using methods described by Goslin and Banker.[4] In brief, rat cortices or hippocampi were dissected and treated with 0.25% trypsin dissolved in calcium- and magnesium-free Hank's solution. Cells were dissociated by trituration through a fire-polished/constricted Pasteur pipette and plated onto polylysine coated culture dishes (3–6 × 105 cells/cm2). Cells were used for biochemical studies 2–4 weeks after plating; by that time, astrocytes had reached confluence.

All pharmacological agents were used in the micromolar range and dissolved in serum-free minimal essential medium. Cells were incubated with the test medium for 1h at 37°C in a 5%CO_2/air incubator. Media were collected, kept at 4°C and applied to Sephadex G-25 columns. Eluates were frozen, dried by vacuum centrifugation and resuspended in an extraction buffer. For SDS PAGE, the volume of sample loaded was corrected for the amount of cell protein per dish. Proteins were transferred to membranes and APPs was detected using mAb22C11 (Boehringer-Mannheim). The protein bands were visualized by a chemiluminescence method (Amersham). Further details on these methods are described by Slack *et al.*[5] When appropriate, the cells were preincubated with L-AP3 or GF109203X for 15 min.

For PI turnover assays, cultured cells were prelabeled with 2μCi myo-[1,2-[^{3}H]inositol per ml of medium. During drug treatments, 10mM LiCl was added to the medium to prevent dephosphorylation of accumulated [^{3}H]-inositol phosphates. Stimulation was terminated by aspirating the medium and adding cold methanol to the cells. Homogenized samples were analyzed for inositol constituents by anion exchange chromatography as described by Berridge *et al.*[6]

RESULTS AND DISCUSSION

Activation of mGluRs by quisqualic acid (QA) or the selective agonist, trans-(1S,3R)-1-aminocyclopentanedicarboxylic acid (ACPD) increased APPs release from both astrocytes and hippocampal neurons. This effect was abolished by pretreatment with the mGluR antagonist L-AP3 (Fig. 1). Phorbol 12-myristate 13 acetate (PMA), which directly stimulates PKC, also increased APPs release in neurons and astrocytes. Inhibition of PKC by GF 109203X[7] blocked the stimulation of APPs release by PMA, QA and ACPD (Fig. 2). These data are consistent with the hypothesis that APPs release in mGluR stimulated cells can be mediated by PKC. Exposure to either QA or ACPD also increased the formation of [^{3}H]-inositol phosphates (Fig. 3). Although we did not measure changes in DAG formation the increases in [3H]-inositol phosphates suggest that the ability of QA and ACPD to stimulate APPs release was mediated by PI hydrolysis.

Stimulation of neuronal ionotropic NMDA receptors with *N*-methyl-D-aspartate did not increase APPs release, nor did the addition of s(-)-5-Fluorowillardiine which activates the AMPA/kainate channels (Fig. 4). These data indicate that APP processing is regulated by second messenger coupled glutamate receptors but is not significantly affected by ligand-gated glutamate receptors.

Although mGluRs have the general property of stimulating PI hydrolysis and mobilizing Ca^{++}, it is clear that ACPD acts on multiple glutamate recep-

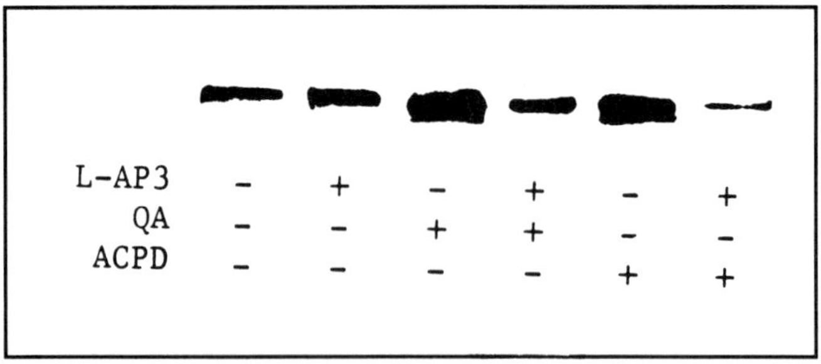

FIGURE 1. Western blot of APP derivatives secreted from astrocytes. mAb22C11 recognizes APPs, a band that corresponds to approximately 100kD. Both QA and ACPD increased APPs release into the medium. This increase was attenuated by the mGluR antagonist, L-AP3. Hippocampal neurons also showed similar responses (data not shown).

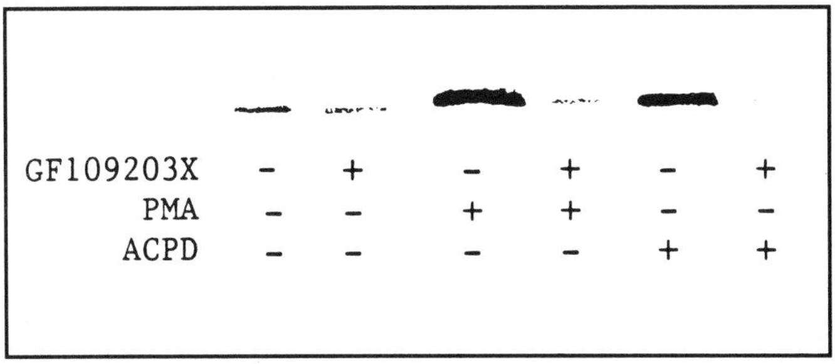

FIGURE 2. Western blot of proteins recognized by mAb22C11. PMA and ACPD stimulated APPs release in hippocampal neurons. This effect was blocked by the PKC inhibitor, GF109203X. Cortical astrocytes also responded in the same manner (data not shown).

tors besides those coupled to PI hydrolysis and initiates multiple signal transduction events.[8,9] Pharmacological and molecular studies have shown that stimulation of mGluRs also induces arachidonic acid release and cAMP formation. To examine the possible roles of cAMP in glutamate-stimulated APP processing we exposed PMA- and ACPD-treated neurons and astrocytes to forskolin or dibutyryl cAMP and found consistent suppression of APPs release in both cell types. Since cAMP does not interfere with PI hydrolysis in hippocampal neurons, it may act downstream of PI turnover or via an independent signal transduction pathway to inhibit APPs release.

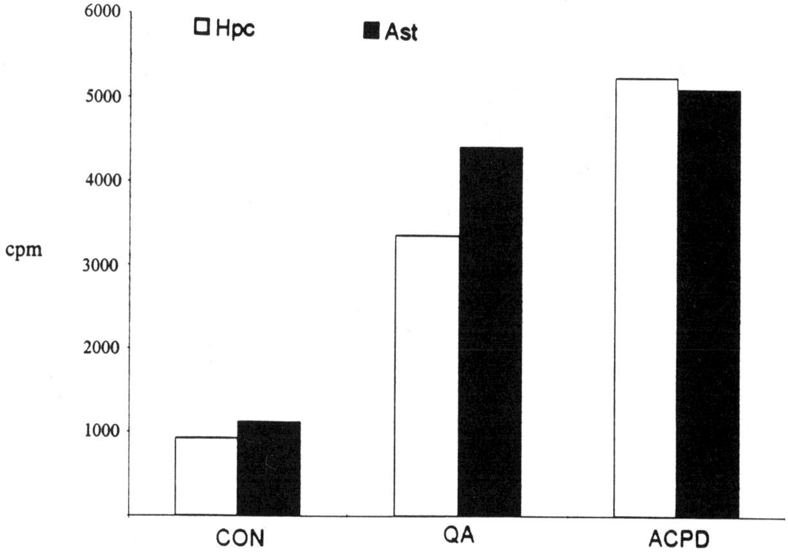

FIGURE 3. PI hydrolysis in hippocampal neurons and astrocytes stimulated by QA and ACPD.

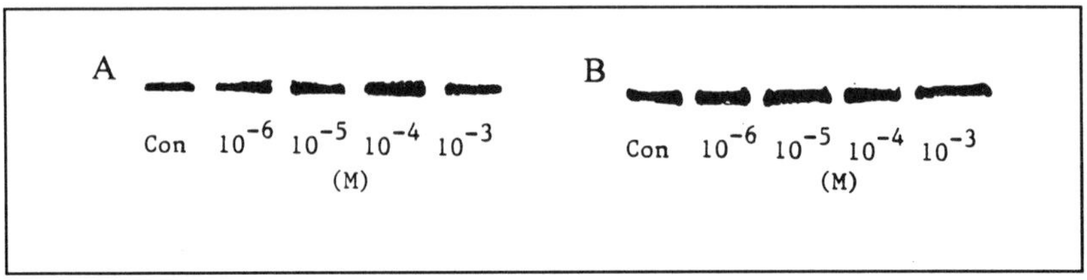

FIGURE 4. Detection of APPs with mAb22C11 after stimulation of (A) astrocytes with the AMPA/kainate agonist, s(-)-5-Fluorowillardiine and (B) hippocampal neuron with *N*-methyl-D-aspartate. None of these ionotropic receptor agonists significantly increased APPs release above baseline levels (Con).

CONCLUSION

The destruction of cortical and hippocampal neurons in AD is associated with a drastic loss of glutamatergic synapses.[10] Such brain lesions not only affect neuronal signaling but also elevate levels of APP in reactive astrocytes.[11] Thus, we sought to determine the extent to which proteolytic processing of APP in neurons and astocytes is disrupted by changes in glutamate signalling.

The data presented here provide the first indication that alterations in glutamate neurotransmission can regulate cellular APP processing, as well as the first evidence that neorotransmitters can modify APPs production in normal brain cells. Both cell types express APP *in vitro* and respond to mGluRs stimulation by upregulating APP metabolism to favor APPs release. It is likely that the widespread disruption in glutamate-mediated excitatory transmission in AD may alter constitutive APP metabolism in ways that promote the formation of amyloidogenic fragments. Our studies suggest that the mGluRs, which modulate synaptic plasticity and diverse signal transduction events, are a viable target for therapeutic strategies aimed at suppressing amyloid formation.

REFERENCES

1. Nitsch, R. M., B. E. Slack, R. J. Wurtman & J. H. Growdon. 1992. Release of Alzheimer amyloid precursor derivatives stimulated by activation of muscarinic acetylcholine receptors. Science **258:**304–307.
2. Buxbaum, J. D., M. Oishi, H. I. Chen, R. Pinkas-Kramarshi, E. A. Jaffe, S. E. Gandy & P. Greengard. 1992. Cholinergic agonists and interleukin 1 regulate processing and secretion of the Alzheimer β/A4 amyloid protein precursor. Proc. Natl. Acad. Sci. USA **89:**10075–10078.
3. McCarthy, K. D. & J. DeVillis. 1980. Preparation of separate astroglial and oligodendroglial cultures from rat cerebral tissue. J. Cell Biol. **85:**890–902.
4. Goslin, K. & G. Banker. 1991. Rat hippocampal neurons in low-density

culture. *In* Culturing Nerve Cells. G. Banker & K. Goslin, Eds.:251–281. Cambridge, MA: MIT.

5. SLACK, B. E., R. M. NITSCH, E. LIVNEH, G. M. KUNZ JR., J. BREU, H. ELDAR & R. J. WURTMAN. 1993. Regulation by phorbol esters of amyloid precursor protein release from Swiss 3T3 fibroblasts overexpressing protein kinase Ca. J. Biol. Chem. **268:**21097–21101.
6. BERRIDGE, M. J., C. P. DOWNS & M. R. HANLEY. 1982. Lithium amplifies agonist-dependent phosphatidylinositiol responses in brain and salivary brands. Biochem. J. **206:**587–595.
7. TOULLEC, D., P. PIANETTI, H. COSTE, P. BELELVERGUE, T. GRAND-PERRET, M. AJAKANE, V. BAUDET, P. BOISSIN, E. BOURSIER, F. LORIOLLE, L. DUHAMEL, D. CHARON & J. KIRILOVSKY. 1991. The bisindolylmaleimide GF109203X is a potent and selective inhibitor of protein kinase C. J. Biol. Chem. **266:**15771–157817.
8. NAKANISHI, S. 1992. Molecular diversity of glutamate receptors and implications for brain function. Science **258:**597–603.
9. SCHOEPP, D. D. & P. J. CONN. 1993. Metabotropic glutamate receptors in brain function and pathology. Trends Pharmacol. Sci. **14:**13–20.
10. FRANCIS, P. T., N. R. SIMS, A. W. PROCTER & D. M. BOWEN. 1993. Cortical pyramidal neurone loss may cause glutamatergic hypoactivity and cognitive impairment in Alzheimer's disease: Investigative and therapeutic perspectives. J. Neurochem. **60:**1589–1604.
11. SIMAN, R., J. P. CARD, R. B. NELSON & L. G. DAVIS. 1989. Expression of β-amyloid precursor protein in reactive astrocytes following neuronal damage. Neuron **3:**275–285.

Development and Characterization of a Monoclonal Antibody 369.2B Specific for the Carboxyl-Terminus of the βA4 Peptide

G. KÖNIG,[a] P. GRAHAM, A. BUSHNELL, S. WEBSTER, D. WUNDERLICH, AND L. S. PERLMUTTER

Institute for Dementia Research, Bayer Corp., Pharmaceutical Division, West Haven, Connecticut USA

ABSTRACT: The β-amyloid deposits in Alzheimer's diseased (AD) brain are made up mainly of βA4 peptides which show both N- and C-terminal heterogeneity. The predominant C-terminal species, identified by peptide sequencing of purified β-amyloid, end either at position 40 or 42 of the βA4 peptide. The distribution of these two βA4 species in postmortem tissue as well as their generation *in vitro* could not be addressed previously due to the lack of specific antibodies that could differentiate them. This report describes the generation of a highly specific monoclonal antibody, MAb 369.2B which was raised against a synthetic peptide consisting of the C-terminus of the 1–42 βA4 species. MAb 369.2B does not recognize the shorter βA4 species 1–40 in solution or in solid phase. Furthermore, both βA4 1–40 and 1–43 were unable to absorb out the antibody when used immunocytochemically. The regional distribution of MAb 369.2B immunoreactivity in AD and non-AD brain samples were compared to MAb 286.8A, an antibody raised against the N-terminal region of βA4. Overall, the staining patterns were very similar. In AD cases with extreme vascular involvement, the N-terminal (MAb 286.8A) antibody was more likely to label the vascular basement membrane of affected vessels, and to label them more uniformly. In addition, preliminary quantitative analyses revealed that the C-terminal antibody labeled fewer, larger plaques than the N-terminal antibody. Qualitative analyses revealed that these smaller plaques were typically subregions of the larger plaques. The data corroborate the biochemical findings of N-terminal raggedness in βA4 plaques. Further studies are required to explain this differential pattern of deposition.

INTRODUCTION

βA4, the 4 kD fragment of the Alzheimer amyloid protein precursor (APP), has recently been shown to be released from primary cell culture and

[a] *Send correspondence to:* Dr. Gerhard König, Institute for Dementia Research/A21, Bayer Corp., Pharmaceutical Division, 400 Morgan Lane, West Haven, CT 06516 USA; Tel: (203) 937-2163; Fax: (203) 937-2075; Email: koenig@wh.bayer.com.

various cell lines.[1,2] Additionally, healthy controls as well as Alzheimer patients contain low nanomolar amounts of βA4 in sera and CSF.[3,4] The hypothesis that the simple release of βA4 was sufficient and therefore responsible for the buildup of β-amyloid plaques in AD brains was therefore no longer tenable. While multiple factors may contribute to the formation of amyloid plaques, the relative amounts of the various subfragments of βA4 may be important.

Certain families with hereditary AD have specific point mutations in and around the βA4 region of the APP gene. This unequivocally demonstrates that the APP gene can be a "disease gene."[5–8] In these families point mutations in the APP gene are sufficient to cause the age of onset at a specific age. Although the vast majority of AD cases are sporadic and probably multifactorial, these familial APP mutations teach us a great deal about amyloidogenesis.

The double mutation at APP codon 670/671 (the "Swedish variant," at the N-terminus of βA4 in APP) causes a 5- to 8-fold higher release of βA4 in cell cultures stably transfected with that mutated APP cDNA.[1,2] Another FAD mutation ("London variant"), which lies C-terminal to βA4 at position 717, decreases the 1-40/1-42 βA4 ratio in AD brain as well as *in vitro*.[9] Although it is unclear how the generation of the transmembrane C-terminus of βA4 occurs the "London mutation" may affect the proteolytic breakdown of APP to βA4. These findings suggest that either increased amounts of βA4, or of a specific form of this peptide, may explain the faster kinetics of βA4 aggregation to β-amyloid plaques in these families.

The present report describes the generation of the monoclonal antibody (MAb) 369.2B, a highly specific antibody which recognizes the C-terminus of βA4 species ending in position 42. The immunocytochemical localization of this epitope in AD and non-AD brain samples was compared to that recognized by MAb 286.8A, an antibody raised against the N-terminal region of βA4. Overall, the staining patterns were very similar. In AD cases with extreme vascular involvement, the N-terminal (MAb 286.8A) antibody was more likely to label the vascular basement membrane of affected vessels and to label them more uniformly. In addition, preliminary quantitative analyses revealed that the C-terminal antibody labeled fewer, larger plaques than the N-terminal antibody. Qualitative analyses revealed that the smaller plaques were typically subregions of the larger plaques. These data corroborate the biochemical findings of N-terminal raggedness in βA4 plaques, and emphasize the difference between these two distinct βA4 species.

MATERIALS AND METHODS

Generation of Monoclonal Antibodies

Balb/c mice were immunized with multiple I.P. inoculations of KLH conjugated peptide #959, a 14 residue synthetic peptide having an N-terminal cysteine attached to a hydrophilic DGDGD spacer and residues 35–42 of the human βA4 (CDGDGDMVGGVVIA). Splenocytes from immunized animals were fused with the mouse myeloma AG8 as described.[10] Supernatants from resultant hybridomas were screened for immunoreactivity to ovalbumin-coupled peptide #958, a 14 residue synthetic peptide having an N-terminal cysteine attached to a GGGGG spacer and residues 35–42 of the human βA4 (CGGGGGMVGGVVIA), using EIA. Hybridomas positive for the expression of immunoreactive MAbs were cloned at least twice by limiting dilution. MAb isotype analysis was performed as described.[10] Purified MAb IgG was prepared from ascites fluid using protein-A affinity chromatography.

Immunoprecipitation/Scintillation Assay (IPSA) for Hybridoma Screening

Approximately 1.5×10^5 DPMs of ^{35}S-methionine-labeled *in vitro*-translated (IVT) βA4 1–42 (Wheat germ translation system, Promega) were added to 10 μl of a 10× immunoprecipitation buffer (150 mM NaCl, 10% NP-40, 5% deoxycholic acid, 1% SDS, 500 mM Tris pH8). Monoclonal cell supernatant (90 μl) from monoclonal fusion 369.2B was added and allowed to react for 2 h at 4°C. Background was determined using 90 μl of a non-reactive clone; the positive control was 90 μl of 286.8A which was made previously against a crude synthetic βA4 (1–42) preparation. After 2 hours, 40 μl of a 10% solution of Omnisorb cells (Calbiochem) equilibrated in 1× immunoprecipitation buffer (RIPA buffer; 150 mM NaCl, 1% NP-40, 0.5% deoxycholic acid, 0.1% SDS, 50 mM Tris pH8) was added and allowed to react for an additional 2 h at 4°C with rocking. The cells were pelleted by centrifugation for 5 min at 4500 × g and 4°C, and washed 3 × with 800 μl cold 1× immunoprecipitation buffer. Pellets were quantitatively transferred to scintillation vials and counted in a Beckman LS6000 scintillation counter in the Auto DPM mode. The percentage of βA4 immunoprecipitated was calculated.

IPSA for Monoclonal Characterization

Approximately 1.5×10^5 DPMs of ^{35}S-methionine-labeled IVT βA4 was added to various amounts of purified MAb 369.2B or 286.8A, in a total volume of 100 μl 1× immunoprecipitation buffer and allowed to react as described above. Immune complexes were precipitated with Omnisorb, washed, and counted as described above.

TABLE 1. Antibody Specifity

Competing Peptide	Staining with N-terminal MAb (286.8A)	Staining with C-terminal (369.2B)
None/buffer	+ + +	+ + +
None/DMSO	+ + +	+ + +
Human 40-1	+ + +	+ + +
Human 1-16	−	+ + +
Mouse 1-16	+ + +	+ + +
Human 1-42	−	−
Human 1-43	−	+ + +
Human 35-42 with linker	+ + +	−

Competition IPSA

Competition IPSAs were performed exactly as above using 1 μg of purified 369.2B in a total volume of 100 μl 1 × immunoprecipitation buffer to which 5 μg of competing peptide were added. Incubations and precipitations were as described above.

Immunocytochemical Localization in Human Brain

Brain tissue was obtained postmortem from 8 AD (age = 73.25 years ± 9.54) and 8 non-AD cases (age = 70.25 years ± 13.16). Neuropathological diagnoses of AD were according to CERAD criteria.[16] Non-AD cases had diagnoses of progressive supranuclear palsy (3), Parkinson's disease (1), Pick's disease (1), alcoholic cirrhosis (1), emphysema (1), and multi-infarct dementia (1). Leptomeningeal vessels were not examined. Samples from Brodmann's area 22 (superior temporal gyrus), a region severely affected by AD lesions, were fixed in 10% neutral buffered formalin, embedded in paraffin and sectioned (10 μm). They were next deparaffinized, treated with hydrogen peroxide, immersed in 88% formic acid for 30 min and incubated with a blocking solution of 10% normal goat serum. MAb 369.2B was used at a concentration of 1:10,000 and MAb 286.8A at 1:8,000 overnight at 4°C. The avidin-biotin-peroxidase kit (Vector), with diaminobenzidine as chromagen, was used. In an additional series of studies, the specificity of the antibodies was determined by preincubating them with a 1000-fold molar excess of a variety of βA4 subfragments (see TABLE 1).

RESULTS

*Identification and Characterization of **βA4 35–42-Specific Monoclonal Antibodies***

The βA4 35–42-specific MAb was identified and characterized using three criteria. IVT βA4 1–42 was used to assess the ability of the MAb 369.2B

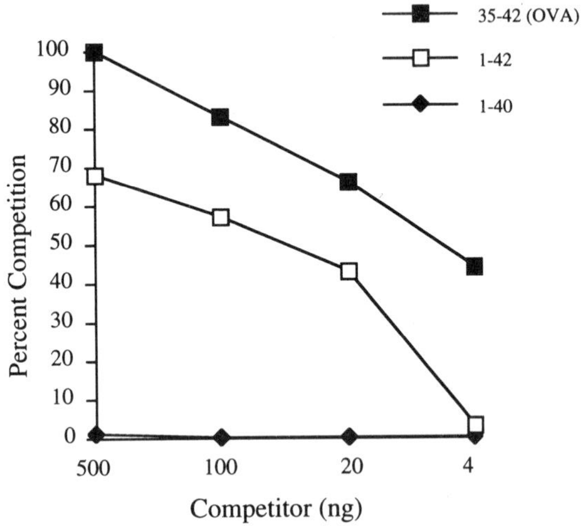

FIGURE 1. Epitope mapping of MAb 369.2 by competitive EIA. C369.2 (50 ng IgG/100 μl) was preincubated with or without synthetic competitor peptides (22°C, 1h) then subjected to EIA against plate-bound Ovalbumin-coupled 35–42 (200 ng/well). Percent competition was calculated relative to MAb binding in the absence of competitor, *i.e.*,

$$\% \text{ competitor} = \frac{(\text{signal w/o competitor}) - [(\text{Signal w competitor}) - \text{Background}]}{(\text{signal w/o competitor})} \times 100$$

to immunoprecipitate βA4. MAb 369.2B (1 μg) immunoprecipitated 19% of total soluble IVT βA4 in an IPSA. Solid-phase EIA demonstrated that MAb 369.2B recognized the βA4 35–42 epitope (peptide #958), as well as βA4 1–42. This signal was competed out by peptides βA4 35–42, βA4 1–42, but not by βA4 1–40 (FIG. 1), demonstrating the specificity of MAb 369.2B. Furthermore, a competition IPSA was performed. MAb 369.2B was added to an approximate 200-fold molar excess of unlabeled competitor peptide at the same time as labeled IVT βA4 1–42. As expected, βA4 1–40, 1–11, 1–28, 12–28, as well as the reverse peptide 40–1 did not compete with the IVT βA4 for immunoprecipitation, whereas the complete 1–42 peptide did (FIG. 2). From these data we conclude that MAb 369.2B is specific for the C-terminal end of the βA4 1–42 peptide.

C-terminal versus N-terminal βA4 Immunoreactivity in AD and non-AD Brain

The immunocytochemical localization of the C-terminal antibody MAb 369.2B was compared with that of MAb 286.8A, which was previously shown to be specific for the N-terminal region of βA4 (epitope βA4 3–8). The specificity of these antibodies, as determined by *in vitro* competition studies, was confirmed in human brain samples using a series of peptide absorptions (TABLE 1). As expected, N-terminal antibody (MAb 286.8A) staining was competed out by human βA4 1–16, 1–40, 1–42, and 1–43, while C-terminal antibody (MAb 369.2B) staining was competed out by human βA4 35–42 and 1–42 (FIG. 3). No competition was observed for the N-terminal antibody absorbed with mouse βA4 1–16, for the C-terminal antibody absorbed with human βA4 1–40 and βA4 1–43.

Brodmann's area 22, a region severely affected in AD, was used in a serial section analysis to compare the immunocytochemical localization of these MAbs. In AD brain, both antibodies labeled numerous plaques of all morphological types, including cored 'doughnuts,' and all sizes of circular dense and amorphous accumulations. Preliminary quantitative analyses established a trend for the C-terminal antibody (MAb 369.2B) to label fewer, larger plaques than the N-terminal antibody (MAb 286.8A). While qualitative analyses revealed that both antibodies tended to label the same plaque population (FIG. 4), the N-terminal antibody frequently labeled multiple, smaller distinct plaques which were identified as one large plaque by the C-terminal antibody. In AD cases with extreme vascular involvement, the N-terminal (MAb 286.8A) antibody was more likely to label the vascular basement membrane of affected vessels, and to label them more uniformly (FIG. 5) No labeling of neuronal cytoplasm, neurites, intracellular tangles or extracellular tangles was seen with either antibody in Brodmann's area 22. In contrast, scattered tangles were labeled in the hippocampal formation by MAB 369.2B and MAb286.8A. This observation was made independent of formic acid pretreatment. In 7/8 non-AD cases, no immunostaining was observed. Only one non-AD case, diagnosed with progressive supranuclear palsy had amorphous accumulations labeled with both antibodies scattered throughout the grey matter of Brodmann's area 22. Serial section analysis demonstrated that these deposits were not thioflavine S-positive, suggesting that they were diffuse plaques.

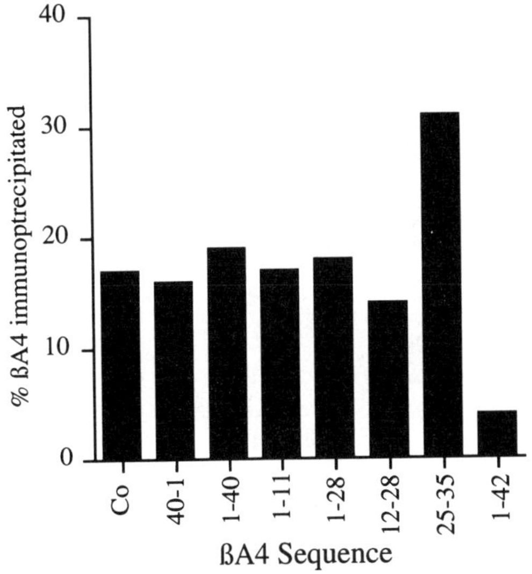

FIGURE 2. Competition IPSA for epitope determination of MAb 369.2. Peptide competitor (5 μg) was mixed with IVT-βA4 ($\sim 1.5 \times 10^5$ DPMs or $\sim$200 pg) then immunoprecipitated with 2 μg 369.2. Percents with a given MAb concentration varied typically only a few percentage points between and within experiments.

Percent βA4 immunoprecipitated

$$= \left[\frac{\text{(dpms with MAb)} - \text{(dpms with non or preimmune control)}}{\text{(total } \beta\text{A4 dpms/reaction)}} \right]$$

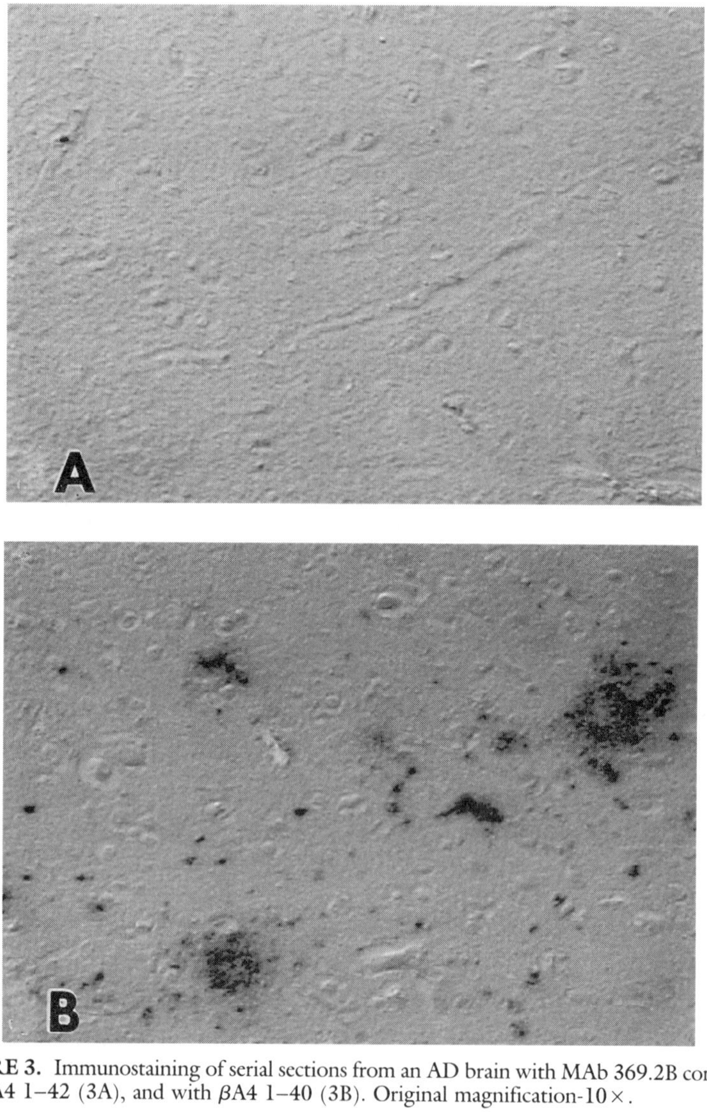

FIGURE 3. Immunostaining of serial sections from an AD brain with MAb 369.2B competed with βA4 1–42 (3A), and with βA4 1–40 (3B). Original magnification-10×.

DISCUSSION

Positions 29 through 42 of the βA4 peptide lie entirely within the putative transmembrane region of APP and are hydrophobic in nature.[11] Synthetic peptides of the C-terminal sequences in this region must overcome the ability of the 34–42 sequence to form an unusually stable β-structure which is virtually insoluble in water and strong denaturants,[12] in order to elicit a good immune response against soluble βA4. In the present study, we designed a hydrophilic spacer five residues long, which would overcome these insolubility problems and also extend the presumed epitope away from the proximity of the carrier. To reduce the likelihood of cross-reactivity with the shorter

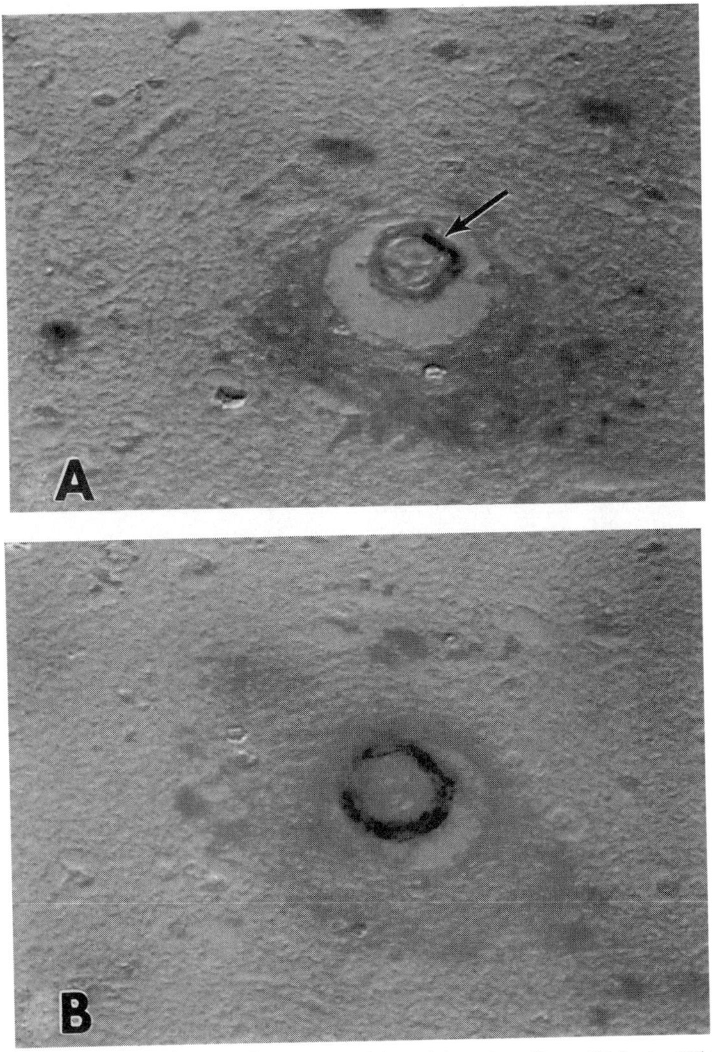

FIGURE 4. Immunostaining of a blood vessel in serial sections from AD brain. While some immunostaining is evident with the C-terminal antibody (arrow, ***A***), the same vessel is uniformly and darkly labeled with the N-terminal antibody (**B**). Original magnification-10×.

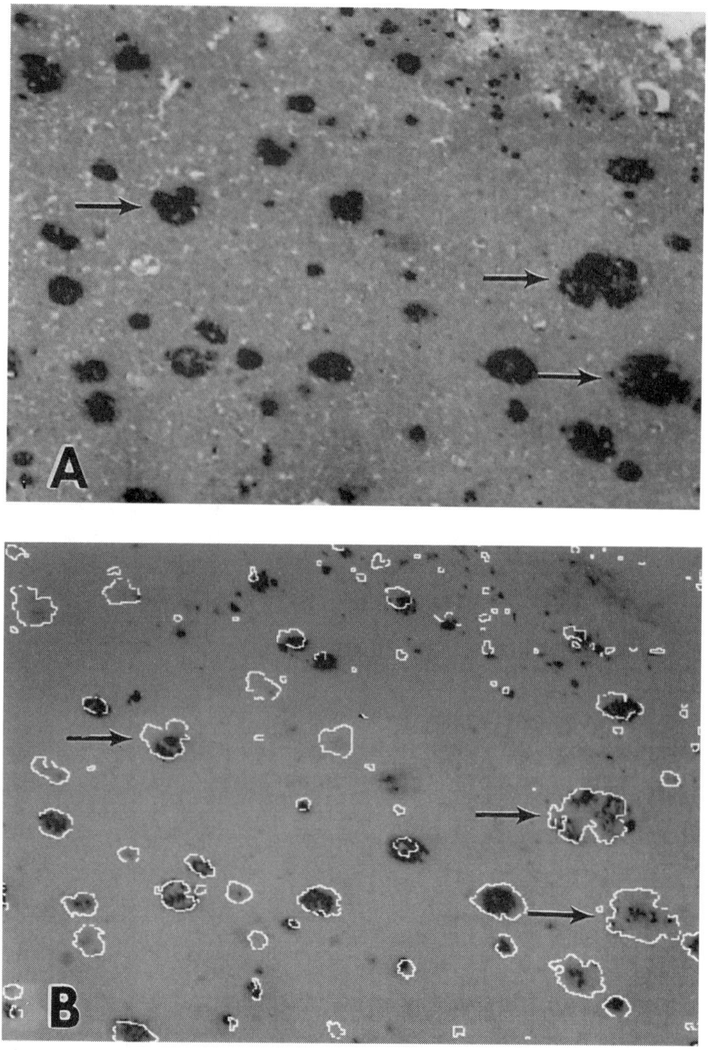

FIGURE 5. Immunolabeling of plaques in serial sections from an AD brain. **A:** The C-terminal antibody (MAb 369.2B) detected fewer, but larger amyloid accumulations. **B:** The boundaries of the C-terminal antibody (MAb 286.8A)-labeled plaques have been superimposed on the plaques recognized by the N-terminal antibody. Qualitative analyses revealed that the N-terminal antibody recognizes multiple subregions of the larger plaques, corroborating the biochemical finding of N-terminal raggedness in plaques. Original magnification-4×.

but major βA4 species, 1–40, we chose a minimal peptidyl epitope of 8 residues corresponding to positions 35–42 of the βA4 sequence. The entire synthetic sequence designed in this way was coupled with a free sulfhydryl group on a N-terminal cyteine residue to KLH. The development of the IPSA allowed a rapid secondary screening of initially identified antibodies by EIA, which were able to immunoprecipitate βA4 effectively.

Both *in vitro* and *in situ* peptide competition studies demonstrated that βA4 sequences other than 1–42 were not effective in inhibiting binding of the C-terminal antibody (MAb 369.2B) to the epitope. The incomplete inhibition by the βA4 1–42 in the competition experiments could be explained by the well-documented aggregation of this peptide leading to a lower effective concentration *in vitro* or simply by the fact that a ~200-fold excess cold competing peptide may not be enough to eliminate the signal. The competition IPSA also demonstrated the enhanced immunoprecipitability of IVT βA4 by the decapeptide 25–35. This phenomenon has also been seen in assays using MAb 286 8A as well as one rabbit polyclonal antisera raised against βA4 1–40 (data not shown). Residues 26–33 are believed to exist as a random coil in aqueous solution[14] and may be able to interact with soluble βA4 in such a way that presents the remaining βA4 molecule in a conformation which enhances its interaction with the MAbs. Interestingly, the βA4 region 25–35 is also the minimal region of βA4 necessary to cause *in vitro* neurotoxicity.[15] In these studies neurotoxicity is correlated with the fibrillar aggregates which can be readily formed by βA4 25–35.

Previously, Murphy *et al.*,[13] using a similar MAb against βA4 1–42, reported the labeling of both intra- and extracellular neurofibrillar tangles in the entorhinal and periamygdaloid cortices. In the present study, however, no labeling of tangles or neuritic structures in Brodmann's area 22 was evident. In the hippocampal formation we observed scattered tangles which were immunopositive with both antibodies. This observation was made independent of the formic acid treatment.

Detailed sequence analysis of βA4 released from cells in culture revealed that the major species end at position 40.[17] This appears to be the major species of the meningovascular form of βA4 as well.[18] The present study, while not specifically examining leptomeninges, did demonstrate an intense labeling of some vessels in a subset of patients with the antibody that recognized βA4 1–40 as well as 1–42 (MAb 286.8A), but less vascular staining with the antibody specific for βA4 1–42 (MAb 369.2B). This is in agreement with previous reports[13,18] and confirms the biochemical finding that βA4 1–40 is the major species deposited at the vasculature.

While the same populations of plaques (diffuse, cored and dense) were labeled by both the C- and N-terminal antibodies, the latter appeared to label multiple subregions of the former. The apparent decreased immunoreactivity is consistent with biochemical demonstrations that parenchymal plaques exhibit an N-terminal raggedness.[3–5] Several hypotheses could account for these findings: different proteases may operate on deposited βA4 in the vascular and parenchymal microenvironments; alternatively, vascular βA4 may derive from different sources (circulating, platelets, cells associated with the vasculature) than the βA4 found in the parenchyma (neurons, glia). Further studies are necessary to explore these hypotheses.

ACKNOWLEDGMENTS

We thank D. Mierz, L. Kelley, and J. Williams for their excellent contributions. Human brain samples were obtained from the Southern California Alzheimers Disease Consortium (NIH 1 P50-AG 05142).

REFERENCES

1. Citron, M., T. Oltersdorf, C. Haass, L. McConlogue, A. Hung, P. Seubert, C. Vigo-Pelfrey, I. Lieberburg & D. Selkoe. 1992. Mutation of the β-amyloid precursor protein in familial Alzheimer's disease increases β-protein production. Nature **360:**672–674.
2. Shoji, M., T. E. Golde, J. Ghiso, T. T. Cheung, S. Estus, L. M. Shaffer, X.-D. Cai, D. M. McKay, R. Tintner, B. Frangione & S. G. Younkin. 1992. Production of the Alzheimer amyloid β protein by normal proteolytic processing. Science **258:**126–129.
3. Busciglio, J., D. H. Gabuzda, P. Matsudaira & B. A. Yankner. 1993. Generation of β-amyloid in the secretory pathway in neuronal and non-neuronal cells. Proc. Natl. Acad. Sci. USA. **90:**2092–2096.
4. Seubert, P., T. Oltersdorf, M. G. Lee, R. Barbour, C. Blomquist, D. L. Davis, K. Bryant, L. C. Fritz, D. Galasko, L. J. Thal, I. Lieberburg & D. B. Schenk. 1993. β-Amyloid precursor protein is cleaved at the amino-terminus of the β-amyloid peptide in an alternative secretory pathway. Nature **359:**325–3.
5. Goate, A. M., M. C. Chartier-Harlin, M. C. Mullan, J. Brown, F. Crawford, L. Fidani, A. Guiffra & A. Haynes. 1991. Segregation of a missense mutation in the amyloid precursor protein gene with familial Alzheimer's disease. Nature **349:**704–706.
6. Carter, D. A., E. Desmarais, M. Bellis, D. Campion, F. Clerget-Darpoux, A. Brice, Y. Agid, A. Jaillard-Serradt & J. Mallet. 1992. More missense in amyloid gene. Nature Genetics **2:**255–256.
7. Murrell, J., M. Farlow, B. Ghetti & M. Benson. 1991. A mutation in the amyloid precursor protein associated with hereditary Alzheimer's disease. Science **254:**97–99.
8. Levy, E., M. D. Carman, I. J. Fernandez-Madrid, M. D. Power, I. Lieberburg, S. G. van Duinen, G. T. A. M. Bots, W. Luyendijk & B. Frangione.

1990. Mutation of the Alzheimer's disease amyloid gene in hereditary cerebral hemorrhage, Dutch type. Science **248:**1124–1126.

9. SUZUKI, N., T. T. CHEUNG, X.-D. CAI, A. ODAKA, L. OTVOS, C. ECKMAN, T. E. GOLDE & S. G. YOUNKIN. 1994. An increased percentage of long amyloid β protein secreted by familial amyloid β protein precursor (βAPP717) mutants. Science **264:**1336–1340.
10. WUNDERLICH, D., A. LEE, R. P. FRACASSO, D. V. MIERZ, R. M. BAYNEY & T. V. RAMABHADRAN. 1992. Use of recombinant fusion proteins for generation and characterization of monoclonal antibodies. J. Immunol. Methods **147:** 1–11.
11. KANG, J., H. G. LEMAIRE, A. UNTERBECK, J. M. SALBAUM, C. L. MASTERS, K. H. GRZESCHIK, G. MULTHAUP, K. BEYREUTHER & B. MÜLLER-HILL. 1987. The precursor of Alzheimer's disease amyloid A4 protein resembles a cell surface receptor. Nature **325:**733–736.
12. BURDICK, D., B. SOREGHAN, M. KWON, J. KOSMOSKI, N. KNAUER, A. HENSCHEN, J. YATES, Z. COTMAN & C. GLABE. 1992. Assembly and aggregation properties of synthetic Alzheimer's A4/β amyloid peptide analogs. J. Biol. Chem. **267:**546–554.
12. HALVERSON, K., P. E. FRASER, D. A. KIRSCHNER & P. T. LANSBURY. 1990. Molecular determinants of amyloid deposition in Alzheimer's disease conformation studies of synthetic β-protein fragments. Biochemistry **29:**2639–2644.
13. MURPHY, G. M., L. S. FORNO, L. HIGGINS, J. M. SCARDINA, L. F. ENG & B. CORDELL. 1994. Development of a monoclonal antibody specific for the COOH-terminal of β-amyloid 1–42 and its immunohistochemical reactivity in Alzheimer's disease and related disorders. Am. J. Pathol. **144:**1082–1088.
14. JARRETT, J. T., E. P. BERGER & P. T. LANSBURY. 1993. The carboxy terminus of the β-amyloid protein is critical for the seeding of amyloid formation: Implications for the pathogenesis of Alzheimer's disease. Biochemistry **32:**4693–4697.
15. YANKNER, B. A., L. R. DAWES, S. FISHER, L. VILLA-KOMAROFF, M. L. OSTER-GRANITE & R. L. NEVE. 1990. Neurotoxicity of a fragment of the amyloid precursor associated with Alzheimer's disease. Science **250:**279–282.
16. MIRRA, S. S., D. W. MCKEEL, B. J. CRAIN, J. P. HUGHES, G. VAN BELLE, A. HEYMAN AND PARTICIPATING NEUROPATHOLOGISTS. 1990. J. Neuropathol. Exp. Neurol. **49:**336.
17. SELKOE, D. 1994. Normal and abnormal biology of the β-amyloid precursor protein. *In* Annual Review of Neuroscience. W. M. Cowan, E. M. Shooter, C. F. Stevens & R. F. Thomson, Eds.
18. MILLER, D. L., I. A. PAPAYANNOPOULOS, J. STYLES, S. A. BOBIN, Y. Y. LIN, K. BIEMANN & K. IQBAL. 1993. Peptide composition of the cerebrovascular and senile plaque core amyloid deposits of Alzheimer's disease. Arch. Biochem. Biophys. **301:**41–52.

Nicotine-induced Protection Against Glutamate Cytotoxicity

Nicotinic Cholinergic Receptor-mediated Inhibition of Nitric Oxide Formation[a]

S. SHIMOHAMA,[b,d] A. AKAIKE,[c] AND J. KIMURA[b]

[b] *Department of Neurology, Faculty of Medicine, Kyoto University, Kyoto, Japan*

[c] *Department of Pharmacology, Faculty of Pharmaceutical Sciences, Kyoto University, Kyoto, Japan*

ABSTRACT: Cortical neurodegeneration in Alzheimer's disease (AD) is suggested to be attributable not only to β-protein but also to glutamate. Although degeneration of cholinergic projection to the cerebral cortex is recognized to be one of the most prominent pathological changes in AD, there is only limited information concerning the cholinergic interaction with the cortical neurodegeneration. This study was performed to examine the protective effect of nicotine against glutamate-induced cytotoxicity using rat cultured cortical neurons. The cell viability was significantly reduced when cultures were briefly exposed to glutamate or *N*-methyl-D-aspartate (NMDA). The simultaneous addition of nicotine did not reduce glutamate cytotoxicity. In contrast, the simultaneous application of NMDA receptor antagonists such as MK-801 reduced glutamate cytotoxicity. Incubating the cultures with nicotine (10 μM) for 0.5-24 h prior to glutamate exposure reduced its cytotoxicity. Neuroprotection by nicotine was dependent on both the concentration and incubation period. In contrast to nicotine, muscarine (10 μM) weakly potentiated glutamate cytotoxicity. The neuroprotective effect of nicotine against glutamate cytotoxicity was antagonized by hexamethonium but not by artopine. Nicotine prevented NMDA cytotoxicity but did not affect cytotoxicity induced by either kainate or α-amino-3-hydroxy-5-methyl-4-isoxazole propionic acid (AMPA). Cell viability was significantly reduced by a brief exposure of cultures to ionomycin, a calcium ionophore. Ionomycin-induced cytotoxicity was abolished by removing Ca^{2+} from incubating medium. S-nitrosocysteine (SNOC), which spontaneously releases nitric oxide (NO), also induced delayed cell death. Nicotine prevented ionomycin-induced cytotoxicity without affecting SNOC-induced cytotoxicity. These results suggest that nicotinic cholinergic receptor stimulation induces neuroprotection against

[a] This work was partly supported by Grants-in-Aid for Scientific Research on Priority Areas (07264221, 07278219) and for Developmental Scientific Research (0755 7051) from the Ministry of Education, Science and Culture, Japan, and by grants from the Ministry of Welfare of Japan and the Japan Medical Association. We also thank Taiho Pharmaceutical Co. Ltd., Japan for financial support.

[d] *Send correspondence to:* Dr. Shun Shimohama, Department of Neurology, Faculty of Medicine, Kyoto University, 54 Shogoinkawaharacho, Sakyo-ku, Kyoto 606, Japan; Tel: 81-75-751-3767; Fax: 81-75-751-9541; Email: i53367@sakura.kudpc.kyoto-u.ac.jp.

glutamate cytotoxicity by its inhibitory action on NO-formation. Therefore, we propose that acetylcholine, acting through nicotinic cholinergic receptors, can function as a putative neuroprotective factor against neurodegeneration caused by the excessive release of glutamate and/or NMDA receptor activation.

INTRODUCTION

Acetylcholine is one of the major neurotransmitters in the CNS. The cerebral cortex contains a dense plexus of cholinergic axon terminals that arise from the cells of the basal forebrain including the nucleus basalis of Meynert.[1] Degeneration of this cholinergic projection is recognized to be one of the most prominent pathological changes in AD.[2]

It has been postulated that glutamate plays an important role in the neurodegeneration observed in hypoxic-ischemic brain injury.[3] Several investigators have also suggested that cortical neurodegeneration in AD is attributable to glutamate.[4] Moreover, a brief glutamate exposure induces delayed cell death in cultured neurons of certain regions including the cerebral cortex, hippocampus and cerebellum. In these brain regions, NMDA subtype of glutamate receptor plays a crucial role in glutamate neurotoxicity. Recent evidence has indicated the existence of NO synthase in the CNS including the cerebral cortex.[5] NMDA receptor stimulation induces Ca^{2+} influx into cells through ligand-gated ion channels, thereby triggering NO formation. Then, NO is suggested to diffuse to adjacent cells, resulting in appropriate physiological responses and/or glutamate-related cell death.

However, there is only limited information concerning the cholinergic interaction with glutamate neurotoxicity. Mattson[6] and Olney *et al.*[7] have demonstrated that stimulation of the muscarinic receptor potentiates neurodegeneration. In the present study, we examine the effects of nicotine on glutamate-induced neurotoxicity using primary cultures of rat cortical neurons.

METHODS

The procedure for the primary culture of cells obtained from the cerebral cortices of fetal rats at 16–18 days of gestation and the quantitative assessment of glutamate-induced neurotoxicity using trypan blue exclusion have been described previously.[8,9] A 10-min exposure to 0.5–1.0 mM glutamate followed by a 1-h incubation with glutamate-free medium were established as the appropriate conditions under which to examine glutamate cytotoxicity and drug-induced protection.[8]

RESULTS

Effects of Nicotine on Glutamate Cytotoxicity

Incubating the cultures with nicotine (10 μM) for 0.5–24 h prior to glutamate exposure reduced its cytotoxicity. The simultaneous addition of nicotine did not reduce glutamate cytotoxicity. In contrast, the simultaneous application of NMDA receptor antagonists such as MK-801 and 3-[(±)2-carboxypiperzin-4-yl]propyl-1-phosphonic acid (CPP) reduced glutamate cytotoxicity. Nicotine-induced neuroprotection was dependent upon both the concentration and incubation period. Nicotine significantly recovered the NMDA cytotoxicity but did not affect that induced by either kainate or AMPA. The neuroprotective effect of nicotine against glutamate cytotoxicity was anatagonized by nicotinic receptor antagonists, mecamylamine and hexamethonium, but not by a muscarinic receptor anatagonist, artopine. Muscarine (10 μM) weakly but significantly potentiated glutamate cytotoxicity. In contrast, carbachol, which acts on both nicotinic and muscarinic receptors, significantly ameliorated glutamate cytotoxicity, although apparently not so potently as nicotine.[10]

Mechanisms of Nicotine-induced Neuroprotection

We next examined the effects of nicotine on the cytotoxicity induced by a brief exposure to ionomycin, a calcium ionophore that induces persistent Ca^{2+} influx. Concentration-dependent cytotoxicity was induced by a brief exposure of cortical cultures to ionomycin at 10 nM–1 μM followed by a 1-h incubation with ionomycin-free medium, though the viability of cultures was not effected immediately after a 10-min exposure to ionomycin. Ionomycin-induced cytotoxicity was mediated by Ca^{2+}, since Ca^{2+} removal from the incubation medium completely abolished the cytotoxicity. NO presumably mediates ionomycin cytotoxicity since ionomycin-induced cytotoxicity was prevented by hemoglobin (20 μM) which traps NO in the incubation medium. Thus, ionomycin directly induces the Ca^{2+}-influx to promote NO formation by skipping the step of NMDA receptor activation. Nicotine (10 μM) but not MK-801 (10 μM) or CPP (10 μM) prevented ionomycin-induced cytotoxicity (FIG. 1). Agents that spontaneously release NO in aqueous solution have been widely used to determine the NO-induced effects on neuronal cells. We employed SNOC as a NO-generating agent to examine the effects of nicotine on NO-induced cytotoxicity. A 10-min exposure to SNOC followed by a 1-h incubation with SNOC-free medium markedly reduced the viability of the cultures. SNOC-induced cytotoxicity was pre-

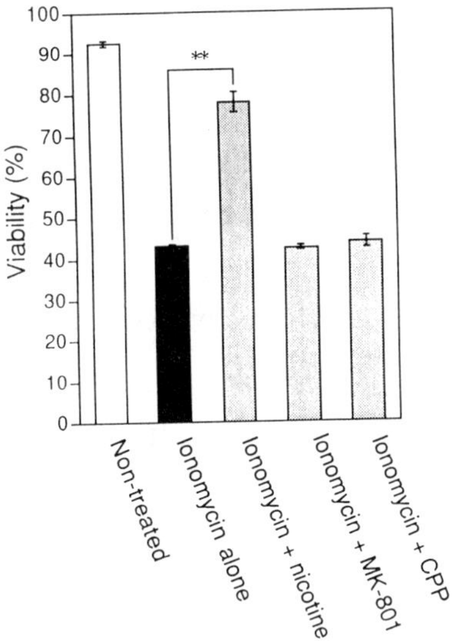

FIGURE 1. Effects of nicotine on ionomycin-induced cytotoxicity in cultured cortical neurons Cultures were exposed to ionomycin (1 μM) for 10 min followed by an incubation with ionomycin-free medium for 1 hour. Nicotine (10 μM) was incubated for 2 h prior to ionomycin exposure. MK-801 (10μM) and CPP (10 μM) were added to both the 10 min- and 1 h-incubation medium. Data were expressed as mean ± SE. $^{**}p < 0.01$, Dunnett's two-tailed test.

vented by hemoglobin (20 μM) but not by Ca^{2+} removal from the incubation medium. Neither nicotine nor MK-801 affected SNOC cytotoxicity (FIG. 2).

DISCUSSION

The present results indicate that the nicotinic cholinergic system is involved in the NMDA receptor-mediated neurodegeneration of the cerebral cortex.

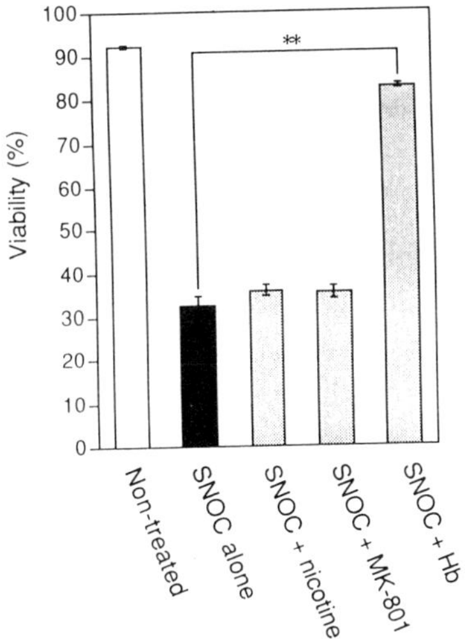

FIGURE 2. Effects of nicotine on SNOC-induced cytotoxicity in cultured cortical neurons. Cultures were exposed to SNOC (300 μM) for 10 min followed by an incubation with SNOC-free medium for 1 hour. Nicotine (10 μM) was incubated for 2 h prior to SNOC exposure. MK-801 (10 μM) was added to both the 10 min- and 1 h-incubation media. Hemoglobin (20 μM) was added only to 10 min-incubation medium. Data were expressed as mean ± SE. $^{**}p < 0.01$, Dunnett's two-tailed test.

Nicotine-induced effects on glutamate cytotoxicity were antagonized by the neuronal nicotinic receptor antagonists, but not by the muscarinic receptor antagonist or NMDA-receptor antagonists, indicating that nicotine induces neuroprotection by its selective action upon nicotinic receptors.

Using the ionomycin and SNOC, we examined the mechanism of nicotine-induced neuroprotection. Ionomycin-induced cytotoxicity was abolished by removing Ca^{2+} from the incubating medium. Both N^{ω}-nitro-L-arginine, NO synthase inhibitor, and hemoglobin, which traps NO, prevented cytotoxicity induced by NMDA or ionomycin. Nicotine but not MK-801 prevented ionomycin-induced cytotoxicity, indicating that nicotine inhibits Ca^{2+}-induced intracellular responses without affecting NMDA receptor activity. SNOC, which spontaneously releases NO, also induced delayed cell death. SNOC-induced cytotoxicity was prevented by hemoglobin but not by Ca^{2+} removal from the incubation medium. Nicotine did not affect SNOC cytotoxicity. These findings suggest that the neuroprotective effects of nicotine are caused by the inhibitory action on NO-formation, but not by reduction of the Ca^{2+} influx nor NO cytotoxicity. It is possible that nicotinic receptor stimulation

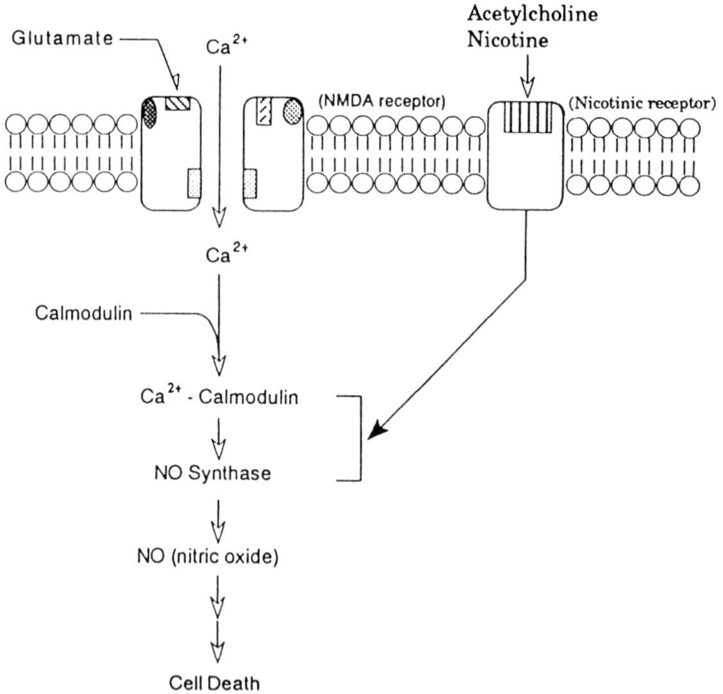

FIGURE 3. Schematic representation of presumed mechanism of nicotine-induced protection against glutamate cytotoxicity.

may cause suppression of a step in NO formation by NO synthase (FIG. 3). However, the precise mechanisms linked with nicotinic receptors are yet to be determined.

Although further study is required to determine the intracellular mechanism of the neuroprotective action of nicotine, this neuroprotective effect of nicotine suggests that certain neurotransmitters or modulators released in the brain protect neurons against the cytotoxic effects of glutamate. Therefore, we propose that acetylcholine acting through nicotinic cholinergic receptors, can function as a putative neuroprotective factor against neurodegeneration caused by the excessive release of glutamate and/or NMDA receptor activation.

REFERENCES

1. MESULAM, M. M., E. J. MUFSON, A. I. LEVEY & B. H. WAINER. 1983. Cholinergic innervation of cortex by the basal forebrain: Cytochemistry and cortical connections of the septal area, diagonal band nuclei, nucleus basalis (substantia innominata) and hypothalamus in the rhesus monkey. J. Comp. Neurol. **214:** 170–197.
2. WHITEHOUSE, P. J., D. L. PRICE, A. W. CLARK, T. T. COYLE & M. DELONG. 1981. Alzheimer's disease evidence for a selective loss of cholinergic neurons in the nucleus basalis. Ann. Neurol. **10:**122–1268.
3. CHOI, D. W. 1988. Calcium-mediated neurotoxicity relationship to specific channel types and role in ischemic damage. Trends Neurosci. **11:**465–469.
4. MARAGOS, W. F., J. T. GREENAMYRE, J. B. PENNEY & A. B. YOUNG. 1986. Glutamate dysfunction in Alzheimer's disease: An hypothesis. Trends Neurosci. **10:**65–68.
5. DAWSON, V. L., T. M. DAWSON, E. D. LONDON, D. S. BREDT & S. H. SNYDER. 1991. Nitric oxide mediates glutamate neurotoxicity in primary cortical cultures. Proc. Natl. Acad. Sci. USA **88:**6368–6371.
6. MATTSON, M. P. 1989. Acetylcholine potentiates glutamate-induced neurodegeneration in cultured hippocampal neurons. Brain Res. **497:**402–406.
7. OLNEY, J. W., J. LABRUYERE, G. WANG, D. F. WOZNIAK, M. T. PRICE & M. A. SESMA. 1991. NMDA antagonist neurotoxicity: Mechanism and prevention. Science **254:**1515–1518.
8. AKAIKE, A., Y. TAMURA, Y. SATO, K. OZAKI, R. MATSUOKA, S. MIURA & T. YOSHINAGA. 1991. Cholecctystokinin-induced protection of cultured cortical neurons against glutamate neurotoxicity. Brain Res. **557:**303–307.
9. SHIMOHANA, S., N. OGAWA, Y. TAMURA, A. AKAIKE, T. TSUKAHARA, H. IWATA & J. KIMURA. 1993. Protective effect of nerve growth factor against glutamate-induced neurotoxicity in cultured cortical neurons. Brain Res. **632:** 296–302.
10. AKAIKE, A., Y. TAMURA, T. YOKOTA, S. SHIMOHAMA & J. KIMURA. 1994. Nicotine-induced protection of cultured cortical neurons against N-methyl-D-aspartate receptor-mediated glutamate cytotoxicity. Brain Res. **644:**181–187.

Free Radicals in the Neurotoxic Actions of β-Amyloid[a]

J. S. RICHARDSON,[b–d] Y. ZHOU,[b] AND U. KUMAR[b]

[b] *Department of Pharmacology and* [c] *Department of Psychiatry, College of Medicine, University of Saskatchewan, Saskatoon, SK, Canada*

ABSTRACT: The fragment of the β-amyloid protein that contains the peptide sequence 25 to 35 of the parent compound, increases cytosolic calcium and is directly cytotoxic to neurons in tissue culture. The cytotoxic action of β-amyloid has been considered to be the primary determinant of the neurodegeneration observed in Alzheimer's disease. The cytotoxic effects of β-amyloid 25–35 or β-amyloid 1–40 added to primary cultures of rat hippocampal or cortical neurons can be reduced or prevented by anti-oxidant lazaroid drugs such as U-74500A, U-78517F, or U-83836E. β-amyloid 25–35 evokes an immediate increase in cytosolic calcium when added to suspensions of PC 12 cells, and is cytotoxic to PC 12 cells in tissue culture. Both the calcium-elevating and the cytotoxic actions of β-amyloid 25–35 on PC 12 cells are reduced or prevented by lazaroid anti-oxidant drugs. Since the lazaroids have been shown to scavenge free radical species, the present results indicate that the actions of β-amyloid 25–35 on calcium regulation and on cell survival are mediated by free radicals.

INTRODUCTION

A free radical is an atom or a molecule that contains an unpaired electron. Generally, unpaired electrons are very unstable, and the free radical will gain stability by donating an electron to, or extracting an electron from, nearby chemicals. This process may turn the nearby chemical into a free radical and this can set off a free radical chain reaction that propagates throughout a chemical network. If that chemical network performs a biological function crucial to the viability of a cell, such as the lipid and protein molecules in the membranes of neurons, or such as the nucleic acid molecules in any cell, the free radicals can produce alterations in cell function ranging from a slight

[a] This research was supported in part by grants from the Alzheimer's Association, U.S.A. (IIRG-91-092), the Upjohn London Neuroscience Program, and the Alzheimer's Society for Lethbridge, Alberta, and by donations from private individuals.

[d] *Send correspondence to:* Dr. J. Steven Richardson, Department of Pharmacology, College of Medicine, University of Saskatchewan, Saskatoon, SK, Canada S7N 5E5; Tel: (306) 966-6301, Fax: (306) 966-6220; Email: RICHARDSON@SASK.USASK.CA.

reduction in cellular efficiency up to cell death. It has been established[1] that free radicals are responsible for the neuronal destruction following cerebral ischemia and reperfusion. There is considerable circumstantial evidence that free radicals are involved in the progressive neurodegeneration of Alzheimer's Disease.[2–4]

Since the demonstration a decade ago of deposits of β-amyloid protein (Aβ) associated with the neuritic plaques and neurofibrillary tangles in the brains of Alzheimer's patients,[5] considerable research has been directed at characterizing Aβ and delineating its actions. In general, Aβ has been considered to be the etiological agent responsible for causing Alzheimer's disease.[6] Aβ is a 39 to 42 peptide molecule that is cleaved from a precursor protein containing around 700 peptides. In turn, the Aβ gives rise to smaller fragment. In certain conditions, Aβ 1–40 has neurotrophic effects. In other conditions, it can be neurotoxic.[7] One of the Aβ fragments, Aβ 25–35, increases cytosolic calcium and potentiates glutamate excitotoxicity,[8,9] and is directly neurotoxic.[10,11] All of these actions have been shown to be mediated by free radicals.

METHODS

Cell Culture

Primary cultures of rat cortical and hippocampal neurons were prepared by standard procedures from E-18 Sprague-Dawley embryos, and were grown on poly-lysine coated tissue culture plates. Cultures of the PC 12 neuronal cell line were grown in collagen-coated flasks and were re-plated onto collagen coated tissue culture plates as needed.

Drugs

The lazaroid free radical scavenger drugs U-74500A, U-78517F, and U-83836E were provided by Dr. John M. McCall, Upjohn Company, Kalamazoo, MI, USA. Aβ 1 to 42, Aβ 25 to 35, and all other chemicals were purchased from commercial sources.

Assays

In cultured cells, growth and morphology were monitored by phase contrast microscopy, and cell viability was determined by the MTT assay. Changes in intracellular cytosolic calcium was quantified by the Fura 2 fluorescence method using dispersed cells harvested from cultures.

Statistics

All data were analyzed statistically using ANOVAs and appropriate post hoc tests contained in the Statview® statistical package for the Macintosh computer.

RESULTS

Mature cultures of rat hippocampal cells exposed to 100 μg/ml of Aβ 1–40 or Aβ 25–35 for 24 hours showed a loss of fine neurites, a decrease in large diameter processes joining cell bodies, a swelling and clumping of cell bodies and the accumulation of cellular debris. The destructive effects of Aβ 25–35 were considerably more marked than those of Aβ 1–40. None of these changes were seen in nontreated control cultures or in cultures exposed to 10 μM U-78517F for 24 hours. Pretreating cultures with 10 μM U-78517F, prevented most of the changes seen in cultures given Aβ 1–40 alone. U-78517F pretreatment reduced but did not prevent the destructive actions of Aβ 25–35.

Aβ 25–35 added in concentrations of up to 500 μM to rat cortical neuron cultures at the time of the initial plating of the cells, reduced cell survival, as determined by the MTT assay, in a dose-dependent manner. The Aβ 25–35 impairment of cell survival at all doses was reduced by the lazaroid free radical scavengers U-74500A and U-83836E, by the antioxidant vitamin E, and by the radical spin trapping agent *N*-t-butyl-α-phenyl nitrone (PBN).

When added to PC 12 cells 24 hours after plating, A-β 25–35 in concentrations up to 10 μM reduced cell survival in a dose dependent manner with 1 μM A-β 25–35 producing a 35% reduction in PC 12 cells 20 hours later. As with neuronal primary cultures, the neurotoxic effects of A-β 25–35 on PC 12 cells were reversed by the lazaroid free radical scavengers U-74500A and U-83836E, and by the anti-oxidant vitamin E. The survival of PC 12 cells is also reduced by treatment with 500 μM buthionine sulfoxime, an irreversible inhibitor of glutathione synthetase, one of the endogenous free radical defensive enzymes. The cytotoxic consequences of glutathione synthetase inhibition are reversed in a dose-dependent manner by U-83836E with concentrations of this lazaroid in the 10 μM range giving complete protection against 500 μM buthionine sulfate.

As determined in PC 12 cells preloaded with Fura-2, A-β 25–35 added in concentrations from 10μM to 50 μM, produces a dose-dependent increase in intracellular cytosolic calcium levels, with 40 μM A-β 25–35 increasing cytosolic calcium by about 60 nM over basal levels. The increase in cytosolic calcium induced by 40 μM A-β 25–35, is not affected by the calcium channel blockers nifedipine (10 μM) or cobalt (500 μM) but is blocked by U-83836E in a dose-dependent manner with a maximal effect attained at doses around 10 μM.

DISCUSSION

The results of these experiments support previous observations that fragments of Aβ are cytotoxic to mature primary cultures of neurons and to PC 12 cells in tissue culture. More importantly, they also demonstrate that the cytotoxic actions of A-β can be blocked or attenuated by various members of the lazaroid family of anti-oxidant compounds. We have demonstrated[12] that the lazaroids U-74500A and U-78517F will scavenge oxygen-centered and nitrogen-centered free radicals. While these drugs might have other pharmacological actions in addition to free radical scavenging, all reported actions of the lazaroids are consistent with antioxidant effects mediated via their free radical scavenging action. The ability of the lazaroids to reduce both the cytotoxicity and the increase in cytosolic calcium produced by Aβ, indicates that these actions of Aβ are mediated by free radicals. Subsequent to the completion of most of the present experiments, Hensley *et al.*[13] reported that Aβ fragments in aqueous solution generate free radicals. Our experiments using electron paramagnetic resonance spectroscopy to demonstrate and identify the free radicals generated by Aβ in cultures of neurons are in progress. However, the present experiments indicate that both the elevation of cytosolic calcium and the cytotoxic actions of Aβ are due to free radicals generated when Aβ 25–35 is added to the experimental preparations.

The observations that free radical mediated neural damage induces the expression of the amyloid precursor protein gene[14–16] suggest that excessive free radical activity could be responsible both for initiating and for maintaining the progressive neurodegeneration of Alzheimer's disease. In this hypothetical scenario, the future Alzheimer's patient experiences a period of time in which free radical activity exceeds the capacity of the free radical defenses, some neural damage results and amyloid precursor protein (APP) formation is induced as part of the endogenous neuro-repair process. However, the excessive free radicals disrupt the normal processing of APP and produce the Aβ fragments that form additional free radicals. This increases the free radical burden of the person and results in more neurodegeneration and more APP production. An amplifying cascade of free radical and APP formation develops that eventually destroys a sufficient number of nerve terminals or entire neurons to disrupt the output of brain areas such as the amygdala, hippocampus, and frontal cortex. When these brain areas no longer have enough nerve terminals or neurons to function normally, the symptoms of Alzheimer's disease appear. Although there is ample *in vitro* evidence implicating free radicals in the pathology of Alzheimer's disease,[3] there is as yet no *in vivo* evidence of abnormal free radical activity in patients with probable Alzheimer's disease. Such *in vivo* evidence could provide the basis for a pre-symptomatic diagnostic test and would allow interventions to block the APP-free radical cascade, to halt the progressive neurodegeneration, and to prevent or postpone the appearance of the symptoms of Alzheimer's disease.

ACKNOWLEDGMENT

The expert stenographic assistance of Mrs. V. Heffner is gratefully acknowledged.

REFERENCES

1. ZHAO, W., J. S. RICHARDSON, M. J. MOMBOURQUETTE, J. A. WEIL, S. IJAZ & A. SHUAIB. Effects of U-78517F or hypothermia on free radicals in the ischemic-reperfused gerbil hippocampus. Submitted.
2. JESBERGER, J. A. & J. S. RICHARDSON. 1991. Oxygen free radicals and brain dysfunction. Int. J. Neurosci. **57:**1–17.
3. RICHARDSON, J. S., U. KUMAR, S. V. KALA, L. CHEN & Y. ZHOU. 1994. *In vitro* evidence for the use of antioxidants in Alzheimer's disease. *In* Alzheimer's Disease: Therapeutic Strategies. E. Giacobini & R. E. Becker, Eds.: 334–342. Boston: Birkhauser.
4. SUBBARAO, K. V., J. S. RICHARDSON & L. C. ANG. 1990. Autopsy samples of Alzheimer's cortex show increased peroxidation *in vitro*. J. Neurochem. **55:** 342–345.
5. GLENNER, G. G. & C. W. WONG. 1984. Alzheimer's disease: Initial report of the purification and characterization of a novel cerebrovascular amyloid protein. Biochem. Biophys. Res. Commun. **120:**885–890.
6. SELKOE, D. J. 1993. Physiological production of the β-amyloid protein and the mechanism of Alzheimer's disease. Trends Neurosci. **16:**403–409.
7. KOH, J.-Y., L. L. YANG & C. W. COTMAN. 1990. β-amyloid protein increases the vulnerability of cultured cortical neurons to excitotoxic damage. Brain Res. **533:**315–320.
8. MATTSON, M. P., B. CHEN, D. DAVIS, K. BRYANT, I. LIEBERBERG & R. F. RYDEL. 1992. β-amyloid peptides destabilize calcium homeostasis and render human cortical neurons vulnerable to excitotoxicity. J. Neurosci. **12:**376–389.
9. MATTSON, M. P., S. W. BARGER, B. CHENG, J. LIEBERBERG, V. L. SMITH-SWINTOWSKY & R. E. RYDEL. 1993. β-Amyloid precursor protein metabolites and loss of neuronal Ca^{++} homeostasis in Alzheimer's disease. Trends Neurosci. **16:**409–414.
10. KUMAR, U., D. M. DUNLOP & J. S. RICHARDSON. 1994. The acute neurotoxic effect of β-amyloid on mature cultures of rat hippocampal neurons is attenuated by the anti-oxidant U-78517F. Int. J. Neurosci. **79:**185–190.
11. LOO, D. T., A. COPANI, C. J. PIKE, E. R. WHITTEMORE, A. J. WALENCEWICZ & C. W. COTMAN. 1993. Apoptosis is induced by β-amyloid in cultured central nervous system neurons. Proc. Natl. Acad. Sci. USA **90:**7951–7955.
12. ZHAO, W., J. S. RICHARDSON, M. J. MOMBOURQUETTE & J. A. WEIL. 1995. An *in vitro* EPR study of the free radical scavenging actions of the lazaroid antioxidants U-74500A and U-78517F. Free Radical Biol. Med. **19:**21–30.
13. HENSLEY, K., J. M. CARNEY, M. P. MATTSON, M. AKSENOVA, M. HARRIS, J. F. WU, R. A. FLOYD & D. A. BUTTERFIELD. 1994. A model for β-amyloid

aggregation and neurotoxicity based on free radical generation by the peptide: Relevance to Alzheimer's disease. Proc. Natl. Acad. Sci. USA **91**:3270–3274.

14. KALARIA, R. N., S. U. BHATTI, W. D. LUST & G. PERRY. 1993. The amyloid precursor protein in ischemic brain injury and chronic hypoperfusion. Ann. N.Y. Acad. Sci. **695**:190–193.
15. KOGURE, K. & H. KATO. 1993. Altered gene expression in cerebral ischemia. Stroke **24**:2121–2127.
16. OOSTVEEN, J. A. & E. D. HALL. 1993. Regional accumulation of amyloid protein precursor following a bilateral chorotid occlusion in the gerbil. Soc. Neurosci. Abstracts **19**:1655.

Molecular Membrane Interactions of a Phospholipid Metabolite

Implications for Alzheimer's Disease Pathophysiology[a]

R. P. MASON,[b,e] M. W. TRUMBORE,[c] AND J. W. PETTEGREW[d]

[b] *Neurosciences Research Center, Medical College of Pennsylvania, Pittsburgh, Pennsylvania USA*

[c] *National Institutes of Health, National Cancer Institute, Bethesda, Maryland, USA*

[d] *Laboratory of Neurophysics, University of Pittsburgh, Pittsburgh, Pennsylvania, USA*

ABSTRACT: Alzheimers's disease is characterized by changes in phospholipid metabolism leading to a perturbation in the levels of phosphomonoesters, including L-Phosphoserine (L-PS). These early changes in lipid metabolism may result in a defect in membrane bilayer structure, leading to increased rates of β-amyloid formation. To investigate the effect of L-PS on membrane lipid bilayers, small angle x-ray diffraction and high resolution differential scanning calorimetry (DSC) approaches were used with liposomes composed of lecithin and cholesterol. A one-dimensional electron density profile of a control dimyristoyl phosphatidylcholine (DMPC)/cholesterol lipid bilayer with a unit cell dimension of 52 Å at 37°C was generated from the x-ray diffraction data. Following incubation with 2.0 mM L-PS, a broad decrease in electron density ±4-12Å from the lipid bilayer center was observed concomitant with an increase in the width of the phospholipid headgroup electron density and a 3Å reduction in lipid bilayer width. The interactions of L-PS with DMPC lipid bilayers were concentration-dependent, highly affected by cholesterol content and reproduced in egg phosphatidylcholine/cholesterol liposomes. DSC analysis showed that millimolar (1.0-5.0 mM) L-PS levels decreased the phase transition cooperative unit size of DMPC liposomes in a highly concentration-dependent manner which was significantly greater in preparations containing 10 mol% cholesterol. These data provide direct evidence that phosphomonoester levels modulate the biophysical properties of the membrane lipid bilayer which may, in turn, lead to altered structure/function relationships in AD.

[a] The research was sponsored by a John A. Hartford Foundation Fellowship and NIA grants AG08371, AG08974 and AG05133.

[e] *Send correspondence to:* Dr. R. P. Mason, Neurosciences Research Center, Allegheny Campus, Medical College of Pennsylvania, 320 East North Avenue, Pittsburgh, PA 15212-4772; Tel: (412) 359-4815; Fax: (412) 359-6390; Email: mason@asri.edu.

INTRODUCTION

The interactions of lipid metabolites with the membrane is of fundamental interest to aging, particularly as it relates to Alzheimer's disease (AD). The central question is whether changes in lipid metabolite levels can alter the molecular structure and physical properties of the lipid bilayer, thereby modulating membrane function. Phosphomonoesters (PME), including L-phosphoserine (L-PS), are metabolites of membrane phospholipids. In autopsy-confirmed AD brain tissue, the levels of PME were reported to be significantly higher than in age-matched controls by 25–50%, as detected by ^{31}P nuclear magnetic resonance.[1,2] The tissue levels of PME were found to be highest in regions of the brain with the least number of senile plaques, suggesting an early role for these metabolites in the disease process.[2] As the extent of pathology progresses, the levels of phosphodiesters (a product of phospholipid metabolism) increased and PME levels decrease, as observed *in vitro* by ^{31}P NMR[2,3] and by chromatographic methods.[4]

To investigate the time-averaged effect of L-PS on lipid bilayer structure, small angle x-ray diffraction was utilized. Changes in the time-averaged location of membrane constituents following addition of phosphomonoester were evaluated by comparing one-dimensional electron density profiles of membranes generated under identical experimental conditions. These experiments provided direct structure information concerning the effect of L-PS on the membrane bilayer. In addition, high resolution differential scanning calorimetry (DSC) was used to test the effect of L-PS on thermodynamic properties of the lipid bilayer.

METHODS

All chemicals used were reagent grade or better and made up in ultra pure deionized water. Dimyristoyl phosphatidylcholine (DMPC), egg phosphatidylcholine (EPC) and cholesterol were purchased from Avanti Polar Lipids, Inc. (Alabaster, AL). L-phosphoserine was obtained from Sigma Chemical Co. (St. Louis, MO).

Differential Scanning Calorimetry

Differential scanning calorimetry (DSC) was used to analyze L-PS containing and control multilamellar vesicles, as previously described.[5] Calorimetry was carried out using a TA Instruments (New Castle, DE) DSC 2910 Differential Scanning Calorimeter and the data were analyzed using a TA Instruments Thermal Analyst 2000 system. The calorimeter was calibrated for baseline, heatflow and temperature using an Indium standard of known mass. The scan rate was 2°C/minute starting at 6°C and ending at 29.5°C. Samples were repeatedly scanned over this range until there was no change in the observed calorimetric trace.

Small Angle X-ray Diffraction

Phospholipid membrane vesicles composed of DMPC or DMPC/cholesterol were prepared for x-ray diffraction experiments as previously described.[5–7] The oriented membrane samples were formed by centrifugation in Lucite sedimentation cells containing an aluminum foil substrate. The "multibilayer" samples were placed in sealed brass canisters in which temperature and humidity were controlled with saturated salts. The membrane samples were placed at near-grazing incidence with respect to the x-ray beam. The radiation source was a collimated, monochromatic x-ray beam (CuK_α x-ray, $\lambda = 1.54$Å) from an Elliot GX-18 rotating anode x-ray generator (Enraf Nonius, Bohemia, NY) operated at 40 kV and 30 mA in the Biomolecular Structure Analysis Center at the University of Connecticut Health Center. A helium tunnel was positioned between the sample and detector to reduce scattering from the air. The experimental method utilized a single Franks' mirror defining a focused, nickel-filtered line source where $K_{\alpha 1}$ and $K_{\alpha 2}$ are unresolved. Bragg's diffraction from the multilamellar samples were recorded on both Kodak DEF-5 film (Eastman Kodak Co., Rochester, NY) and a Braun position-sensitive 1-D detector (Innovative Technologies, Inc., Newburyport, MA). In addition to direct calibration of the detector system, mica, lead stearate and cholesterol were used to verify the calibration and sample to detector distances. Relative intensities for the diffraction orders were obtained directly from digitized computer plots of the detector data using an integration routine.

RESULTS AND DISCUSSION

L-PS alters membrane thermotropic properties in a concentration-dependent manner that is highly dependent on cholesterol content. Differential scanning calorimetry (DSC) demonstrated that L-PS altered the thermotropic behavior of DMPC and DMPC/cholesterol samples in a concentration-dependent manner. In the presence and absence of cholesterol, L-PS caused an increase in the thermal phase transition temperature and a decrease in cooperative unit size associated with the transition without affecting the enthalpy of transition.

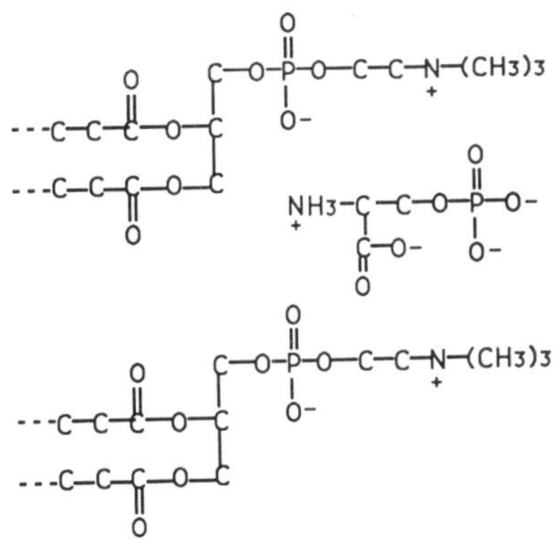

FIGURE 1. Membrane monolayer. Molecular model for the interaction of L-PS with the membrane lipid bilayer. Based on the results of x-ray diffraction, DSC, and consideration of the chemical structure of L-PS, the following model is proposed. L-PS electrostatically binds to the phospholipid headgroup, interacting with oppositely charged groups (*i.e.*, phosphate anionic oxygen). L-PS insertion into this region of the membrane would then alter the packing of phospholipid head-groups, resulting in the observed experimental changes in membrane width and hydrocarbon core electron density.

L-PS also reduced the temperature and altered the enthalpy at which the Lβ-Pβ transition (pretransition) occurred in pure DMPC. These DSC results, supported by the x-ray diffraction data, suggest that L-PS inserts into the headgroup region of liquid-crystalline DMPC resulting in an increase in area per lipid molecule, leading to a decrease in the cooperative unit size (CUS) of the main phase transition (FIG. 1).

In DMPC/cholesterol membranes, L-PS had a much greater effect on membrane CUS but a reduced ability to raise the membrane transition temperature. The endotherm width at half-maximum doubled at 5.0 mM and 1.25 mM L-PS, respectively for DMPC and DMPC/cholesterol samples. Again, the presence of cholesterol would lead to an ordered liquid-crystalline membrane, as evidenced by an increase in the control DMPC/cholesterol CUS, in which the phospholipid molecules are more sensitive to the perturbing effect of L-PS during the thermal melt.

L-PS alters membrane lipid bilayer width and electron density distribution. Small angle x-ray diffraction was used to determine molecular interactions of the phosphomonoester, L-PS, with a phospholipid bilayer. One-dimensional electron density profiles (7-12Å resolution) of the lipid bilayer and determination of membrane d-space (distance from the center of one membrane to the next, including water, at 0.4Å resolution) were ascertained from the x-ray scattering data. The results of the study demonstrated that the phospholipid metabolite, L-PS, perturbed the structure of the DMPC membrane bilayer in a significant and concentration-dependent manner. Although L-PS is charged and readily water soluble, incubation with the membrane resulted in changes throughout the lipid bilayer. In the DMPC/cholesterol lipid bilayer system, the decrease in both hydrocarbon core width by 3-4 Å and electron density ±4–12 Å from the center of the membrane were similar to a reduction in cholesterol content by one-third (from 0.6:1 to 0.4:1 cholesterol: phospholipid mole ratio, data not shown).

As suggested above from the DSC data, the mechanism by which L-PS modulates hydrocarbon core electron density may be related to its incorporation into the headgroup region of the lipid bilayer. Intercalation into the headgroups would effect a lateral displacement of the phospholipid molecules resulting in increased volume for acyl chain movement. This hypothesis is supported by data from DMPC/cholesterol samples which showed a 23% increase in phosphate headgroup peak width at half-maximum compared to control, concomitant with a broad decrease in electron density through a region of the hydrocarbon core ±4–12 Å from the center of the membrane.

Membrane biophysical changes with L-PS may play a role in AD pathophysiology. The effects of L-PS on membrane structure and lipid organization may have important implications for AD pathophysiology. Independent of its role in phospholipid metabolism, there is evidence that elevated L-PS modulates the activity of various neuropharmacological and electrophysiological processes (for review, see ref. 8). These findings suggest a possible nonspecfic mechanism of action for L-PS which may involve perturbation of the lipid bilayer as supported by the results of this study. A decrease in membrane hydrocarbon core width resulting from elevated L-PS could force displacement of amino acids associated with multiple transmembrane segments, thereby altering the integral membrane protein's overall conformation and structure. These changes in membrane structure and physical properties may play a role in abnormal processing of amyloid β-protein from its plasma membrane precursor, β-amyloid precursor protein or APP. Normal APP cleavage by "secretase" is not amino acid sequence-specific but dependent on the protein's tertiary structure and spatial relationship to the membrane surface.[9,10] The data from this study support the hypothesis that elevation of a phosphomonoester, such as L-PS, may fundamentally alter the membrane environment for APP, leading to a change in its conformation, and potentially, its catabolism.

REFERENCES

1. Brown, G. G., S. R. Levine, *et al.* 1989. In vivo ^{31}P NMR profiles of Alzheimer's disease and multiple subcortical infarct dementia. Neurology **39:** 1423–1427.
2. Pettegrew, J. W., J. Moossy, G. Withers, D. McKeag & K. Panchalingam. 1988. ^{31}P nuclear magnetic resonance study of the brain in Alzheimer's disease. J. Neuropathol. Exp. Neurol. **47:**235–248.
3. Miatto, O., R. G. Gonzalez, F. Buonanno & J. H. Growdon. 1986. In vitro ^{31}P NMR spectroscopy detects altered phospholipid metabolism in Alzheimer's disease. Can. J. Neurol. Sci. **13:**535–539.
4. Nitsch, R. M., J. K. Blusztajn, A. G. Pittas, B. E. Slack, J. H. Growdon & R. J. Wurtman. 1992. Evidence for a membrane defect in Alzheimer disease brain. Proc. Natl. Acad. Sci. USA **89:**1671–1675.

5. MASON, R. P., M. W. TRUMBORE & J. W. PETTEGREW. 1995. Membrane interactions of a phosphomonoester elevated early in Alzheimer's disease. Neurobiol. Aging. **16:**531–539.
6. BANGHAM, A. D., M. M. STANDISH & J. C. WATKINS. 1965. Diffusion of univalent ions across the lamellae of swollen phospholipids. J. Mol. Biol. **13:** 238–252.
7. MASON, R. P., W. J. SHOEMAKER, L. SHAJENKO, T. E. CHAMBERS & L. G. HERBETTE. 1992. Structure changes in Alzheimer's disease brain membranes mediated by alteration in cholesterol. Neurobiol. Aging **13:**413–419.
8. KLUNK, W. E., R. J. MCCLURE & J. W. PETTEGREW. 1991. Possible roles of L-phosphoserine in the pathogenesis of Alzheimer's disease. Mol. Chem. Neuropathol. **15:**51–73.
9. MARUYAMA, K., F. KAMETANI, M. USAMI, W. YAMAO-HARIGAYA & K. TANAKA. 1991. "Secretase", Alzheimer amyloid protein precursor secreting enzyme is not sequence-specific. Biochem. Biophys. Res. Commun. **179:**1670–1676.
10. SISODIA, S. S. 1992. β-Amyloid precursor protein cleavage by a membrane-bound protease. Proc. Natl. Acad. Sci. USA **89:**6075–6079.

Brain Glucose Metabolism Is Controlled by Amplification and Desensitization of the Neuronal Insulin Receptor[a]

S. HOYER,[b,d] N. HENNEBERG,[b] S. KNAPP,[c] H. LANNERT,[b] AND E. MARTIN[c]

[b] *Brain Metabolism Group, Department of Pathochemistry & General Neurochemistry,* [c] *Clinic for Anesthesiology, University of Heidelberg, Heidelberg, Germany*

ABSTRACT: Glucose metabolism is essential for brain function and structure. Glucose contributes to the formation of neurotransmitters and is normally the only source for energy formation. There is increasing evidence that brain glucose metabolism is under control of the neuronal insulin/insulin receptor signal transduction. The present data clearly show that intracerebroventricularly administered insulin exerts anabolic effects on cerebral glucose/energy metabolism (amplification of the neuronal insulin receptor complex) whereas cortisol (corticosterone) acts antagonistically (desensitization of the neuronal insulin receptor complex). It is also shown that short-term cortisol (corticosterone) enhanced energy turnover in temporoparietal cortex and hippocampus. In contrast, long-term cortisol (corticosterone) reduced energy turnover in both brain structures studied. This metabolic pattern is reminiscent of that found in very old age. Therefore, it is assumed that long-term cortisol accelerates the aging process in the brain and thus the risk for age-related disorders such as dementia.

INTRODUCTION

For the normal adult mammalian brain, glucose is the major nutrient whose metabolism contributes to the synthesis of the neurotransmitters acetylcholine, glutamate, aspartate, gamma-aminobutyric acid and glycine (for review, see ref. 1). The main product of glucose metabolism is ATP, necessary for both function and structure of nervous cells.[2] Additionally, ATP acts extracellularly as an ultrarapid neurotransmitter[3] and as a mitogen.[4]

[a] The work of N.H. was supported by a grant from the Graduate Program in Molecular and Cellular Neurobiology, University of Heidelberg.

[d] *Send correspondence to:* Dr. Siegfried Hoyer, Brain Metabolism Group, Department of Pathochemistry & General Neurochemistry, University of Heidelberg, In Neuenheimer Feld 220-221, 69120 Heidelberg, Germany; Tel: x6221-566218; Fax: x6221-563328.

In non-nervous tissue, glucose metabolism is under control of the insulin/insulin receptor complex. Insulin amplifies the insulin receptor function via activation of tyrosine kinase.[5] The receptor is desensitized by the activation of phosphotyrosine phosphatase and serine kinase mediated by glucocorticoids[6] and catecholamines.[7] The existence of both insulin and insulin receptors in the CNS is now well established. The latter are dispersed throughout the brain following a highly specific pattern, with the highest density detected in olfactory bulb, hypothalamus, cerebral cortex, and hippocampus.[8] Interestingly, insulin receptor substrate 1 and the insulin receptor are co-expressed in discrete populations of neurons in olfactory bulb and hippocampus.[9]

Insulin gene expression and insulin synthesis was demonstrated to occur in a highly specific pattern, with highest density in hippocampal pyramidal cells, and high densities in medial prefrontal, entorhinal, perirhinal and olfactory cortices.[10] It is, however, not clear as yet whether or not the neuronal insulin receptor is controlled antagonistically by insulin and cortisol as is the case in non-nervous tissue. Therefore, the goal of the present study was to amplify and to densensitize the neuronal insulin receptor and to investigate the effect of amplification and desensitization on brain glucose/energy metabolism.

MATERIAL AND METHODS

The study was conducted on 1-year-old and 2-year-old male Wistar rats. The numbers of animals per group is given in the tables. The final experiment was performed under anesthesia of 0.5 vol% halothane and nitrous oxide/oxygen 70:30 under physiological steady state conditions. The brains were frozen *in situ* by liquid nitrogen. The metabolites and enzyme activities of the cerebral glucose/energy metabolism were determined spectrophotometrically and by HPLC. Statistical significance was assumed at $p \leq 0.05$.

Amplification Studies

Investigations were on 1- and 2-year-old animals: a) acute—intracerebroventricular (icv) single dose administration of insulin (80 mU); b) short-term and long-term icv infusion of insulin (60 mU/day) over 1, 7, and 21 days.

Desensitization Studies

Studies were on 1- and 2-year-old animals: a) icv administration of streptozotocin (STZ) to inhibit tyrosine kinase activity of the insulin receptor;[11] b) short-term administration of corticosterone; day 1 and day 2:20 mg/kg b.w. s.c.; day 3: 1 mg/kg b.w. i.p.; and c) long-term administration of corticosterone; 3 mg/ml sesame oil/rat s.c. (~500 g b.w.) daily over 6 weeks.

RESULTS AND DISCUSSION

Amplification Studies

Icv administration of 80 mU of insulin (single dose) led to significant increases in the activities of the glycolytic key enzymes hexokinase and phosphofructokinase in cerebral temporoparietal cortex.[12] Short-term or long-term icv infusion of insulin (60 mU/day) significantly increased the concentration of creatine phosphate (CrP) in the hippocampus, whereas in cerebral temporoparietal cortex the tissue concentrations of ATP, ADP, CrP, ATP/ADP ratio and energy charge potential were all unaltered[13] (TABLE 1).

Desensitization Studies

Icv STZ caused significant decreases in local glucose utilization in frontal, parietal sensory motor, auditory and entorhinal cerebral cortices, and in all hippocampal subfields.[14] The activities of the glycolytic key enzymes hexokinase, phosphofructokinase, and pyruvate kinase dropped significantly in cerebral temporoparietal cortex and in hippocampus.[15] The concentrations of ATP and CrP fell significantly in cerebral temporoparietal cortex. The ADP concentration and ATP turnover were increased[16] (TABLE 2b).

Neither short-term nor long-term administration of corticosterone varied the size of the energy pool in cerebral temporoparietal cortex and in hippocampus. However, energy turnover expressed as ATP/ADP ratio and GTP/

TABLE 1. Energy Pool and Energy Turnover in Temporoparietal Cortex and Hippocampus after icv Infusion of Insulin

	ATP	ADP	CrP	ATP/ADP
Cortex				
Control	2.38	0.482	4.89	5.77
Icv insulin	2.48	0.471	4.92	5.84
Hippocampus				
Control	2.41	0.547	4.88	4.61
Icv insulin	2.46	0.802	5.42*	4.89

Since neither duration of the experiment nor age of the experimental animals exhibited any influence on response to insulin treatment, data were analyzed together (n = 60/group) (after Henneberg & Hoyer[13]). * $p \leq 0.05$

TABLE 2. a) Energy Pool and Energy Turnover in Temporoparietal Cortex and Hippocampus after Short-term and Long-term Administration of Corticosterosterone b) Energy Pool and Energy Turnover in Temporoparietal Cortex after Icv Injection of Streptozotocin

	ATP	ADP	CrP	GTP	GDP	ATP/ADP	GTP/GDP
				μmol/g ww			
a) Cortex							
Control	2.60	0.483	5.08	0.663	0.392	5.50	1.71
Corticosterone short-term	2.54	0.495	5.02	0.664	0.391	5.40	1.78
Corticosterone long-term	2.53	0.406*	5.34	0.652	0.275*	6.24	2.39*
Hippocampus							
Control	2.21	0.503	4.91	0.605	0.372	4.50	1.71
Corticosterone short-term	2.27	0.700*	5.00	0.616	0.454	3.44	1.38*
Corticosterone long-term	2.25	0.457	4.95	0.595	0.350	5.03	1.74
b) Cortex							
Control	2.68	0.470	4.71			5.70	
Icv streptozotocin	2.28*	0.669*	3.77*			3.42*	

[a] n = 7/group; * $p \leq 0.05$ vs. control.
[b] n = 20/group; after Nitsch & Hoyer[16]; * $p \leq 0.05$ vs. control.

GDP ratio increased after short-term corticosterone in the hippocampus. In contrast, energy turnover was reduced after long-term corticosterone in temporoparietal cortex (TABLE 2a). The latter abnormality is reminiscent of the diminution of energy turnover during old age and mental activity,[17] which is obviously caused by a reduction of the activity of adenine nucleotide translocase.[18]

From the present data, the tentative conclusion is drawn that insulin exhibits a discrete anabolic effect on hippocampal energy metabolism, *i.e.* in a brain area which shows a particular insulin sensitivity.[10,19] Icv STZ treatment caused long-term abnormalities in energy production and utilization. The glucocorticoid corticosterone acts antagonistically to insulin on energy metabolism in temporoparietal cortex and hippocampus. Short-term corticosterone-induced stress is responded by an activation of energy turnover in hippocampus. In contrast, this compensation mechanism fails after long-term corticosterone-induced stress. This metabolic pattern is reminiscent of changes in energy metabolism in very old age. Therefore, the enhancement of cortisol concentration in CSF of aged human beings[20] may be assumed to accelerate the aging process of the brain which may become risky for dementia.[21] The functional deficit of available energy may have drastic impacts on energy-dependent cellular and molecular processes.

REFERENCES

1. HOYER, S. 1990. Brain glucose and energy metabolism during normal aging. Aging **2**:245–258.
2. ERECINSKA, M. & I. A. SILVER. 1989. ATP and brain function. J. Cereb. Blood Flow Metab. **9**:2–19.
3. EDWARDS, F. A., A. J. GIBB & D. COLQUHOUN. 1992. ATP receptor-mediated synaptic currents in the central nervous system. Nature **359**:144–147.
4. NEARY, J. T., S. R. WHITTEMORE, Q. ZHU & M. D. NORENBERG. 1994. Synergistic activation of DNA synthesis in astrocytes by fibroblasts growth factors and extracellular ATP. J. Neurochem. **63**:490–494.
5. WHITE, M. F. & C. R. KAHN. 1994. The insulin signalling system. J. Biol. Chem. **269**:1–4.
6. GIORGINO, F., A. ALMAHFOUZ, L. J. GOODYEAR & R. J. SMITH. 1993. Glucocorticoid regulation of insulin receptor and substrate IRS-1 tyrosine phosphorylation in rat skeletal muscle in vivo. J. Clin. Invest. **91**:2020–2030.
7. HÄRING, H. U. 1991. The insulin receptor: Signalling mechanisms and contribution to the pathogenesis of insulin resistance. Diabetologia **34**:848–861.
8. UNGER, J. W., J. N. LIVINGSTON & A. M. MOSS. 1991. Insulin receptors in the central nervous system: Localization, signalling mechanisms and functional aspects. Progr. Neurobiol. **36**:343–362.
9. BASKIN, D. G., M. W. SCHWARTZ, A. J. SIPOLS, D. A. D'ALAESSIO, B. J.

GOLDSTEIN & M. F. WHITE. 1994. Insulin receptor substrate-1 (IRS-1) expression in rat brain. Endocrinology **134:**1952–1955.

10. DEVASKAR, S. U., S. J. GIDDINGS, P. A. RAJAKUMAR, L. R. CARNAGHI, R. K. MENON & D. S. ZAHM. 1994. Insulin gene expression and insulin synthesis in mammalian neuronal cells. J. Biol. Chem. **269:**8445–8454.
11. KADOWAKI, T., M. KASUGA, Y. AKANUMA, O. EZAKI & F. TAKAKU. 1984. Decreased autophosphorylation of the insulin receptor-kinase in streptozotocin diabetic rats. J. Biol. Chem. **259:**14208–14216.
12. HOYER, S., L. PREM, S. SORBI & L. AMADUCCI. 1993. Stimulation of glycolytic key enzymes in cerebral cortex by insulin. NeuroReport **4:**991–993.
13. HENNEBERG, N. & S. HOYER. 1994. Short-term or long-term intracerebroventricular (i.c.v.) infusion of insulin exhibits a discrete anabolic effect on cerebral energy metabolism in the rat. Neurosci. Lett. **175:**153–156.
14. DUELLI, R., H. SCHRÖCK, W. KUSCHINSKY & S. HOYER. 1994. Intracerebroventricular injection of streptozotocin induces discrete local changes in cerebral glucose utilization in rats. Int. J. Dev. Neurosci. **12:**737–743.
15. PLASCHKE, K. & S. HOYER. 1993. Action of the diabetogenic drug streptozotocin on glycolytic and glycogenolytic metabolism in adult rat brain cortex and hippocampus. Int. J. Dev. Neurosci. **11:**477–483.
16. NITSCH, R. & S. HOYER. 1991. Local action of the diabetogenic drug, streptozotocin, on glucose and energy metabolism in rat brain cortex. Neurosci. Lett. **128:**199–202.
17. DUTSCHKE, K., R. M. NITSCH & S. HOYER. 1994. Short-term mental activation accelerates the age-related decline of high-energy phosphates in rat cerebral cortex. Arch. Gerontol. Geriatr. **19:**43–51.
18. NOHL, H. & R. KRÄMER. 1980. Molecular basis of age-dependent changes in the activity of adenine nucleotide translocase. Mech. Ageing Develop. **14:** 137–144.
19. ZHU, C. Z. & R. N. AUER. 1994. Intraventricular administration of insulin and IGF-I in transient forebrain ischemia. J. Cereb. Blood Flow Metab. **14:** 237–242.
20. SWAAB, D. F., F. C. RAADSHEER, E. ENDERT, M. A. HOFMAN, W. KAMPHORST & R. RAVID. 1994. Increased cortisol levels in aging and Alzheimer's disease in postmortem cerebrospinal fluid. J. Neuroendocrinol. **6:**681–687.
21. HOYER, S. 1994. Age as risk factor for sporadic dementia of the Alzheimer type? Ann. N.Y. Acad. Sci. **719:**248–256.

Cellular Mechanisms of Brain Energy Metabolism

Relevance to Functional Brain Imaging and to Neurodegenerative Disorders[a]

P. J. MAGISTRETTI[b] AND L. PELLERIN

Institut de Physiologie, Universite de Lausanne, Lausanne, Switzerland

ABSTRACT: Astrocyte end-feet surround intraparenchymal microvessels and represent therefore the first cellular barrier for glucose entering the brain. As such, they are a likely site of prevalent glucose uptake. Astrocytic processes are also wrapped around synaptic contacts, implying that they are ideally positioned to sense and be functionally coupled to increased synaptic activity. We have recently demonstrated that glutamate, the main excitatory neurotransmitter, stimulates in a concentration-dependent manner 2-DG uptake and phosphorylation by astrocytes in primary culture.[1] The effect is not receptor-mediated but rather proceeds via one of the recently cloned glutamate transporter. The mechanism involves an activation of the Na^+/K^+ ATPase. Concomitant to the stimulation of glucose uptake, glutamate causes a concentration-dependent increase in lactate efflux. These observations suggest that glutamate uptake is coupled to aerobic glycolysis in astrocytes. In addition, since glutamate release occurs following the modality-specific activation of a brain region, the glutamate-evoked uptake of glucose into astrocytes provides a simple mechanism to couple neuronal activity to energy metabolism. These data also suggest that modality-specific activation visualized using 2DG-based autoradiography or PET may primarily reflect glutamate-mediated uptake of 2DG into astrocytes.

INTRODUCTION

Studies at the organ and regional levels, while providing a focus on synaptic terminals as the principal sites of activation-dependent increases in energy metabolism, lack the cellular resolution required to determine the relative contribution of neurons and glial cells as well as the nature of metabolic exchanges occurring between these two cell types. Studies in cellularly homogeneous preparations in conjunction with others on well compartmentalized and simple nervous tissues have provided strong indications for a prevalent

[a] Research in PJM's laboratory is supported by FNRS Grant 31-40'565.94.

[b] *Send correspondence to:* Dr. Pierre J. Magistretti, Institut de Physiologie, 7 Rue du Bugnon, 1005 Lausanne, Switzerland; Tel: 021 692 5542; Fax: 021 692 5505.

glial localization of the activation-induced increase in glucose utilization. We present data here which lend support to this idea and we describe a mechanism which provides a cellular and molecular basis for PET- and fMRI-based functional brain imaging. In addition, the model proposed might have important implications for certain neurodegenerative diseases considering the possible involvement of metabolic impairments in such disorders.

METHODS

Preparation of Primary Cultures

Primary cultures of mouse cerebral cortical astrocytes were prepared from Swiss albino newborn mice (1–2 days old) as described.[2] This procedure yields cultures that are >95% immunoreactive for glial fibrillary acidic protein. Primary cultures of cerebral cortical neurons were prepared from Swiss albino mice embryos (16–17 days old) as described.[3] This procedure yields cultures which show no immunoreactivity for glial (GFAP), oligodendroglial (GalC), and microglial (Mac1) markers, and >95% immunoreactivity to neuronal markers (Neuron Specific Enolase and Neurofilament-H).

[^{3}H]2DG Uptake

[^{3}H]2DG Uptake was determined as described.[1] For estimation of basal glucose utilization, various concentrations (0.05–10 mM) of cold 2-deoxyglucose were present in the medium and the incubations were carried for 5 min. Apparent Vmax was calculated with Sigmaplot software after generating the best fit curve for the data.

Lactate Release

Lactate release into the medium was measured enzymatically as described.[1]

RESULTS

Basal Glucose Utilization

Basal rate of glucose utilization can be estimated in cellularly homogeneous preparations of astrocytes or neurons using the glucose analog 2-deoxyglucose (2-DG). The basal rate of glucose utilization in astrocytes was found to be higher than in neurons, with a rate in astrocytes of 22.3 nmol/mg prot/min and in neurons of 6.3 nmol/mg prot/min at saturating concentrations of glucose. These values are of the same order as those determined *in vivo* for cortical grey matter, with the 2-DG autoradiographic technique.[4] Thus, in view of this difference and of the quantitative preponderance of astrocytes compared to neurons in the grey matter, these data suggest a significant contribution of astrocytes to basal glucose utilization as determined by 2-DG autoradiography or ^{18}F-2DG PET.

Glucose Utilization during Activation

Since during activation of a given cortical area the concentration of glutamate in the extracellular space increases considerably due to its release from the axon terminals of activated pathways, the effect of glutamate application on glucose utilization by astrocytes and neurons in culture was examined. As shown in FIGURE 1, L-glutamate stimulates 2-DG uptake and phosphorylation by astrocytes in a concentration-dependent manner, with an EC_{50} of 80 μM.[1] This effect is not mediated by specific glutamate receptors since it is not inhibited by any of the specific antagonists tested nor is it mimicked by agonists specific for each receptor subtype such as NMDA, AMPA, quisqualate or *t*-ACPD (TABLE 1). The effect of glutamate is stereospecific, with only the L-isomer being active, and dependent on the presence of extracellular sodium. This pharmacological profile is characteristic of the glutamate transporter. Indeed, the increase in glucose utilization evoked by glutamate is

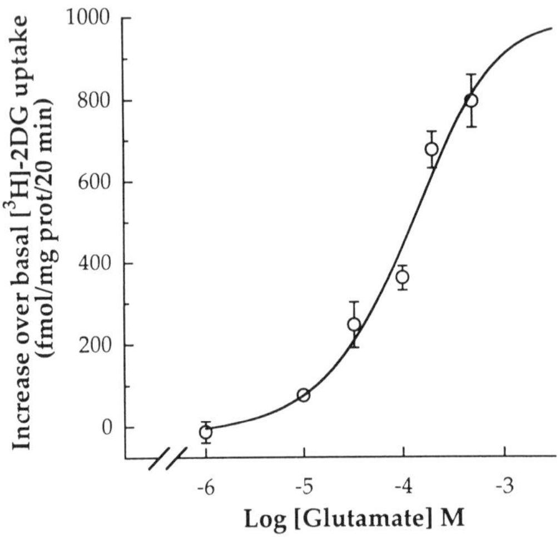

FIGURE 1. Concentration-response curve of the stimulation by glutamate of [^{3}H]2DG uptake by astrocytes. The EC_{50} of this effect is 80 μM (taken from Pellerin and Magistretti[1]).

TABLE 1. Effect of Glutamate Receptor Agonists on [^{3}H]-2DG Uptake by Astrocytes

Treatment	[^{3}]H-2DG Uptake (fmol/mg prot)
None	871 ± 35
Glutamate 500 μM	1581 ± 45
NMDA 500 μM	793 ± 9
Kainate 500 μM	1081 ± 24
Quisqualate 500 μM	918 ± 59
AMPA 500 μM	917 ± 37
t-ACPD 500 μM	840 ± 29

Results are the means ± SEM of at least three separate determinations. Incubation period was 20 minutes.

completely abolished by preincubation of the cultures with the potent glutamate transporter inhibitor DL-*threo*-β-hydroxyaspartate (THA).[1] These results clearly indicate a tight coupling between Na^+-dependent glutamate uptake and glucose utilization by astrocytes. A critical involvement of the Na^+/K^+-ATPase is likely since ouabain completely inhibits the glutamate-evoked 2-DG uptake by astrocytes.[1] Application of glutamate to cultured astrocytes also caused a concentration-dependent release of lactate (FIG. 2), and to a lesser degree pyruvate.[1] This glutamate-evoked lactate release shows the same pharmacology and time-course as the glutamate-evoked glucose utilization thus indicating that glutamate stimulates the processing of glucose through glycolysis.[1]

In contrast to what is observed in astrocytes, glutamate does not stimulate 2-DG uptake in our cultured neurons despite the fact that glutamate can elicit

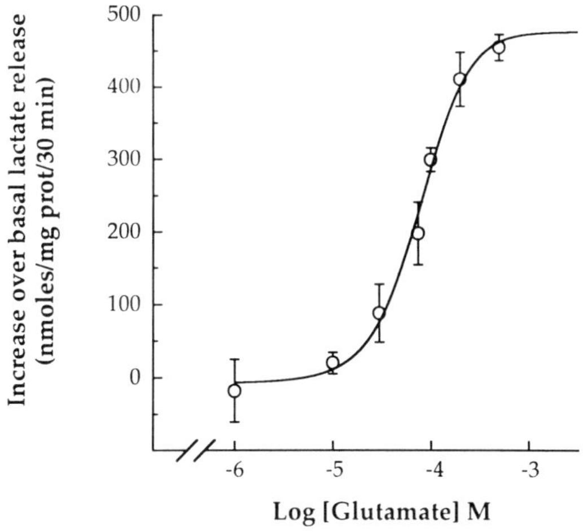

FIGURE 2. Concentration response curve of the stimulation by glutamate of lactate release by astrocytes. The EC_{50} of this effect is 78 μM.

functional responses in these cells which are mediated by ionotropic and metabotropic receptors. Examples of functional responses in these neuronal cultures, which are totally devoid of glial markers, are the NMDA and AMPA/kainate-mediated increases in intracellular calcium,[3] stimulation of arachidonic acid formation[3] and induction of c-fos expression,[5] as well as the metabotropic receptor-mediated stimulation of phospholipase C.[3]

DISCUSSION

Based on our observations, we have proposed a model of cell-specific metabolic regulation (ref. 1 and FIG. 3). In particular, when we consider glutamate-

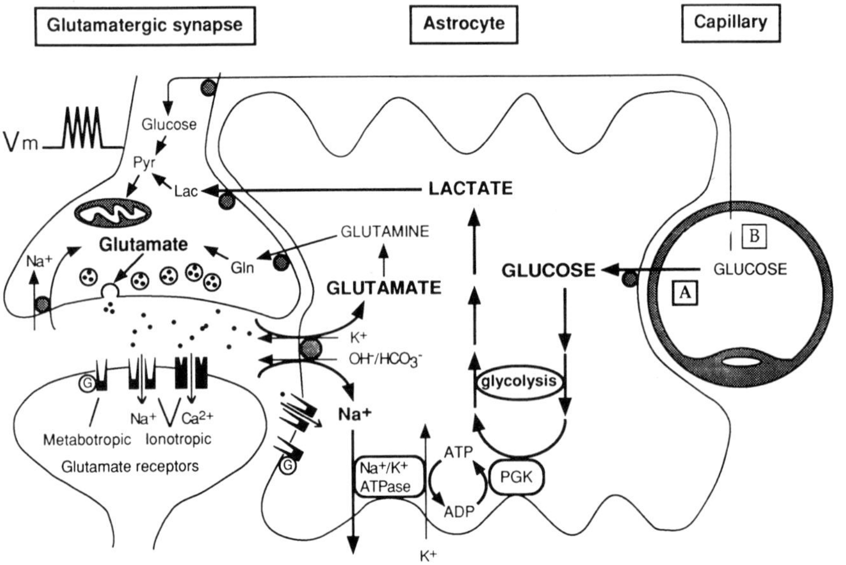

FIGURE 3. Schematic representation of the mechanism for glutamate-induced glycolysis in astrocytes during physiological activation. At glutamatergic synapses, glutamate depolarizes neurons by acting at specific receptor subtypes. The action of glutamate is terminated by an efficient glutamate uptake system located primarily in astrocytes. Glutamate is cotransported with Na^+, resulting in an increase in the intracellular concentration of Na^+, leading to an activation of the Na^+/K^+ ATPase. The pump, fueled by ATP provided by membrane-bound glycolytic enzymes (possibly phosphoglycerate kinase, PGK), activates glycolysis, *i.e.* glucose utilization and lactate production, in astrocytes. Lactate, once released, can be taken up by neurons and serve as an adequate energy substrate. For graphic clarity only lactate uptake into presynaptic terminals is indicated. However, this process could also occur at the postsynaptic neuron. This model, which summarizes *in vitro* experimental evidences indicating glutamate-induced glycolysis, is taken to reflect cellular and molecular events occurring during *activation* of a given cortical area (arrow labeled A: activation). Direct glucose uptake into neurons under basal conditions is also shown (arrow labeled B: basal conditions). Pyr: pyruvate; Lac: lactate; Gln: glutamine; G: G-protein (taken from Pellerin and Magistretti[1]).

induced glycolysis, this model is taken to reflect cellular and molecular events occurring during *activation* of a given cortical area. Such a metabolic compartmentation, whereby glucose taken up by astrocytes is metabolized glycolytically to lactate, which is then released in the extracellular space to be utilized by neurons, is consistent with a number of biochemical and electrophysiological observations. For example, studies in the well-compartmentalized honeybee drone retina and in isolated preparations of guinea pig retina containing photoreceptors attached to Müller cells corroborate such metabolic fluxes between glia and neurons.[6,7]

Correlations with Functional Brain Imaging

These observations and the model described in FIGURE 3 are consistent with some now well-established, but originally unexpected findings obtained in functional brain-imaging studies, for which they provide in fact some cellular and molecular bases. PET studies in which blood flow, oxygen consumption and glucose utilization were determined in the same subject have now clearly established that focal physiological cortical activation results in a metabolic uncoupling whereby the increases in blood flow and glucose utilization are not matched by a commensurate increase in oxygen consumption, indicating a prevalent nonoxidative glucose utilization.[8,9] These *in vivo* observations are consistent with the model of metabolic fluxes summarized in FIGURE 3. Moreover, the glutamate-induced glycolysis in astrocytes we report provides a cellular basis for the metabolic uncoupling demonstrated in PET studies which has led to the notion of activation-induced glycolysis.[9]

Evidence for glycolysis as the principal metabolic pathway operational during activation of a given brain area has also been provided by *in vivo* ^{1}H-NMR spectroscopy studies.[10,11] Such analyses, which are consistent with evidences obtained in microdialysis studies,[12] show increases in the lactate signal in primary visual cortex following physiological activation.[10,11] These *in vivo* ^{1}H-NMR spectroscopy studies taken together with the PET data indicating a metabolic uncoupling, reveal a previously unrecognized prevalence of glycolysis over oxidative phosphorylation during activation. More precisely, it indicates that the mammalian brain shifts to glycolysis as a source of energy during brief increases in neuronal activity.

Implications of Metabolic Impairment for Neurodegenerative Disorders

There have been several reports of hypometabolism or metabolic impairment in a variety of neurodegenerative disorders, and notably Alzheimer's disease, Huntington's disease and amyotrophic lateral sclerosis.[13–15] It is evident that any disturbance in the mechanism we have described could have dramatic consequences for neuronal survival. For example, it has been shown that glutamate uptake is critically dependent on glycolysis.[16] Impairment of energy metabolism which would result in decreased glutamate uptake could lead to excitotoxicity and neuronal cell death. Indeed, deficits in glutamate uptake have been found in certain neurodegenerative diseases.[17,18] Moreover, *in vitro* models have been developed by blocking glutamate uptake which reproduced features of these diseases.[19] In conclusion, it might be well worth to focus further attention on a possible role of energy metabolism impairments in neurodegeneration.

REFERENCES

1. PELLERIN, L. & P. J. MAGISTRETTI. 1994. Glutamate uptake into astrocytes stimulates aerobic glycolysis: A mechanism coupling neuronal activity to glucose utilization. Proc. Natl. Acad. Sci. USA **91**:10625–10629.
2. SORG, O. & P. J. MAGISTRETTI. 1991. Characterization of the glycogenolysis elicited by vasoactive intestinal peptide, noradrenaline and adenosine in primary cultures of mouse cerebral cortical astrocytes. Brain Res. **563**:227–233.
3. STELLA, N., L. PELLERIN & P. J. MAGISTRETTI. 1995. Modulation of the glutamate-evoked release of arachidonic acid from mouse cortical neurons: Involvement of a pH-sensitive membrane phospholipase A2. J. Neurosci. **15**: 3307–3317.
4. SOKOLOFF, L., M. REIVICH, C. KENNEDY, M. H. DES ROSIERS, C. S. PATLAK, K. D. PETTIGREW, O. SAKURADA & M. SHINOHARA. 1977. The [^{14}C]deoxyglucose method for the measurement of local cerebral glucose utilization: theory, procedure, and normal values in the conscious and anesthetized albino rat. J. Neurochem. **28**:897–916.
5. PELLEGRI, G., P. J. MAGISTRETTI & J.-L. MARTIN. 1994. Regulation of gene expression by VIP and glutamate in cultured cortical neurons. Soc. Neurosci. Abstr. **20**:732.
6. TSACOPOULOS, M., A.-L. VEUTHEY, S. G. SARAVELOS, P. PERROTTET & G. TSOUPRAS. 1994. Glial cells transform glucose to alanine, which fuels the neurons in the honeybee retina. J. Neurosci. **14**:1339–1351.
7. POITRY-YAMATE, C. L. & M. TSACOPOULOS. 1992. Glucose metabolism in freshly isolated Müller glial cells from a mammalian retina. J. Comp. Neurol. **320**:257–266.
8. FOX, P. T. & M. E. RAICHLE. 1986. Focal physiological uncoupling of cerebral blood flow and oxidative metabolism during somatosensory stimulation in human subjects. Proc. Natl. Acad. Sci. USA **83**:1140–1144.
9. FOX, P. T., M. E. RAICHLE, M. A. MINTUN & C. DENCE. 1988. Nonoxidative glucose consumption during focal physiologic neural activity. Science **241**: 462–464.
10. PRICHARD, J., D. ROTHMAN, E. NOVOTNY, O. PETROFF, T. KUWABARA, M. AVISON, A. HOWSEMAN, C. HANSTOCK & R. SHULMAN. 1991. Lactate rise

detected by ^{1}H NMR in human visual cortex during physiologic stimulation. Med. Sci. **88:**5829–5831.

11. SAPPEY-MARINIER, D., G. CALABRESE, G. FEIN, J. W. HUGG, C. BIGGINS & M. W. WEINER. 1992. Effect of photic stimulation on human visual cortex lactate and phosphates using ^{1}H and ^{31}P magnetic resonance spectroscopy. J. Cereb. Blood Flow Metab. **12:**584–592.
12. FELLOWS, L. K., M. G. BOUTELLE & M. FILLENZ. 1993. Physiological stimulation increases nonoxidative glucose metabolism in the brain of the freely moving rat. J. Neurochem. **60:**1258–1263.
13. SMITH, G. S., M. J. DE LEON, A. E. GEORGE, A. KLUGER, N. D. VOLKOW, T. MCRAE, J. GOLOMB, S. H. FERRIS, B. REISBERG, J. CIARAVINO & M. E. LA REGINA. 1992. Topography of cross-sectional and longitudinal glucose metabolic deficits in Alzheimer's disease. Arch. Neurol. **49:**1142–1150.
14. BEAL, M. F., B. T. HYMAN & W. KOROSHETZ. 1993. Do defects in mitochondrial energy metabolism underlie the pathology of neurodegenerative diseases? Trends Neurosci. **16:**125–131.
15. DALAKAS, M. C., J. HATAZAWA, R. A. BROOKS & G. DI CHIRO. 1987. Lowered cerebral glucose utilization in amyotrophic lateral sclerosis. Ann. Neurol. **22:** 580–586.
16. SWANSON, R. A. 1992. Astrocyte glutamate uptake during chemical hypoxia in vitro. Neurosci. Lett. **147:**143–146.
17. PERRY, T. L. & S. HANSEN. 1990. What excitotoxin kills striatal neurons in Huntington's disease? Clues from neurochemical studies. Neurology **40:**20–24.
18. PLAITAKIS, A., E. CONSTANTAKAKIS & J. SMITH. 1988. The neuroexcitotoxic amino acids glutamate and aspartate are altered in the spinal cord and brain in amyotrophic lateral sclerosis. Ann. Neurol. **24:**446–449.
19. ROTHSTEIN, J. D., L. JIN, M. DYKES-HOBERG & R. W. KUNCL. 1993. Chronic inhibition of glutamate uptake produces a model of slow neurotoxicity. Proc. Natl. Acad. Sci. USA **90:**6591–6595.

β-Amyloidosis in Normal Aging and Transmitter Signaling in Human Temporal Lobe[a]

E. K. PERRY,[b,e] J. A. COURT,[b] S. LLOYD,[b] M. JOHNSON,[b] M. H. GRIFFITHS,[b,c] D. SPURDEN,[b] M. A. PIGGOTT,[b] J. TURNER,[c] AND R. H. PERRY[d]

[b] *Medical Research Council Neurochemical Pathology Unit, Newcastle General Hospital, Newcastle upon Tyne, United Kingdom*
[c] *Research Laboratories of Schering AG, 13353 Berlin, Germany*
[d] *Department of Neuropathology, Newcastle General Hospital, Newcastle upon Tyne, United Kingdom*

ABSTRACT: Interactions between abnormal amyloid precursor protein metabolism and cholinergic dysfunction are increasingly apparent. Both of these major features of Alzheimer's disease occur in restricted loci in normal aging—a potential model for early Alzheimer type pathology. Entorhinal cortex is particularly vulnerable to β-amyloidosis and compared with other cortical areas is remarkable for the relatively high density of nicotinic (^{3}H-nicotine) but not other cholinergic or glutamate receptor binding. With increasing age, post-maturity, there is a persistent decline in nicotinic receptor binding in entorhinal cortex whereas muscarinic M1 and non-M1, glutamate NMDA and non-NMDA receptors are spared. Normal elderly individuals, distinguished by the absence of βA4 immunoreactive plaques in this area, are differentiated from those with plaques by higher nicotine binding. Amongst individuals with an established history of smoking tobacco, nicotinic receptor binding and hippocampal choline acetyltransferase were elevated compared with non-smokers and preliminary evidence indicates a reduced density of cortical plaques. These findings are consistent with the hypothesis that down regulation of the nicotinic cholinergic receptor—a ligand gated calcium channel known to control the expression of neurotrophins—plays a role in the evolution of Alzheimer-type pathology.

INTRODUCTION

Accumulation of βA4 and defective cholinergic transmission are integral features of Alzheimer's disease which have been linked in various ways[1] in-

[a] This research was partially supported by Schering AG, Berlin and Molecular Geriatrics Inc, Illinois.

[e] *Send correspondence to:* Dr. Elaine K. Perry, Medical Resarch Council Neurochemical Pathology Unit, Newcastle General Hospital, Westgate Road, Newcastle upon Tyne, NE4 6BE, UK; Tel: 0191 273 5251; Fax: 0191 272 5291.

cluding: cholinesterase reactivity at early stages of plaque formation; increasing density of senile plaques related to declining cholinergic activity; and stimulation of normal as opposed to amyloidogenic APP metabolism by M1 and M3 muscarinic receptor activation. Recently gene sequences for rat synuclein, a component of cholinergic synapses, have been reported to bear an overall 90% identity to the human gene coding a non-amyloid component peptide, tightly associated with the amyloid of Alzheimer's disease.[2] This provides a further molecular link between degeneration of cholinergic nerve terminals and plaque formation.

In the normal aging brain the entorhinal and adjacent perirhinal cortex are early sites of plaque formation. This area is associated with relatively high nicotinic receptor binding in the human cortex.[3] Situated at the origin of the perforant pathway the receptor may be important in the integration and transfer of information from diverse cortical areas to the hippocampus. In the present investigation nicotinic and other transmitter receptors have been examined in the normal aging human parahippocampal gyrus and related to the evolution of β-amyloidosis and declining cholinergic activity. Individuals with established histories of tobacco smoking and non-smoking have been compared to further evaluate the role of the nicotinic receptor in the evolution of Alzheimer-type neuropathology.

METHODS

The series of normal cases with no neurological or psychiatric disease, ranging in age from 24 gestational weeks to 95 years, is as previously described.[4] Hippocampus, at the level of the geniculate nucleus, and adjacent gyrus, snap-frozen and maintained for up to 10 years at −70°C, was cryostat sectioned at 20 μm. Nicotinic, muscarinic M1, NMDA, AMPA and kainate receptor autoradiography were performed using [^{3}H]-nicotine, pirenzepine displaced [^{3}H]-*N*-methylscopolamine, [^{3}H]-MK801, [^{3}H]-CNQX and [^{3}H]-kainate.[3] Non-M1 muscarinic receptor binding represented that remaining in the presence of 10^{-6}M pirenzepine. βA4 immunoreactivity was assessed in adjacent formalin fixed sections using a purified polyclonal antibody raised against $\beta A4_{1-42}$. In cases over the age of 60 years, reactive deposits and/or neuritic plaques were evident in 7 and absent in 8 (mean ages 84 ± 11 and 76 ± 16 y). In an extended normal series, examined for nicotinic receptor binding and choline acetyltransferase, clearcut evidence of no, current or previous cigarette smoking was obtained in 24 cases: 8 non-smokers, 8 ex-smokers and 8 smokers (mean ages 80 ± 8, 73 ± 10, 72 ± 11 years).

RESULTS

FIGURE 1, A–F illustrate the patterns of choline acetyltransferase, nicotinic, muscarinic and glutamate receptor binding during development and aging

RESULTS

FIGURE 1, A–F illustrate the patterns of choline acetyltransferase, nicotinic, muscarinic and glutamate receptor binding during development and aging in the parahippocampal gyrus. Nicotinic receptor binding decreased progressively following birth. Choline acetyltransferase peaked in middle age from low neonatal levels and declined about 30% from middle to old age. Muscarinic M1 and non-M1 subtypes (the latter not illustrated) and glutamate NMDA receptor binding varied little across the age span, remaining constant from adult to old age. AMPA binding rose 10-fold from fetus to adulthood and kainate binding declined 3-fold from neonate to adult, both remaining constant from middle to old age. Compared with the other receptors, only the nicotinic altered significantly with aging post maturity declining about 50% from adulthood to old age—this change occurring earlier than and being greater than that for the cholinergic enzyme. In cases over 60 years, nicotine binding was lower in cases with as opposed to without β-amyloidosis (FIG. 1G).

In a further series with complete smoking histories, the nicotinic receptor was higher in smokers compared to non-smokers and ex-smokers (FIG. 1H). Hippocampal choline acetyltransferase was higher in smokers and ex-smokers compared with non-smokers (FIG. 1).

DISCUSSION

The substantial loss of nicotinic receptor binding during aging was not evident for the other cholinergic and glutamate receptors investigated and was confined to entorhinal and perirhinal cortex (compared to hippocampus, dentate gyrus and other neocortical regions, unpublished observations). This regional selectivity of receptor loss is similar to the general pattern of plaque formation in normal elderly individuals. Since β-amyloid plaques rarely appear before the 7th decade the earlier decline in nicotinic receptor binding suggests, as does the lower binding in cases with amyloid, that the receptor loss precedes plaque formation.

The higher receptor binding in smokers compared to non-smokers is consistent with results of experimental administration of nicotine in rodent brain and with previous analysis of human brain tissue in tobacco users.[5] Although in smokers the receptor is significantly increased in entorhinal cortex and hippocampus, choline acetyltransferase was only elevated in hippocampus. This may reflect activation of the perforant pathway projecting to the hippocampus or of other pathways to this area which are stimulated by nicotine (*e.g.* dopaminergic). If substantiated, elevated cholinergic activity in smokers may reflect neuroprotective effects of nicotinic receptor stimulation in offsetting the age-associated decline in hippocampal choline acetyltransferase.[4] Preliminary evidence of reduced β-amyloidosis in tobacco smokers[1] is consistent with accumulating epidemiological evidence that smoking reduces the risk of developing Alzheimer's disease by around 50%.[6] It is possible that abnor-

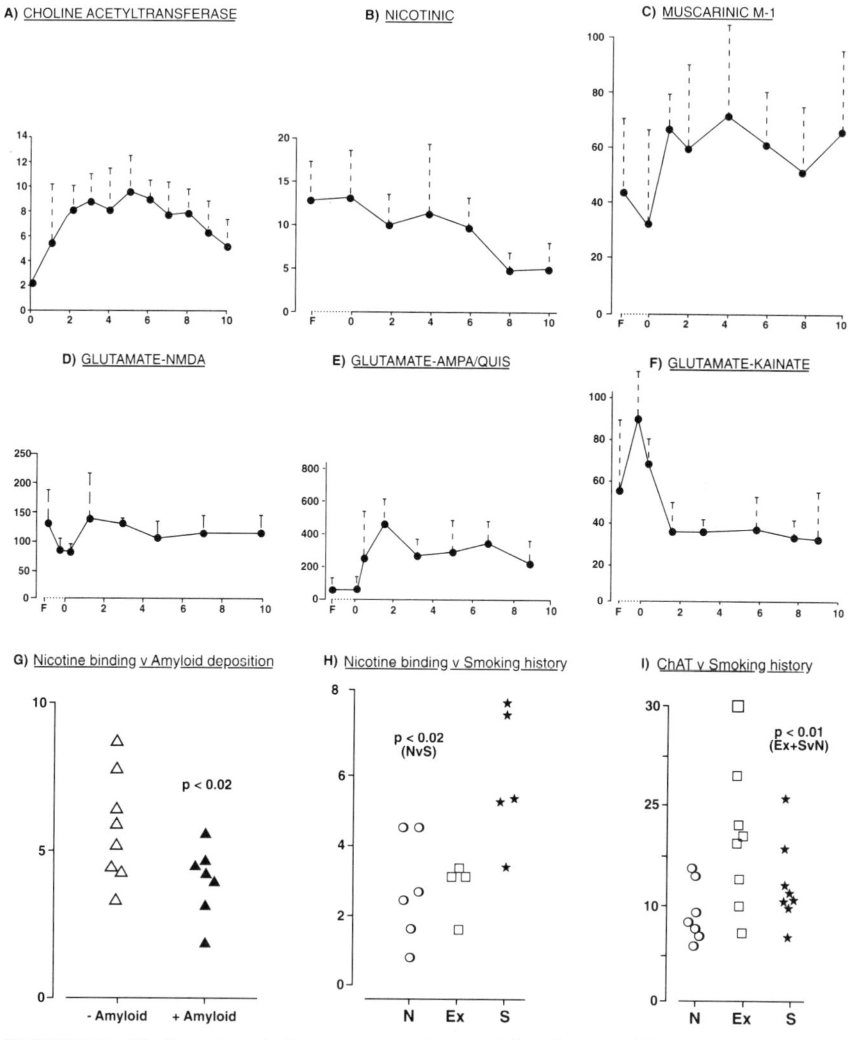

FIGURE 1. Cholinergic and glutamate transmitter activities in normal human entorhinal cortex. **A–F:** alteration in choline acetyltransferase (ChAT), cholinergic and glutamate receptor subtypes during development and aging. Horizontal axis gives age in decades including fetal (F) and neonate (N) groups; points are mean ± SD for 4–10 cases. From adulthood (20–40 y) to old age (up to 100 y) the only significant correlations with age were for ChAT ($r = -0.43$, $p < 0.02$) and nicotinic receptor binding ($r = -0.41, p < 0.05$). **G–I:** differences in nicotinic binding (entorhinal cortex) and ChAT (hippocampus) between age-matched normal individuals with and without entorhinal β-amyloid plaques and between non-smokers (N), ex-smokers (Ex) and smokers (S). P values significant (Student's *t*-test). Units for all receptor activities as fmol/mg and for ChAT as nmol/mg protein/h.

mally low expression of cortical nicotinic receptors, either congenital or acquired, is one of several risk factors for developing Alzheimer's disease. This could be associated with a reduced urge to smoke in individuals destined to develop the disease.

Distinguishing cause from effect amongst the complex range of changes in the normal aging and diseased brain is a continuing challenge. The present data on the normal aging process, considered to be a valuable model of early Alzheimer's disease implicate early nicotinic cholinergic receptor dysfunction in the evolution of Alzheimer-type pathology.

Although apolipoprotein E genotype (the presence of the E4 allele) has been correlated with loss of choline acetyltransferase and nicotinic receptor binding in Alzheimer's disease (Beffert and Poirier, this volume), no such relation has been identified in normal aging. Neither the cholinergic enzyme nor receptor was lower in age-matched normal group with, compared to without, a single E4 allele (hippocampal choline acetyltransferase was 14.4 $\pm$ 5.5 and 12.8 $\pm$ 1.9 nmol/h/mg protein in 17 without compared to 4 with E4; entorhinal cortex nicotine binding was 4.4 $\pm$ 1.1 and 6.9 $\pm$ 2.1 fmol/mg in 6 without compared to 4 with E4). A similar lack of correlation was evident in elderly cases of Alzheimer's disease or Lewy body dementia.[7] Cholinergic dysfunction may then be a consequence of aging and disease which occurs independently of any specific genotype.

REFERENCES

1. Perry, E. K. & R. H. Perry. 1993. Neurochemical pathology and therapeutic strategies in degenerative dementia. Int. Rev. Psychiat. **5**:363–380.
2. Brookes, A. J. & D. St. Clair. 1994. Synuclein proteins and Alzheimer's disease. Trends Neurosci. **17**:404–405.
3. Perry, E. K., J. A. Court, M. Johnson, C. J. Smith, V. James, A. V. Cheng, J. M. Kerwin, C. M. Morris, M. A. Piggott, J. A. Edwardson, N. J. M. Birdsall, J. T. Turner & R. H. Perry. 1993. Autoradiographic comparison of cholinergic and other transmitter receptors in the normal human hippocampus. Hippocampus **3**:307–315.
4. Perry, E. K., M. Johnson, J. M. Kerwin, M. A. Piggott, J. A. Court, P. J. Shaw, P. G. Ince, A. Brown & R. H. Perry. 1992. Convergent cholinergic activities in aging and Alzheimer's disease. Neurobiol. Aging **13**:393–400.
5. Benwell, M. E. M., D. J. K. Balfour & J. M. Anderson. 1988. Evidence that tobacco smoking increases the density of (^{3}H)nicotine binding sites in human brain. J. Neurochem. **50**:1243–1262.
6. Lee, P. N. 1994. Smoking and Alzheimer's disease: A review of the epidemiological evidence. Neuroepidemiology **13**:131–144.
7. Morris, C. H., H. M. Massey, R. Benjamin, A. Leake, C. Broadbent, M. Griffiths, H. Lamb, A. Brown, P. Ince, S. Tyrer, P. Thompson, I. G. Mekeith, J. A. Edwardson, R. H. Perry & E. K. Perry. 1996. Molecular biology of ApoE alleles in Alzheimer's and non-Alzheimer's dementias. J. Neural. Transm. Suppl. In press.

The Effect of Cholinesterase Inhibitors on the Secretion of APPS from Rat Brain Cortex[a]

E. GIACOBINI,[b] F. MORI, AND C-C. LAI

Department of Pharmacology, Southern Illinois University School of Medicine, Springfield, Illinois 62794-9230 USA

ABSTRACT: In this study we examined the question whether cholinesterase inhibitors (ChEI) could alter the release of amyloid precursor protein (APP) from superfused brain cortical slices of the rat following electrical as well as pharmacological stimulation with bethanechol (BETHA). Three ChEI, both reversible and irreversible were tested for their ability to enhance the release of non-amyloidogenic soluble derivatives (APPs). These included physostigmine (PHY), heptyl-physostigmine (HEP) and 2,2-dichlorovinyldimethyl phosphate (DDVP), at the concentrations producing cholinesterase (ChE) inhibition ranging from 5% to 95%. All three ChEI elevated APPs release significantly above control levels. Electrical field stimulation significantly increased the release of APPs within 50 min. Similar increase was observed after muscarinic receptor stimulation with BETHA. Tetrodotoxin (TTX) completely blocked the effect of electrical stimulation. These findings suggest that long-term administration of ChEI to Alzheimer's disease (AD) patients may have a neuroprotective effect by activating normal APP processing and decreasing the formation of amyloidogenic APP products.

INTRODUCTION

Therapeutic strategies designed to treat AD may either: a) produce an attenuation of symptoms improving cognitive function and activity of daily living; or b) slow down deterioration by inhibiting progression of the disease. In the treatment of AD, ChEI have shown the most promising effect than any other drug (*cf.* ref. 1). Cholinesterase inhibitors aim to preserve and enhance the physiological effect of acetylcholine (ACh) by increasing its synaptic levels to close to normal concentrations. The best demonstrated clinical efficacy of ChEI relates to the numerous studies on several thousand patients

[a] Supported in part by Mediolanum Farmaceutici Inc., Milan, Italy.
[b] *Send correspondence to:* Prof. Ezio Giacobini, Institutions Univiversitaires de Geriatrie de Geneve, Route de Mon-Idée, 1226 Thonex-Geneva, Switzerland.

with THA (tacrine, tetrahydroaminoacridine) (cf. ref. 1). With regard to long-term (2–3 years) THA treatment, the question has been raised whether the effect is symptomatic or neuroprotective.[2] A possible mechanism to explain a slower deterioration of the AD patient resulting from ChE inhibition in the CNS can be found in experimental data indicating that cholinergic agonists, ACh or ChEI may regulate processing and secretion of the βA4 amyloid protein through its precursor, APP. Such an effect could presumably slow down or even preclude formation of amyloidogenic APP fragments.

Using *in vitro* non-neuronal cell populations, Nitsch *et al.*[3] and Buxbaum *et al.*[4] demonstrated that the release of APPs is increased by stimulation of muscarinic receptor subtypes m1 and m3 through activation of protein kinase C (PKC).[5] Subsequently, Nitsch *et al.*[6] demonstrated that electrical stimulation of superfused rat hippocampal slices caused a rapid increase in the release of these amino-terminal APP cleavage products. The release of APPs increased with increasing stimulation frequency and was inhibited by the sodium-channel antagonist TTX.[6] These results suggested that neuronal activity may regulate APP processing in mammalian brain.[7]

We suggested the hypothesis that long-term inhibition of ChE having the effect of increasing the level of synaptic ACh may result in the activation of normal APP processing in AD brain[8] (FIG. 1). At the same time, it could slow down the formation of amyloidogenic APP fragments. To determine whether ChEI could alter the release of APP we used superfused brain cortical slices of the rat following the method described by Nitsch *et al.*[6] The irreversible organophosphate DDVP, short-acting reversible ChEI PHY, and its longer-acting carbamate derivative HEP were used in these experiments.[9–11]

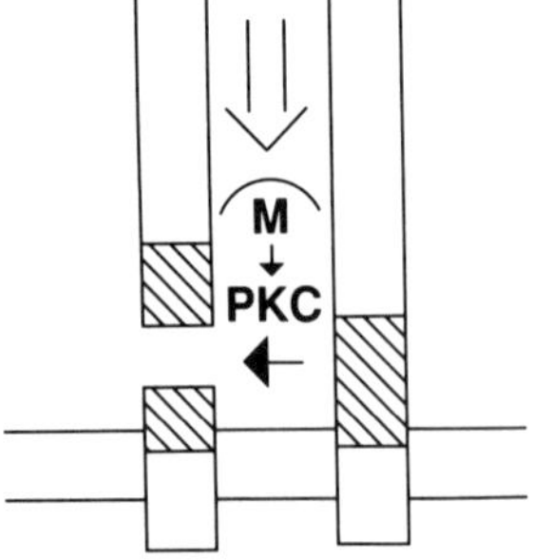

FIGURE 1. Postulated mechanism of action of cholinesterase inhibitors on normal APP processing through muscarinic (M_1 and M_3) receptor stimulation and protein kinase C (PKC) activation.

The concentration of the inhibitors was selected in order to produce a ChE inhibition ranging from 5 to 95%.[12]

MATERIALS AND METHODS

We followed the method described by Nitsch *et al.*[6] with some modifications.[12] Cortical slices (300 μm) were cut from rat brain using a McIlwain tissue chopper, washed three times, randomly chosen, placed (3 slices from each hemisphere per chamber) into superfusion chambers. The drugs were added to the perfusion buffer during equilibration and continuously perfused throughout the collection. The collection time was 50 min. Following incubation with the appropriate drug under the same conditions, time and concentration as for the release experiments, brain slices were homogenized and acetylcholinesterase (AChE) activity was assayed in brain slices and compared to controls using the radiometric method of Johnson and Russell[13] as described by Hallak and Giacobini.[11]

RESULTS AND DISCUSSION

Electrical field stimulation (30 Hz) significantly increased the release of APPs over basal levels within 50 min. Densitometric quantitation indicated a 160% increase (basal release equal to 100%) (TABLE 1). Similar increase was observed in the brain slices after stimulation with 1 μM BETHA (148%) or 100 μM BETHA (153%). Bethanechol did not further increase the effect of electrical stimulation. Tetrodotoxin (1 μM) completely blocked the effect of electrical stimulation indicating that neuronal activity may regulate APP processing. Previous experiments have shown that electrical depolarization enhances the release of APPs from hippocampal rat brain slices.[6] We confirmed these results using slices of rat cortex.[12] In our experiments, as in the

TABLE 1. Drug-stimulated Increase of Basal APPs Release from Rat Brain Cortical Slices[a]

Drug	Conc. μM	Increase (% of basal)	ChE Act. (% Inhib.)
Bethanechol	1	48	0
	100	53	0
Physostigmine	.1	48	25
Heptyl-physostigmine	.1	41	61
DDVP	.02	33	95

[a] Mori *et al.*[12]

previous study of Nitsch *et al.*[6] blockade of action potential with TTX inhibited the depolarization-induced APPs release. These results suggest that APPs release *in vitro* may be a function of neuronal activation and neurotransmitter-dependent. In order to investigate whether cholinergic neurotransmission may modify APPs processing, we used ChEI to enhance ACh levels in our preparation. Three concentrations of PHY were tested (0.01, 0.1 and 1 μM) to demonstrate cholinergic stimulation of APPs release (TABLE 1). Acetylcholinesterase inhibition in the slices was 5%, 25% and 64%, respectively (TABLE 1). The two other concentrations of PHY both elevated APPs release but only the 0.1 μM concentration reached a statistical significance (148%) ($p < 0.05$) (TABLE 1). The enhanced release of APPs evoked by electrical stimulation was not further modified by any concentration of PHY. Both HEP (0.1 μM) and DDVP (20 nM) increased significantly APP release by 141% and 133%, respectively (TABLE 1). Acetylcholinesterase inhibition in the slices was 61% after HEP (0.1 μM) and 95% after DDVP (20 nM) (TABLE 1). The presence of these ChEI did not amplify the effect of electrical stimulation.

This study is the first to demonstrate that ChEI can indeed enhance the release of APPs derivatives in intact brain cortical slices *in vitro.*[12] The concentration of the inhibitors used by us produce an AChE inhibition ranging from a minimum of 5% for PHY (0.01 μM) to 61% for HEP (0.1 μM), to a maximum of 95% for DDVP (20 nM). We have shown that comparable *in vivo* concentrations of the inhibitors and similar levels of AChE inhibition produce maximal increases of ACh levels in rat frontal cortex ranging from 1800% for DDVP to 2500% for PHY and 4000% for HEP.[1] Although no data are available with regard to the actual ACh increase in cortical slices following ChE inhibition we can assume that ACh levels would also be increased in these *in vitro* preparations.

CONCLUSION

Our results showing a relationship among ChE inhibition, ACh increase and APPs processing in rat brain cortex, raise two important questions. First, loss of synapses in human cortex including cholinergic ones, represents the major correlate of cognitive impairment in AD.[14,15] As a result, cortical cholinergic denervation may also impair the neurotransmitter signal (ACh) which normally stimulates the release of APPs through muscarinic receptor activation. One direct consequence of cholinergic denervation could be an accelerated amyloid formation in the brain. Secondly, pharmacological activation of cholinergic muscarinic receptors subsequent to ChEI-induced increase of ACh levels may increase secretion of APPs. This effect would not only enhance the trophic action of APP but also inhibit amyloid deposition and cellular damage resulting in a slower deterioration of the AD patient treated with ChEI.

ACKNOWLEDGMENTS

The valuable help of Drs. S. Farber and R. Nitsch (MIT, Cambridge, MA) in setting up the methodology used in this investigation is acknowledged. The authors thank Elizabeth Williams for technical assistance, and Diana Smith for typing and editing the manuscipt.

REFERENCES

1. GIACOBINI, E. & G. CUADRA. 1994. Second and third generation cholinesterase inhibitors: From preclinical studies to clinical efficacy. *In* Alzheimer Disease: Therapeutic Strategies. E. Giacobini & R. Becker, Eds. 155–171. Cambridge: Birkhauser Boston.
2. GIACOBINI, E. 1994. Therapy of Alzheimer disease symptomatic or neuroprotective? *In* Neuroprotection in Neurodegeneration. P. Riederer, J. Fritze & M. B. H. Youdim eds. Neural Transm. **43**(Suppl.):211–217. New York: Springer-Verlag.
3. NITSCH, R. M., B. E. SLACK, R. J. WURTMAN & J. H. GROWDON. 1992. Release of Alzheimer amyloid precursor derivatives stimulated by activation of muscarinic acetylcholine receptors. Science **258:**304–307.
4. BUXBAUM, J. D., M. OISHI, H. I. CHEN, R. PINKAS-KRAMARSKI, E. A. JAFFE, S. E. GANDY & P. GREENGARD. 1992. Cholinergic agonists and interleukin 1 regulate processing and secretion of the Alzheimer β/A4 amyloid protein precursor. Proc. Natl. Acad. Sci. USA **89:**10075–10078.
5. NISHIZUKA, Y. 1992. Intracellular signaling by hydrolysis of phospholipids and activation of protein kinase C. Science **258:**597–603.
6. NITSCH, R. M., S. A. FARBER, J. H. GROWDON & R. J. WURTMAN. 1993. Release of amyloid β-protein precursor derivatives by electrical depolarization of rat hippocampal slices. Proc. Natl. Acad. Sci. USA **90:**5191–5193.
7. NITSCH, R. M. & J. H. GROWDON. 1994. Role of neurotransmission in the regulation of amyloid β-protein precursor processing. Biochem. Pharmacol. **47**(8):1275–1284.
8. GIACOBINI, E. 1994. Cholinomimetic therapy of Alzheimer disease: Does it slow down deterioration? *In* Recent Advances in the Treatment of Neurodegenerative Disorders and Cognitive Dysfunction, Int. Acad. Biomed. Drug Res., Vol. 7. G. Racagni, N. Brunello & S. Z. Langer, eds. 51–57. Basel: Karger.
9. CUADRA, G. R., K. L. SUMMERS & E. GIACOBINI. 1994. Cholinesterase inhibitor effects on neurotransmitters in rat cortex *in vivo*. J. Pharmacol. Exp. Thera. **270**(1):277–284.
10. DESARNO, P., M. POMPONI, E. GIACOBINI, X.-C. TANG & E. WILLIAMS. 1989.

The effect of heptylphysostigmine, a new cholinesterase inhibitor, on the central cholinergic system of the rat. Neurochem. Res. **14**(10):971–977.

11. HALLAK, M. & E. GIACOBINI. 1987. A comparison of the effects of two inhibitors on brain cholinesterase. Neuropharmacology **26**(6):521–530.
12. MORI, F., C.-C. LAI, F. FUSI & E. GIACOBINI. 1995. Cholinesterase inhibitors increase secretion of APPs in rat brain cortex. NeuroReport **6**(47):633–636.
13. JOHNSON, C. D. & R. L. RUSSELL. 1975. A rapid, simple radiometric assay for cholinesterase, suitable for multiple determinations. Analyt. Biochem. **64:** 229–238.
14. DEKOSKY, S. T., S. D. STYREN & M. E. O'MALLEY. 1994. Cholinergic changes and synaptic alterations in Alzheimer's disease. *In* Alzheimer Disease: Therapeutic Strategies. E. Giacobini & R. Becker, Eds. 93–98. Cambridge: Birkhauser Boston.
15. TERRY, R. D., E. MASLIAH, D. P. SALMON, N. BUTTERS, R. DETERESA, R. HILL, L. A. HANSEN & R. KATZMAN. 1991. Physical basis of cognitive alterations in Alzheimer's disease: synapse loss is the major correlate of cognitive impairment. Ann. Neurol. **30:**572–580.

Therapeutic Effects of CDP-Choline in Alzheimer's Disease

Cognition, Brain Mapping, Cerebrovascular Hemodynamics, and Immune Factors[a]

R. CACABELOS,[b] J. CAAMAÑO, M. J. GÓMEZ,
L. FERNÁNDEZ-NOVOA, A. FRANCO-MASIDE,
AND X. A. ALVAREZ

Institute for CNS Disorders, Basic and Clinical Neurosciences Research Center, La Coruña, Spain
Neurogerontology Unit, Department of Human Physiology, Complutense University Medical School, Madrid, Spain

ABSTRACT: CDP-choline was given to patients with Alzheimer's disease (AD) at a daily dose of 1000 mg/day p.o. for one month. This compound slightly improved mental performance, tended to reduce theta activity in fronto-temporal regions, increasing alpha power in occipital areas, and enhanced cerebrovascular perfusion by increasing blood flow velocity and reducing pulsatility and resistance indexes. In addition, CDP-choline diminished histamine and interleukin-1 levels in blood and serum, respectively, and increased plasma TNF.

INTRODUCTION

CDP-Choline (cytidine-5-diphosphate-choline) is a choline donor involved in the biosynthesis of acetylcholine and in brain phospholipid metabolism.[1] CDP-choline acts as an endogenous intermediate in a biosynthetic pathway incorporating free choline into phosphatidylcholine and choline plasmalogens in brain tissue. This compound has been extensively used in stroke, cerebrovascular disorders, psycho-organic syndromes, memory deficits, and agerelated cognitive disturbances. CDP-choline may contribute to membrane repair in AD, where potential defects have been demonstrated.[2]

[a] This research was sponsored by Ferrer International and EUROESPES Foundation.
[b] *Send correspondence to:* Prof. Dr. Ramón Cacabelos, Institute for CNS Disorders, Basic and Clinical Neurosciences Research Center, P.O. Box 733, 15080-La Coruña, Spain; Tel: 34-81-780505; Fax: 34-81-780505.

In patients with cerebrovascular disorders, CDP-choline reduces slow delta frequencies and increases alpha activity in spectral EEG analysis. In previous studies we have found that CDP-choline improved cognitive function in control subjects and in patients with AD or vascular dementia (VD).[3] In addition, CDP-choline also shows a moderate antidepressant effect, and reduces serum interleukin-1β (IL-1β) levels particularly in patients with early-onset AD who currently have a higher concentration of serum IL-1 than control subjects or other subtypes of AD.[4]

In the present study we have investigated the effects of CDP-choline on cognition, brain mapping, cerebrovascular hemodynamics, and cytokines level in AD patients.

METHODS

Nineteen patients were included (10 males/9 females, age = 66.21 ± 6.45 years): a) early-onset AD (EOAD; N = 13, age = 62.85 ± 4.37 years, range: 57–72 years), and b) late-onset AD (LOAD, N = 6, age = 73.50 ± 3.27 years, range: 70–78 years). All patients met DSM-III-R/NINCDS-ADRDA criteria for primary degenerative dementia and/or probable AD.

Mental performance and cognitive status were assessed with the following psychometric tools: a) Mini-Mental State Examination (MMSE) of Folstein, modified for the Spanish population (maximum score = 35); b) modified models of the Global Deterioration Scale (GDS), brief Cognitive Rating Scale (BCRS), and Functional Assessment Staging (FAST) of Reisberg; and c) Hachinski scale (HACH).

All patients were drug-free for at least 7 days before the clinical trial and none had been taking long-term medication. The patients received CDP-choline 1000 mg/day p.o. after breakfast for 1 month. Clinical evaluation, biomedical monitoring, EEG and brain mapping exploration, brain hemodynamic evaluation, neuropsychological assessment, and cytokine and histamine analysis were carried out prior to and 1 month after daily treatment.

EEG and brain mapping recordings were made in resting conditions and with closed eyes by 19 scalp electrodes located according to the International 10–20 system. The EEG signal was analog-filtered with a band pass of 0.5–25 Hz and digitalized and stored on magnetic disks for further analysis of free-artifact EEG epochs of 2 sec selected for spectral analysis with a Fast Fourier Transform (FFT) using a Brain Atlas v2.345. FFT data were normalized and relative power (%) and spectral amplitude (μ) were used as standard parameters.

Brain hemodynamic exploration was carried out with a TC-2000S multifrequency transcranial Doppler ultrasonograph with automatic processing and a pulsed ultrasound at a frequency of 2 MHz with a cylindrical transducer.

The mean (Mv), systolic (Sv), and diastolic (Dv) blood flow velocities, as well as Gosling's pulsatility index (PI = (Sv-DV)/Mv), and Pourcelot's resistance index (RI = (Sv-Dv)/Sv) of the right (MCA-R) and left (MCA-L) middle cerebral arteries were measured in basal conditions (B) and after 1 month of treatment with CDP-choline (C).

IL-1β and TNF (Tumor Necrosis Factor) were measured in serum and plasma, respectively, by ELISA; and histamine levels were determined by HPLC in whole blood.

RESULTS

Cognitive Effects

After 1 month of treatment with CDP-choline, MMSE scores showed a tendency to increase from 26.9 ± 5.4 to 27.6 ± 5.3 in the total group of AD patients, and from 25.5 ± 5.0 to 26.8 ± 4.5 ($p < 0.005$) in EOAD. However, MMSE scores decreased from 29.5 ± 5.5 to 28.9 ± 6.7 in LOAD. The most significant improvement was detected in the MMSE item of orientation.

Changes in Brain Mapping

CDP-choline induced a significant decrease in spectral amplitude of the theta band, mainly in fronto-temporal areas on the right hemisphere (F8, $p < 0.001$; F4, $p < 0.05$; T4, $p < 0.05$), and in spectral amplitude of beta bands in bilateral frontal electrodes (F7, $p < 0.05$; F8, $p < 0.05$), with no apparent changes in the relative power of any other frequency bands. In LOAD, CDP-choline induced a moderate increase in relative alpha power in occipital regions (O1, O2).

In basal conditions, MMSE scores positively correlated with relative alpha power in T3 ($r = 0.4362$, $p = 0.031$). Orientation scores negatively correlated with relative delta power in C3 ($r = 0.4243$, $p = 0.035$) and P3 ($r = 0.4618$, $p = 0.023$), and positively correlated with relative alpha power in C3 ($r = 0.5305$, $p = 0.01$), P3 ($r = 0.3975$, $p = 0.046$), and P4 ($r = 0.4264$, $p = 0.034$), and with C3 relative beta power ($r = 0.5694$, $p = 0.022$). The MMSE memory item scores negatively correlated with relative delta power in T3 ($r = 0.5694$, $p = 0.005$), P3 ($r = 0.5694$, $p = 0.027$) and T5 ($r = 0.4643$, $p = 0.023$). After treatment, the MMSE scores negatively correlated with relative delta power in T3 ($r = 0.3901$, $p = 0.049$) and T5 ($r = 0.4175$, $p = 0.038$), and positively correlated with relative alpha power in T3, P3, P4, and T5. A negative correlation was found between relative delta power and the orientation and memory scores in the left parietal region (P3), and with language scores in central areas (C3). The relative alpha power positively correlated with orientation scores in T3 and T5, with memory scores in left temporal regions (T3, T5), and with language scores in T3, T4, C3, C4, P3 and P4.

Changes in Cerebrovascular Hemodynamics

By transcranial Doppler ultrasonography significant changes were found in most hemodynamic parameters after one month of treatment with CDP-choline in both MCA-L and MCA-R, including an increase in mean velocity (MvB = 39.16 ± 6.91 vs. MvC = 41.88 ± 7.91 cm/sec, $p < 0.02$) and diastolic velocity (DvB = 24.50 ± 5.59 vs. DvC = 26.55 ± 6.56 cm/sec, $p < 0.02$), and a decrease in pulsatility index (PIB = 1.008 ± 0.30 vs. PIC = 0.96 ± 0.30, $p < 0.05$) and resistance index (RIB = 0.606 ± 0.086 vs. RIC = 0.5690 ± 0.084, $p < 0.05$).

Changes in Cytokines (IL-1β, TNF) and Histamine (HA)

CDP-choline tends to reduce serum IL-1β levels rom 211.2 ± 31.1 to 186.4 ± 32.7 pg/ml, increasing plasma TNF-alpha from 3.6 ± 2.0 to 4.4 ± 2.3 pg/ml.

Basal blood HA levels are higher in EOAD (0.53 ± 0.18 nmol/ml) than in LOAD (0.34 ± 0.14 nmol/ml, $p < 0.05$). After one month of treatment, blood levels of HA were diminished in EOAD (0.30 ± 0.18 nmol/ml, $p < 0.05$), but not in LOAD (0.25 ± 0.13 nmol/ml).

DISCUSSION

According to the present results, CDP-choline slightly improves cognitive function in AD, in agreement with data previously reported.[3] However, this is the first time that clear-cut bioelectrical changes and brain hemodynamic modifications are demonstrated after treatment with CDP-choline in AD patients. CDP-choline significantly decreases spectral amplitude in the theta band in fronto-temporal regions, increasing alpha power in occipital regions, particularly in LOAD. In contrast, cognitive performance was better in EOAD than in LOAD after CDP-choline treatment, this fact leading us to postulate that the effect of CDP-choline might be mediated through a cerebrovascular and/or immunogenic mechanism, since both brain hemodynamic changes and cytokine modifications (IL-1, TNF) seem to be more prominent in EOAD than in LOAD. Furthermore, blood HA levels are also dramatically reduced in EOAD but not in LOAD. It is now well-known that HA levels are markedly increased in brain tissue[5] and serum[6] from patients with AD, as well as IL-1.[7] On the contrary, serum TNF levels tend to be reduced in AD. HA and IL-1 regulate each other at the central level, and both neuroimmune factors might be involved in microglia-associated senile plaque formation and microvascular regulation. Therefore, it might be postulated that compounds with neuroimmunotrophic properties could provide some benefit in AD. In the case of CDP-choline, besides its potential effect as a choline donor and membrane restorative product, its influence on brain hemodynamics, HA and IL-1 may contribute to slow-down neurodegeneration in AD and tissue destruction in VD.

REFERENCES

1. WURTMAN, R. J., J. K. BLUSZTAJN, I. H. ULUS, *et al.* 1990. Choline metabolism in cholinergic neurons: Implications for the pathogenesis of neurodegenerative disease. Adv. Neurol. **51**:117–125.
2. NITSCH, R. M., J. K. BLUSZTAJN, A. G. PITTAS, B. E. SLACK, J. H. GROWDON & R. J. WURTMAN. 1992. Evidence for a membrane defect in Alzheimer disease brain. Proc. Natl. Acad. Sci. USA **89**:1671–1675.
3. CACABELOS, R., X. A. ALVAREZ, A. FRANCO, L. FERNÁNDEZ-NOVOA, J. CAAMAÑO & F. VALLE-INCLÁN. 1992. Therapeutic effects of CDP-choline in Alzheimer's disease and multi-infarct dementia: Psychometric assessment and immune function. Ann. Psychiat. **3**:233–245.
4. CACABELOS, R., A. FRANCO-MASIDE & X. A. ALVAREZ. 1991. Interleukin-1 in Alzheimer's disease and multi-infarct dementia: Neuropsychological correlations. Meth. Find. Exp. Clin. Pharmacol. **13**:703–708.
5. CACABELOS, R., A. YAMATODANI, H. NIIGAWA, *et al.* 1989. Brain histamine in Alzheimer's disease. Meth. Find. Exp. Clin. Pharmacol. **11**:353–360.
6. CACABELOS, R., L. FERNÁNDEZ-NOVOA, A. FRANCO-MASIDE & X. A. ALVAREZ. 1992. Neuroimmune function of brain histamine: Implications for neurotrophic activity and neurotoxicity. Ann. Psychiat. **3**:147–200.
7. CACABELOS, R., X. A. ALVAREZ, L. FERNÁNDEZ-NOVOA, A. FRANCO, R. MANGUES, A. PELLICER & T. NISHIMURA. 1994. Brain interleukin-1β in Alzheimer's disease and vascular dementia. Meth. Find. Exp. Clin. Pharmacol. **16**: 141–151.

HWA 285 (Propentofylline)—A New Compound for the Treatment of Both Vascular Dementia and Dementia of the Alzheimer Type

M. ROTHER,[a] B. KITTNER, K. RUDOLPHI, M. RÖßNER, AND K.-H. LABS

HOECHST AG, HR PGU Cardiovascular Agents, Wiesbaden, Germany

ABSTRACT: The pharmacological profile of HWA 285 favors its use in patients with both Alzheimer's disease (PDD) and/or vascular dementia (MID). Clinical trials showed clinically relevant, statistically significant efficacy in the domains of cognitive function, global function and activities of daily living (ADL) in both PDD and MID. HWA 285 had a prolonged symptomatic effect for at least 12 months, although therapeutic effects were seen already after the first 3 months of treatment. HWA 285 was very well tolerated for at least 1 year.

INTRODUCTION

The pathophysiologies of Alzheimer's disease (also known as primary degenerative dementia for PDD) and vascular dementia (VaD: also known as multi-infarct dementia or MID) appear to be quite different.[1] However, common elements underlying Alzheimer's disease and vascular dementia, such as inflammation and neuronal death, may lead to some common therapeutic approaches.[2] Under ischemic conditions, neuronal death is associated with excess release of the amino acid neurotransmitter glutamate and an influx of calcium ions.[3] Although the causes of neuronal death in Alzheimer's disease are not known, a pathological role has been proposed for glutamate in the progressive neurodegeneration.[4] The loss of cholinergic neurons in the basal forebrain has been implicated in the cognitive decline associated with Alzheimer's disease,[5] and possibly vascular dementia. Inflammatory processes and the consequent generation of cytokines, such as interleukin-1 β(IL-1β), tumor necrosis factor (TNF-α), and neurotoxic oxygen free radicals, may also

[a] *Send correspondence to*: Dr. Matthias Rother, HOECHST AG, Department of Clinical Angiology, PGU Cardiovascular agents, Rheingaustr. 190-196, D-65203 Wiesbaden, Germany; Tel: +49/611/962 5787; Fax: +49/611/962 9271.

contribute to the pathophysiology of both Alzheimer's disease and vascular dementia. Free-radical-induced aggregation of β-amyloid peptide may be an essential part of the pathogenesis of amyloid plaques in Alzheimer's disease.[6] By interfering with some of these common elements in the neurodegenerative process, a single neuroactive therapy may attenuate the cognitive decline in both Alzheimer's disease and vascular dementia.

MODE OF ACTION

The molecular mechanism of action of HWA 285 is not yet fully elucidated. HWA 285 blocks adenosine transport, inhibits a cAMP and cGMP phosphodiesterases (PDE), and is also a weak antagonist of the adenosine A_1-receptor. The reinforcement of the effects of the homeostatic neuromodulator adenosine and the second messenger cAMP may be important for the pharmacodynamic effects of HWA 285.[7] The effects of adenosine are also partly mediated by an increase in tissue cAMP, which acts as a second messenger of physiological processes such as vasodilation, inhibition of platelet aggregation, and metabolism.

PHARMACOLOGICAL ACTIONS IN MODELS OF CEREBRAL ISCHEMIA AND ALZHEIMER'S DISEASE

Pre- or post-lesion treatment with HWA 285 reduces ischemic nerve cell death in the brain, in models of global and focal ischemia.[7] HWA 285 also decreases brain edema,[8] intraneuronal calcium accumulation,[9] and reactive astrocytosis;[10] ameliorates mitochondrial function and the energy state of the ischemic brain;[11] and prevents the delayed ischemia-induced loss of acetylcholine receptors (muscarinic-1) in the hippocampus.[12]

In rats with experimental basal forebrain lesions, an Alzheimer's-type dementia model, chronic treatment with HWA 285 improves learning and memory, reverses the deficit in hippocampal choline acetyltransferase activity, and restores the changes of the muscarinic receptors in the cortex.[13] It also improves performance in other models of impaired cognition in rats and mice.[14] HWA 285 stimulates long-term potentiation (LTP) in the guinea pig hippocampus;[15] this may be an electrophysiological correlate of memory formation.

In view of a) recent evidence that excessive microglial and immunological markers are associated with the plaque pathology of Alzheimer's disease, b) the inhibitory effects of HWA 285 on the formation of oxygen free radicals by reactive microglia, and c) the HWA 285-induced decrease in immunostaining for reactive microglia, it seems that HWA 285 may interfere with the disease process itself.[16,17] With regard to possible attenuation of the neurodegenerative process, HWA 285 has been shown to increase the release of nerve growth factor (NGF) from astrocyte cultures,[18] and prevent the age-associated decline of NGF-immunoreactivity in aged rats.[19]

CLINICAL EXPERIENCE

Early clinical studies, in which HWA 285 was given with food, failed to demonstrate efficacy. The most probable explanation for these results came from a later pharmacokinetic study, which indicated a strong food-drug interaction.

A subsequent 12-week double-blind, placebo-controlled phase II study in patients with memory and intellectual deficits[20] was designed to assess whether 300 mg HWA 285 given three times daily (orally, 1 h before meals) was efficacious when the interaction with food was avoided. The Gottfries-Bråne-Steen (GBS) Scale, the Clinical Global Impressions (CGI), and the Mini Mental State Examination (MMSE), which were chosen to investigate clinical efficacy, showed statistically significant differences in favor of HWA 285.

The neuroprotective effect of HWA 285 was examined in a double-blind, placebo-controlled, clinical trial in stroke patients in which positron emission tomography of glucose metabolism was used as an efficacy criterion.[21] After 3 months of treatment, the HWA 285 group showed a significant increase of glucose metabolism in the infarct areas. Additional analysis of the data showed a significant beneficial effect particularly in the penumbra region.

The phase III program included 2 studies with similar protocols and treatment periods of 6 months and 12 months, respectively. Patients with mild to moderate dementia according to DSM-III-R (PDD and/or MID) and with a MMSE score between 15 and 25 were eligible. The proportions of PDD, MID and mixed were 59%, 26%, and 15% respectively.

Study protocols were agreed and studies were started before the current EC draft guidelines for clinical trials in dementia were published.[22] These guidelines require appropriate tests covering three domains; cognitive function, global function and activities of daily living (ADL). All the tests and scales used in the phase III studies could be allocated to these domains, and the results are discussed in the context of the domains in order to comply with the guidelines, 492 patients were included in the domain analysis.

Three tests covered the domain cognitive function; the SKT (Syndrome Short Test for german: Syndrom Kurz Test), the MMSE orientation subscale and the Digit Symbol Substitution Test (DSST) SKT results have been reported to correlate well with those of the Alzheimer Disease Assessment Scale (ADAS) and MMSE.[23] Since the three tests assess different aspects of cognitive function, a composite score was created, with weighting according to the importance of the tests. The results of the composite score showed differences in favour of HWA 285 ($p < 0.001$).

Global assessments measure the patients overall clinical condition (or global function). Two global assessments were used; the GBS Scale, a comprehensive assessment, and the clinical global improvement with its subscales severity (item I) and change (item II). A composite score of the global measures was created giving each score equal weight. The results of the composite score showed differences in favor of HWA 285 ($p < 0.001$).

The ADL assessments are of particular relevance as they reflect the most tangible consequences of dementia, *i.e.* the patients inability to perform normal daily activities. However, ADL tests are generally insensitive to change in mild to moderate dementia. Assessment of ADL was based on two scales from the Nuremberg gerontopsychological inventory: the Nürnberger-Alters-Beobachtungs-Skala (assessment of the patient by a carer, friend, or relative) and the Nürnberger-Alters-Alltags-Skala (patient self-assessment). A composite score of the two test scores was made in which the variables were weighted according to their clinical importance. The result of the composite score showed differences in favor of WW17285 ($p > 0.005$).

The results of the Befindlichkeitsskala (BfS Scale) showed no difference between HWA 285 and placebo. There was therefore no reason to suggest that the efficacy results described in the previous sections were a consequence of HWA 285 acting primarily as a mood-altering drug.

Efficacy was evaluated in PDD, MID and mixed patients. The analyses of the composite scores for the three domains indicated efficacy in both PDD and MID patients in each domain. HWA 285 may, however, be more effective in patients in whom the vascular component of the disease is more pronounced.

In all three double-blind studies 371 patients received HWA 285, of whom 52% reported adverse events. The spectrum of individual adverse events reported was typical for an elderly and demented population. Of the 372 patients who received placebo, 40% reported adverse events. The most commonly reported adverse events in both HWA 285 and placebo patients were those affecting the nervous system, digestive system, cardiovascular system and body as a whole. Of these, the frequencies of adverse events affecting the nervous and digestive systems were notably higher in HWA 285 patients compared to placebo patients (nervous system 26% vs 17%, digestive system 19% vs 9%). There was a low incidence of serious adverse events, and safety problems led to discontinuation of the active treatment in only a low number of patients. HWA 285 was equally well tolerated in all patient subgroups investigated.

REFERENCES

1. KORCZYN A. D. 1991. The clinical differential diagnosis of dementia: Concept and methodology. Psychiatr. Clin. North Am. **14**(2):237–206.

2. BOWEN, D. M. & A. N. DAVISON. 1986. Can the pathophysiology of dementia lead to rational therapy? *In* Treatment Development Strategies for Alzheimer's Disease. T. Crook, R. Bartus, S. Ferris & S. Gershon, Eds. :36–66. Madison, CT: Mark Powley Associates.
3. BEAL, M. & M. FLINT. 1991. Mechanisms of excitotoxicity in neurologic diseases. FASEB J. **6**:3338–3344.
4. FOSTER, A. C. 1990. Physiology and Pathophysiology of Excitatory Amino Acid Neurotransmitter Systems in Relation to Alzheimer's Disease. R. J. Wurtman, *et al.*, Eds. Neurology; **51**:97–102. New York: Raven Press.
5. BARTUS, R. T. *et al.* 1982. The cholinergic hypothesis of geriatric memory dysfunction. Science **217**:408–417.
6. FRIEDLICH, A. L. & L. L. BUTCHER. 1994. Involvement of free oxygen radicals in amyloidosis: An hypothesis. Neurobiol. Aging; **15**:443–455.
7. PARKINSON, F. E., K. A. RUDOLPHI & B. B. FREDHOLM. 1994. HWA 285: A nucleoside transport inhibitor with neuroprotective effects in cerebral ischemia. Gen. Pharmac. **25**:1053–1058.
8. MRSULJA, B. B., D. V. MICIC & V. STEFANOVICH. 1985. HWA 285 and post-ischaemic brain edema: Relation to Na^+-K^+-ATPase activity. Drug Dev. Res. **6**:339–344.
9. KADOYA, F., A. MITANI, T. ARAI & K. KATAOKA. 1992. Effects of HWA 285 on hypoxia-hypoglycaemia-induced calcium accumulation in gerbil hippocampal slices. J. Cereb. Blood Flow Metab. **12/2**:301–305.
10. DELEO, J., L. TOTH, P. SCHUBERT, K. RUDOLPHI & G. W. KECUTZBERY. 1987. Ischemia-induced neuronal cell death, calcium accumulation, and glial response in the hippocampus of the mongolian gerbil and protection by propentofylline (HWA285). J. Cereb. Blood Flow Metab. 7:745–751.
11. SASAKI, M., H. NARITOMI, M. KANASHIRO, H. NISHIMURA & T. SAWADA. 1989. Effects of HWA 285 on energy metabolism of the ischemic brain studied by in vivo ^{31}P nuclear magnetic resonance spectroscopy. Arzneim.-Forsch./Drug Res. **39**:886–889.
12. HIRATA, H., N. OGAWA, K. HABA, M. ASANUMA, H. CHOU & A. MORI. 1992. Effect of HWA 285 on ischaemia-induced loss of muscarinic cholinergic receptor binding in the gerbil hippocampus. Res. Commun. Chem. Pathol. Pharmacol. **75/3**:365–368.
13. FUJI, K., M. HIRAMATSU, S. HAYASHI, T. KAMEYAMA & T. NABESHIMA. 1993. Effects of HWA 285, a NGF synthesis stimulator, on alterations in muscarinic cholinergic receptors induced by basal forebrain lesions in rats. Neurosci. Lett. **150**:99–102.
14. GOTO, M., N. DEMURA & T. SAKAGUCHI. 1987. Effects of HWA 285 on disorder of learning and memory in rodents. Jpn. J. Pharmacol. **45**:373–378.
15. YAMADA, Y., H. NAKAMURA & Y. OKADA. 1994. HWA 285 enhances the formation of long-term potentiation in guinea pig hippocampal slices. Neurosci. Lett. **176**:189–192.
16. BANATI, R. B., P. SCHUBERT, G. ROTHE, J. GEHRMANN, K. RUDOLPHI, G. VALET & G. W. KREUTZBERG. 1994. Modulation of intracellular formation of

reactive oxygen intermediates in peritoneal macrophages and microglia/brain macrophages by HWA 285. J. Cereb. Blood Flow Metab. **14:**145–149.

17. McRae, A., K. A. Rudolphi & P. Schubert. 1994. HWA 285 depresses amyloid and Alzheimer's CSF microglial antigens after ischemia. Neuroreport **5:**1193–1196.
18. Shinoda, I., Y. Furukawa & S. Furukawa. 1990. Stimulation of nerve growth factor synthesis/secretion by HWA 285 in cultured mouse astroglial cells. Biochem. Pharmacol. **39:**1813–1816.
19. Nabeshima, T., A. Nitta & T. Hasegawa. 1993. Impairment of learning and memory and the accessory symptoms in aged rats as senile dementia model (3): Oral administration of HWA 285 produces recovery of reduced NGF content in the brain of aged rat. Jpn. J. Psychopharmacol. **13:**89–95.
20. Möller, H. J., I. Maurer & B. Saletu. 1994. Placebo-controlled trial of the xanthine derivative HWA 285 in dementia. Pharmacopsychiatry **27/4:** 159–165.
21. Huber, M., B. Kittner, C. Hoyer, G. R. Fink, M. Neveling & W. D. Heiss. 1993. Effect of HWA 285 on regional cerebral glucose metabolism in acute ischemic stroke. J. Cereb. Blood Flow Metab. **13:**526–530.
22. Working Party on Efficacy of Medicinal Products, Note for Guidance: Antidementia Medicinal Products. November 1992.
23. Kim, Y. S., D. W. Nibbelink & J. E. Overall. 1993. Factor structure and scoring and the SKT test battery. J. Clin. Psychol. **49:**61–71.

Preliminary Evaluation of Besipirdine for the Treatment of Alzheimer's Disease[a]

F. J. HUFF[b]

THE BESIPIRDINE STUDY GROUP[c]

Hoechst-Roussel Pharmaceuticals Inc., Somerville, New Jersey 08876 and Department of Neurology, University of Medicine and Dentistry of New Jersey, Piscataway, New Jersey 08855 USA

ABSTRACT: Besipirdine hydrochloride (HP 749) is an indole-substituted analog of 4-aminopyridine. Besipirdine enhances both cholinergic and adrenergic neurotransmission in the central nervous system, and may have greater efficacy than purely cholinergic agents in treating dementia due to Alzheimer's disease.[1] The present study examined the efficacy and tolerability of two doses of besipirdine (5 and 20 mg BID) in 275 patients with Alzheimer's disease during 3 months of treatment and during 3 months after withdrawl of treatment.

Besipirdine was generally well tolerated. The level of performance on a cognitive test was sustained during 3 months of treatment with besipirdine, whereas the performance of patients treated with placebo deteriorated over the same time period. The results suggest a dose-response relationship, with greater efficacy after 3 months of treatment and longer persistence after treatment withdrawal for besipirdine 20 mg BID than for 5 mg BID. A clinical global rating did not detect a besipirdine treatment effect. The full efficacy after 3 months of treatment did not persist after withdrawal of treatment, suggesting that the benefit is primarily symptomatic. Treatment with higher doses and for longer periods may enhance efficacy on both cognitive and clinical global assessments.

INTRODUCTION

The study design is represented in FIGURE 1. Increasing dementia severity scale scores are reflected on the ordinate. The abscissa reflects intervals of

[a] Supported by Hoechst-Roussel Pharmaceuticals, Inc.

[b] *Send correspondence to*: Dr. F. Jacob Huff, B-103, Hoechst-Roussel Pharmaceuticals Inc., Route 202-206 N, Somerville, NJ 08876; Tel: (908) 231-2960; Fax: (908) 231-3324.

[c] The Besipirdine Study Group comprised the following investigators. M. Alter, Philadelphia, PA; P. Antuono, Milwaukee, WI; S. Cohen and C. Hirsch, St. Petersburg, FL; N. Cutler, Beverly Hills, CA; J. Delagandara and A. Pedraza, Pompano Beach, FL; G. Gottlieb, Philadelphia, PA; R. Green and S. Hersch, Atlanta, GA; R. Holub, Albany, NY; J. Little, Austin, TX; P. McDonald and J. Tomlinson, Wichita Falls, TX; W. Reichman, Piscataway, NJ; M. Rosenthal, San Diego, CA; M. Rymer, Kansas City, MO; R. Strub, New Orleans, LA; and F. Zemlan and M. Keys, Cincinnati, OH.

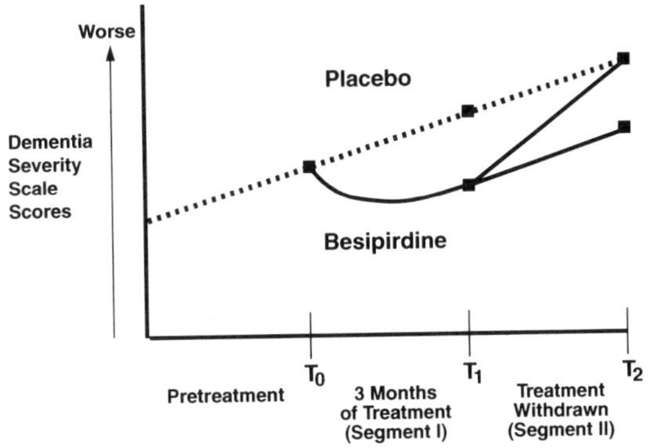

FIGURE 1. Design and hypothetical out-comes of besipirdine treatment study.

time before, during, and after treatment with besipirdine. The slope of the line through the pre-treatment interval describes the rate of progression in the absence of treatment. For the placebo group, the same rate of progression continues through the treatment and post-treatment intervals (dashed line in the figure). The solid lines in the figure represent possible outcomes in patients receiving besipirdine.

The hypothesis that besipirdine is efficacious during the treatment interval was tested by computing the change in dementia severity scale scores between T_0 and T_1. The change in scores may be negative (patient improved), zero, or positive (patient deteriorated). The efficacy hypothesis predicted that the change would be less positive (less deteriorated) for the besipirdine group than for the placebo group. The efficacy hypothesis was the primary study hypothesis.

The anticipated efficacy associated with besipirdine at T_1 may be either a reversible effect, illustrated in the figure by the upper solid line intersecting the placebo line at T_2, or a persistent effect as described by the lower line in the post-treatment interval. The hypothesis that besipirdine has a persistent effect was tested by computing change in dementia severity scores between T_0 and T_2, and testing the prediction that the change was less positive (less deteriorated) in the group that received besipirdine during the treatment interval than in the group that received placebo. If a difference at the end of Segment I persists at the end of Segment II, an effect on dementia progression (neuroprotective effect) may be inferred.

METHODS

Patients were randomized to one of two doses of besipirdine (5 mg BID, 20 mg BID) or placebo. Efficacy was tested during 3 months of double-blind treatment (Segment I). Patients were followed for 3 additional months on placebo (Segment II), during which time the persistence of efficacy was evaluated. One of every 12 of patients on active treatment in Segment I remained on it in Segment II in order to maintain blinding. A total of 275 patients with Alzheimer's disease by NINCDS-ADRDA criteria[2] enrolled in 16 clinical centers. The primary efficacy measures were the Alzheimer's Disease Assessment Scale (ADAS) Cognitive Subscale[3] and the Clinician's Interview Based Impression of Change (CIBIC) global rating (P. Leber, unpublished letter). Secondary measures included the Cornell Scale for Depression in Dementia.[4]

RESULTS

There were no statistically significant differences among the three treatment groups with regard to age and other demographic variables, or pretreatment baseline ADAS Cognitive scores. Besipirdine was generally well tolerated. Of 184 patients assigned to besipirdine (5 mg BID or 20 mg (BID), 162 (88%) completed the last Segment I assessment. Among 91 placebo patients, 86 (95%) completed Segment I.

The primary efficacy analysis included all subjects (intent-to-treat analysis with last observation carried forward for incomplete cases) through the endpoint of Segment I, the double-blind treatment segment of the study. Progressive deterioration over the 12-week period was observed in placebo treated patients, whereas besipirdine-treated patients (both dosage groups combined) were essentially unchanged from baseline over the same period. In testing the primary study hypothesis, the difference in mean change scores between the placebo and besipirdine treatment groups approached but did not reach statistical significance (p = .067). The magnitude of the difference at the endpoint of treatment was 1.25 points on the ADAS scale. When the same comparison was computed only for patients who completed all assessments in Segment I, the magnitude of the treatment difference at endpoint was 1.41 and was statistically significant ($p < .05$).

Secondary analyses involved separate comparisons of the besipirdine 5 mg and 20 mg BID treatment groups with the placebo group. In the intent-to-treat analyses, the magnitude of the mean change from baseline was essentially the same for both the 5 mg and 20 mg BID dosing regimens. In the analyses of Segment I completers, the treatment difference compared with placebo was 1.23 points for the 5 mg BID group and 1.58 points for the 20 mg BID group, suggesting a positive dose-response relationship.

The mean change from baseline at the Segment II endpoint in the combined besipirdine treatment groups was 0.53 ADAS points less than the mean change for the placebo group in the intent-to-treat analysis. This difference indicates that only about half of the efficacy (1.25 points) apparent after 12 weeks of treatment persisted after a further 12 weeks without treatment. It is noteworthy that the treatment difference for the besipirdine 5 mg BID group compared with placebo was 0.16 ADAS points, whereas the difference was 0.87 points for the 20 mg BID group at the Segment II endpoint. The effect of besipirdine 20 mg BID treatment thus appears to subside more slowly than the effect of 5 mg BID after treatment is withdrawn.

Similar analyses were performed for the CIBIC global rating. A rating of 4 on this scale reflects no change from baseline, a rating of 1 reflects marked improvement, and a rating of 7 indicates the patient is markedly worse. More than 50% of patients were rated as unchanged at weeks 4, 8, and 12 of treatment. Mean ratings by treatment group across all Segment I and II assessments spanned the narrow range of 3.9 to 4.1, indicating that substantial differences among treatment conditions were not discerned by the clinician raters over this observation period.

Because the effect of besipirdine on norepinephrine uptake and release indicates a potential for modifying mood, the Cornell Scale for Depression in Dementia was used as a secondary outcome measure in the study. Patients with clinically important depression were excluded from the study, and scores on the Cornell scale during the study reflected mood status well below the threshold for diagnosis of major depression. The mean at endpoint indicated a slight improvement in the placebo group and slight deterioration in the combined besipirdine treatment groups, the 0.75 point difference being statistically significant ($p < .05$). An important scale item contributing to this difference was increased "agitation" in patients receiving besipirdine.

DISCUSSION

The results of this Phase II efficacy study suggest that besipirdine has an effect on the cognitive deficit in patients with Alzheimer's Disease. Whereas patients treated with placebo deteriorated over 3 months (12 weeks), patients treated with besipirdine remained essentially unchanged. Changes observed on the ADAS scale were not detected on the CIBIC global rating. It is possible that the magnitude of the cognitive effect was not sufficiently large for the clinicians to discern it in their global ratings. If higher doses or longer treatment increase the magnitude of the cognitive effect, the CIBIC ratings may reflect the treatment effect. Another factor potentially affecting the CIBIC ratings is that there apparently is a clinically small, but statistically significant, "dysphoric" effect measured by the Cornell scale over the 12 weeks of treatment with besipirdine. This effect may in part reflect increased agitation due to the adrenergic activity of besipirdine.

The results of the withdrawal of treatment from weeks 12 through 24 indicate that the effect of besipirdine on cognition is reversible after withdrawal of treatment, and suggest that its efficacy is primarily symptomatic rather than an effect on dementia progression. Disappearance of the treatment effect was slower in the 20 mg BID treatment group than in the 5 mg BID treatment group, and these results do not exclude the possibility of some neuroprotective effect. Treatment with higher doses and for longer periods may enhance the efficacy of besipirdine on both cognitive and clinical global assessments.

REFERENCES

1. HUFF, F. J. 1995. Multiple Neurotransmitter Strategies for treatment of Alzheimer's disease. *In* Alzheimer's Disease: Clinical and Treatment Aspects. N. R. Cutler, C. G. Gottfries & K. R. Siegfried, Eds.: 151–162. UK: John Wiley & Sons.
2. MCKHANN, G., D. DRACHMAN, M. FOLSTEIN, R. KATZMAN & E. M. STADLAN. 1984. Clinical diagnosis of Alzheimer's Disease: Report of the NINCDS-ADRDA work group under the auspices of the Department of Health and Human Services Task Force on Alzheimer's Disease. Neurology **34:**939–944.
3. ROSEN, W. G., R. C. MOHS & K. L. DAVIS. 1984. A new rating scale for Alzheimer's disease. Am. J. Psychiatry **141:**1356–1364.
4. ALEXOPOULOS, G. S., R. C. ABRAMS, R. C. YOUNG & C. A. SHAMOIAN. 1988. Cornell Scale for Depression in Dementia. Biol. Psychiatry **23:**271–284.

Immortalized Neural Cells from Trisomy 16 Mice as Models for Alzheimer's Disease

K. FREDERIKSEN,[a,e] A. THORPE,[b] S. J. RICHARDS,[b] J. WATERS,[c] S. B. DUNNETT,[d] AND B. E. B. SANDBERG[a]

[a] *Pharmacological Research, H. Lundbeck A/S, Copenhagen-Valby, Denmark*
[b] *Department of Medicine, University of Cambridge, Addenbrooke's Hospital, Cambridge, United Kingdom*
[c] *Regional Cytogenetics Unit, East Birmingham Hospital, Birmingham, United Kingdom*
[d] *MRC Brain Repair Centre, Cambridge, United Kingdom*

ABSTRACT: The trisomy 16 mouse (Ts16) is a general accepted animal model for both Downs syndrome (DS) and Alzheimer's Disease (AD). However, the efficacy of this model is severely hampered by the fact that Ts16 is lethal after about 18–20 days of gestation. Chimeras, long-term tissue culture and neural transplantation of Ts16 material have previously been used to overcome this limitation presented by death *in utero* of the Ts16. In this paper we describe a new strategy to overcome this limitation, *i.e.* immortalization of primary cells from Ts16 mice with retrovirus-mediated gene transfer of a temperature sensitive immortalizing oncogene. By this method we have obtained a total of 21 stable cell lines from Ts16 hippocampus, Ts16 cortex, normal hippocampus, and normal cortex. So far, two of the cell lines have been karyotyped and as expected, the cell line immortalized from Ts16 embryos has retained three copies of chromosome 16. We are currently characterizing these cell lines with respect to expression of APP, T-antigen, Nestin, GFAP, NF and Map-2. Moreover, the processing and secretion of APP fragments are being investigated by immunoblotting.

In summary, we have immortalized CNS cells from Ts16 mice and we expect that these cell lines will be useful as *in vitro* and *in vivo* models for studying various aspects of the pathology of Alzheimer's disease.

INTRODUCTION

Downs syndrome (DS, trisomy 21) is an important high risk factor for the predisposition to early onset Alzheimer's disease. Nearly all individuals

[e] *Send correspondence to*: Dr. Kristen Frederiksen, Pharmacological Research, H. Lundbeck A/S, DK-2500 Copenhagen-Valby, Denmark; Tel: +45 36 30 37 47 ext. 2516; Fax: +45 36 30 37 47.

with DS develop in their fourth decade neuropathological alterations typical of Alzheimer's disease (AD).[1,2] The trisomy 16 mouse exhibits a close genotypic and phenotypic homology with human trisomy 21.[3–5] It is notable that the gene coding for the Amyloid Precursor Protein (APP) has been located to human chromosome 21 and to mouse chromosome 16 (FIG. 1). However, the use of trisomy 16 mice as a model for Alzheimer's disease is severely hampered by death of the trisomic embryo *in utero*, occurring 18–20 days after gestation. Hence, until now any such model had to be based on the utilization of embryonic tissue, or primary cultures derived from such tissue.

To overcome these limitations, a project aiming at the immortalization of various CNS cells from Ts16 mice was initiated. Previous work has shown that primary CNS cells from rat embryos can be immortalized through retrovirus-mediated gene transfer of different oncogenes.[6] Hence, the same technique was adopted to immortalize hippocampal and neocortical primary cells derived from embryos of Ts16 mice.[7]

METHODS

A special breeding scheme was established between male mice with balanced bilateral Robertsonian translocations of chromosome 16 (*i.e.*, fusion of chromosome 16 with another chromosome) and normal females (FIG. 1). Such mating may result in embryos with three copies of chromosome 16 (Trisomy 16, Ts16) in up to 30% of the total litter.[8] Of a total of about 60 embryos from 12 pregnant mice, we obtained six Ts16 embryos. Hippocampus and cortex were dissected from the 6 trisomic embryos and from 6 euploid littermates. The tissues were trypsinized, triturated and the resulting cell suspension was seeded on tissue culture dishes and cultured at 37°C in DMEM + 5 % FCS or optimum with B27 as supplement. We used a previously described psi2-derived producer cell line to obtain infectious retrovirus particles carrying the temperature sensitive mutant form of SV40 large T-antigen (tsA58) as the immortalizing oncogene and a neomycin gene as the selection resistant marker for G418. Both genes were inserted into the shuttle vector pZipNeo SV(X)1.[6] This producer cell line was cultured in DMEM + 10% FCS and the supernatant containing retrovirus particles was harvested and used to infect the primary cultures from Ts16, and from euploid littermates. The infection procedure was essentially as described by Frederiksen *et al.* 1988.[6] Briefly, the primary cell cultures were exposed to virus containing supernatant in the presence of 8 μg/ml polybreen (Aldrich). Two to 3 hours later, the supernatant containing the virus particles was replaced by normal growth media and moved to 33°C for further growth. Two to 4 days later G418 was added to the culture medium and the cells were grown for 3–4 weeks under G418 selection pressure to select for infected cells. The surviving

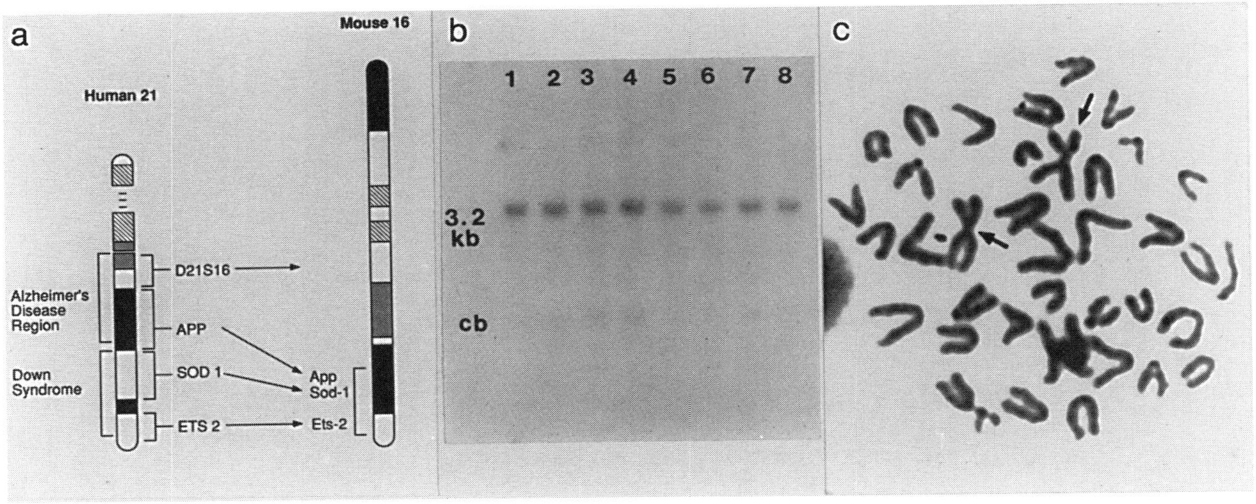

FIGURE 1. **a:** Homology between human chromosome 21 and mouse chromosome 16; **b:** Southern blot analysis of gene dosage of APP from Ts16 (lanes 1–4) and normal tissues (lanes 5–8); **c:** Karyotype from Ts16 mouse tissue confirming the trisomic nature. Arrows indicate the Robertsonian translocations (adapted from Richards *et al.*[20]).

G418 resistant colonies were picked by using cloning rings and transferred to 96-well plates. The colonies were subsequently expanded into cell lines and frozen in liquid nitrogen.

RESULTS

Out of approximately 60 selected colonies we obtained 21 stable immortalized cell lines *i.e.* 10 cell lines from Ts16 hippocampus, 4 cell lines from Ts16 cortex, 3 cell lines from normal hippocampus, and 4 cell lines from normal cortex. Southern blot analysis of the cell lines was performed to ensure that the cell lines had only one copy of the viral insert, indicating that the cell lines are monoclonal.[6] So far we have confirmed that ten of the immortalized cell lines are derived from single clones. Two of the cell lines have been karyotyped, one from normal, and one from trisomic, hippocampus. As expected, the cell line immortalized from Ts16 embryos has three copies of chromosome 16, whereas the cell line drived from normal euploid embryo has only two copies of chromosome 16. In order to further characterize the cell lines, staining with different cell type specific antibodies was performed. Most of the cell lines tested so far at the permissive temperature (33°C) are positive for Nestin, but are negative for Map-2. As Nestin (recognized by the monoclonal antibody Rat-401 or a polyclonal serum) has previously been shown to be transiently expressed in all CNS precursor cells in the rat,[9,10] we expect that these Nestin positive cell lines are multipotential CNS precursor cells which at 39°C may differentiate into either neurons or astrocytes.[6]

DISCUSSION

A number of recent studies points to significant differences between primary cultured cells from Ts16 embryos and cells from normal euploid littermates, *i.e.* fewer surviving cholinergic neurons in culture from Ts16 basal forebrain than control,[11] enhanced stimulated superoxide production in Ts16 microglia,[12] higher production of interleukin-1 (IL-1) in Ts16 microglia,[13] differential sensitivity of Ts16 and control neurons to NGF,[14] higher intracellular Ca^{++} in Ts16 astrocytes,[15] accelerated death of Ts16 hippocampal neurons,[16] dysregulation of neuropeptide Y containing neurons caused by Ts16 glia cells,[17] abnormal action potentials in Ts16 hippocampal neurons.[18] Some of these differences may relate to AD. Moreover, transplantation of Ts16 tissue into the brain of normal mice has been employed in an attempt to generate an *in vivo* model for Alzheimer's disease.[5,19,20] However, these and other studies have been hampered by the fact that Ts16 embryos do not survive *in utero*. Thus, we anticipate that access to immortalized cell-lines derived from Ts16 cortex and hippocampus may open new avenues to the development of pathophysiological *in vivo* and *in vitro* models applicable to the study of Alzheimer's disease. No doubt, these cell lines constitute interesting tools for the screening of novel drug candidates aimed at therapeutic targets for the treatment of Alzheimer's disease.

ACKNOWLEDGMENT

Dr. P. S. Jat, Ludwig Institute for Cancer Research, London, UK, is thanked for the generous gift of the pZipNeoSVtsA58 producer cell line.

REFERENCES

1. BALL, M. J., M. B. SCHAPIRO & S. I. RAPOPORT. 1989. Neuropathological relationships between Down syndrome and senile dementia of the Alzheimer's type. *In* The Neurobiology of Down Syndrome. C. J. Epstein, Ed. :35–58. New York: Raven Press.
2. WISNIEWSKI, K. E., H. M. WISNIEWSKI & G. Y. WEN. 1985. Occurrence of neuropathological changes and dementia of Alzheimer's disease in Down syndrome. Ann. Neurol. **17:**278–382.
3. COX, D. R. & C. J. EPSTEIN. 1985. Comparative gene mapping of human chromosome 21 and mouse chromosome 16. Ann. N.Y. Acad. Sci. **450:** 169–177.
4. REEVES, R. H., N. K. ROBAKIS, M. L. OSTER-GRANITE, H. M. WISNIEWSKI, J. T. COYLE & J. D. GEARHART. 1987. Genetic linkage in the mouse of genes involved in Down syndrome and Alzheimer's disease in man. Mol. Brain Res. **2:**215–221.
5. RICHARDS, S. J. 1991. The neuropathology of Alzheimer's disease investigated by transplantation of mouse Trisomy 16 hippocampal tissues. TINS **14**(8): 334–338.
6. FREDERIKSEN, K., P. S. JAT, N. VALTZ, D. LEVY & R. MCKAY. 1988. Immortalization of precursor cells from the mammalian CNS. Neuron **1:**439–448.
7. THORPE, A., K. FREDERIKSEN & S-J. RICHARDS. 1994. Immortalization of hippocampus and neocortex cells from the trisomy 16 mouse. Neurobiol. Aging **5**(Suppl. 1):Abstract S11,45.
8. GEARHART, J. D., M. T. DAVISSON & M. L. OSTER-GRANITE. 1986. Autosomal aneuploidy in mice: Generation and developmental consequences. Brain Res. Bull. **16:**789–801.
9. FREDERIKSEN, K. & R. MCKAY. 1988. Proliferation and differentiation of neuroepithelial precursor cells *in vivo*. J. Neurosci. **8:**1144–1151.
10. LENDAHL, U., L. B. ZIMMERMAN & D. G. MCKAY. 1990. CNS stem cells express a new class of intermediate filament protein. Cell **60:**585–595.
11. CORSI, P. & J. T. COYLE. 1991. Nerve growth factor corrects developmental

impairments of basal forebrain cholinergic neurones in the trisomy 16 mouse. Proc. Natl. Acad. Sci. USA **88:**1793–1797.

12. COLTON, C. A., J. YAO, D. GILBER & M.-L. OSTER-GRANITE. 1990. Enhanced production of superoxide anion by microglia from trisomy 16 mice. Brain Res. **519:**236–242.
13. COLTON, C. A., J. YAO, R. E. TAFFS, J. E. KERI & M.-L. OSTER-GRANITE. 1991. Abnormal production of interleukin-1 by microglia from trisomy 16 mice. Neuroscience Letters **132:**270–274.
14. COAN, E. J., Z. GALDZICKI & S. I. RAPOPORT. 1994. Effects of nerve growth factor on whole-cell currents and other electrical membrane properties in cultured dorsal root ganglion neurons from normal and trisomy 16 mice. Brain Res. **650:**161–165.
15. GOLOVINA, V. A., I. L. BAMBRICK, P. J. YAROWSKY & B. K. KRUEGER. 1994. Altered regulation of intracellular Ca^{++} in astrocytes from the trisomy 16 mouse. Soc. Neurosci. Abstr. **20:**343.9.
16. BAMBRICK, L. L., D. BROUGHER, P. J. YAROWSKY & B. K. KRUEGER. 1994. Hippocampal neurons from the Trisomy 16 mouse: Accelerated death *in vitro*. Soc. Neurosci. Abstr. **20:**290.16.
17. CASERTA, M. T. 1994. Neuropeptide Y immunoreactive neurons in murine trisomy 16 cortical cultures plasticity of expression and differentiation. Molecular and chemical neuropathology. **22**(3):197–210.
18. GALDZICKI, Z., E. COAN & S. I. RAPOPORT. 1993. Cultured hippocampal neurons from trisomy 16 mouse, a model for Down's syndrome, have an abnormal action potential due to a reduced inward sodium current. Brain. Res. **604**(1–2):69–78.
19. HOLTZMAN, D. M., Y. LI, S. J. DEARMOND, M. P. MCKINLEY, F. H. GAGE, C. J. EPSTEIN & W. C. MOBLEY. 1992. Mouse model of neurodegeneration: Atrophy of basal forebrain cholinergic neurones in trisomy 16 transplants. Proc. Natl. Acad. Sci. USA **89:**1383–1387.
20. RICHARDS, S. J., J. J. WATERS, D. C. ROGERS, F. L. MARTEL, D. R. SPARKMAN, C. L. WHITE, K. BEYREUTHER, C. L. MASTERS & S. B. DUNNETT. 1990. Hippocampal grafts derived from embryonic trisomy 16 mice exhibit amyloid (A4) and neurofibrillary pathology. Prog. Brain Res. **82:**215–223.

Mice Deficient for the Amyloid Precursor Protein Gene

H. ZHENG,[a,f] M. JIANG,[a] M. E. TRUMBAUER,[a] R. HOPKINS,[b] D. J. S. SIRINATHSINGHJI,[b] K. A. STEVENS,[c] M. W. CONNER,[d] H. H. SLUNT,[e] S. S. SISODIA,[e] H. Y. CHEN,[a] AND L. H. T. VAN DER PLOEG[a]

[a] *Department of Genetics & Molecular Biology, Merck Research Laboratories, Merck & Co. Inc., Rahway, New Jersey 07065 USA*

[b] *Merck Sharp & Dohme Research Laboratories, Neuroscience Research Center, Terlings Park, Eastwick Road, Essex, CM20 2QR, United Kingdom*

[c] *Laboratory Animal Resources, Merck Research Laboratories, Merck & Co. Inc., Rahway, New Jersey 07065 USA*

[d] *Safety Assessment, Merck Research Laboratories, Merck & Co., Inc. West Point, Pennsylvania 19486 USA*

[e] *Neuropathology Laboratory, The Johns Hopkins University School of Medicine, Baltimore, Maryland 21205 USA*

ABSTRACT: To understand the *in vivo* function of the amyloid precursor protein (APP) we generated an APP null mutation in mice by homologous recombination in embryonic stem (ES) cells. We show here that homozygous APP deficient mice were produced at expected frequencies. Neither APP mRNA nor protein could be detected in these animals. Yet the homozygous APP mutant mice are fertile and do not show overt abnormalities at up to 12 weeks of age. Neuroanatomical studies of the brain did not reveal significant differences in the knockout mice as compared to the wild-type controls. These results argue against an essential function of APP in mouse embryonic and early neuronal development.

INTRODUCTION

Alzheimer's disease (AD) is a neurodegenerative disorder characterized by the deposition of β-amyloid plaques, the presence of neurofibrillary tangles (NFT) and neuronal cell loss. The risk of developing AD increases exponentially with age. The β-amyloid peptide (Aβ), the major component of the neuritic plaques characterizing AD, is a 39 to 43 amino acid peptide derived

[f] *Send correspondence to*: Dr. Hui Zheng, Department of Genetics and Molecular Biology, Merck Research Laboratories, P.O. Box 2000, Rahway, NJ 07065-0900; Tel: (908) 594-6145; Fax: (908) 594-5700.

from proteolytic cleavage of a larger amyloid precursor protein (APP). APP is a gene of approximately 400 kilobase (kb) and contains at least 18 exons.[1,2] It undergoes alternative splicing to generate several isoforms, encoding proteins that range from 365 to 770 amino acid residues. Among these, APP695 is expressed predominantly in neurons and APP751 and APP770 can be detected in all tissues examined.[3] Mutations in APP gene have been identified which cause familial, early onset Alzheimer's disease, making it the strongest evidence that APP metabolism is a central event in Alzheimer's disease progression. APP is one of the most abundant proteins in the brain and Aβ is produced not only in AD patients, but also in cerebrospinal fluid (CSF) of normal individuals. APP has been implicated in numerous activities based on *in vitro* studies, such as regulations of cell adhesion, neurite outgrowth, intraneuronal calcium and protein half-life.[4–8] However, the *in vivo* function of APP remains unclear.

We wish to inactivate the APP gene in mice in order to address its role in mouse development and its function in the central nervous system (CNS). In this report, we show that we have generated a complete deficiency of APP in mice by homologous recombination in embryonic stem (ES) cells and these animals fail to exhibit significant anatomical deficits.

METHODS

Construction of APP Targeting Vector and Gene Targeting in Murine ES Cells

The targeting vector was constructed by ligating a 1.4 kb Bgl II-Xho I fragment upstream of APP promoter and a 7.1 kb Xho I-Bgl II fragment from the 3′ portion of exon 1. In doing so, a 3.8 kb fragment containing the APP promoter, the first exon and part of the first intron was deleted. A 1.5 kb Xho I-Sal I phosphoglycerate kinase promoter driving the neomycin phosphotransferase gene (PGKneo) cassette was inserted at the site of deletion. The final vector was constructed by ligating the 2 kb Xho I fragment of a pMC1-thymidine kinase (HSV-TK) cassette to the 3′ end of the long arm.

The vector was linearized by digestion with Not I, leaving the pKS backbone attached to the end of the TK gene prior to electroporation. Electroporation and selection of targeted clones were carried out as described.[9]

Gross and Microscopic Analysis

Mice were anesthetized and subjected to diagnostic necropsies. All tissues were fixed in 10% buffered formalin. Bones were decalcified using formic acid. Sections of tissues were prepared by routine methods and stained with hemotoxylin and eosin (see ref. 10 for a detailed list of tissues examined).

RESULTS

To target the APP gene in murine ES cells, a positive-negative selection strategy was used. The targeting vector (FIG. 1) contained a total of 8.5 kb of homology with the APP gene. A 3.8 kb sequence containing the APP promoter,[11] the first exon and part of the first intron was deleted and replaced at the site of the deletion with a positive selectable PGKneo marker. An HSV-TK cassette was inserted at the end of the vector for negative selection. Correct homologous recombination between the targeting vector and one of the APP alleles in ES cells results in a deletion of the APP promoter and exon 1 encoding the signal peptide.

The targeting vector was electroporated into AB2.1 ES cells. G418 and FIAU resistant clones were screened by a mini-Southern protocol.[9] A 4.4-fold enrichment was achieved by selecting the cells with FIAU. Six targeted clones were identified, and the frequency of targeted recombination versus random integration at the APP locus was 1/140 (data not shown). Of the four clones injected to generate chimeric mice, two transmitted the targeted APP allele into the offspring. Heterozygous matings were set up to produce mice homozygous for the disrupted APP gene.

Homozygous APP knockout mice were generated at expected frequencies (data not shown). These mice appeared normal and healthy up to 12 weeks of age. Northern blot analysis of both brain and kidney RNA using APP695 cDNA as a probe showed that APP mRNA was not produced in any size class in the homozygous mice. APP mRNA levels were reduced by approximately 50% in the heterozygous mice when compared to wild-type controls.[10] The APP null mutation was further confirmed by Western blot analysis which failed to detect any APP protein.[10] Since APP is expressed at the highest level in the brain, neuroanatomical studies were performed on 6 and 14 week old mice. No significant differences were observed in the knockout mice (data not shown).

DISCUSSION

APP is conserved among mammalian species.[12] APP homologues have been identified in human, mouse, *Drosophila melanogaster* and *Caenorhabditis elegans*.[13–16] These proteins may be functionally related since the behavioral deficit observed in flies deleted for the Drosophila APP homologue (Appl)

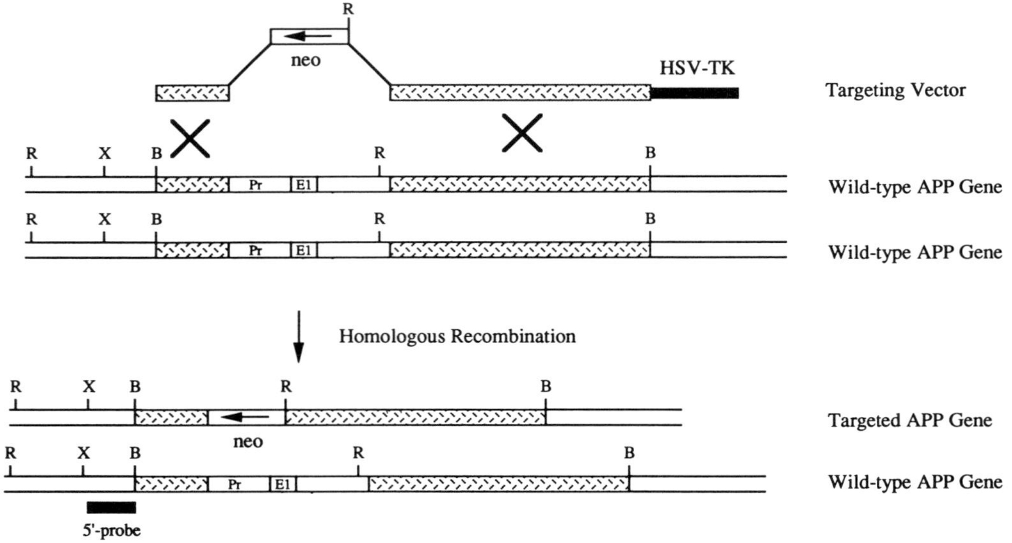

	EcoRI
Wild-type APP Gene	9.0 kb
Targeted APP Gene	6.5 kb

FIGURE 1. Schematic representation of gene targeting at the APP locus. The targeting vector contains from left to right, 1.4 kb of 5 homology, a PGKneo expression cassette inserted in the opposite orientation as the APP gene, 7.1 kb of 3′ APP homology and a HSV-TK gene. A 3.8 kb sequence containing the APP promoter (Pr), the first exon (E1) and part of the first intron was deleted. The homologous sequences are represented as shaded rectangles. Homologous recombination between the vector and the wild-type APP locus results in the deletion of the 3.8 kb APP sequence followed by replacement of the 1.5 kb neo sequence. The probe used for Southern blot analysis was the 1.0 kb Xba I-Bgl II fragment (5′-probe). Eco RI was used to differentiate the wild-type and the targeted APP alleles, which generates a 9.0 kb and 6.5 kb fragments. R: Eco RI, X: Xba I, B: Bgl II.

can be partially rescued by the human APP695[17] and behavioral defects were also observed in mice with a modified APP.[18]

Since none of the APP homologues contains the Aβ peptide, unidentified APP-like genes encoding an Aβ sequence may exist in *Drosophila* or *C. elegans* or the Aβ may serve specific functions in vertebrates. In human and mouse, the APP homologue APLP2 is expressed in a very similar pattern to that of the APP, which indicates that the expression of APLP2 may compensate for the loss of APP. However, Northern analysis did not reveal any differences in APLP2 expression when compared between the wild-type mice and the heterozygous and homozygous APP deficient mice (data not shown).

Since APP overexpression or mutation is likely to be one of the causes of Down syndrome (DS) and Alzheimer's disease, transgenic mice have been generated that express various forms of the human APP gene.[19] However, pathology resembling human Alzheimer's disease could not yet be detected in these mice. One of the factors contributing to the failure to produce a mouse model of AD might be due to three amino acid differences in the Aβ region of mouse as compared to human Aβ and the murine sequence may prevent human Aβ from forming plaques in the mouse. The APP deficient mice may therefore, serve as a valuable recipient to insert the human APP gene in an effort to generate a mouse model of Alzheimer's disease.

REFERENCES

1. LAMB, B. T., S. SISODIA, A. LAWLER, H. SLUNT, C. KITT, W. KEARNS, P. PEARSON, D. PRICE & J. GEARHART. 1993. Introduction and expression of the 400 kilobase precursor amyloid protein gene in transgenic mice. Nature Genet. **5**:22–30.
2. YOSHIKAI, S.-I., H. SASAKI, K. DOH-URA, H. FURUYA & Y. SAKAKI. 1990. Genomic organization of the human amyloid β-protein precursor gene. Gene **87**:257–263.
3. KOSIK, K. S. 1992. Alzheimer's disease: A cell biological perspective. Science **256**:780–783.
4. MATTSON, M. P., B. CHENG, A. R. CULWELL, F. S. ESCH, I. LIEBERBURG & E. R. RUSSEL. 1993. Evidence for excitoprotective and intraneuronal calcium-regulating roles for secreted forms of the β-amyloid precursor protein. Neuron **10**:243–254.
5. NILWARD, E. A., R. PAPADOPOULOS, S. J. FULLER, R. D. MOIR, D. SMALL, D. BEYREUTHER & C. L. MASTERS. 1992. The amyloid protein precursor of Alzheimer's disease is a mediator of the effects of nerve growth factor on neurite outgrowth. Neuron **9**:129–137.
6. PONTE, P., P. GONZALEZ-DEWHITT, J. SCHILLING, J. MILLER, D. HSU, B. GREENBERG, K. DAVIS, W. WALLACE, I. LIEBERBURG, F. FULLER & B. CORDELL. 1988. A new A4 amyloid mRNA contain a domain homologous to serine proteinase inhibitors. Nature **331**:525–527.

7. SIROH, R., M. SUNDAMO, J.-M. ROCH, N. KIMURA, G. COLE, D. SCHUBER, T. OLTERSDORF & D. SCHENK. 1989. Secreted form of amyloid β protein precursor is involved in the growth regulation of fibroblasts. Cell **58:**615–622.
8. SCHUBERT, D., L.-W. JIN, T. SAITOH & G. COLE. 1989. The regulation of amyloid β protein precursor secretion and its modulatory role in cell adhesion. Neuron **3:**689–694.
9. RAMIREZ-SOLIS, R., J. RIVERA-PEREZ, J. D. WALLACE, M. WIMS, H. ZHENG & A. BRADLEY. 1992. Genomic DNA microextraction: A method to screen numerous samples. Anal. Biochem. **201:**331–335.
10. ZHENG, H., M. JIANG, M. TRUMBAUER, D. SIRINATHSINGHJI, R. HOPKINS, D. SMITH, R. HEAVENS, G. DAWSON, S. BOYCE, M. CONNER, K. STEVENS, H. SLUNT, S. SISODIA, H. CHEN & L. VAN DER PLOEG. 1995. β-amyloid precursor protein-deficient mice show reactive gliosis and decreased locomotor activity. Cell **81**:525–531.
11. IZUMI, R., T. YAMADA, S.-I. YOSHIKAI, H. SASAKI, M. HATTORI & Y. SAKAKI. 1992. Positive and negative regulatory elements for the expression of the Alzheimer's disease amyloid precursor-encoding gene in mouse. Gene **112:**189–195.
12. YAMADA, T., H. SAKASI, H. FURNYA, T. MIYATA, I. GOTO & Y. SAKAKI. 1987. Complementary DNA for the mouse homolog of the human amyloid beta protein precursor. Biochem. Biophys. Res. Commun. **149:**665–671.
13. DAIGLE, I. & C. LI. 1993. Apl-1, a *Caenorhabditis elegans* gene encoding a protein related to the human β-amyloid protein precursor. Proc. Natl. Acad. Sci. USA **90:**12045–12049.
14. ROSEN, D. R., L. MARTIN-MORRIS, L. LUO & K. WHITE. 1989. A Drosophila gene encoding a protein resembling the human beta-amyloid protein precursor. Proc. Natl. Acad. Sci. USA **86:**2478–2482.
15. SLUNT, H. H., G. THINAKARAN, C. VAN KOCH, A. C. Y. LO, R. E. TANZI & S. S. SISODIA. 1994. Expression of a ubiquitous, cross-reactive homologue of the mouse β-amyloid precursor protein (APP). J. Biol. Chem. **269:**2637–2644.
16. WASCO, W., K. BUPP, M. MAGENDANTZ, J. GUSELLA, R. TANZI & F. SOLOMON. 1992. Identification of a mouse brain cDNA that encodes a protein related to the Alzheimer disease-associated amyloid β protein precursor. Proc. Natl. Acad. Sci. USA **89:**10758–10762.
17. LUO, L., T. TULLY & K. WHITE. 1992. Human amyloid precursor protein ameliorates behavioral deficit of flies deleted for Appl gene. Neuron **9:**595–605.
18. MÜLLER, U., N. CRISTINA, Z. W. LI, *et al.* 1994. Behavioral and anatomical deficits in mice homozygous for a modified β-amyloid precursor protein gene. Cell **79:**755–765.
19. PRICE, D. L. & S. S. SISODIA. 1994. Cellular and molecular biology of Alzheimer's disease and animal models. Annu. Rev. Med. **45:**435–447.

The D_1 Agonist Dihydrexidine Releases Acetylcholine and Improves Cognition in Rats

T. D. STEELE, D. B. HODGES, JR., T. R. LEVESQUE, K. W. LOCKE, AND B. W. SANDAGE, JR.[a]

Interneuron Pharmaceuticals, Inc., Lexington, Massachusetts 02173 USA

ABSTRACT: Neurochemical and behavioral studies have elucidated extensive interactions between dopaminergic and cholinergic systems in brain areas associated with movement and cognition. The initial goal of these studies was to evaluate the effect of the anti-Parkinson drug dihydrexidine (DHX), a dopamine D_1 full-efficacy agonist, on brain acetylcholine (ACh) release using *in vivo* microdialysis techniques. Moderate doses of DHX (3 and 10 mg/kg) produced approximately a 50% increase in striatal ACh release that was blocked by the D_1 agonist SCH23390 (0.3 mg/kg). A higher dose of DHX (17.5 mg/kg) was less effective in raising striatal ACh, possibly due to D_2 receptor activation. In frontal cortex, DHX (10 mg/kg) evoked a more robust increase in ACh release (+200%) that was blocked by SCH23390 (0.3 mg/kg). Since elevations in brain ACh are associated with cognitive improvement, the effectiveness of DHX in a passive avoidance model of learning and memory was also evaluated. These studies revealed a significant improvement in performance by 0.3 mg/kg DHX in scopolamine-induced amnestic rats. These results provided support for the hypothesis that DHX improves cognitive performance as a consequence of ACh release in relevant brain regions. Further, D_1 agonists may have novel therapeutic potential in the treatment of dementia.

INTRODUCTION

Dopamine (DA) receptors have classicially been divided into two major subclasses, D_1 and D_2, based on pharmacological, biochemical and behavioral data.[1] A major breakthrough in the characterization of D_1 receptors was the synthesis and pharmacological evaluation of dihydrexidine (trans-10,11-dihydroxy-5,6,6a,7,8,12b-hexahydrobenzo[a]phenanthridine, DHX) as a potent, selective, full-efficacy D_1 agonist.[2,3] DHX has displayed a significant therapeutic potential for the treatment of Parkinson's disease in MPTP-treated primates.[4] The D_1 antagonist SCH23390, but not the D_2 antagonist

[a] *Send correspondence to:* Dr. Bobby W. Sandage, Jr., Interneuron Pharmaceuticals, Inc., 99 Hayden Avenue, Suite 340, Lexington, MA 02173; Tel: (617) 861-8444; Fax: (617) 861-3830.

remoxipride, blocked the improvement produced by DHX indicating that D_1 receptor activation alleviated the Parkinsonian syndrome.[5]

Recent neurochemical studies have shown reciprocal control by DA on striatal acetylcholine (ACh) release depending on which receptor subtype is activated; D_1 receptor activation stimulates and D_2 activation inhibits release.[6] Thus, the initial goal of the studies presented here sought to characterize D_1 versus D_2 activities of DHX using *in vivo* striatal ACh release as the model. In other brain regions, acetylcholine-releasing effects of D_1 agonists may have additional therapeutic implications because of the association of the dementia of Alzheimer's[7,8] and Parkinson's disease[9] with central cholinergic deficiencies. To evaluate this possibility with DHX, *in vivo* microdialysis studies were conducted to assess its effects on ACh release in rat frontal cortex. Once significant effects were identified, the cognition-enhancing potential of DHX was evaluated in scopolamine-induced amnestic rats using a passive avoidance procedure.

METHODS

In Vivo *Microdialysis*

Male Sprague-Dawley rats were anesthetized with ketamine/xylazine (87/13 mg/kg) for stereotaxic implantation of a guide cannula (Bioanalytical Systems Carnegie-Medicin, West Lafayette, IN) for dialysis probe placement into the striatum (A: 1.5, L: 2.6, V: 5.5) or frontal cortex (A: 3.2, L: 1.3, V: 1.0). Following recovery from surgery (24–48 h), a dialysis probe was inserted into the guide and perfused at a flow rate of 1.5 μl/min with artificial CSF. Dialysate samples (20 μl) were collected every 20 min. No experimental manipulations were invoked until three consecutive samples were collected which did not vary in composition of acetylcholine by more than 10%. Dialysate ACh was analyzed using an analytical column and acetylcholinesterase/choline oxidase post-column enzyme reactor.

Passive Avoidance

On training (acquisition) day, groups of 8 male Sprague-Dawley rats were injected with scopolamine (3.0 mg/kg, i.p.) or vehicle 30 min prior to the training session. Ten min prior to the session, animals were injected with either dihydrexidine (0.1, 0.3 or 1.0 mg/kg, i.p.) or vehicle. The sessions were initiated by placing the rats in the light compartment of a two compartment chamber facing away from the opening between the two compartments. When a rat entered the dark compartment completely, a 1 mA, 3 sec scrambled shock was delivered to the entire grid floor. The animal was allowed to remain in the dark compartment or escape to the light compartment during the shock period. Twenty-four hours after training, each rat was tested for retention of the task. The latency for animals to enter the darkened side (*i.e.*, step-through latency) was recorded up to a maximum of 600 sec.

RESULTS AND DISCUSSION

Neurochemical studies in the striatum showed that doses of 3 and 10 mg/kg of DHX stimulated acetylcholine release. Blockade of the effect with the D_1 antagonist SCH23390 (0.3 mg/kg) suggests D_1 receptor involvement. The maximal level of stimulation (150–160% of baseline) achieved with these doses of DHX approximates that reported with other direct D_1 agonists.[10,11] The highest test dose of DHX (17.5 mg/kg) had a negligible effect on extracellular striatal acetylcholine levels. This result may be due to the loss of receptor subtype selectivity of DHX in this dosage range. Support of this hypothesis comes from the observation that the D_2 agonist quinpirole (2.0 mg/kg) significantly reduced striatal acetylcholine release. Thus, activation of D_2 receptors by high doses of DHX may inhibit acetylcholine release counteracting its stimulatory effect through D_1 receptors.

In frontal cortex, DHX produced a robust increase in extracellular concentrations of cortical acetylcholine to 200–300% of baseline values. This effect was completely blocked by SCH23390 (0.3 mg/kg), supporting a D_1 site of action. In contrast to striatal acetylcholine release, D_2 receptor activation also stimulated cortical acetylcholine release. The level of stimulation due to D_2 activation appears to be relatively modest compared to the D_1 effect since a large, behaviorally active dose of the D_2 agonist quinpirole (2 mg/kg, s.c.) elevated extracellular concentrations of cortical acetylcholine to only 150% of baseline.

These studies provide neurochemical evidence that the D_1 receptor agonist DHX increases extracellular acetylcholine levels in rat striatum and frontal cortex. Passive avoidance studies were conducted which suggested that DHX (0.3 mg/kg) improves cognitive performance in scopolamine-induced amnestic rats. The observed improvement by DHX is consistent with other rodent behavioral studies which showed that dopamine D_1 receptor agonists improve performance in a radial-arm maze task.[12] In addition, DHX improves cognitive performance in monkeys treated chronically with low doses of MPTP[13] and in young, intact monkeys[14] by SCH23390-sensitive mechanisms. Together these findings support the hypothesis that elevation of central cholinergic neurotransmission by DHX is responsible for its cognition-improving effects.

REFERENCES

1. KEBABIAN, J. W. & D. B. CALNE. 1979. Nature **277:**93–96.
2. MOTTOLA, D. M., W. K. BREWSTER, L. L. COOK, D. E. NICHOLS & R. B. MAILMAN. 1992. J. Pharmacol. Exp. Therap, **262:**383–393.
3. BREWSTER, W. K., D. E. NICHOLS, R. M. RIGGS, D. M. MOTTOLA, T. W. LOVENBERG, M. H. LEWIS & R. B. MAILMAN. 1990. J. Med. Chem. **33:** 1756–1764.
4. TAYLOR, J. R., M. S. LAWRENCE, D. E. REDMOND, J. D. ELSWORTH, R. H. ROTH, D. E. NICHOLS & R. B. MAILMAN. 1991. Eur. J. Pharmacol. **199:** 389–391.
5. SPICER, M. A., J. D. ELSWORTH, M. S. LAWRENCE, R. H. ROTH, J. R. TAYLOR, R. B. MAILMAN, D. E. NICHOLS & D. E. REDMOND. 1993. 1st International Meeting on Dopamine Agonists (abstract).
6. BERTORELLI, R. & S. CONSOLO. 1990. J. Neurochem. **54:**2145–2148.
7. WHITEHOUSE, P. J., D. L. PRICE, R. G. STRUBLE, A. W. CLARK, J. T. COYLE & M. R. DELONG. 1982. Science **215:**1237–1239.
8. BARTUS, R. T., R. L. DEAN, B. BEER, & A. S. LIPPA. 1982. Science **217:** 408–417.
9. WHITEHOUSE, P. J. & J. R. UNGERSTADT. 1988. Eur. Neurol. **28**(Suppl. 1): 36–41.
10. DAMSMA, G., G. S. ROBERTSON, C. S. THAM & H. C. FIBIGER. 1990. Eur. J. Pharmacol. **186:**335–338.
11. DAMSMA, G., G. S. ROBERTSON, C. S. THAM & H. C. FIBIGER. 1991. J. Pharmacol. Exp. Therap. **259:**1064–1072.
12. LEVIN, E. D. & J. E. ROSE. 1991. Pharmacol. Biochem. Behav. **38:**243–246.
13. SCHNEIDER, J. S., Z. Q. SUN & D. P. ROELTGEN. 1994. Brain Res. **663:** 140–141.
14. ARNSTEN, A. F. T., X. T. CAI, B. L. MURPHY & P. S. GOLDMAN-RAKIC. 1994. Psychopharmacology **116:**143–151.

Subject Index

Index of Contributors